Handbook of Dynamics and Probability

Peter Müller

Handbook of Dynamics and Probability

Springer

Peter Müller
Department of Oceanography
University of Hawaii at Manoa
Honolulu, HI, USA

ISBN 978-3-030-88485-7 ISBN 978-3-030-88486-4 (eBook)
https://doi.org/10.1007/978-3-030-88486-4

This Springer imprint is published by the registered company Springer Nature Switzerland AG
The registered company address is: Gewerbestrasse 11, 6330 Cham, Switzerland

Foreword

Computer models pervade in all fields of science. For the study of complex systems, such as the climate system, they are indispensable. The output of such computer models is difficult to assess, for many reasons, and often misunderstood and misinterpreted. Important questions are How does one determine the fidelity of computer models? Does an effect seen in the model corresponds to an effect in the real world or is it due to the approximations, parametrizations, and discretizations in the model? How does one interpret probability statements? What is one to make of the statement that global warming is man-made at the 95% confidence level? It surely does not mean: had we 100 climate systems 95 of them would show man-made global warming. Proper answers to these and other questions are essential for extracting useful and valid information from computer models, and for establishing public trust and confidence in their results.

The basic theories to frame and answer these questions are dynamical systems theory and probability theory. Computer models integrate the equations of motion that govern the dynamical evolution of the system. Dynamical systems theory identifies and studies the common mathematical structure of these equations, independent of the specific context in which they arise. When constructing a computer model one needs to specify which processes to explicitly resolve, which processes to neglect or parameterize, and what numerical algorithm to apply. Though different choices are made, resulting in different models for the same system, there is general agreement about which model best suits a specific purpose and what constitutes an improvement. This part of computer modeling is fairly objective.

The situation is quite different when uncertainties need to be quantified. Though probability theory unambiguously provides the rules of how to carry out calculations with given probabilities, it is silent about the assignment and interpretation of probabilities. Subjective choices come into play. Care must be taken, and there is genuine disagreement among researchers. Their underlying assumptions need to be understood if actions are to be taken based on probabilistic forecasts. A much advocated approach is Bayesian inference, an updating procedure, mimicking the way we learn. Bayesian inference also makes explicit the fact that there is no inference without prior assumptions, either openly stated or hidden behind some professional jargon.

In every case, dynamical systems theory and probability theory are the basic theories that need to be understood to answer questions like the ones above and to use computer models effectively.

Here is a book that will surely help with this task. It covers, in a concise and unified approach, the basic concepts and statements of dynamical systems and probability theory. In addition, it combines these theories to discuss the behavior of broadened initial conditions and of stochastic dynamical systems. It carefully elucidates the role that objective probability theory and subjective assumptions play in statistical inference. As a prime application it revisits statistical mechanics, a much studied subject but here treated as an exercise in the application of the principle of maximum entropy. The book is accessible to the applied scientist, self-contained, with many appendices covering technical details, it even has a uniform notation over its broad range of subjects.

It is a valuable book needed at this time. It will help the applied scientist to not become overwhelmed by the technicalities of computer models but instead to focus on extracting highly useful and valid information from them.

Hamburg, Germany Klaus Hasselmann
October 2021

Preface

Why another book about dynamics and probability? Our time is characterized by an explosive growth of information and access to it. Huge amounts of information are available on the World Wide Web and elsewhere. Assimilation of knowledge requires a conceptual framework. This book is to provide such a framework. It presents an unifying overview of the basic concepts and statements of dynamical systems theory and probability theory and their interrelations. These concepts are mathematical since the language of science is mathematics.

The book emphasizes precise definitions and solid statements. It takes the epistemological view that mathematical theories provide useful models for real-world systems, independent of whether or not the mathematical constructs actually exist in the real world. The book tries to remain as simple as possible without sacrificing too much mathematical rigor. As an overview the book also tries to be as general as possible. There are, of course, many examples. Some of these illustrate how dynamical systems and probability theory might be used to model quantitative aspects of systems in the real world, but no systematic analysis of such system is attempted since this would require a detailed explication of the to-be-modeled system. The book requires some familiarity with analysis, linear algebra, and physics.

The book is aimed at students, post-docs, and researchers in the applied sciences such as the earth sciences and engineering who are interested in better understanding the conceptual and mathematical underpinnings of the theories and models that they use in their respective fields. Despite the peculiarities of any of these applied sciences, dynamics and probability remain the basis for a reduced and effective description of the complicated systems encountered in these fields.

Overall, the aim of the book is to define and explicate the dynamical and probabilistic concepts that an applied scientist with a mathematical inclination might wonder about. I try to present these concepts in a transparent, lucid, clear and light way, making the book hopefully an easy and illuminating read.

The book grew out of tutorials and lectures and eventually a course on dynamics and probability that I gave at the University of Hawaii. I owe thanks to the students and post-docs who encouraged and inspired me to write this book. I also would like to thank Hajo Leschke of the University of Erlangen who was an early companion in the conception of this book and Andrei Natarov of the Pacific Tsumami Warning Center for his intellectual and technical support over many years.

Honolulu, USA Peter Müller

Contents

Chapter 1
Introduction

Dynamical systems theory and probability theory are basic tools of science. They are mathematical theories. They are used to describe, understand, model, and predict systems in quantitative sciences such as physics, biology, and economics. The time evolution of many of these systems is governed by dynamical laws that can be cast into a common mathematical form that makes them a dynamical system. Dynamical systems theory studies the properties of these systems independent of the context in which they arise. Science also has to deal with uncertainty. We do not know exactly that an event will occur under certain circumstances or as the outcome of a certain experiment. Mathematicians have developed the concept of probability to quantify uncertainty. Uncertainty may be intrinsic, quantum mechanics being the prime example; or it may be epistemic, due to our lack of complete knowledge or control. Though probability theory grew out of the analysis of games of chances it is also an axiomatic theory independent of any specific context. Part I and Part II of this book review dynamical systems theory and probability theory.

A dynamical system consists of two parts: a state space whose points completely specify the state of the system and a family of maps that uniquely determines the time evolution of states. To qualify as a dynamical system the family of maps has to be consistent with the additive structure of time. Mathematically, dynamical systems are most often realized by iterative (or recursive) maps and by ordinary or partial differential equations.

To study the properties and behavior of dynamical systems additional concepts need to be introduced. Observables, quantities of interest that depend on the state of the system, and orbits or trajectories, the curves that a state traces out in state space as time evolves, are two such additional concepts. They are purely set theoretical concepts. For other concepts one needs to endow the state space with additional structure. Distinguishing between linear and nonlinear systems requires the state space to be a vector space. Comparing the behavior of neighboring points and studying the long-time behavior requires the state space to be at least a topological space so that notions such as neighborhoods and convergence are defined. Determining what fraction of the state space exhibits a certain behavior requires the state space to

P. Müller, *Handbook of Dynamics and Probability*,
https://doi.org/10.1007/978-3-030-88486-4_1

be a measure space. Of course, in many applications the state space is endowed with all of these structures but for a systematic discussion it is important to identify the underlying mathematical structure that is required for a certain concept. Indeed, this book uses these different structures of the state space as an organizing principle, and discusses linear and nonlinear systems, the stability and long-time behavior, and the invariant and ergodic measures of dynamical systems.

The dynamical systems considered in the first part of this book are deterministic. Though deterministic, these systems may exhibit very complicated behavior. To analyze such complicated behavior dynamical systems theory has developed many unique and innovative concepts such as attractors, chaos, Lyapunov exponents, state space portraits, and bifurcations. One important discovery of dynamical systems theory is that simple dynamical prescriptions can lead to very complicated behavior. This discovery was aided by the use of digital computers. There was no incentive and no one had the stamina to iterate a simple quadratic map for a few thousand times for different initial conditions and parameter values. It were computer calculations which unveiled the intricate structure of this map. A second important discovery was that dynamical systems may exhibit sensitivity to initial conditions. Two neighboring points in state space may diverge from each other exponentially as time progresses. Thus, when initial conditions cannot be measured or represented accurately the system becomes unpredictable. It was also realized that the new tools of dynamical systems theory can only be explicated for low-dimensional systems. Thus most of dynamical systems theory has evolved around the analysis of low-dimensional system where it has produced many fascinating results, and beautiful pictures. Sometimes low-dimensional systems exhibit essential features of higher dimensional more realistic systems. The sensitivity to initial conditions in the three-dimensional Lorenz system has been used to rationalize the unpredictability of the weather beyond a few days. In general, it is, however, unclear to what extent results for low-dimensional systems carry over to higher dimensional systems. Despite these limitations dynamical system theory provides a useful intellectual framework for the analysis of arbitrary dynamical systems, and this is the attitude taken in this book, but the explicit calculation of key quantities is usually only possible for low-dimensional systems.

Only few dynamical systems allow for analytic solutions. Most of them must be solved numerically on a computer. A digital computer can only manipulate a finite amount of information. It has a finite word length. Real numbers can only be represented to about fifteen significant decimal digits, implying a round-off error. If such a number is iterated for many times by a map the numerical result given back might deviate grossly from the correct result. The problem is exacerbated when differential equations, instead of iterative maps, are considered. Then derivatives must be replaced by finite differences, an operation which introduces truncation errors. Some of the numerical algorithms used for solving first-order ordinary and second-order partial differential equations will be discussed in order to give the reader a sense of the issues and potential pitfalls of numerical algorithms. The important point is that the original dynamical system and its numerical realization are different dynamical systems. One must be careful and diligent to infer properties of the original system from the numerical system. This assessment is not meant to discourage the use of numerical computations. They are a powerful tool. As with any tool numerical computation can be a source of valuable information and significant insight in the hands of the able scientist but a source of delusion and misconception in the hands of dilettantes.

It is one of the ironies of history that computer modeling has reversed one of the major achievements in the history of science: calculus. Calculus allowed scientists to quantify the concept of change. The finite difference quotient became the derivative. Most of the physical laws are formulated as differential equations. To solve these differential equations one now has to undo this process and approximate derivatives by finite differences.

Life is uncertain, death and taxes not withstanding. Even science has to deal with uncertainty. Mathematicians have developed the concept of probability to quantify uncertainty. It was put on an axiomatic basis by Kolmogorov. The basis of probability theory is a probability space. A probability space consists of three parts: a set of possible outcomes, a family of subsets of this set whose members are called events, and a function that assigns to each event a value between zero and one, called the probability of the event. The probability of an event is supposed to characterize its uncertainty. Outcomes may be the face value of a card drawn from a deck or the reading of an analog measuring device. The family of events must be closed under countable set operations and the probabilities of events must satisfy Kolmogorov's axioms. The distinction between outcomes and events allows probability theory to include real numbers as outcomes. Probability may be viewed as a normalized measure and probability theory as a sub-discipline of measure theory.

Probability theory introduces two key concepts: random variables and statistical dependence. A random variable is a (measurable) function that assigns to each outcome a real number. This number is called a realization of the random variable. The set of all possible realizations is called an ensemble. A random variable induces a probability space where the outcomes are real numbers. This induced probability space is the starting point for most of probability theory. We only consider discrete and continuous random variables. They are discrete if the support of the induced probability measure consists of an at most countable set of real numbers and continuous if the support is the real line or an interval. Discrete random variable can be characterized by a probability mass function, continuous random variables can often be characterized by a probability density function. Composing a given random variables with a real-valued function generates a new random variable. Much of probability theory consists of deriving the probability distributions of the given random variable from the underlying probability space or the probability distribution of the new random variable from the distribution of the given random variable.

The second concept is statistical dependence. It is a concept usually not considered in measure theory. Statistical dependence describes the fact that the occurrence of one event may tell you something about the occurrence of another event. Such statistical dependence is quantified by conditional probabilities. Statistical dependence of two random variables is quantified by covariance matrices and correlation coefficients. Statistical dependence of two events or two random variables does not imply any causal relationship between the two events or random variables. The adjective 'statistical' distinguishes this kind of dependency from other dependencies such as linear dependence of vectors.

Once random variables are introduced one can consider infinite sequences of random variables. One interpretation of such an infinite sequence of random variables is that it represents the outcome of an experiment that is repeated infinitely often. An important question is then whether or not these sequences converge and in what sense. Different kinds of convergences have to be distinguished. One important result is the strong law of large numbers which states that the arithmetic mean of a single realization of a sequence of independent, identically distributed random variables of finite mean equals the ensemble mean almost surely. This probabilistic law of large numbers and its generalization, the ergodic law of large numbers, are central to the interpretation and application of probability concepts. It links the world of possible outcomes, of parallel universes, to the one realization, to the one world we live in.

When the sequence of random variables is supposed to represent the non-deterministic evolution of a system it is called a discrete stochastic process. Important properties are whether or not these stochastic processes are stationary, reversible, or ergodic. The simplest discrete stochastic processes are independent identically distributed processes, where all random variables in the sequence are mutually independent and have the same distribution. The most prominent example is the Bernoulli process which can only take two different values with prescribed probabilities. Next in line are the Markov processes where the probability of an outcome only depends on the outcome at the previous time. The corresponding conditional probability is called the transition probability. The future only depends on the present and not on the past history, in close analogy to dynamical systems. Such Markov processes are uniquely determined by a transition probability. The most prominent examples are random walks on a discrete lattice or on a continuum.

Families of random variables where the index set represents continuous time are called continuous stochastic processes. There are some issues that arise from the fact that probability theory only provides means for the manipulation of countable sets of events whereas uncountable sets occur in the analysis of continuous processes. These issues can be resolved by restricting the analysis to separable processes, which is not much of a restriction. Their properties are almost surely determined on a countable dense subset of time, and applying probability does not meet with principal problems. Again, Markov processes are the most studied processes. There are jump processes whose transition probabilities evolve according to an integro-differential equation, the master equation, and processes with continuous sample paths whose transition probabilities evolve according to a partial differential equation, the Fokker–Planck equation. Random walks where a particle moves on a discrete grid or in a continuous space are again the most important applications. The most important continuous independent process is white noise, a non-differentiable process with infinite variance.

It is emphasized that there is nothing random in a random variable. It is just a function. What will later in applications become random in the colloquial use of the word is the argument of the random variable and hence its value. It will be uncertain which value will occur. The values will be random. This randomness is characterized by the probability measure. There are different degrees of uncertainty or randomness. One probability measure might be more random than another. To quantify the degree of randomness one introduces the concept of information entropy which might be interpreted as the amount of missing information. Information entropy is a key concept in information theory and a property of the probability measure. The more information is missing the larger the information entropy. The probability distribution of a fair die has a higher information entropy than that of a loaded die. Whereas probability characterizes the uncertainty of an event information entropy characterizes the degree of randomness of a probability measure. An important principle will be the principle of maximum entropy which determines the probability measure that has the maximal information entropy given some constraints. Information entropy is an important and central concept of this book. The information entropy must be distinguished from Clausius' entropy in thermodynamics and is related to Boltzmann's and Gibbs' entropy in statistical mechanics.

Part III of this book applies probabilistic concepts to deterministic dynamical systems. Probability theory offers itself when something is or must be regarded as uncertain in the dynamical system. One important case is that the initial conditions must be regarded as uncertain, either because we do not know them exactly or, more importantly, because we cannot control them. As in ergodic theory one endows the state space with a probability measure, but this probability measure now reflects the uncertainty in the initial condition, and not 'volumes' of the state space. We consider the special case where this probability measure is given by a probability density function. One calls this density a broadened initial state, as opposed to a point in the state space which is called a pure state in this context. The dynamics uniquely determines the time evolution of this probability density function, in a linear manner. One thus turns a generally nonlinear dynamical system for the evolution of points into a linear deterministic dynamical system for the evolution of densities. If, for example, the original dynamical system is governed by a (nonlinear) ordinary differential equation then the new system is governed by the Liouville equation, a linear partial differential equation. An important result is that these broadened states often can do things that pure states cannot do. Even when points endlessly wander around in state space, densities might approach a stationary density. There might be attractors for densities though there were none for points. Though the incorporation of random initial conditions formally converts a dynamical system into a stochastic process, this process is of a trivial kind. Each realization is a trajectory that evolves deterministically according to the dynamical law. From each initial condition there only emanates one trajectory. There is no loss of information over time and the entropy of the broadened state does not change over time (if the trajectories do not merge). Broadened initial conditions alone are not the root of irreversibility and the second law of thermodynamics.

A more interesting behavior and loss of information might occur when one considers the evolution of observables. Observables are generally many-to-one functions. The initial value of an observable then does not, in general, determine a single trajectory. Many trajectories might be consistent with the initial value and might result in different values of the observable at future times. When the initial conditions are random one can then determine the fraction of trajectories that will end up at a particular future value of the observable. This fraction constitutes a transition probability and the evolution of the observable becomes a genuine stochastic process. Thus, observables evolving according to a deterministic dynamical system represent stochastic processes when random initial conditions are assumed. This is indeed the most common way to construct realizations of stochastic processes. Since one does not know exactly which future value of the observable will be obtained there is loss of information and the information entropy will increase. However, considering observables does not break any time-symmetry of the underlying dynamical system. Considering observables is also not the root of irreversibility and the second law of thermodynamics.

Two important applications are coarse graining and the elimination of variables. Coarse grain the state space into a countable number of boxes. The index function that assigns to every point of the state space the number of the box that the state occupies is then an observable. When eliminating variables the projection onto the retained variables constitutes an observable.

The random initial conditions are imposed onto the deterministic dynamics (by the investigator). One may ask whether a deterministic system might intrinsically generate some probabilistic structure. The answer is yes. A system consisting of two subsystems with largely different time scales will do so, when one eliminates the fast variable. The effect of the fast subsystem on the slow subsystem is given by a time integral. In the limit that the ratio of the time scales is infinite the integral extends to infinity and the time integral can be replaced by the mean with respect to an invariant measure, if the fast subsystem is ergodic. This invariant measure is 'chosen' by the dynamics and not imposed by the investigator. The underlying physical reason is that very high frequency forcing becomes indistinguishable from stochastic forcing.

An important generalization of deterministic dynamical systems is stochastic dynamical systems. They are motivated by the desire to add noise or some other random or non-deterministic aspect to a dynamical system. Formally, this is accomplished by assuming that a parameter of the dynamical system is a random variable or stochastic process. We consider autoregressive and multiplicative processes where either the forcing term or the growth rate of a linear iterative map is assumed to be random. The continuous analogue of the autoregressive process is the Langevin equation which is a linear differential equation with a random forcing term. It was heuristically derived to describe the random impact of water molecules on ink particles in Brownian motion. The Langevin equation is a special case of a stochastic differential equation. Stochastic differential equations require a new calculus to account for the non-differentiability of stochastic processes.

Part IV of the book considers the use of probability theory in scientific reasoning. There is first the problem that probability theory, and dynamical systems theory, are mathematical theories. They belong to the Platonic world if one distinguishes between the Platonic, physical, and mental world, the latter two often combined and called the real world. Mathematics studies the objects of the Platonic world. Ontology studies the physical world, what objects or systems *are*. Epistemology studies the mental processes, how we *know* about objects and systems. To apply these mathematical theories to the real world one must identify entities in the real world that correspond to the mathematical concepts and structures. This identification raises deep philosophical questions which have been studied by many great thinkers and are still largely unresolved. This identification is especially difficult in the case of the mathematical concept of probability and goes under the name of "interpretation of probability". What does it mean to say that an event in the real world has a certain probability? Is probability an objective property of the real world or a subjective property of our mental world? There are at least four different interpretations of probability:

– The physicalist places probability squarely into the physical world. Probability is an objective property of a system. This interpretation often bases its arguments on ergodic theory.

– The frequentist regards probability as a property of finite or infinite sequences of events. The probability of a specific event is the relative frequency with which this event occurs in the sequence of events.

– The logicist views probability as a logical relation between propositions. Probability theory becomes an extension of propositional logic to the case where there is insufficient knowledge.

– The subjectivist interprets probability as a subjective degree of belief. This view has its roots in decision-making and betting strategies which do not make you a 'sure looser'.

This book does not take any of these views and does not attempt to answer other related philosophical questions such as do we create or discover mathematics or is the world deterministic or what is and causes randomness? Rather it takes a pragmatic or utilitarian view. It takes mathematical concepts, structures, and theorems as tools, used to describe, comprehend, model, and predict the behavior of real world systems. If the tools work, fine; if they do not, discard them. 'Work' means that the tools pass the epistemological loop.

The epistemological loop consists of the following steps. First, one uses observations to infer a mathematical structure. This step might be the construction of a mathematical model or a theory or even a law. Second, one explores by logical reasoning or computer analysis the implications of this mathematical structure and formulates hypotheses which, in the third step, are to be tested against *new* observations. The comparison of the prediction with the new data might either confirm the assumed mathematical structure, or lead to modifications or to outright rejection. This epistemological loop is the standard practice of science. This loop can be formalized in various ways.

In this utilitarian view, mathematical structures are tools to be used rather than ontological statements to be assessed for truth or falsity. A probabilistic model might be useful, independent of whether or not one agrees with Einstein that "He does not throw dice". Bohr's model of the atom with electrons circling around the nucleus is useful even though it is not a correct quantum mechanical description. Whether or not a mathematical structure works also depends on the questions being asked, on the context, and on other contingencies. Euclidean geometry works for some problems and not for others. Given a time series one might either regard it as the deterministic trajectory of a chaotic dynamical system or as a realization of a stochastic process, whatever works. The distinction between these two interpretations is indeed muted by the fact that realizations of stochastic processes can be constructed from deterministic algorithms. Ever noticed that some random number generators on computers always give the same sequence of 'random' numbers? What the time series truly is, is hard to tell, other than it is a sequence of numbers.

The above epistemological attitude might be scolded as relativism. Everything goes as long as it works. Where is scientific truth? The pragmatic attitude is taken here because the author simply does not know whether the world is deterministic or probabilistic and whether or not these are even the right concepts to describe the world.

Aside from these philosophical problems there is the problem that probability theory does not tell us how to assign probabilities. It only tells us how to do calculations with probabilities. The task of assigning probabilities falls to a discipline called statistics. Statistics has a bad reputation. There is the famous quote by Benjamin Disraeli: "There are three lies - lies, damn lies and statistics". The main reason for this bad reputation is the unreflected and uncritical use of statistical methods. Most generally, statistics addresses the problem of how to reason or make inferences or decisions under uncertainty. To do this properly the statistician has to carefully evaluate the epistemological situation. What is the data or evidence at hand? What can reasonably be assumed? The statistician must also clearly formulate the questions to be answered or the goals to be achieved and the risks that wrong conclusions might entail. This process involves many subjective choices by the statistician, and not everyone makes the same choices.

One of the basic tasks of statistics is parameter estimation. Given a sample, interpreted as a realization of a random variable, one would like to obtain estimates of parameters of the underlying probability measure. The standard approach considers functions from the sample space into the parameter space. These functions are also called statistics. Substituting the observed sample as an argument into this function produces an estimate of the parameter. When one composes this function with the random variable of which the sample is a realization one obtains a random variable, called the estimator. The properties of different estimators can be compared and an estimator be selected that has desirable properties though what is or is not desirable is in the eye of the statistician. It is important to realize that this standard approach tells you something about the particular estimator chosen and not too much, actually nothing, about a particular estimate. If one were to repeatedly apply the same estimator then one can say that a certain fraction of the estimates would fall into certain ranges. A second basic task of statistics is the testing of hypotheses about the underlying probability measure. Such testing is enabled by binary functions on the sample space, the two outcomes being either rejecting or accepting the hypothesis. The bottom line is the same as for parameter estimation. The test is chosen according to the preferences of the statistician. When one repeatedly applies the test one can say something about its performance. This kind of parameter estimation and hypothesis testing is inherently based on a frequentist's interpretation of probability. It cannot be applied to single or individual events.

Bayesian inference can be applied to single events. It makes use of Bayes theorem for conditional probabilities which reflects its premise that all inference is conditional. Bayesian inference is an updating algorithm. A given prior probability distribution is updated to a new posterior distribution when new data or evidence becomes available. This approach is very flexible and can be applied to a variety of problems such as backward probabilities, not just the estimation of parameters or the testing of hypotheses. Bayesian inference formalizes how we learn and make inferences in everyday life. It shows that one has to have a somewhat open mind to get maximal benefit from new information. If the prior probability distribution is too narrow, meaning one is too narrow-minded, then new evidence will hardly modify it, one sticks to one's opinion, no matter what. If, on the other hand, the prior distribution is too broad, one does not have an opinion, then the new evidence determines the posterior distribution, one's opinion is tossed around by each new piece of information.

The major subjective element in Bayesian inference is the specification of the prior distribution. The prior probabilities must be specified by the statistician. Often the prior distribution is the posterior distribution of a previous inference (or experiment). Or one may employ Laplace's principle of insufficient reason and attempts to be maximally non-judgmental with respect to missing information. This latter attempt is often formalized by the principle of maximum entropy. One chooses as the prior probability distribution that distribution that maximizes the entropy, given what one already knows. The dependence on a subjectively chosen prior distribution is somewhat softened by a convergence property. When one starts out with different prior distributions and repeatedly updates the distribution as new evidence becomes available one should eventually arrive at the same posterior distribution. If only politicians and scientists would agree to using Bayesian inference. Eventually all wrangling would stop. The author advocates the Bayesian inference

Statistical inference is an exercise in obtaining knowledge from data. In real life gaining knowledge is often not enough. One must take action and make decisions based on this knowledge. Knowing all the odds, shall I or shall I not buy the lottery ticket? Shall I gamble or shall I not? The author's advice will be not to gamble but to buy casino stocks. To assess the consequences of decisions one introduces loss (or utility) and risk functions. These loss and risks functions are not only subjective but might also be different for different stakeholders. The patient, the doctor, the insurance company, and the national health service all face different risks.

The comparison of a model prediction with observational data, the third step in our epistemological loop, may be formulated as a statistical test, with the two decisions being either to accept or to reject the model. As a statistical test this comparison is subjected to all the pitfalls and subjective choices that are inherent in statistical inference.

Though the author shied away from stating what probability exactly is in the real world he advocates the use of Bayesian inference. It realizes a key point about statistics namely that there is no inference without assumptions. Bayesian inference forces you to explicate these often professionally and psychologically ingrained assumptions, which constitutes good science.

Part V applies probability concepts to physics, to statistical mechanics and quantum physics. Statistical mechanics is the discipline that attempts to derive the macroscopic properties of matter as formulated in equilibrium and non-equilibrium thermodynamics from the microscopic dynamics of the constituent molecules. The most elegant approach is due to Gibbs and relies on five basic assertions: The first assertion is that macro-variables correspond to observables of the microscopic system. In general, many micro-states are compatible with a given value of a macro-variable. The second assertion is that the macroscopic state is specified by a broadened initial condition, by a probability density function on the microscopic state space. Such broadened initial conditions represent our limited control of the microscopic state. The third assertion is that the probability density function is the one that maximizes the entropy, subject to the constraints of the experimental set-up. The fourth assertion is that a measurement of a macro-variable produces the mean value of the corresponding observable. Finally, when considering macroscopic system in the thermodynamic limit the laws of large number takes effect and the fluctuations around the mean value diminish and the predictions become deterministic or as B.V. Gnedenko and A.N. Kolmogorov state: "In fact, all epistemological value of the theory of probability is based on this: that large-scale random phenomena in their collective action create strict, non-random regularity".

These assertions provide recipes that work extremely well. Statistical mechanics, both of equilibrium and non-equilibrium systems, is an extremely successful theory. It demonstrates that the combination of probability and dynamics can lead to an effective reduced description of complicated systems. Why statistical mechanics works is, however, still not fully understood, especially for non-equilibrium systems. There are still many unresolved foundational issues. In a situation where the macroscopic variables evolve from one macro-state to a second macro-state according to some macroscopic dynamical law one must minimally require that the overwhelming majority of the micro-states that are compatible with the first macro-state evolve into states that are compatible with the second macro-state. This minimal requirement for statistical mechanics to work cannot explicitly be proven for systems other than simple toy systems. The situation is even worse when the macroscopic system evolves from the first to the second state irreversibly. Then we must require additionally that there are much more micro-states compatible with the second macro-state than with the first macro-state, an even harder requirement to prove. It is also unclear why an ensemble approach should work at all. Why should an average over an ensemble, over many independent parallel universes, should have anything to do with what happens in this unique universe that we live in.

Statistical mechanics has been covered in numerous textbooks. Here, we reformulate it as the result of a Bayesian inference process: choosing a probabilistic approach with the probability distribution determined by the principle of maximum entropy is simply the best that we can do given the circumstances.

Statistical mechanics shares the property of not having solid foundations with another immensely successful theory, namely quantum mechanics which also has

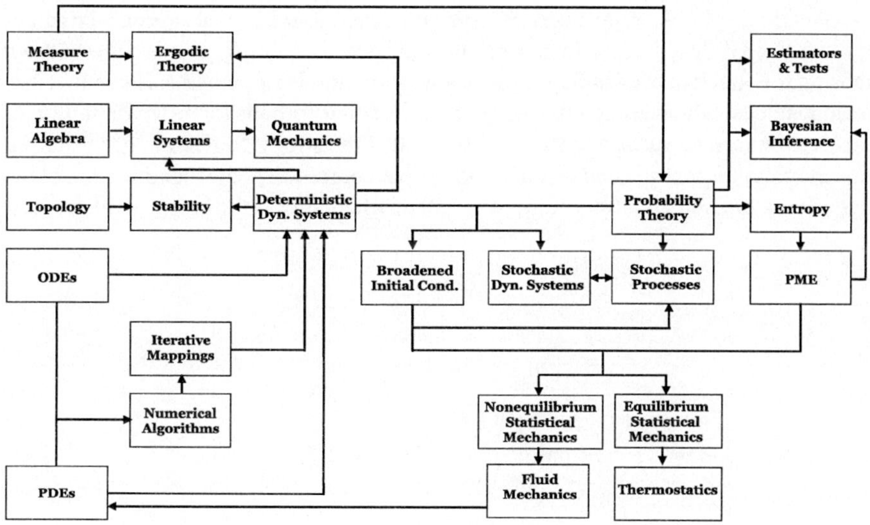

Fig. 1.1 Logical connections between the different chapters and sections of this book

a host of unresolved foundational issues and which, perhaps not accidentally, also combines dynamic and probabilistic concepts, albeit in a different way. The chapter on quantum mechanics highlights these differences, which can be done, fortunately, without requiring the complete quantum mechanical apparatus. Probability enters quantum mechanics already at the interpretation of the state. The state space of a quantum mechanical system is a Hilbert space and observables are self-adjoint operators on this Hilbert space. The possible values of an observable are the eigenvalues of the operator. If the system is in a specific state then the probability of obtaining any of the possible eigenvalues in a measurement is given by a rule that generalizes the theorem of Pythagoras. Probability in quantum mechanics is thus intrinsic and arises from Pythagoras theorem rather than from measure theory. This difference manifests itself in that quantum mechanics obeys the rules of quantum logic rather than classical propositional logic. To this Pythagorean probability structure one can add a measure theoretical probability structure to describe broadened initial states and arrives at the concept of the density matrix. Standard statistical interpretations are also used when quantum-mechanical probabilities are determined by repeatedly performing measurements on identical systems. Though observables have an intrinsic probability structure in quantum mechanics the dynamical time evolution of a quantum mechanical state is actually quite simple. It is given by the deterministic Schrödinger equation which describes a uniform rotation in state space. From a dynamical system's point of view quantum mechanics is a linear deterministic dynamical system.

The Fig. 1.1 shows the logical connections between the different chapters and sections of this book.

Overall, this book emphasizes the conceptual and foundational aspects of dynamics and probability. It tries to present these aspects as simply as possible without loosing too much mathematical rigor. Some of required mathematics is summarized in appendices, which are to provide quick access to a perhaps forgotten definition or theorem but are no substitute for a textbook. It is the strong conviction of the author that the clear disposition and explanation of concepts requires the identification of the underlying mathematical structure. After all, mathematics is the language of science.

Part I
Dynamics

This part presents the basic concepts of dynamical systems theory. What is a system and what is dynamics?

A system is an abstraction of an object in the real world whose properties the investigator wants to study. The object becomes a system once it is described by a specific set of state variables, comprising the state space. The object might be a gas in a container or the financial market. A given object may give rise to different systems, depending on the investigator's interests. In case of the gas the researcher might be interested in the microscopic positions and momenta of all the molecules that make up the gas. These positions and momenta then become the state variables and the gas becomes a N-particle system. Or the researcher might just be interested in the macroscopic velocity, pressure and temperature fields. These fields then become the state variables and the gas becomes a fluid system. The same physical object, the gas in a container, becomes a different system depending on the interests of the researcher.

Everything changes. One can merely describe these changes. This would be kinematics. Dynamics goes beyond mere description. It identifies and quantifies the causes of the changes, the laws that govern the changes. Classical mechanics is the science where dynamics was first established. There, dynamics studies the movement of bodies under the action of forces. Forces cause accelerations, as formulated in Newton's famous second law. Since then the word "dynamics" has been extended to other quantitative sciences. It now applies to any kind of time evolution that is governed by some causal or other law. Dynamics encodes these laws mathematically.

This first part considers deterministic dynamical systems where the evolution is uniquely and exactly determined by the dynamical law. Though many deterministic dynamical systems have their roots in causal laws, the definition of a deterministic dynamical system does not make any references to causes and effects. It merely states that future (and perhaps past) states are determined by the present state, for whatever reason. Dynamical systems theory does not address the profound philosophical issues surrounding the concept of causality. A deterministic dynamical system is then a system characterized by a set of state variables together with a mathematical dynamical law that determines the time evolution of these state variables.

Stochastic dynamical systems where the evolution is not precisely and uniquely determined are discussed in Part III. It is also stressed that determinism is different from predictability. Determinism is a property of the system. Determinism says that the future is determined. In philosophical terms, determinism is an ontological concept. Predictability is our ability to predict the future. Predictability is an epistemological concept. The future might be determined but we might not be able to predict it.

Dynamical systems theory abstracts from the peculiarities of a specific science or object. While classical mechanics always keeps a close connection to the real physical world when it studies the nature of space and time and introduces concepts like force, momentum and energy, dynamical systems theory abstracts from the meaning that state variables might have and studies their time evolution independent of the specific contexts from which they arise. Dynamical systems theory studies the mathematical structure of the dynamical laws, which might be identical for the motion of a star in a galaxy or the motion of a molecule in a gas. Though dynamical systems theory studies dynamical systems independent of any specific contexts we elucidate concepts and results with specific examples, mostly drawn from the fields of classical mechanics and fluid dynamics. A very brief and superficial introduction into these fields is provided in appendices, hopefully sufficient for the uninitiated to put our examples into context.

The first chapter gives the formal definition of a dynamical system, identifies iterative maps and ordinary and partial differential equations as the main mathematical structures that give rise to a dynamical system, and introduces the important concepts of observables and of invariant sets and orbits. These are all purely set-theoretical concepts.

There are many questions that can be asked about dynamical systems such as:
– Can the dynamical evolution be expressed analytically or in closed form?
– Is the dynamical evolution predictable and computable?
– Can systems be decomposed into parts that evolve independently?
– Is there a simple classification of dynamical systems?
– Are there equilibrium states and are they stable or unstable?
– Do systems evolve towards a particular limiting behavior?
– How does the evolution change when initial states are slightly changed?
– How does the evolution change when parameters are slightly changed?
– What fraction of the state space exhibits a certain behavior?

Answers to such questions require that the state space is endowed with additional structures. These additional structures are the organizing principle of this part. In the chapter “Linear systems" we assume that the state space constitutes a vector space so that the notion of linearity and the related notion of integrability can be defined. In the chapter “Stability and long-time behaviour” we assume that the state space has a topology so that one can introduce the notions of neighborhoods and convergence, compare the behavior of neighboring states, and study limiting behavior. In the chapter “Ergodic theory” we assume the state to be a measure space so that we can quantify the fraction of the state space that exhibits a certain behavior.

Dynamical systems theory arrives at quite general and powerful statements such as: invertible dynamical systems only exhibit three types of orbits: fixed points, periodic orbits, and non-intersecting non-periodic orbits; linear systems only exhibit exponential growth or decay, or periodic motions, or a superposition of these; discrete-time systems can exhibit chaos and sensitivity to initial conditions already in one-dimensional spaces whereas continuous-time systems can do so only in three- and higher-dimensional spaces; the behavior of systems can change drastically when parameters are slightly altered; every set of finite measure in a finite measure space will be revisited by almost all its points infinitely often.

While such general statements are important one would also like to assess the specific behavior of systems. This behavior can only rarely be expressed in analytic or closed form. More often one has to resort to numerical algorithms to be executed on digital computers. Such algorithms replace the original system by a new discrete-time system and the errors and distortions introduced by this replacement are discussed in the final chapter "Numerical algorithms".

Chapter 2
Deterministic Dynamical Systems

2.1 Introduction

In this chapter, we first give the general and abstract definition of a deterministic dynamical system. This definition has as its basic ingredients a state space whose points characterize the state of the system and a family of maps that determines the time evolution of states according to a dynamical law. The state space is comprised of the minimal set of variables that determines all properties of interest. To qualify as a dynamical system the family of maps must satisfy a composition rule to conform with the additive structure of time.

Deterministic dynamical systems are most often realized by iterative maps or by ordinary or partial differential equations. Establishing this connection is straightforward for iterative maps, includes some fine points for ordinary differential equations, and presents a major challenge in the case of partial differential equations.

Besides the variables that determine the state of the system there are additional variables that are of interest or relevance. If these additional variables depend on the state they are called observables. If they are independent of the state they are called parameters. Dynamical systems theory studies the behavior of dynamical systems and of their observables and the dependence of this behavior on parameters.

Two key concepts in the analysis of dynamical systems are invariant sets, which are sets comprised of states that will not leave the set during time evolution, and orbits which are the sets underlying the curves that an initial state traces out in state space during its time evolution. These orbits are the preferred tool to visualize the behavior of dynamical systems.

The definition of a dynamical system and of its observables and invariant sets only requires the mathematical concepts of sets and maps, nothing more.

P. Müller, *Handbook of Dynamics and Probability*,
https://doi.org/10.1007/978-3-030-88486-4_2

2.2 States and Their Time Evolution

The defining elements of a dynamical system are a state space, a law or rule that determines the time evolution of states and the ability to compose time evolutions. These elements are formalized in the following definition of a dynamical system:

State space. Assume that the state of the system can be characterized by some set of variables, denoted by x. The set of all possible states comprises the state space X. States are thus points in the state space, $x \in X$. For the general definition, one does not need to specify the state space any further. The state space may be the set of integers, real numbers, vectors, matrices, functions, or any other set. It is only in later chapters that we will need to define the precise nature of the state space and endow it with certain structures. Nevertheless, it is often helpful to envision the state x as the position of a particle. We will often employ this 'particle picture' since it makes many statements more vivid and intuitively accessible. In classical mechanics and equilibrium statistical mechanics, the state space is called the *phase space* and often denoted by the symbol Γ. The actual choice of the state space depends on both the object that one studies and the purpose of the study.

Family of maps. The dynamical law that governs the time evolution of the state is assumed to be given by a two-parameter family of maps:

$$
\begin{aligned}
S_{t_2,t_1}\colon X &\mapsto X \\
x &\mapsto S_{t_2,t_1}(x)
\end{aligned}
\tag{2.1}
$$

where $(t_2, t_1) \in T \times T$ and the index set T denotes time. If the system is in a state x at time t_1 it will be in the state $S_{t_2,t_1}(x)$ at time t_2. Note that future states at time t_2 are solely determined by the initial state at time t_1, no matter how the system reached the initial state. This requirement imposes an important restriction on the dynamical law. It generally excludes dynamical laws with memory and delay effects where future states do not only depend on the present state but on past states as well, though such dynamical laws can sometimes be cast into the above form by expanding the state space to a history space.

Composition rule. To qualify as a dynamical system, the composition of maps must satisfy the rule

$$
\begin{aligned}
S_{t_3,t_2} \circ S_{t_2,t_1} &= S_{t_3,t_1} \\
S_{t,t} &= I
\end{aligned}
\tag{2.2}
$$

where the circle denotes the composition of maps and I the identity map on X ($I(x) = x$). Since time spans can be added this composition rule assures that the evolution of the system conforms with the additive structure of time.

A deterministic dynamical system determines the dynamical evolution of the system for all times and for all points of the state space. This determinism is different from predictability. Determinism is a property of the system. It states that the future (or past) is determined. Predictability is our ability to predict the future. The future might be determined but we might not be able to predict it. One main reason for this inability, namely sensitivity to initial conditions, will be discussed in detail in Sect. 4.8.

Autonomous dynamical systems. A dynamical system is invariant under a translation in time or does not depend explicitly on time if $S_{t_1+t,t_2+t} = S_{t_1,t_2}$ for all $t \in T$. Such systems are called autonomous systems and can be characterized by the one-parameter family $S_t := S_{t,0}$ which maps every state x at the arbitrarily chosen initial time $t = 0$ into the state $S_t(x)$ at time t

$$\begin{aligned} S_t : X &\mapsto X \\ x &\mapsto S_t(x) \end{aligned} \tag{2.3}$$

This one-parameter family satisfies the composition rule

$$\begin{aligned} S_t \circ S_{t'} &= S_{t+t'} \\ S_0 &= I \end{aligned} \tag{2.4}$$

The original two-parameter family can be recovered by $S_{t_2,t_1}(x) = S_{t_2-t_1}(x)$. The family of maps $(S_t)_{t\in T}$ forms a group under the operation of composition. The index set T that represents time also forms a group, under the operation of addition. The composition rule (2.4) states that these two groups are isomorphic.

One generally studies autonomous systems. Non-autonomous maps can be turned into autonomous maps by an expansion of the state space. Given a two-parameter family S_{t_2,t_1} on X, one can define a one-parameter family $\tilde{S}_t$ on $X \times T$ by

$$\begin{aligned} \tilde{S}_t : X \times T &\mapsto X \times T \\ (x, t') &\mapsto (S_{t'+t,t'}(x), t' + t) \end{aligned} \tag{2.5}$$

It is easy to show that this one-parameter family satisfies the composition rule (2.4). By expanding the state space by one dimension one can reduce the index set by one dimension. This construction is just a formal one. Nothing is really gained by it. It might even complicate matters. The state space X might be endowed with some structure that cannot be extended to the state space $X \times T$. The relativistic space-time continuum notwithstanding one should not mix states and times. In Sect. 3.3, we give an example of the perils of such a construction.

Simple examples. If the state space X is a complex vector space (so that one can multiply states by complex numbers) then

$$S_t(x) = e^{-i\omega t}x \tag{2.6}$$

with $\omega \in \mathbb{C}$ is an autonomous deterministic dynamical system. It satisfies the composition rule (2.4) because of the properties of the exponential function. If $X = \mathbb{C}$ and $\omega \in \mathbb{R}$ then this system represents a harmonic oscillator. If X is a Hilbert space and $\omega = H/\hbar$ where H is a self-adjoint operator on X, called the Hamiltonian, and $\hbar$ is Planck's constant then this deterministic dynamical system becomes $S_t(x) = e^{-itH/\hbar}x$ and describes the evolution of quantum mechanical states $x \in X$ according to the Schrödinger equation.

Another simple example of an autonomous dynamical system is

$$S_t(x) = x + u_0 t \tag{2.7}$$

since it also satisfies the composition rule (2.4). It may represent the position of a particle that moves with constant velocity u_0.

Equivalent dynamical systems. Dynamical systems can be represented in different equivalent ways. Two dynamical systems $\{X, (S_t)_{t\in T}\}$ and $\{\tilde{X}, (\tilde{S}_t)_{t\in T}\}$ are called dynamically equivalent if there exists a bijection $f : X \mapsto \tilde{X}$ such that

$$\tilde{S}_t = f \circ S_t \circ f^{-1} \tag{2.8}$$

for all $t \in T$. Examples of equivalent systems will be given in the next section. In proper mathematical language, one should thus talk about equivalence classes of dynamical systems. Once we endow the state space with additional structure we will require that this structure will be preserved under the bijection, leading to more restrictive conditions for equivalence, such that the bijection should be a coordinate transformation, a continuous function, or a measurable function.

A first classification of dynamical systems. Dynamical systems can be classified according to the properties of the index set T, of the state space X, and of the maps $(S_t)_{t\in T}$. Let us start with the index set T which denotes time. Time is an ordered set. We distinguish between the past $t < 0$, the present $t = 0$, and the future $t > 0$. Time evolves from the past to the future. This is the arrow of time. Time might be viewed as either discrete or continuous. The dynamical law might determine only future or future and past states. One thus distinguishes between discrete-time and continuous-time dynamical systems and between invertible (both the past and the future are determined) and non-invertible (only the future is determined) systems, with the index set being

$T = \mathbb{N}$	discrete-time, non-invertible systems
$T = \mathbb{Z}$	discrete-time, invertible systems
$T = \mathbb{R}_+$	continuous-time, non-invertible systems
$T = \mathbb{R}$	continuous-time, invertible systems

Invertible systems satisfy $(S_t \circ S_{-t})(x) = S_0(x) = x$. Invertible systems thus map the state space onto the state space in a one-to-one manner. They are bijections. S_{-t} is the inverse of S_t. Non-invertible deterministic dynamical systems are also called semi-deterministic dynamical systems.

Invertibility must be distinguished from concepts such as invariance under time-reversal or reversibility. Invertibility only states that given the present state both future and past states are determined. Invariance under time-reversal is a more stringent symmetry of dynamical systems occurring in physics. Reversibility is a concept that applies to stochastic and thermodynamic processes. These concepts will be introduced and compared later in the book although it is already stressed here:

Time has a direction. Invertibility, reversibility, and time-reversal invariance are properties of the dynamical system, not of time.

To avoid the somewhat awkward adjectives 'discrete-time' and 'continuous-time' we might just call such dynamical systems discrete or continuous dynamical systems but emphasize that these adjectives just state the nature of the index set T and not of the state space.

Recapitulating, an (autonomous) deterministic dynamical system is a pair $\{X, (S_t)_{t \in T}\}$ where T represents time, X the state space and $(S_t)_{t \in T}$ a family of maps that satisfies the composition rule (2.4).

The key property of a dynamical system is that the future state is determined by the present state, not by the way the system arrived at the present state. This property seems to exclude all systems that have a memory and where the evolution depends on its history but such systems can often be transformed into dynamical systems by enlarging the state space to a history space, or by some other manipulations. We will encounter a few examples of such transformations in this book. The composition rule might also strike one as a very restricting requirement but it is merely a statement about time, namely that time evolves unaffected by the state and has an additive structure. Any reasonable description of the deterministic evolution of a system should thus honor this requirement. Next we consider the most common ways to actually construct dynamical systems.

2.3 Discrete Systems and Iterative Maps

Discrete-time dynamical systems have a discrete index set T which is either $\mathbb{N}$ or $\mathbb{Z}$. Such discrete dynamical systems can be realized by iterative (recursive) maps. We first consider simple iterative maps and then higher order iterative maps.

Iterative maps. Any function $F : X \mapsto X$ generates an iterative map

$$x_{n+1} = F(x_n) \tag{2.9}$$

where $n \in \mathbb{N}$. Successive application with initial condition x_0 gives

$$x_n = F^n(x_0) \tag{2.10}$$

where F^n is the n-fold composition of F with itself. Thus, every function F defines a discrete dynamical system

$$S_n : X \mapsto X$$
$$x \mapsto S_n(x) := F^n(x) \tag{2.11}$$

for $n \in \mathbb{N}$. The composition rule (2.4) is satisfied since $F^n \circ F^m = F^{n+m}$. If the inverse F^{-1} exists, $x_{n-1} = F^{-1}(x_n)$ and the dynamical system becomes invertible. If F depends explicitly on time then the iteration gives rise to a non-autonomous dynamical system.

Example: Turning of a coin. The state space $X = \{0, 1\}$ consists of just the two numbers 0 (head) and 1 (tail) and the map is given by

$$F(x) = \begin{cases} 0 \text{ if } x = 1 \\ 1 \text{ if } x = 0 \end{cases} \tag{2.12}$$

This map is invertible and generates the sequence $(..., 0, 1, 0, 1, ...)$. This is admittedly a rather trivial example but will serve some purpose later in the book.

Example: Logistic map. The logistic map first arose in the context of population dynamics. If a population x grows by a fixed increment Δx at each time step

$$x_{n+1} = x_n + \Delta x \tag{2.13}$$

then the population grows linearly, $x_n = x_0 + n\Delta x$. If the increment is assumed to be proportional to the current value, $\Delta x = \gamma x$, as is customarily assumed for the number of births, then

$$x_{n+1} = (1 + \gamma)x_n \tag{2.14}$$

and the population grows geometrically, $x_n = (1 + \gamma)^n x_0$. Finally, if the growth rate is assumed to decrease as $1 + \gamma = r(x_c - x)$ when approaching a carrying capacity x_c, then one obtains

$$x_{n+1} = r(x_c - x_n)x_n \tag{2.15}$$

The special case $x_c = 1$ leads to the logistic map

$$F(x) = r(1 - x)x \tag{2.16}$$

which is defined on the interval $[0, 1[$. The parameter r can take any value in the interval $[0, 4]$. Graphs of the function F for a few values of r are shown in Fig. 2.1, together with the geometric procedure to perform the iterations. The logistic map can exhibit very complicated behavior, including, as we will demonstrate later, chaotic behavior for certain values of the parameter r. The logistic map demonstrates that simple prescription can lead to complicated behavior. For these theoretical reason and for many fruitful applications, the logistic map is a much studied system.

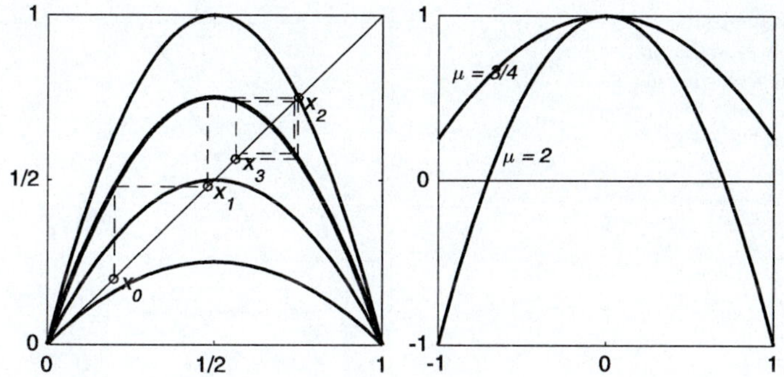

Fig. 2.1 Left: Graphs of the logistic map F for parameter values $r = 1, 2, 3, 4$ and the first three iterations for the $r = 3$ map starting at x_0. Right: Graphs of the logistic map $\tilde{F}$ for the parameter values $\mu = 2$ (Ulam map) and $\mu = 3/4$

Example: Ulam map. The Ulam map is defined on the interval $[-1, 1[$ by

$$\tilde{F}(x) = 1 - 2x^2 \tag{2.17}$$

It is a special case of a general class of maps

$$\tilde{F}(x) = 1 - \mu x^2 \tag{2.18}$$

with $\mu \in [0, 2]$, which are also called logistic maps. These logistic maps are also shown in Fig. 2.1 for two-parameter values. The common name reflects the fact that the maps F in (2.16) for values $r \in]2, 4]$ and restricted to the interval $[1/2 - (r - 2)/4, 1/2 + (r - 2)/4[$ are dynamically equivalent to the maps $\tilde{F}$ in (2.18) with values $\mu \in]0, 2]$, where $\mu = r(r - 2)/4$. Recall, that two dynamical systems F and $\tilde{F}$ are called equivalent when there exists a coordinate transformation f such that $\tilde{F} = f \circ F \circ f^{-1}$, the coordinate transformation being in this case

$$f(x) = \left(x - \frac{1}{2}\right)\frac{4}{r - 2} \tag{2.19}$$

Example: The doubling map. The doubling map is defined on the interval $[0, 1[$ by

$$F(x) = \begin{cases} 2x & \text{for } x \in [0, \frac{1}{2}[\\ 2x - 1 & \text{for } x \in [\frac{1}{2}, 1[\end{cases} \tag{2.20}$$

or $F(x) = 2x \; mod \, 1$. A graph of the function F together with the geometric construction of successive iterations is shown in Fig. 2.2.

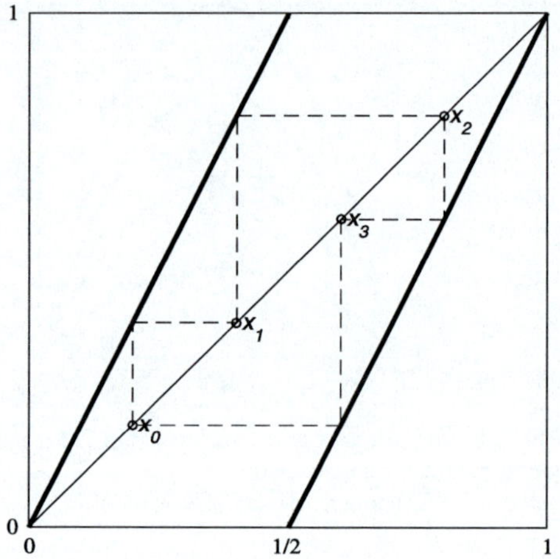

Fig. 2.2 Graph of the doubling map and the first four iterations starting at $x_0 = 1/5$, leading to a periodic map

Bernoulli shift. Take an infinite sequences of 0 s and 1 s. Delete the first number and shift the remaining numbers to the left. This is the Bernoulli shift or binary shift map, formally given by

$$\begin{aligned} T_1 : \{0, 1\}^{\mathsf{N}} &\mapsto \{0, 1\}^{\mathsf{N}} \\ (k_1, k_2, k_3, \ldots) &\mapsto (k_2, k_3, k_4, \ldots) \end{aligned} \tag{2.21}$$

This binary shift map T_1 is a special case of a more general shift map that acts on sequences of integers or real numbers and is discussed in Sect. 5.10.

The Bernoulli shift is dynamically equivalent to the doubling map. Write the initial real number x_0 in its binary representation

$$x_0 = \sum_{n=1}^{\infty} k_n 2^{-n} \tag{2.22}$$

where k_n takes the values 0 or 1 then the first iterate of the doubling map becomes

$$x_1 = \sum_{n=1}^{\infty} k_{n+1} 2^{-n} \tag{2.23}$$

Each iteration of the doubling map thus consists of deleting the first digit of the binary representation and shifting the remaining digits to the left. Choosing a rational number as the initial state results in a periodic map. Formally, this equivalence is accomplished by the bijection

$$f : \{0,1\}^{\mathsf{N}} \mapsto [0,1[$$
$$(k_1, k_2, k_3, \dots) \quad \mapsto \quad \sum_{n=1}^{\infty} k_n 2^{-n} \tag{2.24}$$

for which one indeed finds $F = f \circ T_1 \circ f^{-1}$.

There are other equivalent representations of the doubling map. Instead of considering it on the interval $[0, 1[$ one might define it on the unit circle C_1 in the complex plane. One then obtains the equivalent system $\tilde{F}(y) = y^2$ for $y \in C_1$, with the bijection between the two systems being $f(x) = e^{i2\pi x}$.

Example: Baker transformation. The baker transformation derives its name from the fact that it mimics the kneading of a piece of dough. It operates on the state space $[0, 1[\times[0, 1[$. Each iteration consists of squashing the square to a rectangle $2 \times 1/2$, cutting it along the line $x_1 = 1$ and placing the right part on top of the left part (see Fig. 2.3). Formally,

$$F(x_1, x_2) = \begin{cases} (2x_1, x_2/2) & \text{for } x_1 \in [0, \frac{1}{2}[\\ (2x_1 - 1, x_2/2 + 1/2) & \text{for } x_1 \in [\frac{1}{2}, 1[\end{cases} \tag{2.25}$$

If one writes x_1 and x_2 in binary notation the baker transformation removes the first digit from x_1 and adds it to x_2

$$\hat{F}(0.i_1 i_2 i_3 ..., 0.j_1 j_2 j_3 ...) = (0.i_2 i_3 ..., 0.i_1 j_1 j_2 ...) \tag{2.26}$$

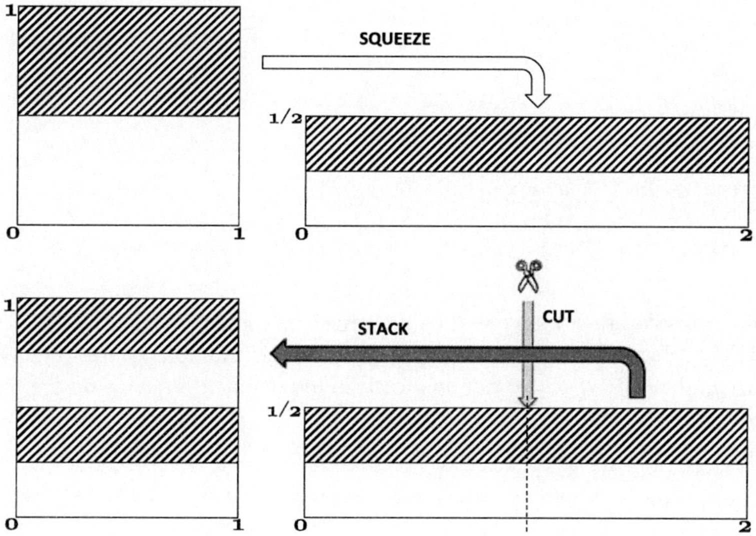

Fig. 2.3 The baker transformation

The baker transformation is invertible with

$$F^{-1}(x_1, x_2) = \begin{cases} (x_1/2, 2x_2) & \text{for } x_2 \in [0, \frac{1}{2}[\\ (x_1/2 + 1/2, 2x_2 - 1) & \text{for } x_2 \in [\frac{1}{2}, 1[\end{cases} \tag{2.27}$$

and it has the further property that $F \circ f = f \circ F^{-1}$ if $f(x_1, x_2) = (x_2, x_1)$.

We will return to these examples later in the book. The logistic, Ulam, and Bernoulli maps are examples of one-dimensional and non-invertible maps. The baker transformation is an example of a two-dimensional invertible map. All these maps are fairly simple prescriptions but lead, as we will see, to complicated behavior.

A concrete physical system that generates an iterative map is billiards. When one registers the positions where a billiard ball encounters the boundary one obtains an iterative map. Billiard tables of different shapes and their associated iterative maps are discussed quite extensively in Berger (2001).

Higher order iterative maps. Higher order iterative maps of the form

$$x_{n+1} = G(x_n, x_{n-1}, \ldots, x_{n-K+1}) \tag{2.28}$$

where K is called the order of the map include memory effects. A particular simple example of such a higher order iterative map is the autoregressive map

$$x_{n+1} = a_1 x_n + a_2 x_{n-1} + \cdots + a_K x_{n-K+1} \tag{2.29}$$

A stochastic generalization of it will be considered in Sect. 13.2.

Consider the K-tuple

$$y_n := (x_n, x_{n-1}, \ldots, x_{n-K+1}) \tag{2.30}$$

It satisfies

$$\begin{aligned} y_{n+1} &= (x_{n+1}, x_n, \ldots, x_{n-K+2}) \\ &= (G(x_n, x_{n-1}, \ldots, x_{n-K+1}), x_n, \ldots, x_{n-K+2}) \\ &=: F(y_n) \end{aligned} \tag{2.31}$$

A higher order iterative map can thus be turned into a simple iterative map on an enlarged state space. This procedure represents a simple example where systems with memory on the state space X can be turned into dynamical systems on the history space Y.

2.4 Continuous Systems and ODEs

Continuous-time dynamical systems have a continuous index set T which is either $\mathbb{R}$ or $\mathbb{R}_+$. Such continuous dynamical systems can be realized by systems of ordinary differential equations.

Systems of ordinary differential equations. Let $u : \mathbb{R} \times T \mapsto \mathbb{R}$, $(x, t) \mapsto u(x, t)$ a prescribed 'velocity field'. Then, an ordinary differential equation (ODE) has the form

$$\frac{d}{dt} x(t) = u\left(x(t), t\right) \tag{2.32}$$

where $x : \mathbb{R}(= T) \mapsto \mathbb{R}, t \mapsto x(t)$ is the unknown function. Equation (2.32) has to be solved subject to the initial condition $x(t = 0) = x_0$. The solution is then denoted by $x(t; x_0)$, indicating the parametric dependence of the solution on the initial value. Note that we follow standard ODE notation where the letter 'x' denotes a function and not an element of the state space as in the previous section. If confusion might arise we include the argument t when we consider the function.

Equation (2.32) also represents a system of coupled ODEs when

$$x(t) = (x_1(t), \ldots, x_N(t))^T \quad \text{and} \quad u(x, t) = (u_1(x, t), \ldots, u_N(x, t))^T \tag{2.33}$$

are regarded as N-dimensional (column) vectors.

Differential equations of second order

$$\ddot{z}(t) = u\left(z(t), \dot{z}(t), t\right) \tag{2.34}$$

can be reduced to first-order ODEs by introducing the two-dimensional vector $x(t) = (z(t), \dot{z}(t))$ which results in a set of two coupled first order ODEs

$$\dot{x}_1(t) = x_2(t) \tag{2.35}$$

$$\dot{x}_2(t) = u\left(x_1(t), x_2(t), t\right) \tag{2.36}$$

Higher order ODEs are amenable to the same procedure. Higher order ODEs can thus be reduced to a set of first ODEs by enlarging the state space.

Two sets of ODEs will be discussed quite extensively in this book: Hamilton's equations which are the backbone of classical mechanics and the Lorenz equations that led to the discovery of chaos.

Example: Hamilton's equations. In classical mechanics, the state of point particles is given by $x = (q, p)$ where q is the coordinate and p the momentum of the point particle. The velocity can be derived from a Hamilton function $H(q, p, t)$ by

$$u_q(q,p,t) = \frac{\partial H(q,p,t)}{\partial p}$$
$$u_p(q,p,t) = -\frac{\partial H(q,p,t)}{\partial q} \tag{2.37}$$

and the equations of motion are

$$\dot{q}(t) = u_q(q(t), p(t), t)$$
$$\dot{p}(t) = u_p(q(t), p(t), t) \tag{2.38}$$

and are called Hamilton's equations. Here, we have used a dot for time differentiation, $\dot{q} := dq/dt$, $\dot{p} := dp/dt$, for compactness of notation. An important example of a Hamiltonian system is the harmonic oscillator whose Hamilton function is given by

$$H(q,p) = \frac{1}{2}\omega p^2 + \frac{1}{2}\omega q^2 \tag{2.39}$$

where ω is the frequency of the oscillator. This Hamiltonian is one of the many possible representations of the harmonic oscillator, other representations can be found in the Appendix A.4.

Example: Lorenz equations. The Lorenz equations (Lorenz 1963) are

$$\dot{x}_1(t) = -\sigma x_1(t) + \sigma x_2(t) \tag{2.40}$$
$$\dot{x}_2(t) = r x_1(t) - x_2(t) - x_1(t)x_3(t) \tag{2.41}$$
$$\dot{x}_3(t) = x_1(t)x_2(t) - b x_3(t) \tag{2.42}$$

with parameters $\sigma, r, b > 0$. The velocity field is convergent, $\nabla_x^T u = -\sigma - 1 - b < 0$. The Lorenz system is a three-dimensional system. It is invertible.

The Lorenz equations are presumed to describe the essential features of Rayleigh-Bénard convection. Convection is a phenomenon that is governed by the equations of fluid dynamics which are a set of coupled partial differential equations (PDEs) describing the evolution of continuous fields like velocity, temperature, and pressure. When one expands the spatial structure of the fields into a (countable) set of basis functions one obtains an infinite set of ODEs for the time-dependent expansion coefficients. Truncation of this infinite set at any order then leads to a finite set of ODEs. This is indeed the way the Lorenz equations are 'derived' from the fluid dynamical equations. The details of this derivation are given in Appendix D.1.

Initial value problem. Given an ODE the initial value problem is the problem of finding a function $x(t)$ that satisfies the differential equation in a time interval I surrounding $t = 0$ and takes a prescribed value $x(t = 0) = x_0$ at time $t = 0$. We denote such a solution by $x(t; x_0)$ to show the parametric dependence of the solution on the initial value. There exists well-established conditions for the existence and uniqueness of such solutions.

A sufficient condition for the existence of such a solution (given by the Peano existence theorem) is that the velocity field $u(x, t)$ is continuous in a neighborhood surrounding the point $(x_0, t = 0)$. The theorem only assures existence of a local solution, in an interval surrounding surrounding $t = 0$.

The uniqueness of a local solution (stated by the Picard-Lindelöf theorem) is assured when the velocity gradient tensor $\nabla_x u^T$ (where $\nabla_x = (\partial/\partial x_1, \ldots, \partial/\partial x_N)^T$ is the gradient vector) is continuous in the neighborhood of $(x_0, t = 0)$ as well. An alternative sufficient condition for uniqueness is that $u(x, t)$ is locally Lipshitz in x, meaning that their exist a neighborhood of $(x_0, t = 0)$ and a constant c such that $|u(x_1, t) - u(x_2, t)| \leq c\,|x_1 - x_2|$ for all x_1, x_2 and t in the neighborhood.

There are well-known examples where these conditions are not met. The Bernoulli equation

$$\dot{x}(t) = 3x^{2/3}(t) \tag{2.43}$$

has the solution

$$x(t; x_0) = \left(t + x_0^{1/3}\right)^3 \tag{2.44}$$

for any initial condition x_0. At the initial condition $x_0 = 0$ the gradient of the velocity field $u(x) = 3x^{2/3}$ is not continuous (nor is the velocity field locally Lipshitz there) and there exists the additional solution $x(t; 0) = 0$ and combinations of the two types of solutions as depicted in Fig. 2.4. The solution is therefore not unique.

When the conditions of the Picard-Lindelöf theorem are met then one can also prove that there exists a maximum time interval over which a unique solution exists. This solution is referred to as the global solution. This maximum interval might still

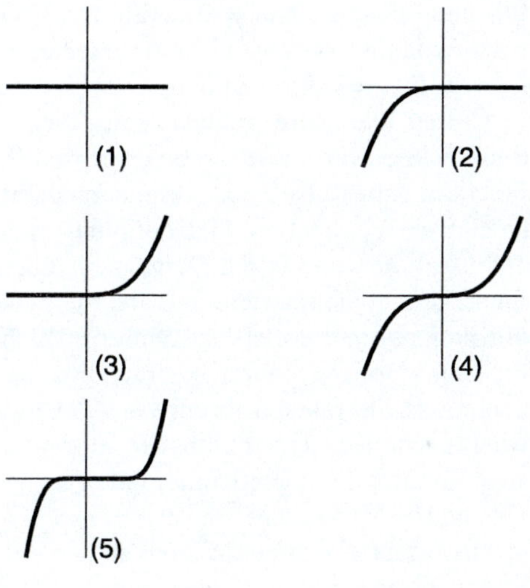

Fig. 2.4 Graphs of the possible solution of the Bernoulli equation with initial condition $x_0 = 0$. There are four different local solutions, (1), ...,(4) and an infinite set of global solutions, such as (5)

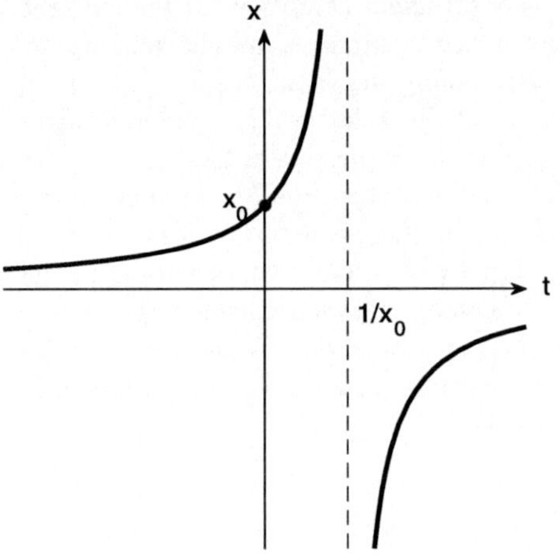

Fig. 2.5 Graph of the solution to the Ricatti equation on the real axis

be smaller than the time range over which the velocity field is continuous and might depend on x_0. A well-known example is the Ricatti equation

$$\dot{x}(t) = x^2(t) \tag{2.45}$$

whose velocity field $u(x) = x^2$ is continuous but whose solution

$$x(t; x_0) = \frac{x_0}{1 - x_0 t} \tag{2.46}$$

for initial condition $x_0 \neq 0$ develops a singularity at time $t = 1/x_0$ (Fig. 2.5). The solution is thus only valid in the interval $I_{max} =]-\infty, 1/x_0[$ for $x_0 > 0$ or in the interval $I_{max} =]1/x_0, \infty]$ if $x_0 < 0$.

Closed form and analytic solutions. A solution of an ODE is called a solution in closed form if it can be expressed in terms of a finite number of elementary functions using only basic arithmetic operations. The question is, of course, what is an elementary function. One definitely counts polynomials, roots, the exponential, and logarithm among the elementary functions but some people include the error function, the gamma function, and the hyper-geometric functions. A broader class of solutions are the analytic solutions which extends the above definition by including infinite series, continued fractions, integrals, and limits among the allowable operations. Despite the absence of a generally accepted definition, it is usually clear what functions and operations are allowed if an author writes down a closed form or analytic solution. A particular kind of analytic solution are the solutions that can be obtained by 'quadrature' and can be expressed as an integral. The name 'quadrature' results from the interpretation of an integral as the 'area under the curve'.

Having an analytic solution, such as $x(t) = x_0 e^t$ for the ODE $\dot{x}(t) = x(t)$, allows one to utilize the distinct properties of the exponential function in the interpretation and further analysis of the solution. Not having an analytic solution requires one to numerically integrate the ODE which generally leads to less transparent solutions. Despite these advantages of analytic solutions, it must be realized that the actual numerical value of e^t is still approximate. We have to use one of the various series expansion of the exponential function and can then approximate the value to any required accuracy by a finite truncation. Indeed, the values of the exponential and other elementary and non-elementary functions used to be tabulated in books like Abramowitz and Stegun (1965). It is also often argued that a solution like $x(t) = x_0e^t$ allows one to calculate the solution at time t without the need to know the solution at intermediate times. When we discuss numerical solutions in Chap. 6 we will see that this assertion is actually not correct.

Relation between ODEs and dynamical systems. If the system of ODEs (2.32) has a unique and well-defined solution for all initial condition $x \in \mathbb{R}^N$ and all $t \in \mathbb{R}$ then it corresponds to the invertible dynamical system

$$S_t : \mathbb{R}^N \mapsto \mathbb{R}^N \tag{2.47}$$

$$x \mapsto S_t(x) \tag{2.48}$$

which is obtained by solving

$$\frac{d}{dt}S_t(x) = u(S_t(x)) \tag{2.49}$$

$$S_0(x) = x \tag{2.50}$$

for all $x \in \mathbb{R}^N$ and all $t \in \mathbb{R}$.

If the dynamical system S_t is given, then Eq. (2.49) can be used to construct the velocity field of the differential equation. If the velocity field depends on time then one arrives at a non-autonomous dynamical system.

When the ODE does not have unique and well-defined solutions for all values of $x \in \mathbb{R}^N$ and all times $t \in \mathbb{R}$ it may still form the basis of a dynamical system. This is the case for the Ricatti equation shown in Fig. 2.5. One may add a point ∞ to the real line where this point represents both ends of the real line and assign the velocity $u(\infty) = \infty$ to this additional point. This extended set $\mathbb{R} \cup \{\infty\}$ can be mapped onto the unit circle by the stereographic projection $\varphi = 2\cot^{-1} x$, resulting in $\dot{\varphi} = -2\cos^2 \varphi/2$. This transformed Ricatti equation on the unit circle shows a more benign behavior and forms the basis of a dynamical system on the unit circle which is depicted in Fig. 2.6. It shows the interesting behavior that the point $x = \infty$ which is equivalent to $\varphi = 0$ can be reached in finite time.

Similarly, the Bernoulli equation of Fig. 2.4 defines a dynamical system if one chooses for the initial condition $x_0 = 0$ the solution (4) among the many solutions depicted in Fig. 2.4.

The major point is that not all ODEs correspond to a dynamical system. If they do, the difference between the differential equation and the dynamical system is that the differential equation describes the local time rate of change of the variable x,

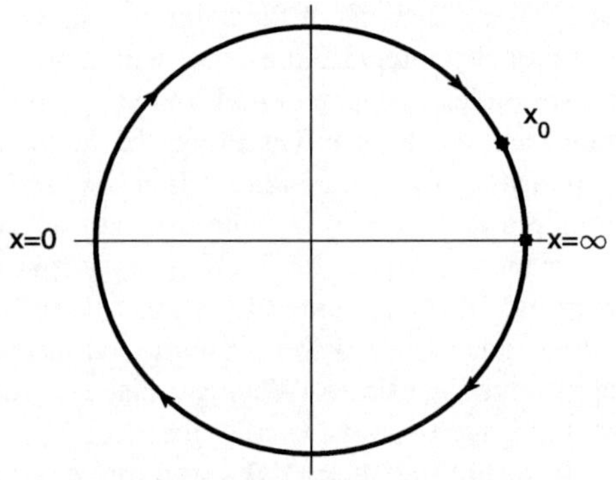

Fig. 2.6 Graph of the solution to the Ricatti equation on the unit circle. The solution reaches the point $x = \infty$ or $\varphi = 0$ at time $t = 1/x_0$

whereas the dynamical system describes the family of all solutions to the differential equation which must be obtained by solving the differential equation for all initial conditions and all times. The theory of ODEs studies properties of the functions $x(t)$, whereas dynamical systems theory studies the properties of the family of functions $S_t(x)$. This difference is also reflected in the notation used by these two communities and here. The solution $x(t)$ of an ODE metamorphoses via $x(t; x_0)$ and $x_t(x_0)$ and $x_t(x)$ into $S_t(x)$.

Integro-differential equations. Dynamical laws with memory and delay effects are often governed by integro-differential or delay equations. Such equations can often be cast into the dynamical systems form by enlarging the state space. A simple example is the integro-differential equation

$$\dot{x}(t) = u(x(t), t) + \int_{-\infty}^{t} dt' v(x(t'), t') \tag{2.51}$$

If one introduces the additional variable

$$y(t) := \int_{-\infty}^{t} dt' v(x(t'), t') \tag{2.52}$$

then the integro-differential equation (2.51) turns into the set of ordinary differential equations

$$\dot{x}(t) = u(x(t), t) + y(t) \tag{2.53}$$

$$\dot{y}(t) = v(x(t), t) \tag{2.54}$$

which corresponds to a dynamical system.

Relation between discrete and continuous dynamical systems. We have made the important distinction between discrete and continuous dynamical systems but, of course, any continuous system $(S_t)_{t\in\mathbb{R}}$ implies discrete dynamical systems $(\hat{S}_n)_{n\in\mathbb{Z}}$ by the definition

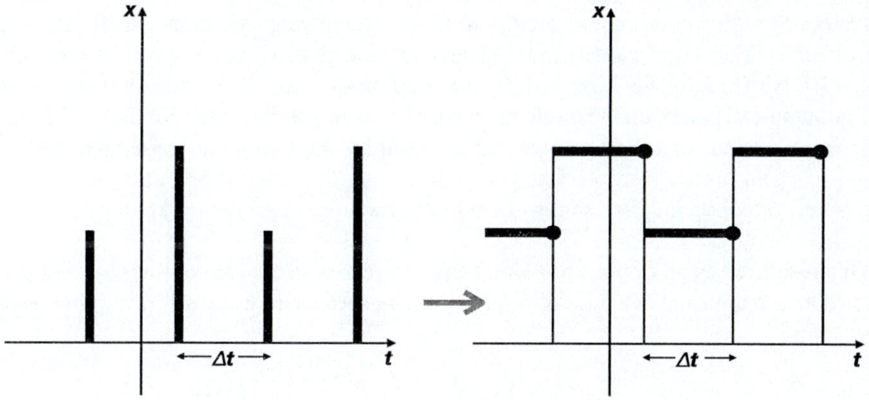

Fig. 2.7 Sketch of a discrete system and its continuous counterpart

$$\hat{S}_n := S_{n\Delta t} \quad n \in \mathbb{Z} \tag{2.55}$$

where $\Delta t > 0$ is any time interval. The discrete system $\hat{S}_n$ is called a stroboscopic version of the continuous system S_t.

Thus, every continuous system implies a whole suit of discrete systems. What about the converse: is every discrete system the stroboscopic version of a continuous system? Consider the iterative map $x_{n+1} = F(x_n)$. One can of course extend the series of 'spikes' to a series of step functions that are right-continuous by the definition (see Fig. 2.7).

$$x(t) := x_n \text{ for } (n-1)\Delta t < t \le n\Delta t \tag{2.56}$$

This series of step functions satisfies the differential equation

$$\dot{x}(t) = [F(x(t)) - x(t)] \sum_{n=-\infty}^{\infty} \delta(t - n\Delta t) \tag{2.57}$$

which is easily proven by integrating this differential equation from $t = n\Delta t - \epsilon$ to $t = (n+1)\Delta t - \epsilon$ and then letting $\epsilon > 0$ approach zero. This result might suggest that any iterative map F implies a time-continuous system, namely (2.57), that has F as its stroboscopic version. The above construction is, however, somewhat superficial. First, Eq. (2.57) is not a proper ODE: it contains delta functions, which are not functions, but distributions. Second, the Eq. (2.57) represents a non-autonomous system. Indeed, all that was done was to take a time-discrete system, extending it to a series of step functions that are right-continuous and to identify the 'ODE' (2.57) as the equation that produces this series of step functions.

If one imposes more stringent requirements on the continuous system then the answer is negative: not every discrete system is the stroboscopic version of a continuous system. For example, there is no ODE $\dot{x} = u(x)$ and associated continuous-time system S_t with a continuous velocity field u that generates the binary shift map T_1, i.e., there is no continuous $u(x)$ and no Δt such that $S_{\Delta t}(x) = 2x \; mod \, 1$ for all $x \in [0, 1[$. Think of the binary shift as a map on the unit circle. For the flow to be continuous all points on the circle must have the same period. This conflicts with the different periods that the binary shift map exhibits. An important consequence is

Discrete dynamical systems may exhibit behavior that continuous systems can not.

This result raises an important point. Suppose real systems are continuous. Are we introducing non-real behavior by approximating such systems by discrete dynamical systems?

2.5 Continuous Systems and PDEs

Continuous dynamical systems can also be realized by partial differential equations.

Partial differential equations. Let $x \in \Omega \subseteq \mathbb{R}^N$, then a partial differential equation has the form

$$P(x, t, \rho, \partial_t \rho, \nabla_x \rho, \partial_t \partial_t \rho, \nabla_x \nabla_x^T \rho, \ldots) = 0 \tag{2.58}$$

Here, we use the standard notation for PDEs, where ρ denotes the dependent variable which is a function of the independent variables x and t. The vector x comprises the spatial coordinate and t the time coordinate, ∂_t denotes the partial derivative with respect to time and $\nabla_x = (\partial_{x_1}, \ldots, \partial_{x_N})$ the gradient operator. If x is one-dimensional then we denote the partial derivative with respect to x by ∂_x. Note that the time variable appears as an argument of ρ, not as a subscript as is our convention for dynamical systems, since a subscript often denotes partial differentiation in the theory of PDEs.

Initial-boundary value problem. The initial-boundary value problem consists of finding a solution for given initial value $\rho(x, t = 0)$ within the domain Ω that satisfies appropriate conditions on the boundary $\partial\Omega$. Studies address infinite, finite, and periodic domains. In the case of an infinite domain, one has to impose conditions on the solution as $|x|$ approaches infinity.

The mathematical analysis of PDEs distinguishes between strong and weak solutions. Strong solutions are solutions that satisfy the PDE. Weak solutions are functions that satisfy the equations when the scalar product with appropriate test functions is taken. Typically, this scalar product consists of multiplying the PDE with the test functions and integrating over the spatial domain. The theory of PDEs also uses the concept of smoothness. A function is called smooth if all its spatial derivatives exist. If the domain is infinite smoothness includes ‘proper’ behavior at infinity.

Whereas there exists a well-established theory for the existence and uniqueness of solutions for ODEs this is not the case for PDEs. Each PDE has to be considered in its own right. Many PDEs have been studied quite extensively. Important results have been obtained but open questions also remain. The Navier–Stokes equations, described in Appendix C.2, are such a well-studied system of PDEs. In two dimensions there exist unique strong and weak solutions for all times for smooth initial velocity fields. In three dimension one can show that weak solutions exist for all times but it is an open question whether or not they are unique. There exist unique strong solutions up to a finite time but it is an open question whether or not they exist for all times. Indeed the latter question is one of the Clay Institute's million dollar question which is formulated as

"Prove or give a counter-example of the following statement: In three space dimensions and time, given an initial smooth velocity field, there exists a vector velocity and a scalar pressure field, which are both smooth and globally defined, that solve the Navier-Stokes equations."

When constructing dynamical systems from PDEs a major problem is the identification of an appropriate function space. Here we consider two examples where this construction is fairly straightforward, namely the Liouville equation and the diffusion equation.

Example: Liouville equation. The PDE

$$\partial_t \rho(x,t) + \nabla_x^T (u(x,t)\rho(x,t)) = 0 \tag{2.59}$$

with the domain being $\mathbb{R}^N$ and $u(x,t)$ being a prescribed velocity field is called the Liouville equation and plays an important role in many fields. If ρ is the density of a 'substance' it describes the conservation of this substance as it is being advected by the velocity field $u(x,t)$. We will later consider the cases where the 'substance' is mass or probability. The solution will obviously depend on the type of velocity field. To arrive at an autonomous dynamical system we assume that the velocity field does not depend on time t and is of a structure that the ODE $\dot{x}(t) = u(x(t))$ corresponds to an invertible dynamical system U_t such that the Jacobian $J_t(x) := \partial(U_t(x))/\partial(x)$ exists. One can then show by direct differentiation that

$$\rho_t(x) = J_{-t}(x)\rho_0(U_{-t}(x)) \tag{2.60}$$

is a solution. Here, we converted to the subscript notation of dynamical systems theory. The major point now is that this explicit solution can be extended to all integrable functions

$$\rho \in \mathcal{L}^1 := \{\rho : \int_{\mathbb{R}^N} dx|\rho(x)| < \infty\} \tag{2.61}$$

independent of whether or not the functions ρ are differentiable, as would be required for a solution of the Liouville equation. The Liouville equation thus implies a dynamical system

$$S_t : \mathcal{L}^1 \mapsto \mathcal{L}^1 \tag{2.62}$$

$$\rho_0(x) \mapsto \rho_t(x) = J_{-t}(x)\rho_0(U_{-t}(x)) \tag{2.63}$$

One has only to prove that $\rho_t(x)$ belongs to $\mathcal{L}^1$ if $\rho_0 \in \mathcal{L}^1$. This is indeed the case since

$$\begin{aligned}\int_{\mathbb{R}^N} dx \rho_t(x) &= \int_{\mathbb{R}^N} dx J_{-t}(x)\rho_0(U_{-t}(x)) \\ &= \int_{\mathbb{R}^N} dy \rho_0(y) < \infty\end{aligned} \tag{2.64}$$

upon substitution of $y = U_{-t}(x)$ in the integral.

Example: Diffusion equation. The PDE

$$\partial_t \rho(x,t) = \partial_x D(x,t) \partial_x \rho(x,t) \tag{2.65}$$

is called the diffusion equation and $D(x,t)$ the diffusion coefficient which may depend on space x and time t. If more than one spatial dimension is considered the diffusion coefficient becomes a diffusion matrix. If the domain is the real line and if the diffusion coefficient is assumed to be a constant $D_0 > 0$ then one can prove by direct differentiation that

$$\rho_t(x) = \int_{\mathbb{R}} dx' (4\pi D_0 t)^{-1/2} \exp\left\{-(x-x')^2/4D_0 t\right\} \rho_0(x') \tag{2.66}$$

is a solution. The solution is represented through an integral kernel. Again one can prove that any $\rho_t(x,t)$ given by this integral representation, independent of whether or not it is differentiable, is a member of $\mathcal{L}^1$ if $\rho_0(x,t) \in \mathcal{L}^1$. The integral representation thus defines a map $S_t : \mathcal{L}^1 \mapsto \mathcal{L}^1$.

The Liouville and diffusion equations are particular simple examples of PDEs since they are both linear equations in ρ. The construction of an associated dynamical system is possible because we happen to have the explicit solutions (2.60) and (2.66). In the chapter on linear dynamical systems we consider a more systematic way to analyze linear PDEs. When nonlinear PDEs are considered the analysis gets much more complicated but there are still cases where solutions and associated dynamical systems can be constructed. One prominent example is the KdV equation whose initial boundary value problem can be converted into a dynamical system when a method called the inverse scattering transform is employed. Details are given in Appendix D.2. Though the inverse scattering transform is fairly involved and has some serendipitous features, it provides a comprehensive solution to the initial value problem of the nonlinear KdV equation by applying a number of linear operations. The inverse scattering transform can also be applied to some other nonlinear PDEs, including the nonlinear Schrödinger equation and the Sine-Gordon equation. There are other serendipitous algorithms to solve PDEs but a general theory is lacking. The important point is

Every PDE has to be considered in its own right. The identification of the appropriate function space constitutes the first major task.

2.6 Observables and Parameters

In addition to variables that determine the state of a system there are other variables that are of interest or relevance. If these additional variables depend on the state of the system they are called observables. They are contingent quantities, contingent on the state. If these additional variables are independent of the state they are called parameters.

Observables. When introducing observables one has to distinguish between two functions. The first function is

$$A : X \mapsto \mathbb{R} \tag{2.67}$$

$$x \mapsto A(x) \tag{2.68}$$

The function A assigns to each point x in the state space the value $A(x) \in \mathbb{R}$. This definition follows the definition in classical physics. In quantum mechanics, an observable is a self-adjoint operator on the Hilbert space and does not fit this definition.

One might easily generalize the above definition to include functions into an arbitrary space Y. One may also include an explicit time dependence of the observable and define more generally

$$A : X \times T \mapsto Y \tag{2.69}$$

$$(x, t) \mapsto A(x, t) \tag{2.70}$$

The time-honored example of an observable from physics is the energy of a Hamiltonian system. In this case the function A is given by the Hamilton function $H(q, p, t)$. It assigns to each point in the state space and each time instant an energy value. Another example is the projection of the state vector onto some of its components, say $A(x_1, \ldots, x_N) = x_1$. Of course, not every conceivable function A corresponds to some observable with a well-defined measuring device. Also note that the concept of an observable does not require any specific structure of the state space.

To motivate the second function consider again the Hamilton function $H(q, p, t)$ and a particle that is initially at the point (q, p). At time t this particle will be at the point $(q_t(q, p), p_t(q, p))$ and will have the energy value $E_t(q, p) =$

$H(q_t(q, p), p_t(q, p), t)$, the value that the Hamilton function assigns to its current position. Thus given the function A and a dynamical system S_t one introduces the second function

$$f_t^A := A \circ S_t \tag{2.71}$$

Its values are given by $f_t^A(x) := A(S_t(x))$ and especially $f_0^A(x) = A(x)$. If the system starts out at x at time $t = 0$ then $f_t^A(x)$ is the value of the observable at time t. We will call both functions A and f_t^A observables and suppress the superscript 'A' when not necessary. If the function A depends on time then $f_t^A := A \circ (S_t \times I)$ with values $f_t^A(x) := A(S_t(x), t)$ and especially $f_0^A(x) = A(x, t = 0)$. It is important to distinguish between the two observables A and f_t^A which is best done by using two different symbols.

When viewed as a function on a function space, $A \mapsto f_t^A$, this function is linear, $f_t^{aA+bB} = af_t^A + bf_t^B$, positive, $f_t^A > 0$ if $A > 0$, and multiplicative, $f_t^{AB}(x) = f_t^A(x) f_T^B(x)$.

If the dynamical systems arises from an ODE with velocity field $u(x, t)$ one may differentiate f_t^A at fixed x which yields

$$\frac{df_t^A}{dt}(x) = \frac{DA}{Dt}(S_t(x), t) \tag{2.72}$$

where

$$\frac{DA}{Dt}(y, t) := (\partial_t + u^T \nabla_y) A(y, t) \tag{2.73}$$

is called the advective (or material or Eulerian) derivative. The notation on the right-hand side of (2.72) is to indicate that the function DA/Dt has to be evaluated at $y = S_t(x)$. There are thus two ways to calculate time changes of the observable. One can time differentiate the function $f_t^A(x)$ at fixed initial position x. This approach corresponds to the Lagrangian description in fluid dynamics and to the Heisenberg representation in quantum mechanics. Alternatively, one can take the advective derivative of the function $A(y, t)$ at $y = S_t(x)$. This approach corresponds to the Eulerian description in fluid dynamics and to the Schrodinger representation in quantum mechanics. In the first approach, the state x is fixed and the observable $f_t^A(x)$ evolves in time. In the second approach, the observable $A(x, t)$ is fixed and the state $S_t(x)$ evolves in time.

Dynamical maps S_t and observables f_t^A constitute the two basic ingredients of dynamical systems whose properties dynamical systems theory studies.

Constants of motion. A most important class of observables are constants of motion for which the value of the observable does not change in time

$$\frac{df_t^A}{dt} = 0 \tag{2.74}$$

In the context of ODEs such constants of motion are called first integrals of motion.

Consider the energy of a Hamiltonian system. Hamilton's equations imply

$$\begin{aligned}\frac{df_t^H}{dt} = \frac{DH}{Dt} &= \partial_t H + \partial_q H \dot{q} + \partial_p H \dot{p} \\ &= \partial_t H + \partial_q H \partial_p H - \partial_p H \partial_q H \\ &= \partial_t H \end{aligned} \tag{2.75}$$

The value of the energy does not change in time if the Hamilton function does not explicitly depend on time. The energy is then a constant of motion and the trajectories are constrained to the $(N-1)$-dimensional energy surface $\Gamma_E := \{(q, p) : H(q, p) = E\}$. The constants of motion of Hamiltonian systems are related to symmetries of the system. Energy conservation reflects the invariance under infinitesimal time translations.

Determining the constants of motion is an important part of the analysis of systems in physics and we have more to say about it in the chapter on integrable systems (3.6).

Flux laws. The Liouville and diffusion equations considered in the section on PDEs are all of the form

$$\partial_t \rho(x, t) + \nabla_x^T j(x, t) = 0 \tag{2.76}$$

with a flux vector $j = u\rho$ for the Liouville equation and $j = -D_0 \partial_x \rho$ for the diffusion equation. Many other PDEs such as the KdV equation and the Navier–Stokes equation can also be cast into the form (2.76). One says that the quantity of which ρ is the density is locally conserved. In many applications, there also appear source or sink terms on the right-hand side of (2.76).

Define the observable

$$A := \int_\Omega dx \rho(x) \tag{2.77}$$

where Ω is the time-independent spatial domain on which the PDE is defined. Then f_t^A changes according to

$$\begin{aligned}\frac{df_t^A}{dt} = \frac{d}{dt}\int_\Omega dx \rho(x, t) &= \int_\Omega dx \partial_t \rho(x, t) \\ &= -\int_\Omega dx \nabla_x^T j(x, t) \\ &= -\int_{\partial\Omega} ds n^T(s) j(s, t) \end{aligned} \tag{2.78}$$

on account of Gauss' theorem. Here $\partial\Omega$ is the (time-independent) surface enclosing the domain Ω and n its outward normal vector. If there is no flux through the enclosing surface, $n^T(s) j(s, t) = 0$, then the observable does not change and the observable is a constant of motion. One says the quantity of which ρ is the density is globally conserved.

Closure problem. In general, observables are many-to-one functions. Many states map onto the same value of the observable. In Hamiltonian dynamics, all states on the energy surface Γ_E map onto the single value E of the energy. The state determines the value of the observable but the value of the observable does not, in general, determine the state. This fact implies that the evolution of the observable is in general not autonomous. The initial value y_0 of an observable does not imply a unique value y_t of the observable at other times. This is because not all initial states in the pre-image $A^{-1}(y_0)$ that are compatible with a certain value y_0 of the observable at time $t = 0$ evolve to states that have the same value y_t at time t (see Fig. 2.8). More formally, introduce the function

$$R_t^A(y) := (A \circ S_t \circ A^{-1})(y) \tag{2.79}$$

for all $y \in Y$. If A is a bijection then R_t^A represents a dynamical system on Y equivalent to S_t. If the observable A is a many-to-one function then R_t^A might become a multi-valued function or, more properly, a function from Y into its power set $\mathcal{P}(Y)$. Only in the case that R_t^A is a single-valued function of y do all states $x \in A^{-1}(y)$ map onto states $S_t(x)$ that have the same value $y_t = A(S_t(x))$, and R_t^A constitutes a dynamical system on $Y = A(X)$. This condition must hold for all $y \in Y$ and all $t \in T$. As the many 'for all' conditions indicate these are quite restrictive conditions but can be met. One trivial example are constants of motion where R_t^A becomes the identity. A less trivial example derives from the baker transformation. The observable $A(x_1, x_2) = x_1$, which maps every state (x_1, x_2) onto its first component x_1, evolves according to the Bernoulli shift, i.e., R_t^A becomes the Bernoulli shift. The evolution of the observable $A(x_1, x_2) = x_2$, on the other hand, does not depend only on x_2 but also on x_1. Note that even if R_t^A constitutes a dynamical system its character might be quite different from the one of S_t. The baker transformation is invertible; the Bernoulli shift is not. If R_t^A is a dynamical system one says that one has a closed equation for the evolution of the observable. In general, however, R_t^A, is a multi-valued function and one faces a closure problem This closure problem is one of the major foundational issues in statistical mechanics. The important result is

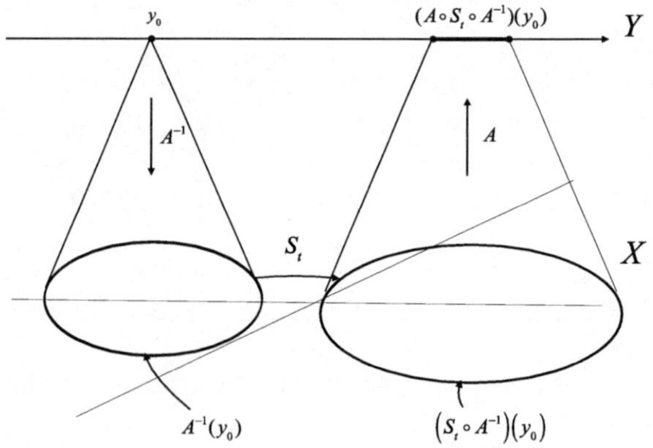

Fig. 2.8 The closure problem. The function $R_t^A(y) := (A \circ S_t \circ A^{-1})(y)$ is generally not a single-valued function

The time evolution of observables does not generally represent a dynamical system.

Symbolic dynamics. An important observable is obtained by coarse-graining the state space. Coarse-graining partitions the state space X into mutually disjoint cells or boxes B_k, $k = 0, 1, 2, \ldots$ such that

$$X = \cup_{k=0}^{\infty} B_k, \tag{2.80}$$

with $B_k \cap B_l = \emptyset$ if $k \neq l$. The index function

$$I(x) := \sum_{k=0}^{\infty} k\chi_{B_k}(x) \tag{2.81}$$

which assigns to each point the index of the box that x occupies is then an observable with associated

$$f_t^I(x) = I(S_t(x)) \tag{2.82}$$

which depends both on the dynamical system and the partition.

The observable generates for each point $x \in X$ and for any ordered sequence of times $t_0 < t_1 < \ldots$ an ordered sequence $i_0, i_1, \ldots,$ of numbers where i_n is the index of the box that $S_{t_n}(x)$ occupies. This is a one-sided sequence. If S_t is invertible one can generate two-sided sequences. This sequence of numbers is called a symbolic dynamics of S_t. The name symbolic dynamics originates from the practice to label the different cells of a finite partition by different symbols, say letters from an alphabet. Though the name includes the word 'dynamics', symbolic dynamics is generally not a dynamical system but an observable.

Studying its symbolic dynamics often provides interesting insights into the nature of a dynamical system. As an illustration consider the doubling map $F(x) = 2x \; mod \; 1$ on the interval $[0, 1[$. Consider the crudest partition of the state space, namely into the two intervals $B_0 = [0, 1/2[$ and $B_1 = [1/2, 1[$. Each initial condition x then generates a sequence $i_0 i_1 i_2 \ldots$ where i_k is either 0 or 1 depending on whether the kth iterate of x falls into B_0 or B_1. The symbolic dynamics generates the binary representation of the number x, with all the insights that go along with this representation.

Parameters. Parameters are variables that do not depend on the state of the system but characterize properties of the system. The frequency ω is a parameter of the harmonic oscillator. The parameter r characterizes the logistic map, the parameters (σ, r, b) the Lorenz equations, the velocity field $u(x, t)$ the Liouville equation and the diffusion coefficient D_0 the diffusion equation. These parameters remain fixed when the evolution of the system is considered. However, we will consider in Sect. 4.12 how the properties of a dynamical system change when its parameters are changed.

2.7 Invariant Sets and Orbits

Invariant sets and orbits are important tools in the analysis and visualization of dynamical systems.

Invariant sets. Graphically, an invariant set is a set that if you are in it and move with the flow you stay in it and if you outside of it and move with the flow you stay outside. More formally, given a dynamical system S_t, a set $A \subseteq X$ is called an invariant set if

$$S_t^{-1}(A) = A \tag{2.83}$$

for all $t \in T$. If A is an invariant set then $S_t(A) \subseteq A$. If $S_t(A) = A$ the set A is called strictly invariant. If $S_t(A) \subset A$ it is called positively invariant. If A is an invariant set then $X \setminus A$ is also an invariant set and hence $S_t(X \setminus A) \subseteq X \setminus A$. The state space X itself is an invariant set. If the system is invertible the state space is a strictly invariant set. If the system is non-invertible the state may be strictly invariant (as in the case of the Bernoulli shift) or only positively invariant (as in the case of the logistic map for $r < 4$ where the state space shrinks as time progresses).

Orbits. Consider the curve in state space that is traced out in time by $S_t(x)$ for fixed x. This curve defines the set

$$O(x) := \{S_t(x); t \in T\} \subseteq X \tag{2.84}$$

and is called the orbit through the point x. For invertible systems orbits are strictly invariant sets, $S_t(O(x)) = O(x)$, and the orbits of x and $S_t(x)$ coincide, $O(S_t(x)) = O(x)$ for all $t \in T$. The state space becomes the union of mutually disjunct orbits. For non-invertible systems the orbits satisfy $S_t(O(x)) \subseteq O(x)$ and also $O(S_t(x)) \subseteq O(x)$. When you are on an orbit you stay on the orbit. The orbits are, however, not invariant sets since they can merge. Points from outside $O(x)$ can enter $O(x)$. The union of all merging orbits becomes, however, an invariant set.

Orbits are sets. When plotting orbits in state space-time information may be retained by arrows indicating the direction of time and by tick marks indicating time intervals. Orbits with these accessories are often called trajectories and are the preferred way to visualize the behavior of dynamical systems, as opposed to plotting $S_t(x)$ as a function of t for fixed x, as is customary in the analysis of ODEs. Consider the harmonic oscillator governed by the ODE

$$\begin{aligned} \dot{q} &= \omega p \\ \dot{p} &= -\omega q \end{aligned} \tag{2.85}$$

and solved by

$$\begin{aligned} q_t &= p_0 \sin \omega t + q_0 \cos \omega t \\ p_t &= -q_0 \sin \omega t + p_0 \cos \omega t \end{aligned} \tag{2.86}$$

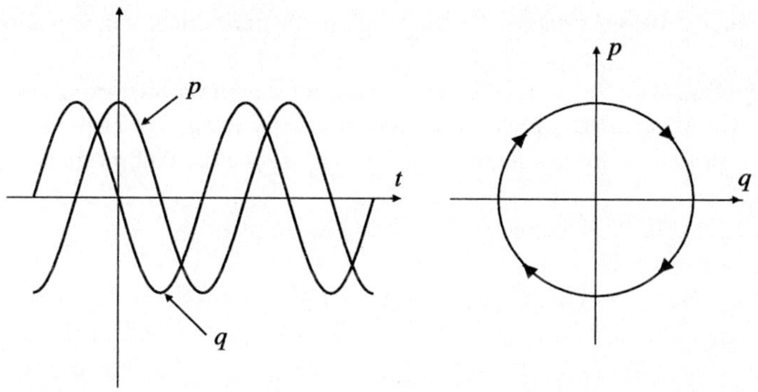

Fig. 2.9 Harmonic oscillator depicted in the traditional view showing the time dependence of the position and momentum and in the dynamical systems view showing the trajectory in state space

Instead of explicating that the position q and momentum p vary sinusoidally with time, dynamical systems theory presents the curve that is traced out by the solution in the (q, p)-plane, which is a circle (see Fig. 2.9) traversed in a clockwise direction. Dynamical systems theory asserts that these circles are the essential features of the harmonic oscillator. This representation of the solution is also called the geometric view of the system and shows, for example, that all initial conditions on the circle move along the same circle.

We made a distinction between a trajectory and an orbit. A trajectory is the curve traced out in state space by the function $S_t(x)$ for fixed x. The trajectory retains time information. The orbit is a set, sometimes invariant. Thus Kepler's first law "The orbit of every planet is an ellipse with the Sun at one of the two focuses" is a statement about an orbit, whereas his second law "A line joining a planet and the Sun sweeps out equal areas during equal intervals of time" is a statement about a trajectory. Our distinction between a trajectory and an orbit is not always made in the literature but we try to adhere to it. The important point is:

Dynamical systems theory aims at understanding dynamical systems by taking the geometric view, studying the orbits and trajectories in state space rather than plotting the evolution of states versus time.

Classification of orbits. What else can be said about orbits (or trajectories)? Consider first invertible continuous systems, $T = \mathbb{R}$, as most often realized by sets of ODEs. The first statement we can then make is that different orbits do not merge and do not intersect. Otherwise, $S_t(x)$ would not have a unique solution at the intersection. We can make another fairly remarkable statement. For fixed x the function $S_t(x)$ can only have one of the following three types of behavior

(1) $S_t(x) = x$ for all $t \in \mathbb{R}$
(2) (i) there exists a pair of times $t_1 \neq t_2$ such that $S_{t_1}(x) = S_{t_2}(x)$, and
    (ii) there exists at least one time t_3 such that $S_{t_3}(x) \neq x$
(3) $S_{t_1}(x) \neq S_{t_2}(x)$ for all $t_1 \neq t_2$

implying three different types of orbits: fixed points, periodic orbits, and non-periodic orbits.

Fixed points. The first type of behavior implies that the orbit consists of the single point $\{x\}$ which is called a fixed point. We generally use a subscript '$*$' to indicate that a point in state space is a fixed point. Thus, if x_* is a fixed point then $S_t(x_*) = x_*$ for all $t \in T$.

Periodic orbits. The second type of behavior implies

$$S_{t+t_2-t_1}(x) = (S_{t-t_1} \circ S_{t_2})(x) = (S_{t-t_1} \circ S_{t_1})(x) = S_t(x) \tag{2.87}$$

for all t. The function S_t is periodic with period $p := t_2 - t_1$. This argument is only valid for invertible systems since $t - t_1$ might not be in T for a non-invertible system. The second condition (ii) excludes the case that these periodic functions represent a fixed point. If p_1 and p_2 are periods then it is easily proven that $-p_1$ and $p_1 \pm p_2$ are also periods. Periods form a group under addition.

Non-periodic orbits. The third type of possible behavior is that $S_{t_1}(x) \neq S_{t_2}(x)$ for all $t_1 \neq t_2$. The orbits are thus curves that do not intersect and have no beginnings and no ends.

These three cases indeed cover all possible cases. The three cases may also be characterized by periods $p_0 = 0$, periods $0 < p_0 < \infty$, and period $p_0 = \infty$.

The above arguments also hold for invertible discrete systems, where the orbits do not trace out a continuous curve in state space but consist of sequences of points. Periodic orbits satisfy $S_{n+p} = S_n$ for all $n \in \mathbb{Z}$ with p being a non-zero integer. The case $p = 1$ corresponds to a fixed point. A periodic orbit of period p may be regarded as a fixed point of the p times iterated map. If the state space is finite then there exist only fixed points and periodic orbits.The important point is:

Invertible dynamical systems only exhibit three types of orbits: fixed points, periodic orbits and non-periodic orbits with no beginning and no end.

The situation becomes more varied for non-invertible systems, as they occur for iterative maps such as the Bernoulli shift or for PDEs such as the diffusion equation. There also exist fixed points, periodic orbits, and non-periodic orbits. However, orbits now have a beginning; they may also have an end; they may merge, but not cross. The Bernoulli shift exhibits these new features. The pre-image of any point of the Bernoulli shift consists of two points. The points of period p are the 2^p points with the binary representation $i_0 i_1 i_2 \ldots i_{p-1} i_0 i_1 i_2 \ldots i_{p-1} \ldots$. All rational starting points will become eventually periodic whereas irrational points are non-periodic.

All these different types of orbits are depicted in Fig. 2.10.

The situation becomes even more varied when observables are considered. The curves traced out by $f_t^A(x)$ in Y can intersect and show other peculiar behavior absent from trajectories.

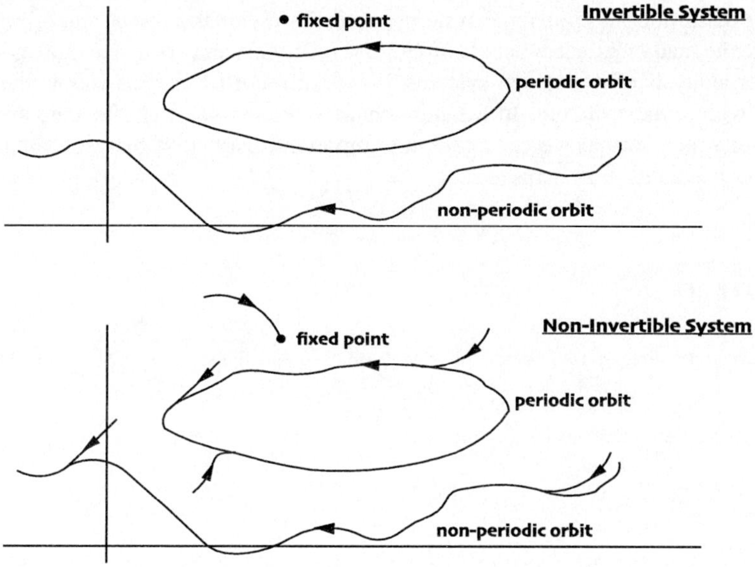

Fig. 2.10 Sketch of the possible orbits for invertible and non-invertible systems

2.8 Summary

In this chapter, we have given the abstract definition of a dynamical system, which consists of a state space and a one-parameter family of maps (in the autonomous case) that satisfies a composition rule making it compatible with the additive structure of time. Depending on the representation of time we distinguished between discrete and continuous and between invertible and non-invertible dynamical systems. There are no restrictions on the state space, as long as it consists of proper mathematical objects. Discrete dynamical systems are most often realized by iterative maps and continuous systems by sets of ordinary or partial differential equations. Most dynamical laws in quantitative sciences are of this type and hence represent dynamical systems.

In addition to the variables that describe the state of the system there are other variables of interest or relevance. If these additional variables depend on the state they are observables, if they do not they are parameters. The evolution of observables is in general not autonomous and does not constitute a dynamical system. Their evolution faces the closure problem. Dynamical maps and observables are the building blocks of dynamical systems theory.

Invariant sets and orbits are the most important tools for the analysis and visualization of dynamical systems. Orbits are invariant sets for invertible systems. Invertible systems exhibit only three types of orbits: fixed points, periodic orbits, and non-periodic orbits. Orbits do not intersect. For non-invertible systems orbits have a beginning and can merge but not intersect.

The definitions of dynamical systems, observables, invariant sets, and orbits only require the mathematical concepts of sets and functions, and are hence quite general. Further analysis of dynamical systems, however, requires that we endow the state space with some structure. In the next chapter, we assume that the state space is a vector space so that we can make the important distinction between linear and nonlinear dynamical systems.

References

Abramowitz, M., Stegun, I.A.: Handbook of Mathematical Functions. Dover Publications, Mineola (1965)

Berger, A.: Chaos and Chance. Walter de Gruyter (2001)

Lorenz, E.N.: Deterministic Nonperiodic Flow. J. Atmos. Sci. **20** (1963)

Chapter 3
Linear and Integrable Systems

3.1 Introduction

This chapter considers linear dynamical systems. The concept of linearity requires the state space to be a vector space. The elements of the state space can thus be added and multiplied by real (complex) numbers. The time evolution of linear systems is represented by a linear operator on this vector space. The prime example of a linear dynamical system is the time evolution of a quantum mechanical system according to the Schrödinger equation. Linear algebra provides powerful tools for the analysis of linear operators and hence of linear systems.

The two main results about linear dynamical systems are: First, linear systems can be decomposed into components which do not interact and the complete system is just a superposition of these components. The whole is just the sum of the parts. The second main result is: The time evolution of each component is extremely simple. Its amplitude varies sinusoidally in time, with a possible exponential decay or growth. A consequence of these two main results is that linear systems are predictable. Mathematically, these two results are obtained by an eigenanalysis where the evolution of the system is projected onto the eigenvectors of the linear operator. The properties of eigenvectors are covered by various spectral theorems. The most prominent spectral theorem is: The eigenvectors of a Hermitian operator on a finite-dimensional vector space form an orthonormal basis and the eigenvalues are real. We will present a systematic analysis of linear systems generated by finite sets of linear ODEs and of linear iterative maps. We also present two examples of linear PDEs which lead to dynamical systems on vector spaces of countably and non-countably infinite dimension.

Linear systems might not be very interesting in themselves but they do provide the background against which the more interesting nonlinear systems must be viewed. Despite their simple formal structure the explicit calculation of eigenvectors and eigenvalues might still be very complicated. Quantum mechanics is not simple, even the hydrogen atom has many subtleties. Oceanic tides which represent the ocean's near linear response to the gravitational pull of the sun and moon are predictable but

P. Müller, *Handbook of Dynamics and Probability*,
https://doi.org/10.1007/978-3-030-88486-4_3

very complicated phenomena due to the complicated topographical structure of the ocean basins.

Linearity is a powerful concept. It pervades much of our thinking. There is a tendency to linearize problems, which is often a sensible approach but care must be taken not to eliminate essential features, or "to throw out the baby with the bathwater". Indeed, the incentive for the development of dynamical systems theory was to provide tools for the understanding of nonlinear systems that cannot be linearized.

Since linear systems are so simple an important question is whether nonlinear systems can be turned into linear systems by a suitable coordinate transformation. There are certainly examples that such transformations exist. The most systematic investigation of this question has been done for Hamiltonian systems where one searches for transformations that turn the Hamilton function into an integrable form.

A second important question is what happens if one slightly perturbs a linear system. Again, the most distinct results have been obtained for Hamiltonian systems. One key result is the KAM theorem which demonstrates the intricate changes that occur when a small nonlinearity is added to a linear Hamiltonian system.

Once the state space is a vector space one can characterize a dynamical system also by the dimension of its state space. The degree of linearity and the dimensionality of the state space are two of the most important characteristics of dynamical systems.

3.2 Vector Spaces

Before starting the analysis of linear systems we recall some key concepts of linear algebra. More details can be found in the appendices.

Real (complex) vector spaces. A set X is called a vector space if its elements $x, y, \ldots$, called vectors, can be added such that $x + y \in X$ and multiplied by a scalar $a \in \mathbb{R}(\mathbb{C})$ such that $ax \in X$. The addition must be associative and commutative. There must exist a null vector 0 and to any vector x a negative vector $-x$ such that $x + (-x) = 0$. The scalar multiplication must be distributive and associative. If the scalar a is a real number the vector space is called a *real* vector space, if it is a complex number it is called a *complex* vector space.

Euclidean (unitary) vector spaces. Euclidean (unitary) vector spaces are real (complex) vector spaces X that are endowed with a scalar (inner) product $\langle x, y\rangle \in \mathbb{R}(\mathbb{C})$. Having a scalar product one can define the norm (or length) of the vector by

$$\| x \| :=(\langle x, x\rangle)^{1/2} \tag{3.1}$$

and the orthogonality of two vectors $x \perp y$ by $\langle x, y\rangle = 0$. A Euclidean (unitarian) vector space is thus also a normed space. A distance between two vectors can then be defined by

$$d(x, y):= \| x - y \| \tag{3.2}$$

which makes the Euclidean (unitarian) vector space also a metric space.

Bases and dimensions. Bases and dimensions are the most important concepts. One has to distinguish between finite, countably infinite, and non-countably infinite dimensions.

Given a finite set of vectors $B = \{e^{(1)}, \ldots, e^{(N)}\}$, one can form the *finite linear combinations*

$$x = \sum_{n=1}^{N} x_n e^{(n)} \tag{3.3}$$

If every vector $x \in X$ can be written as such a finite linear combination the set B of vectors is called *complete.* The set B of vectors is called linearly independent if $x = 0$ implies $x_n = 0$ for all $n = 1, \ldots, N$. If the set of vectors B is complete and linearly independent then the set B is called a *basis.* The representation (3.3) is then unique and the expansion coefficients x_n are called the *components* of the vector x. All bases have the same number of vectors, called the *dimension* of the vector space. A basis is called orthonormal if $\langle e^{(i)}, e^{(j)} \rangle = \delta_{ij}$. Such basis vectors have length one and are mutually orthogonal. For such finite-dimensional vector spaces all (nontrivial) scalar products and hence metrics and norms are equivalent in that they generate the same topology. One thus generally takes the canonical scalar product

$$\langle x, y \rangle = \sum_{i=1}^{N} x_i^* y_i = x^\dagger y \tag{3.4}$$

Given a countable set of vectors $B = \{e^{(1)}, e^{(2)}, \ldots\}$ one can form the infinite linear combinations

$$x = \sum_{n=1}^{\infty} x_n e^{(n)} \tag{3.5}$$

If X is a Euclidean (unitary) vector space it is endowed with a norm and one can assess whether or not these infinite linear combinations converge and hence are well defined. The definitions of completeness, independence, and basis then carry over to these infinite linear combinations. The dimension is countably infinite.

A typical example of a vector space with a countably infinite dimension is the set of all piecewise smooth complex-valued functions on the interval $[-\pi, \pi]$ of the real line. Such functions can be represented by a Fourier series, a countably infinite sum of sine and cosine functions. These sine and cosine functions form an orthonormal basis.

Progressing to vector spaces of even higher cardinality one considers sets $B = \{e^{(\omega)} | \omega \in \mathbb{R}\}$ and linear combinations of the form

$$x = \int_{\mathbb{R}} d\omega x(\omega) e^{(\omega)} \tag{3.6}$$

where the vector space must be endowed with a structure that such integrals can be defined. A typical example is the Fourier integral which represents all integrable functions $\mathbb{R} \mapsto \mathbb{C}$ by such an integral with the basis functions $e^{(\omega)} = e^{i\omega t}$. For such infinite-dimensional vector spaces there might exist non-equivalent topologies and a limit might exist in one topology but not in another. For an alternative definition of a basis in infinite-dimensional vector spaces see Appendix E.4.

Linear operators. A function or map $\tilde{A} : X \mapsto X$ is called a *linear operator* on the vector space X if

$$\tilde{A}(x+y) = \tilde{A}(x) + \tilde{A}(y) \tag{3.7}$$

for all $x, y \in X$ and

$$\tilde{A}(cx) = c\tilde{A}(x) \tag{3.8}$$

for all $c \in \mathbb{C}(\mathbb{R})$. If $\tilde{A}$ and $\tilde{B}$ are linear operators then the sum, $(\tilde{A}+\tilde{B})(x){:=}\tilde{A}(x) + \tilde{B}(x)$, and the product, $\tilde{A}\tilde{B}(x) = \tilde{A}(\tilde{B}(x))$ are also linear operators. The linear operators on a vector space form a (generally non-commutative) algebra.

If X is a finite-dimensional vector space and $(e^{(1)}, \ldots, e^{(N)})$ an orthonormal basis then a linear operator $\tilde{A}$ can be represented by a square matrix A such that $y = \tilde{A}(x)$ becomes

$$y_i = \sum_{j=1}^{N} A_{ij} x_j \tag{3.9}$$

where the matrix A has components

$$A_{ij} = \langle e^{(i)}i, \tilde{A}(e^{(j)})\rangle \tag{3.10}$$

and y_i and x_i are the components of the vectors y and x in that basis. The matrix representation of a linear operator depends on the choice of basis. In the following we do not always distinguish clearly between the operator and its matrix representation. Operators on vector spaces of countably infinite dimensions can similarly be represented by infinite matrices.

3.3 Linear Dynamical Systems

Here, we give the formal definition of linear dynamical systems, state some of their basic properties, and note the slightly different use of the term 'linear' in the theory of ODEs.

Linear dynamical systems. An autonomous dynamical system $\{X, (S_t)_{t \in T}\}$ is called a linear dynamical system if X is a vector space and S_t a linear operator for all $t \in T$. Thus, S_t must satisfy the superposition principle

$$S_t\left(x + x'\right) = S_t\left(x\right) + S_t\left(x'\right) \tag{3.11}$$

for all $x, x' \in X$ and all $t \in T$ and the homogeneity condition

$$S_t(cx) = cS_t(x) \tag{3.12}$$

for all $c \in \mathbb{R}(\mathbb{C})$, $x \in X$, and $t \in T$. Otherwise the system is called nonlinear. The above two conditions can be combined into the single requirement $S_t\left(cx + c'x'\right) = cS_t(x) + c'S_t\left(x'\right)$ for all $x, x' \in X$, all $t \in T$ and all $c, c' \in \mathbb{R}(\mathbb{C})$. Similarly, a non-autonomous dynamical system is called linear if $S_{t_2,t_1}\left(cx + c'x'\right) = cS_{t_2,t_1}(x) + c'S_{t_2,t_1}\left(x'\right)$.

The origin $x = 0$ is a fixed point of a linear system. Linear systems can have multiple fixed points but these cannot be isolated. Indeed, if $x_1 \neq x_2$ are two different fixed points then $c_1x_1 + c_2x_2$ are also fixed points for all $c_1, c_2 \in \mathbb{R}(\mathbb{C})$. Similarly, linear systems can have multiple periodic orbits but these cannot be isolated since for every periodic orbit $x(t)$ there exist neighboring periodic orbits $cx(t)$ for any $c \neq 0$.

If the state space has a finite dimension linear operators can be represented by a matrix A. Assuming this matrix to be time-independent one arrives at the linear dynamical system

$$S_n(x) = A^n x \tag{3.13}$$

in the discrete-time case, which corresponds to the iterative map $x_{n+1} = Ax_n$, and at the linear dynamical system

$$S_t(x) = e^{At}x \tag{3.14}$$

in the continuous-time case which correspond to the ODE $\dot{x}(t) = Ax(t)$.

Examples. Important examples of linear systems are the linear oscillator which is governed by the ODE

$$\begin{pmatrix} \dot{q}(t) \\ \dot{p}(t) \end{pmatrix} = \begin{pmatrix} 0 & \omega \\ -\omega & 0 \end{pmatrix} \begin{pmatrix} q(t) \\ p(t) \end{pmatrix} \tag{3.15}$$

where ω is the frequency of the oscillator and quantum systems which are governed by the Schrödinger equation

$$i\hbar\partial_t x(t) = H(x(t)) \tag{3.16}$$

where H is a self-adjoint operator called the Hamiltonian, $\hbar$ Planck's constant, and x an element of a Hilbert space. We also discussed the linear systems

$$S_t : \rho_0(x) \mapsto J_{-t}(x)\rho_0(U_{-t}(x)) \tag{3.17}$$

arising from the Liouville equation and

$$S_t : \rho_0(x) \mapsto \int_{\mathbb{R}} dx'(4\pi D_0 t)^{-1/2} \exp\left\{-(x - x')^2/4D_0 t\right\} \rho_0(x') \tag{3.18}$$

arising from the diffusion equation. All our other examples, the logistic map, the Ulam map, the doubling map, the Bernoulli shift, the baker transformation, and the Lorenz equations represent nonlinear systems.

Linear ODEs. The language of dynamical systems theory differs slightly from that of the theory of ODEs. There a set of ODEs is called linear if it is of the form

$$\dot{y}(t) = A(t)y(t) + f(t) \tag{3.19}$$

where $A(t)$ is a prescribed matrix and $f(t)$ a prescribed forcing vector. The ODE is called homogeneous if $f(t) = 0$, otherwise inhomogeneous. If the matrix A is independent of time the ODE is called an ODE with constant coefficients, otherwise an ODE with variable coefficients.

A solution of (3.19) can be obtained by 'variation of the constant' and results in

$$y(t_2) = e^{\int_{t_1}^{t_2} dt' A(t')} y(t_1) + e^{\int_{t_1}^{t_2} dt' A(t')} \int_{t_1}^{t_2} dt' e^{-\int_{t_1}^{t'} dt'' A(t'')} f(t') \tag{3.20}$$

The solution of the inhomogeneous problem consists of the general solution of the homogeneous problem (the first term) plus a particular solution $y_*(t)$ of the inhomogeneous problem (the second term). The function $x(t):=y(t) - y_*(t)$ thus satisfies the homogeneous ODE $\dot{x}(t) = A(t)x(t)$. If one considers this function or if the forcing term vanishes one thus obtains the solution manifold

$$x(t_2) = e^{\int_{t_1}^{t_2} dt' A(t')} x(t_1) \tag{3.21}$$

which corresponds to the linear non-autonomous dynamical system

$$S_{t_2,t_1}(x) = e^{\int_{t_1}^{t_2} dt' A(t')} x \tag{3.22}$$

Non-autonomous systems can be turned into an autonomous system by adding another variable $z = t$ with $\dot{z} = 1$. This transformation has its price. Here, it turns a set of linear ODEs with variable coefficients into a set of nonlinear ODEs with constant coefficients. If the coefficients of the matrix are constant we recover the linear autonomous dynamical system (3.14).

In the above discussion we have casually introduced the exponential of a matrix or linear operator. Formally, such exponentials are defined by

$$e^A := \sum_{n=0}^{\infty} \frac{A^n}{n!} \tag{3.23}$$

Though formally correct, this definition is not very insightful. The very essence of linear systems is obtained when one performs an eigenanalysis, when one expands the solution with respect to the eigenvectors of the linear operator.

3.4 Eigenanalysis

Linear dynamical systems are defined by a linear operator A. Explicit solutions for such linear dynamical systems can be obtained by an eigenanalysis or spectral analysis which expresses the solutions in terms of the eigenvectors of the linear operator. Here, we give a fairly general and detailed discussion of the eigenanalysis for finite dimensions where A is a constant $N \times N$ matrix which enters the linear ODE $\dot{x}(t) = Ax(t)$ or the linear iterative map $x_{n+1} = Ax_n$. Here one must distinguish between non-defective and defective matrices. For infinite dimensions we sketch the eigenanalysis of the diffusion equation.

Eigenvectors and eigenvalues. Eigenvectors $e^{(s)}$ and eigenvalues $\lambda^{(s)}$ of a linear operator A are the solutions of the equation

$$Ae^{(s)} = \lambda^{(s)} e^{(s)} \tag{3.24}$$

The set of all eigenvalues is called the spectrum of the operator. In case of finite dimensions, the eigenvalues are the zeros of the characteristic polynomial $det(A - \lambda I) = 0$ where I is the identity matrix. The eigenvectors $\hat{e}^{(s)}$ of the adjoint operator $A^\dagger$ are solutions of

$$A^\dagger \hat{e}^{(s)} = \kappa^{(s)} \hat{e}^{(s)} \tag{3.25}$$

which can also be written

$$\hat{e}^{(s)\dagger} A = \kappa^{(s)*} \hat{e}^{(s)\dagger} \tag{3.26}$$

For this reason the eigenvectors $\hat{e}^{(s)}$ are also called the left eigenvectors and the eigenvectors $e^{(s)}$ the right eigenvectors of the operator A. Their eigenvalues are related by $\lambda^{(s)*} = \kappa^{(s)}$.

Non-defective matrices. A matrix A is called non-defective if it has a complete set of independent right eigenvectors $e^{(s)}$, $s = 1, \ldots, N$ with eigenvalues $\lambda^{(s)}$. The left eigenvectors are then also complete and right and left eigenvectors can be normalized to satisfy the completeness relation

$$I = \sum_{s=1}^{N} e^{(s)} \hat{e}^{(s)\dagger} \tag{3.27}$$

The matrix A can then be written in its spectral representation

$$A = \sum_{s=1}^{N} \lambda^{(s)} e^{(s)} \hat{e}^{(s)\dagger} \tag{3.28}$$

Multiplying this completeness relation by x we obtain the expansion

$$x = \sum_{s=1}^{N} e^{(s)} \hat{e}^{(s)\dagger} x = \sum_{s=1}^{N} e^{(s)} a^{(s)} \tag{3.29}$$

where the components or amplitudes

$$a^{(s)} = \hat{e}^{(s)\dagger} x \tag{3.30}$$

are the projection of the vector x onto the left eigenvectors $\hat{e}^{(s)}$.

Linear ODEs. The solution of a set of coupled linear ODEs $\dot{x}(t) = Ax(t)$ with a non-defective matrix A can be obtained in various equivalent ways. One way is to substitute the expansion

$$x(t) = \sum_{s=1}^{N} a^{(s)}(t) e^{(s)} \tag{3.31}$$

into the ODE which results in

$$\sum_{s=1}^{N} (\dot{a}^{(s)}(t) - \lambda^{(s)} a^{(s)}(t)) e^{(s)} = 0 \tag{3.32}$$

Since the eigenvectors are independent this expression factors into

$$\dot{a}^{(s)}(t) - \lambda^{(s)} a^{(s)}(t) = 0 \quad s = 1, \ldots, N \tag{3.33}$$

with solution

$$a^{(s)}(t) = e^{\lambda^{(s)} t} a^{(s)}(0) \tag{3.34}$$

and hence

$$x(t) = \sum_{s=1}^{N} e^{\lambda^{(s)} t} a^{(s)}(0) e^{(s)} = \sum_{s=1}^{N} e^{\lambda^{(s)} t} e^{(s)} \hat{e}^{(s)\dagger} x(0) \tag{3.35}$$

The solution is just a linear superposition of N independent components where each component varies exponentially in time, an extremely simple result. The solution (3.35) also shows that the exponential of the matrix A can be written in the form

$$e^{At} = \sum_{s=1}^{N} e^{\lambda^{(s)} t} e^{(s)} \hat{e}^{(s)\dagger} \tag{3.36}$$

rather than the infinite sum (3.23).

The above derivation only requires the completeness of the eigenvectors. The eigenvectors of non-defective matrices also satisfy the orthogonality relation

$$\hat{e}^{(s)\dagger} e^{(s')} = \delta_{ss'} \tag{3.37}$$

One can thus derive the above result more economically by differentiating the amplitude with respect to time which gives

$$\begin{aligned}
\dot{a}^{(s)}(t) &= \hat{e}^{(s)\dagger} A x(t) \\
&= \lambda^{(s)} \hat{e}^{(s)\dagger} x(t) \\
&= \lambda^{(s)} \hat{e}^{(s)\dagger} \sum_{s'=1}^{N} e^{(s')} a^{(s')}(t) \\
&= \lambda^{(s)} \sum_{s'=1}^{N} a^{(s')}(t) \hat{e}^{(s)\dagger} e^{(s')} = \lambda^{(s)} a^{(s)}(t)
\end{aligned} \tag{3.38}$$

where both completeness and orthogonality have been invoked.

Yet another derivation makes use of the fact that the matrix E whose columns are the eigenvectors of the matrix A is an invertible matrix. This matrix induces an invertible coordinate transformation such that

$$D = E^{-1} A E \tag{3.39}$$

is a diagonal matrix with the eigenvalues being the diagonal elements. Thus, a set of coupled linear ODEs with a non-defective matrix can be diagonalized which decomposes the problem into a set of N uncoupled ODEs. This formulation is a special case of a more general result called singular value decomposition.

The above analysis simplifies for normal matrices, matrices that satisfy $A^\dagger A = AA^\dagger$, for which the right and left eigenvectors are identical and the matrix E becomes a unitary matrix. If the matrix A is Hermitian then the eigenvalues are real.

There are two technical points:

1. For ease of notation we assumed that all eigenvalues are non-degenerate. If an eigenvalue $\lambda^{(s)}$ is degenerate with degeneracy or algebraic multiplicity $d(s)$ then there exist $d(s)$ independent eigenvectors $(e^{(s,1)}, \ldots, e^{(s,d(s))})$ with this same eigenvalue. The expansion must then be written

$$x(t) = \sum_{s=1}^{M} \sum_{i=1}^{d(s)} a^{(s,i)}(t) e^{(s,i)} \tag{3.40}$$

and the analysis carried through with this expansion.

2. Even if the matrix A is real the eigenvalues may be complex. However, the eigenvalues, eigenvectors, and amplitudes then form complex conjugate pairs, say

$\lambda^{(1)} = \lambda^{(2)*}$, $e^{(1)} = e^{(2)*}$, $\hat{e}^{(1)} = \hat{e}^{(2)*}$ and $a^{(1)} = a^{(2)*}$. The real (subscript r) and imaginary (subscript i) parts of the amplitude then evolve according to

$$\begin{pmatrix} a_r^{(1)}(t) \\ a_i^{(1)}(t) \end{pmatrix} = e^{\lambda_r^{(1)} t} \begin{pmatrix} \cos \lambda_i^{(1)} t & -\sin \lambda_i^{(1)} t \\ \sin \lambda_i^{(1)} t & \cos \lambda_i^{(1)} t \end{pmatrix} \begin{pmatrix} a_r^{(1)}(0) \\ a_i^{(1)}(0) \end{pmatrix} \tag{3.41}$$

The exponential describes a growth or decay and the matrix a rotation by angle $\lambda_i^{(1)} t$ in the plane spanned by the two amplitudes $a_r^{(1)}$ and $a_i^{(1)}$. The contribution of such a complex conjugate pair to the vector x is given by

$$x(t) = e^{\lambda_r^{(1)} t} \cos \lambda_i^{(1)} t (e^{(1)} \hat{e}^{(1)\dagger} + e^{(1)*} \hat{e}^{(1)T}) x(0) + e^{\lambda_r^{(1)} t} \sin \lambda_i^{(1)} t (i e^{(1)} \hat{e}^{(1)\dagger} - i e^{(1)*} \hat{e}^{(1)T}) x(0) \tag{3.42}$$

The harmonic oscillator (3.15) is such a real problem with complex eigenvalues $\lambda^{(1,2)} = \pm i\omega$ and evolution

$$\begin{pmatrix} q(t) \\ p(t) \end{pmatrix} = \begin{pmatrix} \cos \omega t & \sin \omega t \\ -\sin \omega t & \cos \omega t \end{pmatrix} \begin{pmatrix} q(0) \\ p(0) \end{pmatrix} \tag{3.43}$$

In summary: In case of non-defective matrices, the components either grow or decay exponentially or spiral exponentially towards zero or infinity or are periodic. No other motion is possible. This result is explicitly demonstrated for two-dimensional flows.

Example: Two-dimensional flows. Two-dimensional flows are governed by the ODE $\dot{x} = Ax(t)$ where the matrix A is a 2×2 matrix. Since the matrix is assumed to be non-defective it has two independent eigenvectors $e^{(1)}$ and $e^{(2)}$ with eigenvalues

$$\lambda^{(1,2)} = \frac{1}{2}[tr(A) \pm \sqrt{tr(A)^2 - 4det(A)}] \tag{3.44}$$

If the eigenvalues are identical then one can choose two independent eigenvectors. If $tr(A)^2 < 4det(A)$ the eigenvalues form a complex conjugate pair. The possible behavior of such two-dimensional flows is shown in Fig. 3.1

Transient growth. The trajectories in Fig. 3.1 are drawn for a non-normal matrix. A subtle phenomena may occur for such non-normal matrices. If both eigenvalues are negative all trajectories will ultimately approach the origin. The distance from the origin might, however, grow for a finite amount of time before it decays to zero. This behavior is explicitly demonstrated in Fig. 3.2 for the non-normal matrix

$$A = \begin{pmatrix} -1 & 1 \\ 0 & a \end{pmatrix} \tag{3.45}$$

with $a \neq -1$. The eigenvalues are $\lambda^{(1)} = -1$ and $\lambda^{(2)} = a$ and the right and left eigenvectors are

$$e^{(1)} = \begin{pmatrix} 1 \\ 0 \end{pmatrix} \quad e^{(2)} = \begin{pmatrix} 1 \\ 1 + a \end{pmatrix} \tag{3.46}$$

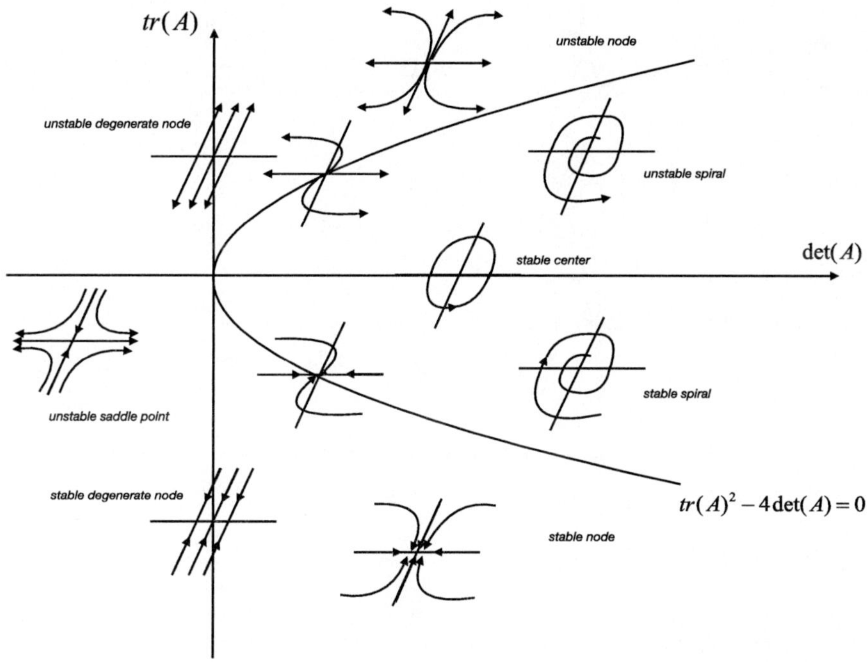

Fig. 3.1 Behavior of two-dimensional flows in the parameter space spanned by the determinant $det(A) = \lambda^{(1)}\lambda^{(2)}$ and the trace $tr(A) = \lambda^{(1)} + \lambda^{(2)}$ of the matrix A which are both real for real matrices. The inlays show typical trajectories in two-dimensional state space at the respective parameter values. If both eigenvalues are positive, all trajectories move away from the origin (zero) to infinity. The origin is called an unstable node. If one eigenvalue is positive and the other negative then the origin represents a saddle point where the trajectories approach the origin from one direction and get repelled to infinity in the other direction. If both eigenvalues are negative then all trajectories move towards the origin which is a stable node. If the eigenvalues form a complex conjugate pair then the trajectories spiral inward towards the origin (stable spiral) or outward towards infinity (unstable spiral) depending on whether the real part is smaller or larger than zero. If the two eigenvalues are purely imaginary then all orbits circle the origin with a fixed period $2\pi\lambda_i^{-1}$ (stable center). If one of the eigenvalues is zero ($det(A) = 0$) the trajectories move towards or away from a straight line of connected fixed points (stable or unstable degenerate node)

$$\hat{e}^{(1)} = \begin{pmatrix} 1 \\ -(1+a)^{-1} \end{pmatrix} \quad \hat{e}^{(2)} = \begin{pmatrix} 0 \\ (1+a)^{-1} \end{pmatrix} \tag{3.47}$$

The matrix thus has two non-orthogonal eigendirections, a fast and a slow eigendirection. All trajectories become parallel to the slow eigendirection as $t \to \infty$ and parallel to the fast eigendirection as $t \to -\infty$. The distance squared from the origin, or the energy in many contexts, changes according to

$$\frac{d}{dt}[(x_1)^2 + (x_2)^2] = -2(x_1)^2 + 2nx_1x_2 - 4(x_2)^2 \tag{3.48}$$

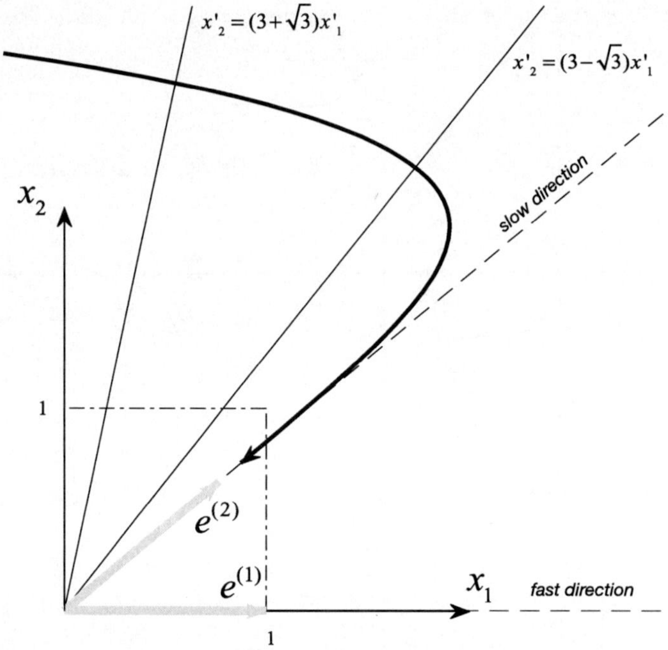

Fig. 3.2 Trajectory in a two-dimensional flow governed by the non-normal matrix $A = \begin{pmatrix} -1 & 1 \\ 0 & a \end{pmatrix}$ with $a = -1/6$. Within the cone the trajectory moves away from the origin

The derivative is zero on the two lines $x_2 = b_{+,-}x_1$ with $b_{+,-} = -(1 \pm \sqrt{1+4a})/(2a)$. If $-1/4 < a < 0$ these two lines delineate a cone in which the distance from the origin grows. Trajectories starting within this cone move away from the origin before they finally converge onto the origin. A trajectory for the case $a = -1/6$ is depicted in the Fig. 3.2. The effect is subtle, hardly discernible, but the distance to the origin is indeed smaller where the trajectory enters the cone than where it leaves the cone. We have chosen this example because it demonstrates the transition to defective matrices.

Defective matrices. If the matrix A is defective, the expansion (3.31) does not work since the set of eigenvectors is not complete, one then has to use generalized eigenvectors. Let us exemplify this approach for the simplest case of a 2×2 matrix which has one eigenvalue λ with algebraic multiplicity 2 and geometric multiplicity 1. Then there exist only one eigenvector $e^{(1)}$. One can of course choose any other independent vector $e^{(2)}$ to form a basis, even one that is perpendicular to $e^{(1)}$, but the solution takes a particular simple form if one chooses the vector $e^{(2)}$ from the null-space of $(A - \lambda I)^2$

$$(A - \lambda I)(A - \lambda I)e^{(2)} = 0 \tag{3.49}$$

and among these the specific vector that satisfies

$$(A - \lambda I)e^{(2)} = e^{(1)} \tag{3.50}$$

Then, the solution takes the form

$$x(t) = a^{(1)}(0)e^{\lambda t}e^{(1)} + a^{(2)}(0)e^{\lambda t}(e^{(2)} + e^{(1)}t) \tag{3.51}$$

Despite the extra term proportional to time t the asymptotic behavior of the solution is still determined by the eigenvalue λ, approaching the origin if $\lambda_r < 0$ and getting repelled to infinity if $\lambda_r > 0$. An example of a 2×2 defective matrix is the case $a = -1$ of (3.45) which is the matrix

$$A = \begin{pmatrix} -1 & 1 \\ 0 & -1 \end{pmatrix} \tag{3.52}$$

It has the eigenvalue $\lambda = -1$ with only one eigenvector $e^{(1)} = (1, 0)^T$. The vector $e^{(2)} = (0, 1)^T$ is then the one that satisfies (3.49) and (3.50) and the solution becomes $x(t) = e^{At}x(0)$ with

$$e^{At} = e^{-t}\begin{pmatrix} 1 & t \\ 0 & 1 \end{pmatrix} \tag{3.53}$$

The difference between the algebraic and geometric multiplicity of an eigenvalue $\lambda^{(s)}$ is called the defect. If the defect K is larger than one there appear terms proportional to $t^k e^{\lambda^{(s)}t}$ with $k = 1, \ldots, K$ in the solution.

The linear case also demonstrates that the character of the flow in state space changes as the parameters of the system, here the coefficients of the matrix, change. These changes are more systematically analyzed in Sect. 4.12 on bifurcations.

Iterative maps. When the eigenanalysis is applied to an iterative map $x_{n+1} = Ax_n$ then one obtains

$$x_n = \sum_{s=1}^{N} \lambda^{(s)n} e^{(s)} \hat{e}^{(s)\dagger} x_0 \tag{3.54}$$

The exponential $e^{\lambda^{(s)}t}$ in (3.35) is replaced by the nth power $\lambda^{(s)n}$. The behavior now depends on whether the real parts of the eigenvalues $\lambda^{(s)}$ are smaller than one or larger than one. Iterative maps with defective matrices require use of generalized eigenvectors, analogously to the time-continuous case.

The continuous and discrete results can be related. The iterative map

$$x_n = (I + \Delta t A)^n x_0 \tag{3.55}$$

converges to $x_t = e^{At}x_0$ in the limit $n \to \infty$, $\Delta t \to 0$ with $n\Delta t = t$, using Euler's formula $\lim_{n\to\infty}(1 + A/n)^n = e^A$.

Eigenanalyses can also be carried out for linear operators in vector spaces of infinite dimensions as they occur for PDEs. Here, we sketch the eigenanalysis for the diffusion equation on an interval and on the real line. The eigenvectors then become eigenfunctions, which are also called normal modes or eigenmodes.

Example: Diffusion equation on an interval. Consider the diffusion equation

$$\partial_t \rho(x,t) = \partial_x D_0 \partial_x \rho(x,t) \tag{3.56}$$

on the interval $[0,\pi]$ with prescribed boundary conditions $\rho(x=0,t)=\rho(x=\pi,t)=0$ for all t and prescribed initial condition $\rho(x,t=0)=\rho_0(x)$ and constant diffusion coefficient D_0. The eigenfunctions of the diffusion operator with these boundary conditions are $\sin nx, n=1,2,\ldots$ with eigenvalues $\lambda^{(n)} == -D_0 n^2$ resulting in the representations

$$\rho(x,t) = \sum_{n=1}^{\infty} c_n(t) \sin nx \tag{3.57}$$

with coefficients

$$c_n(t) = \frac{1}{\pi}\int_0^{\pi} dx\, \rho(x,t) \sin nx \tag{3.58}$$

Note that this eigendecomposition is not a Fourier series expansion which would result in a representation

$$\rho(x,t) = \frac{a_0(t)}{2} + \sum_{n=1}^{\infty} (a_n(t)\cos 2nx + b_n(t)\sin 2nx) \tag{3.59}$$

where the boundary conditions would impose the constraint $0 = a_0/2 + \sum_{n=1}^{\infty} a_n$ on the coefficients.

The time derivative of the coefficients is given by

$$\begin{aligned}
\dot{c}_n(t) &= \frac{1}{\pi}\int_0^{\pi} dx \partial_t \rho(x,t) \sin nx \\
&= \frac{1}{\pi}\int_0^{\pi} dx\, D_0 \partial_x \partial_x \rho(x,t) \sin nx \\
&= -\frac{1}{\pi}\int_0^{\pi} dx \sum_{n'}^{\infty} c_{n'}(t) D_0 n'^2 \sin n'x \sin nx \\
&= -D_0 n^2 c_n(t)
\end{aligned} \tag{3.60}$$

where the last line is obtained by using the orthogonality of the sine functions. Thus, by using an expansion into eigenfunctions, the PDE is turned into a countably infinite set of ODEs for the coefficients $c_n(t)$. This set of ODEs is easily solved by

$$c_n(t) = c_n(0)e^{-D_0 n^2 t} \tag{3.61}$$

and substitution into the eigenexpansion (3.57) leads to

$$\rho(x,t) = \sum_{n=1}^{\infty} c_n(0)e^{-D_0 n^2 t} \sin nx \tag{3.62}$$

Square-integrable functions

$$\rho \in \mathcal{L}^2 := \left\{ \rho : \int_0^{\pi} dx |\rho(x)|^2 < \infty \right\} \tag{3.63}$$

also have a representation as in (3.57) even if they are not twice differentiable. One can therefore define a dynamical system on $\mathcal{L}_2$, now in dynamical systems notation

$$\begin{aligned} S_t &: \mathcal{L}^2 \mapsto \mathcal{L}^2 \\ \rho_0(x) &\mapsto \rho_t(x) = \sum_{n=1}^{\infty} c_n(0)e^{-D_0 n^2 t} \sin nx \end{aligned} \tag{3.64}$$

What needs to be shown is that S_t is a map from $\mathcal{L}^2$ into $\mathcal{L}^2$ or, equivalently, that $\rho_t \in \mathcal{L}^2$ when $\rho_0 \in \mathcal{L}^2$. Since

$$\int_0^{\pi} dx |\rho(x,t)|^2 = \frac{2}{\pi} \sum_{n=1}^{\infty} |c_n(t)|^2 = \frac{2}{\pi} \sum_{n=1}^{\infty} |c_n(0)|^2 e^{-2D_0 n^2 t} \tag{3.65}$$

one has for $t > 0$ the upper bound

$$\int_0^{\pi} dx |\rho(x,t)|^2 < e^{-2D_0 t} \frac{2}{\pi} \sum_{n=1}^{\infty} |c_n(0)|^2 \tag{3.66}$$

and $\rho_t \in \mathcal{L}^2$ if $\rho_0 \in \mathcal{L}^2$. For $t < 0$ no such bound exist and the sum might diverge for finite negative times. Indeed, using a step function as initial condition there is no solution for negative times at all. Physically, diffusion ‘smooths’ the function in the forward time direction and ‘sharpens’ it in the backward time direction. A step function cannot be sharpened any further.

Example: Diffusion equation on the real line. In Sect. 2.5 we considered the diffusion equation

$$\partial_t \rho(x,t) = \partial_x D_0 \partial_x \rho(x,t) \tag{3.67}$$

on the real line and found that it induces a dynamical system $S_t : \mathcal{L}_1 \mapsto \mathcal{L}_1$. This system can also be analyzed in terms of the eigenfunctions of the operator $D_0 \partial_x \partial_x$ which are e^{ikx}, $k \in \mathbb{R}$ with eigenvalues $-D_0 k^2$. Expansion of any element $\rho \in \mathcal{L}_1$

with respect to these eigenfunctions is equivalent to the Fourier transform and results in the pair

$$\rho_t(x) = \frac{1}{2\pi}\int_{-\infty}^{\infty} dk \rho_t(k) e^{ikx} \tag{3.68}$$

$$\rho_t(k) = \int_{-\infty}^{\infty} dx \rho_t(x) e^{-ikx} \tag{3.69}$$

The time evolution of the Fourier amplitudes $\rho_t(k)$ is therefore governed by

$$\begin{aligned}
\partial_t \rho_t(k) &= \int_{-\infty}^{\infty} dx \partial_t \rho_t(x) e^{-ikx} \\
&= \int_{-\infty}^{\infty} dx D_0 [\partial_x \partial_x \rho_t(x)] e^{-ikx} \\
&= \int_{-\infty}^{\infty} dx D_0 \left[\partial_x \partial_x \frac{1}{2\pi} \int_{-\infty}^{\infty} dk' \rho_t(k') e^{ik'x} \right] e^{-ikx} \\
&= \int_{-\infty}^{\infty} dx [-D_0 k'^2] \frac{1}{2\pi} \int_{-\infty}^{\infty} dk' \rho_t(k') e^{ik'x} e^{-ikx} \\
&= -D_0 k^2 \rho_t(k)
\end{aligned} \tag{3.70}$$

The last equality sign is obtained through an integration of $e^{i(k'-k)x}$ over x which results in the δ-function $2\pi\delta(k'-k)$ and then a final integration over k'. The solution is

$$\rho_t(k) = e^{-D_0 k^2 t} \rho_0(k) \tag{3.71}$$

Again, the Fourier amplitudes evolve independently from each other.

Recombining, one obtains

$$\begin{aligned}
\rho_t(x) &= \frac{1}{2\pi}\int_{-\infty}^{\infty} dk \rho_t(k) e^{ikx} \\
&= \frac{1}{2\pi}\int_{-\infty}^{\infty} dk e^{-D_0 k^2 t} \rho_0(k) e^{ikx}
\end{aligned} \tag{3.72}$$

Substitution of

$$\rho_0(k) = \int_{-\infty}^{\partial} dx' \rho_0(x') e^{-ikx'} \tag{3.73}$$

leads to

$$\begin{aligned}
\rho_t(x) &= \int_{-\infty}^{\infty} dx' \frac{1}{2\pi} \int_{-\infty}^{\infty} dk e^{-D_0 k^2 t} e^{ikx} e^{-ikx'} \rho_0(x') \\
&= \int_{-\infty}^{\infty} dx' \frac{1}{\sqrt{4\pi D_0 t}} e^{-\frac{(x-x')^2}{4D_0 t}}
\end{aligned} \tag{3.74}$$

and we recover the result (2.66).
The construction of the solution by expansion into eigenmodes (=eigenvectors or eigenfunctions) is conceptually straightforward and easy to prove. The actual calculation of eigenmodes and eigenvalues and the projection of the initial state onto these eigenmodes might, however, be quite a computational challenge. Also, superposing all the eigenmodes might result in a fairly complicated time evolution. But there remains the important point:

Linear systems can be decomposed into independent parts, the eigenmodes of the system.

It is also stressed:

The independent parts of a linear dynamical system are the eigenmodes of the linear operator, not any parts of the physical system that is being represented. The independent parts of the diffusion equation are the sinusoidal eigenmodes, not parts of the diffusing substance.

3.5 Nonlinear Interactions

Eigenanalysis might also be useful for nonlinear problems if the problem contains an appropriate linear operator. The eigenvectors or eigenfunctions of the linear operator can then be used to represent the solution of the nonlinear system. Add a quadratic term to the diffusion equation

$$\partial_t \rho(x,t) = \partial_x D_0 \partial_x \rho(x,t) + \epsilon \rho^2(x,t) \tag{3.75}$$

on the real line. Performing the same operations as in (3.70) we find for the nonlinear term

$$\begin{aligned}
\partial_t \rho(k,t) + D_0 k^2 \rho(k,t) &= \epsilon \int_{-\infty}^{\infty} dx \rho^2(x,t) e^{-ikx} \\
&= \epsilon \int_{-\infty}^{\infty} dx \frac{1}{2\pi} \int_{-\infty}^{\infty} dk' \rho(k',t) e^{ik'x} \frac{1}{2\pi} \int_{-\infty}^{\infty} dk'' \rho(k'',t) e^{ik''x} e^{-ikx} \\
&= \epsilon \frac{1}{2\pi} \int_{-\infty}^{\infty} dk' \rho(k',t) \int_{-\infty}^{\infty} dk'' \rho(k'',t) \delta(k'+k''-k) \\
&= \epsilon \frac{1}{2\pi} \int_{-\infty}^{\infty} dk' \rho(k',t) \rho(k-k',t)
\end{aligned} \tag{3.76}$$

Fourier amplitudes at wavenumber k' interact with the Fourier amplitudes at $k'' = k - k'$ to affect the Fourier amplitude at k. To interact the wavenumber must thus satisfy the resonance condition $k = k' + k''$. A third order nonlinearity would imply that Fourier components k' and k'' interact with amplitudes $k''' = k - k' - k''$ to affect the Fourier amplitude k. The important point apart from these technical details is: For nonlinear systems the amplitudes do not evolve independently but interact.

3.6 Integrable Systems

The simplicity of linear systems raises the important question of whether or not a nonlinear system can be transformed into a linear system by a suitable coordinate transformation. There certainly exist examples that this is indeed possible. When discussing ordinary differential equations we already noticed that the inhomogeneous linear ODE $\dot{y} = Ay + f$ can be turned into a homogeneous and hence proper linear system by subtracting out a particular solution. A slightly less trivial example from the theory of ODEs is the equation

$$\dot{y}(t) = A(t)y(t) + f(t)y^n(t) \tag{3.77}$$

which becomes a linear ODE

$$\dot{x}(t) = (1-n)[A(t)x(t) + f(t)] \tag{3.78}$$

if one substitutes $x = y^{1-n}$.

The question of whether systems can be linearized by suitable coordinate transformations has most intensely been studied for Hamiltonian systems. There one searches for transformation that renders the Hamilton function 'integrable'. The concept of integrability is easily exemplified by a set of N uncoupled harmonic oscillators.

Uncoupled harmonic oscillators. A set of N uncoupled harmonic oscillators is a linear system governed by the Hamilton function

$$H(q_1, \ldots, q_N, p_1, \ldots, p_N) = \frac{1}{2}\sum_{n=1}^{N}(p_n^2 + \omega_n^2 q_n^2) \tag{3.79}$$

with frequencies $\omega_n \geq 0$. Introduce action $J_n \geq 0$ and angle θ_n variables by the transformation

$$q_n = \sqrt{\frac{2J_n}{\omega_n}} \sin\theta_n \tag{3.80}$$

$$p_n = \sqrt{2\omega_n J_n} \cos\theta_n \tag{3.81}$$

which mimic polar coordinates and inverse transformation

$$J_n = \frac{1}{2\omega_n}(p_n^2 + \omega_n^2 q_n^2) \tag{3.82}$$

$$\theta_n = \arctan \frac{\omega_n q_n}{p_n} \tag{3.83}$$

Under this (canonical) transformation the Hamilton function transforms into

$$H(\theta_1, \ldots, \theta_N, J_1, \ldots, J_N) = \sum_{n=1}^{N} \omega_n J_n \tag{3.84}$$

and Hamilton's equations become

$$\dot{\theta}_n = \partial H/\partial J_n = \omega_n \tag{3.85}$$

$$\dot{J}_n = -\partial H/\partial \theta_n = 0 \tag{3.86}$$

which are explicitly solved by

$$J_n(t) = J_n(0) \tag{3.87}$$

$$\theta_n(t) = \omega_n t + \theta_n(0) \tag{3.88}$$

The actions are constant. The angles evolve linearly in time. An extremely simple result. This transformation is possible because the set of N uncoupled harmonic oscillators has N constants of motions, the energies $E_n = \omega_n J_n$ of each harmonic oscillator.

Integrable systems. With this example in mind, one calls a Hamiltonian system $H(q, p)$ with N degrees of freedom integrable if there exists a (single-valued, analytic) canonical transformation to a set of N angular variables θ_n and N action variables J_n such that the Hamiltonian becomes a function of the action variables J_n only, $H(J_1, \ldots, J_N)$. Frequencies are then defined by $\omega_n{:=}\partial H/\partial J_n$. The set of uncoupled harmonic oscillators is a degenerate case in the sense that the frequencies are constants, whereas in the general case the frequencies may depend on the values of the action variables, $\partial(\omega)/\partial(J) \neq 0$. Also note that the transformation from the (q, p) variables to the (J, θ) variables turns a set of linear homogeneous differential equations into a set of linear inhomogeneous equation.

Integrable systems have N (single-valued, analytic, global) constants of motion, restricting their motion to a N-dimensional subspace of the $2N$-dimensional phase space. The qualifiers single-valued, analytic, and global are necessary since every Hamiltonian system with N degrees of freedom has $2N - 1$ constants of motion. They can be obtained by eliminating time t from the $2N$ solutions $q_t(q, p)$ and $p_t(q, p)$. The intersection of these $2N - 1$ constants of motion defines the orbit of the solution in state space. However, most of these constants of motion, except the energy, are generally multivalued, pathological quantities which do not constrain the motion in any sensible way. This is already the case for our set of N uncoupled

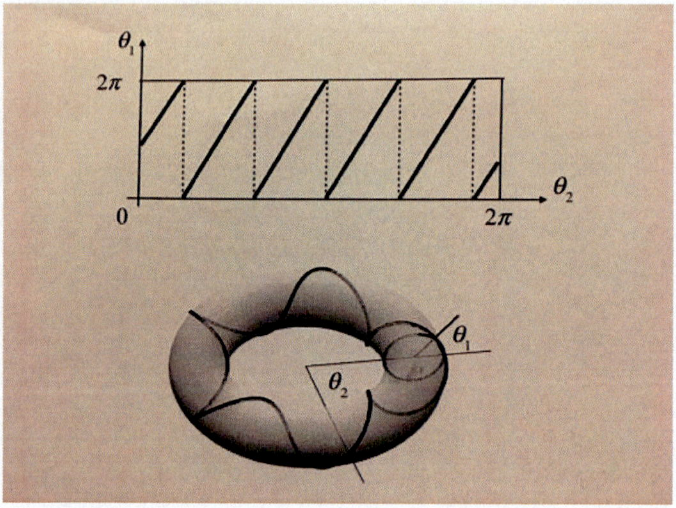

Fig. 3.3 A trajectory of two uncoupled harmonic oscillators on the plane and on the torus

harmonic oscillators. Eliminating t from the solution for the angle variables gives the $N-1$ additional constants of motion

$$A_n(J,\theta) = \theta_n - \frac{\omega_n}{\omega_1}\theta_1 \tag{3.89}$$

Due to the arctan-function in the definition (3.83) of θ_n these additional constants of motion are, however, multivalued functions and can only be implemented locally but not globally.

If the motion is spatially bounded then the angular variables are indeed angles and the frequencies are angular frequencies. If $N = 1$ the motion proceeds along a circle. If $N = 2$ the trajectories can be represented in the (θ_1, θ_2)-plane, as depicted at the top of Fig. 3.3. The slope of the trajectory is given by ω_2/ω_1. The periodicity of the motion is, however, more elegantly accounted for by representing the motion on the torus obtained by first joining the top and bottom of the square to form a tube and then joining the two ends of the tube to form a torus, depicted at the bottom of Fig. 3.3. Formally, this torus represents the Cartesian product of two circles. If the ratio of the two frequencies is a rational number then all trajectories on the torus are closed or periodic; otherwise the trajectories are quasi-periodic and each trajectory is (topologically) dense on the torus. This fact is often referred to as: 'Each trajectory fills out the torus', though this language is not correct because a line cannot fill out a plane. If one views the torus as a wheel with a tire, one may interpret the radius of the wheel as the total energy and the radius of tire as the energy of the first oscillator. Varying the energy of the first oscillator while keeping the total energy constant then leads to a succession of tires within tires.

These simple results carry over to $N > 2$. The motion is then confined to a torus of dimension N, which is the Cartesian product of N circles, which we cannot visualize anymore. The trajectories on the torus are multiply periodic. More precisely: the trajectories are quasi-periodic if $\omega^T k \neq 0$ for all vectors of integers $k = (k_1, \ldots, k_N)^T$ and the trajectories are closed or periodic if there exist at least one vector $k \neq 0$ such that $\omega^T k = 0$, i.e., if the frequencies are linearly dependent. If the integrable system represents a set of N uncoupled oscillators with given frequencies then all trajectories are either periodic or non-periodic. This is a peculiarity of the system of uncoupled oscillators. In general the frequencies depend on the angular variables, $\partial(\omega)/\partial(J) \neq 0$, and one encounters the more generic situation that in the neighborhood of every torus with periodic orbits there lies a torus with non-periodic orbits.

For a long time, physicists have searched for the magic transformation that would turn the Hamilton function into its integrable form. At the end of the nineteenth century they realized that 'most' interesting systems, starting from the celebrated three-body problem, are not integrable. The reason that these Hamiltonian systems are not integrable is that they generally have only one (single-valued, analytic, global) constant of motion, namely the energy. The motion is thus constrained to a $2N-1$-dimensional surface in phase space. Furthermore, the trajectories of ergodic systems, which we will discuss later, 'fill out' the energy surface. Filling out the $2N-1$-dimensional energy surface (as ergodic systems do) is quite different from being constrained to a N-dimensional subspace (as integrable systems are), especially for $N = O(10^{23})$.

The preceding paragraph is just a qualitative description. There exists theories and theorems that quantify the described behavior. An example is a theorem by Siegel (1941). It considers Hamilton functions that can be Taylor-expanded about the equilibrium point ($q = 0$, $p = 0$) in the following way

$$H(q, p) = \frac{1}{2}\sum_{n=1}^{N}(p_n^2 + \omega_n^2 q_n^2) + \sum_{k_1,\ldots,k_{2N}}^{\infty} c_{k_1,\ldots,k_N,k_{N+1},\ldots,k_{2N}} q_1^{k_1} \cdot \ldots q_N^{k_N} \cdot p_1^{k_{N+1}} \cdot \ldots p_N^{k_{2N}} \tag{3.90}$$

where the first sum represents a set of uncoupled oscillators and the second sum quadratic and higher order interaction terms. Such Hamilton functions correspond to points in the infinite-dimensional space $\mathcal{C}$ spanned by all expansion coefficients c_k where the index k comprises $2N$ integers that each run from zero to infinity. One can introduce a topology in this space and then prove the theorem: In every neighborhood of a Hamilton function (whether integrable or not), there exist a non-integrable Hamilton function. The set of non-integrable Hamiltonians is dense in the space $\mathcal{C}$.

3.7 The KAM Theorem

An important question is what happens if a small perturbation is added to an integrable system. Fermi, Pasta, and Ulam (1989) addressed this question.

The FPU system. Fermi, Pasta, and Ulam considered a one-dimensional chain of $N + 1$ particles with nonlinear interactions between nearest neighbors (see Fig. 3.4). The Hamilton function of their α-model is given by

$$H(q, p) = \sum_{n=1}^{N-1} \frac{1}{2} p_n^2 + \sum_{n=0}^{N-1} \frac{1}{2} (q_{n+1} - q_n)^2 + \frac{\alpha}{3} \sum_{n=0}^{N-1} (q_{n+1} - q_n)^3 \tag{3.91}$$

where $q_n(t)$ is the displacement of the nth particle from its equilibrium position. The second term describes a linear restoring force and the cubic term a quadratic restoring force. The boundaries are fixed, $q_0(t) = q_N(t) = 0$ and $p_0(t) = p_N(t) = 0$ for all times. Only $N - 1$ particles are moving. The mass and linear spring constant are set to one since it is the dependence on the quadratic force, characterized by the parameter α, that is of interest. There is also a β-model which has cubic instead of quadratic forces.

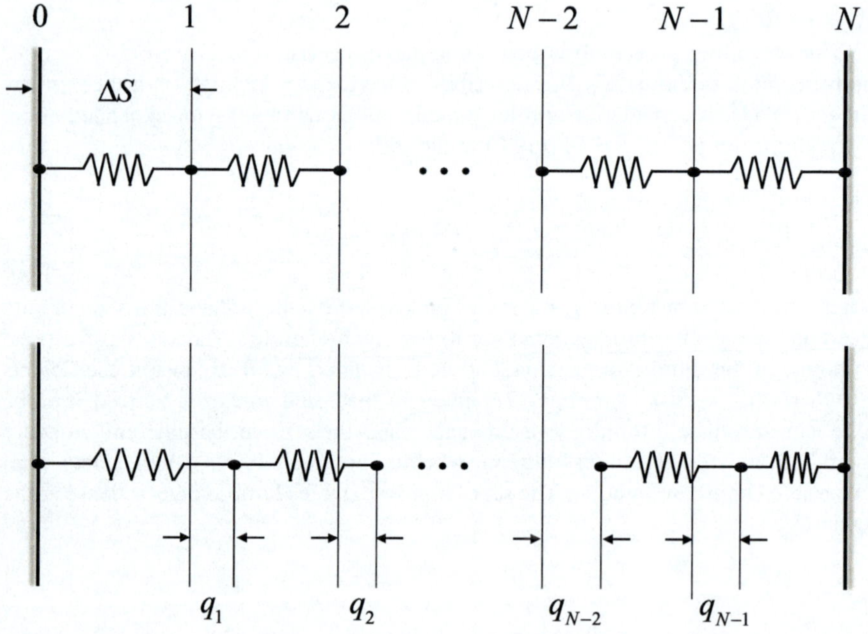

Fig. 3.4 Sketch of the FPU system consisting of $N - 1$ moving particles with nonlinear forces acting between neighboring particles

Instead of the displacements q_n and their corresponding momenta p_n FPU found it useful to expand these variables with respect to the $N-1$ eigenvectors or normal modes of the linear problem, resulting in the representation

$$q_n = \sum_{s=1}^{N-1} Q^{(s)} \sqrt{\frac{2}{N}} \sin \frac{sn\pi}{N} \tag{3.92}$$

$$p_n = \sum_{s=1}^{N-1} P^{(s)} \sqrt{\frac{2}{N}} \sin \frac{sn\pi}{N} \tag{3.93}$$

where $Q^{(s)}$ and $P^{(s)}$ are the normal mode amplitudes. The superscript s counts the eigenvectors. The Hamilton function then takes the form

$$H(Q, P) = \sum_{s=1}^{N-1} \frac{1}{2}[P^{(s)2} + \omega^{(s)2} Q^{(s)2}] + \alpha H_3(Q^{(1)}, \ldots, Q^{(N-1)}) \tag{3.94}$$

The first term represents a set of $N-1$ uncoupled harmonic oscillators with frequencies

$$\omega^{(s)} = 2 \sin \frac{s}{2} \quad s = 1, \ldots, N-1 \tag{3.95}$$

and the second term H_3 the quadratic interactions. Its precise form is not relevant for what follows. The expansion into normal modes transforms the problem from particles interacting with nearest neighbors to interacting modes (phonons) that do not have a characteristic range in normal mode space. Without the quadratic force the system would be just a set of $N-1$ uncoupled oscillators.

Fermi, Ulam and Pasta solved the equations for the time evolution of the normal mode amplitudes numerically for $N = 32$. They started out with most of the energy in just the gravest mode $Q^{(1)}(0) = 1$ and $P^{(1)}(0) = 0$. As time developed energy spread into all the other modes and after a while the energy became thoroughly mixed, as in thermal equilibrium. This was the expected result, but they continued their numerical integration nevertheless, only to find that after some more time the system returned to nearly its initial state. The nonlinear FPU system exhibits quasi-periodic behavior, in much the same way as the linear system does. This was quite a surprising result.

The KAM theorem. The counter-intuitive result of Fermi, Ulam, and Pasta later found its theoretical explanation in the KAM theorem (Kolmogorov, Arnold and Moser; see Arnold 1981, pp. 399–415) which is one of the backbones of contemporary dynamical systems theory. The KAM theorem considers systems with N degrees of freedom that are governed by a Hamilton function of the form

$$H(J, \theta) = H_0(J, \theta) + \epsilon \Delta H(J.\theta) \tag{3.96}$$

The action and angle variables are N-dimensional vectors. The Hamilton function is supposed to satisfy the following conditions:

1. The unperturbed Hamilton function is integrable, $H_0(J, \theta) = H_0(J)$. It has frequencies $\omega = \partial H/\partial J$.
2. The unperturbed Hamilton function satisfies the non-degeneracy condition $\partial(\omega)/\partial(J) \neq 0$.
3. The perturbation Hamilton function is periodic in the angle variable, $\Delta H(J, \theta + 2\pi) = \Delta H(J, \theta)$
4. The parameter ϵ is sufficiently small.

Note that the Hamilton function of the FPU system is of the required form, with the parameter ϵ corresponding to the interaction coefficient α but that it does not satisfy the non-degeneracy condition.

The motion of the unperturbed system is confined to an N-dimensional torus since the energy of each normal mode is conserved. When the interactions are switched on the state space expands to $2N - 1$ dimensions since only the total energy is conserved. The expanded state space is made up of all the tori of possible linear systems that have the same total energy. In the $N = 2$ example of Fig. 3.3, this expanded state space is the interior and surface of the tire that touches the origin. The KAM theorem states: when the system is perturbed then some of these tori survive (get only distorted), whereas others are destroyed. The ones that survive are those with quasi-periodic motion; the ones that get destroyed are those with periodic motions. The motion on the distorted tori is also quasi-periodic motion albeit with different frequencies due to the non-degeneracy condition (2). As an aside: it is in general a nontrivial problem to assess whether or not a set of frequencies is linearly dependent and leads to periodic or quasi-periodic motions. Are there any values of N such that the frequencies $\omega^{(s)} = 2 \sin s/2$, $s = 1, \ldots, N - 1$ of the FPU system lead to a periodic motion?

The tori have interesting topological and measure-theoretical properties. The tori that survive form a set that is nowhere dense in state space but their union carries most of the Lebesgue measure of the state space. The complementary set of destroyed tori is dense but its measure goes to zero as $\epsilon \to 0$. The KAM theorem also specifies how small ϵ has to be for the above statements to be true.

A nowhere dense set of finite measure. To demonstrate the intricacies of the KAM theorem we note that the distinction between destroyed and distorted tori is not simply analogous to the difference between rational and irrational numbers. Though the irrational numbers carry the Lebesgue measure, whereas the rational numbers have zero Lebesgue measure both are dense. To arrive at an analogy one has to do a more involved construction such as: Enumerate all rationals in the unit interval $[0, 1]$ as $r_n, n = 1, \ldots$. Around each such rational number choose an open interval of length $1/3^n$. Subtract all such intervals from the unit interval. The remaining set K is nowhere dense since it is closed and has an empty interior. It does not contain any open intervals. It has, however, at least the length $1/2$ since the subtracted intervals of length $1/3^n$ have at most the length $1/2$, at most, because of the possible overlap.

The KAM theorem shows what happens to a (Hamiltonian) system when its dynamics changes from linear to nonlinear. The intricate changes found gave rise to many concepts of modern dynamical systems theory. This historical development is traced in Ford (1992). The KAM theorem also enters our discussion of the ergodic justification of equilibrium statistical mechanics.

3.8 Dimensions of Dynamical Systems

Once the state space is a vector space one can not only classify dynamical systems as to whether they are linear or nonlinear but also as to the dimension of the state space. Figure 3.5 shows this classification for the examples considered so far.

In classical physics, the dimensionality of the state space is related to the number of degrees of freedom which is the number of independent generalized coordinates. A particle that can move freely in three-dimensional physical space has therefore 3 degrees of freedom. The state space consists of the three space coordinates and of the three momentum components and is 6-dimensional.

In physics, the motion is often constrained by conservation laws (or integrals of motion). This fact can be used to reduce the dimensionality of the problem. Consider the classical Kepler or two-body problem. The original two-body problem is a 12-dimensional problem. First, one separates the relative motion of the two bodies

Dimension of state space	linear	nonlinear
1		Doubling map Bernoulli shift Logistic map
2	Harmonic oscillator	Baker transformation
3		Lorenz equations
64		FPU system
$O(10^{23})$		N-particle system
∞	Liouville equation Diffusion equation Schrödinger equation	Navier-Stokes equations

Fig. 3.5 Classification of dynamical systems

from the (trivial) uniform translation of the complete system. This step reduces the dimensionality to 6 and one arrives at the Kepler problem proper, the motion of a mass point in a prescribed gravitational field. One may then impose the conservation of angular momentum. Conservation of its direction (two constraints) then reduces the problem to four dimensions, to the motion in a plane perpendicular to the specified angular momentum vector. Conservation of total energy (one constraint) and of the magnitude of the angular momentum (one constraint) then gives a further reduction by two dimension and one arrives at a two-dimensional problem, usually formulated as a set of two first-order ODEs for the polar coordinates of the orbit. There exists one more conservation law, namely the conservation of the Runge-Lenz or eccentricity vector which reduces the problem by yet another dimension. It is up to the researcher to define the state space in which he or she wants to study the Kepler problem, according to computational efficiency, physical insight or some other criteria.

Because of the existence of conservation laws one should explicitly define the state space of the dynamical system. Given a time-independent Hamilton function $H(q, p)$ for a N-particle system one should explicitly state whether one considers the corresponding dynamical system on a 6N-dimensional state space or on a (6N-1)-dimensional hyper-surface defined by a specific value of the (conserved) energy. This distinction is not always made and can lead to some confusion. The system on the (6N-1)-dimensional state space might have properties that the system on the 6N-dimensional state space does not have.

Note also that, for any given physical problem, the number of conservation laws is not automatically given, but that these laws need to be discovered. The conservation of potential vorticity in hydrodynamics was actually discovered 200 years after the foundations of fluid dynamics were established. The reason for this late discovery is the fact that conservation of potential vorticity corresponds to the fluid system being invariant under infinitesimal particle relabeling transformations which can be applied to continuous systems but has no counterpart in discrete particle systems.

Dimension and degree of nonlinearity require the state space to be a vector space. If it is not one might still assign a number to the system that is the number of independent variables that are necessary to define the state of the system. This number might serve as a substitute for dimension. The important point is

If defined, the dimension and the degree of nonlinearity are two of the most important properties of a dynamical system.

Linear systems are conceptually straightforward though actual calculation may become complicated for high-dimensional systems. Such is the case for quantum mechanical systems. Note that we assigned the dimensionality infinity to quantum mechanical systems in the table, though there are of course also finite-dimensional quantum mechanical systems such as spin systems. Oceanic tides are another example. They represent the ocean's response to the gravitational pull of the sun and moon and are hard to calculate because of the complicated shape of the ocean basins though they represent a (nearly) linear response, as can be seen by the superposition of the lunar and solar tides.

Dynamical systems theory aims at describing the complete flow in state space rather than specific trajectories. This aim requires the solution for every possible initial state, a tall order for high dimensional and nonlinear systems. From a dynamical systems point of view, the $N \gg 1$ particle Hamiltonian systems and fluid dynamics are thus the most challenging systems. For these reasons, most of dynamical systems theory has focused on low-dimensional systems. Indeed, most of systems that have been studied have a dimension smaller than four. Although explicit results are mostly obtained for such low-dimensional systems, the conceptual approach taken by dynamical systems theory is useful for higher dimensional systems as well.

3.9 Summary

This chapter introduced the important concept of a linear dynamical system. This concept requires the state space to be a vector space. Linear systems are then characterized by a linear operator. Linear systems are somewhat uninteresting from a dynamical systems point of view. Their evolution can be projected onto the eigenvectors of the linear operator. For each projection the amplitudes exhibit a very limited kind of behavior, they vary sinusoidally in time with a possible added exponential growth or decay. These projections can then be recombined. This is the superposition principle for linear systems. The whole is just the sum of its parts. Technically, these results are obtained by an eigenanalysis, as we demonstrated explicitly for finite sets of ODEs and iterative maps and the diffusion equation. The actual calculation of eigenvalues and eigenvectors might be a rather cumbersome task in higher dimensions, as every student of quantum mechanics can attest to. Linear systems are predictable. Tides, the linear response to the suns and moon's gravitational pull are predictable for a long time; the weather, the atmosphere's nonlinear response to the sun's radiation is only predictable for a few days. The degree of nonlinearity and the dimension of the state space are two of the most important characteristics of a dynamical system.

As we will discuss in the next chapters nonlinear system can exhibit much more varied behavior. The components of nonlinear systems interact with each other and everything depends on everything else. The whole is more than its parts.

References

Ford, J.: The Fermi-Pasta-Ulam problem: paradox turns discovery. Phys. Rep. **213**(5), 271–310 (1992)

Siegel, C.I.: On the integrals of canonical systems. Ann. Math. **42**, 806–822 (1941)

Fermi, E., Pasta, J., Ulam, S.: Studies in nonlinear problems, I. Los Alamos Report LA-1940, (1955). Article is reproduced in: Newell, A.C., (ed.) Nonlinear Wave Motion, Am. Mathematical Soc., (1974) and in: Tabor, M.: Chaos and Integrability in Nonlinear Dynamics: An Introduction, The FUP Experiment p. 280. John Wiley and Sons, New York (1989)

Chapter 4
Stability and Long-Time Behavior

4.1 Introduction

Dynamical systems theory aims at an understanding and comprehension of the complete dynamical system not just at documenting the properties of a particular solution. Essential elements of such a comprehension are the stability of orbits and limiting sets. The stability of orbits addresses the question of whether or not two orbits that start out close together stay close together or eventually diverge from each other. Limiting sets address the question of the long-term behavior of non-periodic orbits.

Answers to these questions require the topological concepts of neighborhoods and convergence. Thus the state space has to be endowed with a topology. Quantitative statements require more than just a topology, namely a metric or a norm. In most of our examples the state space is even a vector space endowed with a scalar product which induces a norm, metric and topology. These mathematical concepts are defined in the Appendices E.

The stability of fixed points can be assessed by a linear stability analysis, which is an exercise in linear algebra and requires the determination of the eigenvalues of the velocity gradient tensor or the displacement gradient tensor at the fixed point. Linear stability analysis is not always conclusive. The stability of fixed points can also be analyzed by the use of a Liapunov function, if such a function can be found which is the case for gradient and Hamiltonian systems. The stability of periodic orbits is harder to assess. One way is to analyze the stability of the corresponding fixed point of the Poincaré map, which however needs first to be constructed. A concept closely related to stability is the sensitivity to initial conditions, asking whether two trajectories initially close to each other will remain so or diverge from each other.

The long-term behavior of trajectories is studied with the help of concepts such as limiting sets and attractors. These are the sets that non-periodic orbits converge to. Many different trajectories might converge on the same attractor. All initial conditions that converge on an attractor constitute the basin or domain of attraction for this attractor.

P. Müller, *Handbook of Dynamics and Probability*,
https://doi.org/10.1007/978-3-030-88486-4_4

Fixed points, periodic orbits, their stability, attractors and their basins of attraction are the basic elements for constructing a state space portrait of the flow. This state space portrait provides a global description of the flow field. Even if only qualitative it might be more illuminating than the study of single trajectories. It is this global geometric description of the flow field that dynamical systems theory strives for.

A state space portrait might exhibit a very irregular structure. One such irregular structure is called chaos. Chaos is a topological concept, is differently defined and encompasses many different kind of irregular behavior.

Having a state space portrait, an important question is how it changes when the parameters of the system change. If a small change in a parameter results in a profound change of the structure of the state space portrait, the systems is said to have gone through a bifurcation. Bifurcation theory is an important part of the study of dynamical systems.

4.2 Topology

To define concepts such as closeness, neighborhoods, continuity, and convergence requires that the state space X is endowed with a topology.

Topology. A topological set is a set X together with a family $\mathcal{T}$ of subsets, called open sets. The empty set $\emptyset$ and X are open sets. Arbitrary unions of open sets are open sets. Finite intersection of open sets are open sets. The complement of an open set is called a closed set. Sets can be open, closed or neither. A set N is called a neighborhood of a point $x \in X$ if there exists an open set T such that $x \in T \subseteq N$. The notions of open and closed sets and neighborhoods together with the notions of interior and exterior points and the interior and closure of a set form the basic tool box of topology. These concepts are briefly reviewed in Appendix E.12. A topological set $\{X, \mathcal{T}\}$ is the most general way to define the concepts that are needed. We also require that the dynamical system is consistent with the topological structure, that is we assume that $S_t(x)$ is a continuous function with respect to the state space variable x and with respect to the time variable t, in the case of continuous-time systems. Quantitative theorems and results require the state to be metric or normed space.

Metric spaces. A metric space is a space X together with a distance $d(x, y)$ between two elements $x, y \in X$ that satisfies the axioms given in the Appendix E.6. For metric spaces the open sets are generated by open balls (spheres)

$$B(x, \epsilon) := \{y : d(x, y) < \epsilon\} \tag{4.1}$$

for all $x \in X$ and all $\epsilon > 0$. A topology defines neighboring points. A metric defines how close these are.

Euclidean (unitary) vector spaces. The state spaces X of our examples are generally vector spaces. A vector space allows the algebraic operations of addition of vectors and multiplication of vectors with a scalar. We require that these vector spaces are topological vector spaces which are vector spaces with a topology such that the operation of addition of vectors and multiplication with a scalar are continuous. The vector spaces that we consider are generally endowed with a scalar (inner) product $\langle x, y \rangle \in \mathbb{R}(\mathbb{C})$ that assigns to any two vectors x, y a real or complex number called their scalar product. A real vector space with a scalar product is called a Euclidean vector space and a complex vector space with a scalar product a unitary vector space. Having a scalar product one can define the norm (or magnitude or length) of the vector by $\| x \| := \langle x \rangle^{1/2}$ and a distance between two vectors by $d(x, y) = \| x - y \|$. A Euclidean (unitary) vector space is thus also a normed space and a metric space. For convenience of notation we generally formulate our results in terms of norms.

Now that the state space is a topological space one needs to refine the notion of equivalent dynamical systems. Two dynamical systems S_t and $\tilde{S}_t$ are called topological conjugates if there exists a *continuous* bijection or homeomorphism f such that $f \circ S_t = \tilde{S}_t \circ f$ for all t.

4.3 Stability

Orbits are a purely set theoretical concept. If the state space is a topological space one can ask how stable these orbits are which is to ask what happens to points in the neighborhood of these sets, or to points which are close to these sets. Since the state space is assumed to be a topological space the notions of 'neighborhoods' and 'close' are well-defined.

Stability. An invariant set A_* is called stable if for every neighborhood Ω of A_* there exist a neighborhood Ω_0 of A_* such that trajectories that originate in Ω_0 stay within the neighborhood Ω. More formally: the image set

$$S_{\mathbb{R}_+}(\Omega_0) := \{S_t(x) : x \in \Omega_0, t \in \mathbb{R}_+\} \subset \Omega \tag{4.2}$$

has to be a subset of Ω. Otherwise the invariant set is called unstable. There exist points arbitrarily close to the invariant set that will eventually diverge from it, or more formally, as an exercise in negation: An invariant set A_* is unstable if there exists a neighborhood Ω of A_* such that in every neighborhood Ω_0 of A_* there exist at least one point $y \notin A_*$ and a time $t > 0$ such that $S_t(y) \not\subset \Omega$. For time-discrete systems $\mathbb{R}_+$ has to be replaced by $\mathbb{N}$. The neighborhoods in the above definitions can be envisioned as spheres around fixed points, tori or tubes around periodic and non-periodic orbits.

If the state space is a normed space and the invariant set a fixed point the stability criterion is usually formulated as follows: A fixed point x_* is stable if for all $\epsilon > 0$ there exists a $\delta(\epsilon) > 0$ such that for all $x(0)$ with $\| x(0) - x_* \| < \delta$ one finds $\| x(t) - x_* \| < \epsilon$ for all $t > 0$. An even more stringent property is asymptotic stability which requires that x_* is stable and in addition there exists a $\delta(\epsilon) > 0$ such that for all

$\| x(0) - x_* \| < \delta$ one finds $\lim_{t\to\infty} \| x(t) - x_* \| = 0$. This formulation demonstrates that the stability of a fixed point is a local property. The challenge is to find a δ however small the ϵ is. You give me an ϵ I give you a δ. All that matters is the behavior of the system in the vicinity of the fixed point. Stability is a local property like continuity or differentiability which can also defined using an (ϵ, δ)-formulation.

The stability or instability can of course be assessed by explicitly calculating trajectories, generally a demanding task. One, of course, tries to get away with less. We will discuss the stability of fixed points, using two methods: linear stability analysis and Liapunov functions. The stability of periodic orbits will be assessed by the behavior of Poincaré sections and the stability of non-periodic orbits will be discussed in the Sect. 4.8: "Sensitivity to initial conditions".

4.4 Linear Stability of Fixed Points

Linear stability analysis linearizes the dynamics at the fixed point and determines the stability of the fixed point with respect to the linear dynamics, which turns out to be just an exercise in linear algebra. We carry out this linear stability analysis for systems arising from finite sets of ODEs and of iterative maps. The question is, of course, whether linear stability, the stability of the fixed point with respect to the linear dynamics, implies (nonlinear) stability, the stability of the fixed point with respect to the nonlinear dynamics of the original system. For finite-dimensional systems it does. PDEs are again more complicated, not all norms are equivalent and linear stability does not necessarily imply stability.

Sets of ODEs. Consider a set of ODEs $\dot{x}(t) = u(x(t))$ with a velocity field $u(x)$ whose derivative exists and is continuous, and an isolated fixed point x_*. Introduce the perturbation $x' :== x - x_*$. The ODE can then be written

$$\begin{aligned} \dot{x} &= u(x_* + x') \\ &= Ax' + Q(x') \end{aligned} \tag{4.3}$$

where we have Taylor-expanded the velocity at the fixed point x_*. Here A is a matrix with components $A_{ij} = \partial u_i / \partial x_j (x = x_*)$. It is the transpose of the velocity gradient tensor. It represents a linear operator and the higher order terms are $Q(x') = O(|x'|^2)$. The idea is that as long as the perturbations x' stay small one can neglect the nonlinear term Q and consider the system

$$\dot{x}' = Ax' \tag{4.4}$$

which is a homogeneous system of linear differential equations with constant coefficients. Formally, the linearized system is the linear tangent map. This map is called linearly stable (unstable) if this linear system is stable or unstable. As we discussed in Sect. 3.3 the solution manifold for non-defective matrices is given by

$$x'(t) = e^{At} = \sum_{s=1}^{N} e^{\lambda^{(s)}t} a^{(s)}(0) e^{(s)} \tag{4.5}$$

where $e^{(s)}, s = 1, \ldots, N$ are the eigenvectors, $\lambda^{(s)}$ the eigenvalues and $a^{(s)}$ the coefficients in the expansion $x'(t) = \sum_{s=1}^{N} a^{(s)}(t)e^{(s)}$. The components $a^{(s)}$ decay, rotate or grow depending on whether the real part of $\lambda^{(s)}$ is smaller than zero, zero or larger than zero. In the case of non-defective matrices the stability of the linearized system thus depends on the eigenvalues of the matrix A. More precisely:

- If the real part of all eigenvalues is smaller than zero, perturbations approach the fixed point. The fixed point is asymptotically stable.
- If at least one of the eigenvalues has a real part larger than zero, the fixed point is unstable.
- If the real part of all eigenvalues is less or equal to zero the fixed point is stable, also called neutrally stable.

Linear stability versus nonlinear stability. The important question is, of course, whether or not the stability (instability) of the fixed point with respect to the dynamics of the linearized system implies stability (instability) with respect to the dynamics of the original nonlinear system. For finite dimensional systems this is the case. Explicitly: if the real part of all eigenvalues is smaller than zero then the fixed point is (nonlinearly) stable. If there exists one eigenvalue with a real part larger than zero then the fixed point is (nonlinearly) unstable. The proof of these assertions is based on two inequalities. First, let $\lambda_1, \ldots, \lambda_N$ be the finite number of eigenvalues of A. If they all have real parts smaller than zero then there exists constants $\lambda_* > 0$ and $c_1 > 0$ such that

$$\| e^{At} \| < c_1 e^{-\lambda_* t} \tag{4.6}$$

for all times $t > 0$. Furthermore, if the velocity field u has a continuous derivative then there exists for sufficiently small $\| x'(t) \|$ another constant $c_2 > 0$ such that

$$\| u(x_* + (x'(t)) \| < c_2 \| x'(t) \|^2 \tag{4.7}$$

The existence of these two bounds imply the assertions. The explicit proof is given for example in Strauss (2012). The only caveat of this positive result is that the linear analysis is inconclusive if there is at least one eigenvalue with a zero real part and all the other eigenvalues have real parts less or equal to zero. This inconclusive case is called neutral stability and requires higher order analysis.

Let us apply these general results to one- and two-dimensional flows.

One-dimensional flows. The one-dimensional case is of course trivial. In this case the matrix is just a number $A = \lambda$. If $\lambda < 0$ the perturbations x' decay. The fixed point is asymptotically stable. If $\lambda > 0$ the perturbations grow and the fixed point is unstable.

A simple example is $\dot{x} = \sin x$ which has fixed points $x_*^{(n)} = n\pi$ for $n \in \mathbb{Z}$. The linear tangent map is $\dot{x}' = x'$ for even n and $\dot{x}' = -x'$ for odd n. The fixed points for even n are thus unstable and those for odd n are asymptotically stable.

If $\lambda = 0$ then we cannot say anything. An example is the Ricatti equation $\dot{x} = -x^2$ which we considered in Sect. 2.4. It has a fixed point $x_* = 0$ and the derivative of the velocity field at the fixed point is zero. So the linear stability analysis is inconclusive. We can, however, infer from the explicit solution in Fig. 2.5 that the fixed point is unstable.

Two-dimensional flows. In the case of two-dimensional flows the matrix A is a 2×2 matrix. The solutions of this two-dimensional linear problem were already given in detail in the previous chapter, distinguishing between non-defective and defective matrices and among the non-defective matrices between normal and non-normal matrices. The principal behavior of the solution was sketched in Fig. 3.1 in a plane spanned by $det(A) = \lambda^{(1)}\lambda^{(2)}$ and $tr(A) = \lambda^{(1)} + \lambda^{(2)}$ which are both real for real matrices. We can just take over this figure and interpret it as a stability diagram, as is done in Fig. 4.1.

In Sect. 3.4 we pointed out a subtle phenomena for non-normal matrices, namely that perturbations can initially grow before they finally approach the fixed point.

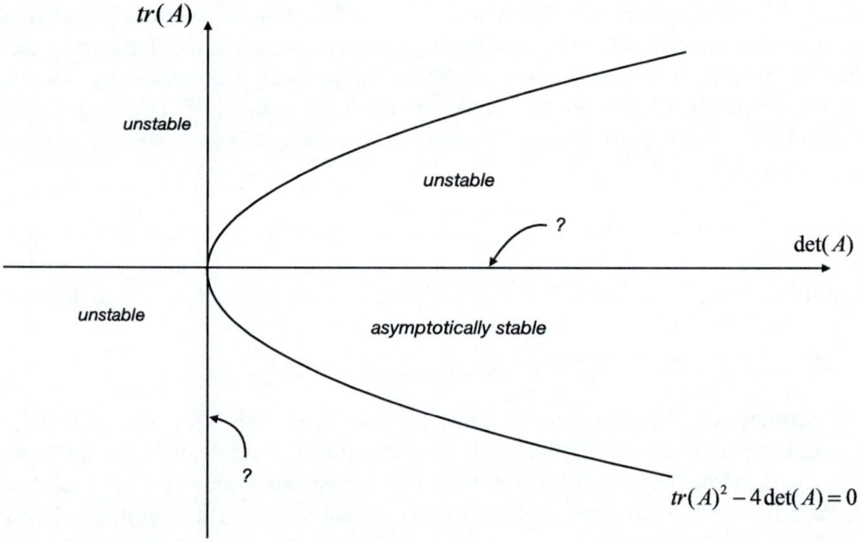

Fig. 4.1 The fixed point is asymptotically stable in the quadrant $det(A) > 0$, $tr(A) < 0$ and unstable in the other three quadrants. Along the line $det(A) = 0$ one of the eigenvalues is zero. For $tr(A) > 0$ the fixed point is unstable since the other eigenvalue is larger than zero; for $tr(A) < 0$ the other eigenvalue is smaller than zero so the stability behavior must be determined by higher order analysis in the direction of the eigenvector with zero eigenvalue. If $tr(A) = 0$ the two eigenvalues are $\lambda^{(1,2)} = \pm 2\sqrt{-det(A)}$. The fixed point is unstable for $det(A) < 0$ and requires higher order analysis for $det(A) > 0$

This finite growth does not affect stability considerations. One can always limit the initial perturbation to a small enough neighborhood to keep the finite growth under the required threshold. The system is still asymptotically stable with respect to infinitesimal perturbations. However, finite perturbations can grow and the system may evolve beyond the limit of the validity of the linearization. This has been a problem in the study of the stability of real fluids.

For defective matrices, the eigenvector expansion does not work since there is only one eigenvector and one has to use the generalized eigenvalue analysis discussed in Sect. 3.4. If $\lambda \neq 0$ then the stability is governed by the sign of the eigenvalue. If $\lambda = 0$ the fixed point is unstable because of the term proportional to t in the solution (3.51).

In summary, the linear stability of fixed points of ODEs is an exercise in linear algebra and requires the determination of the eigenvalues of a matrix.

Iterative maps. Iterative maps $x_{n+1} = F(x_n)$ can be treated similarly. The perturbation equation becomes

$$x'_n = Ax'_{n-1} = A^n x'_0 \tag{4.8}$$

where now A is a matrix with components $A_{ij} = \partial F_i/\partial x_j(x_*)$. It is the transpose of the displacement gradient tensor. If the matrix is non-defective eigenvector expansion leads to $x'_n = A^n x_0$ with

$$A^n = \sum_{s=1}^{N} (\lambda^{(s)})^n e^{(s)} \hat{e}^{(s)\dagger} \tag{4.9}$$

The stability now depend on whether the magnitudes of the eigenvalues $\lambda^{(s)}$ are smaller than one or larger than one. The fixed point is asymptotically stable if all eigenvalues have a magnitude smaller than one. If at least one of the eigenvalues has a magnitude larger than one the fixed point is unstable. Again, if the matrix A is non-normal trajectories might exhibit intermediate growth.

For a two-dimensional state space we find the stability diagram shown in Fig. 4.2. If the two-dimensional matrix A is defective then one again finds instability in the case that the eigenvalue has a magnitude one. An example is the defective matrix

$$A = \begin{pmatrix} 1 & 1 \\ 0 & 1 \end{pmatrix} \tag{4.10}$$

which has the eigenvalue $\lambda = 1$ but its nth power is

$$A^n = \begin{pmatrix} 1 & n \\ 0 & 1 \end{pmatrix} \tag{4.11}$$

and the fixed point is unstable. Analogous results hold for iterative maps in higher but finite dimensions.

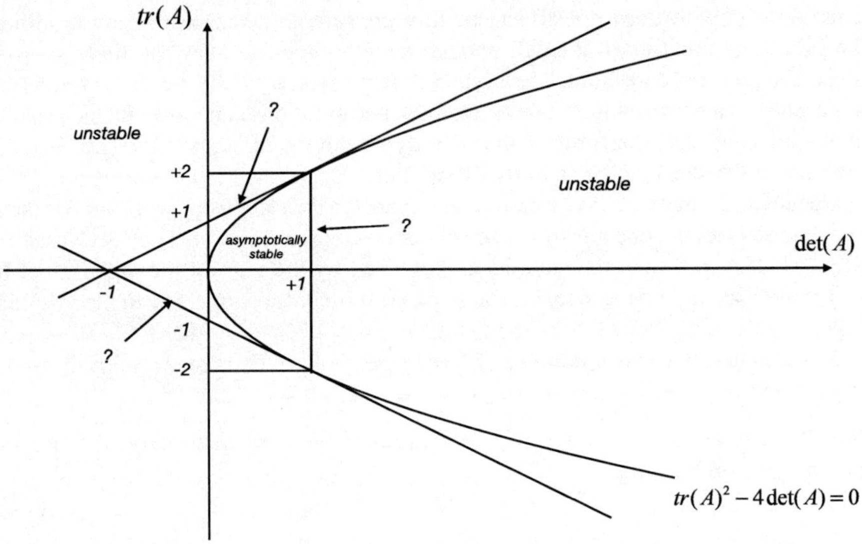

Fig. 4.2 Stability diagram for two-dimensional iterative maps. The fixed point is stable inside the triangle formed by the points $(-1, 0)$, $(1, 2)$, $(1 - 2)$ and unstable outside

For finite dimensions linear stability analysis thus suffices to determine the (non-linear) stability, unless the fixed point is neutrally stable. The situation is different for infinite dimensions which we demonstrate for PDEs.

PDEs. The functions on which the differential operators are acting are generally elements of a function space, sometimes of countably infinite dimension but often also of higher cardinality. Typical examples are the flows in fluid dynamics where instability is associated with the onset of turbulence. The analysis of the stability of such fluid flows becomes much more complex and no general systematic analysis exists. Here we just list a few of the new aspects that arise in the analysis of fluid flows and of PDEs in general:

1. As we already have pointed out, the spectrum, the set of all eigenvalues, of linear operators in infinite dimensions may have a complicated structure, including continuous parts. It becomes much harder to determine bounds on $\| e^{At} \|$ as we could in (4.6) for finite dimensions.

2. There exist different non-equivalent norms in the infinite-dimensional case whereas all norms are equivalent in finite dimensions. What is the appropriate norm? We demonstrate this point with the two-dimensional Euler equations

$$\begin{aligned} \partial_t u_h + (u_h^T \nabla_h) u_h &= -\nabla_h p \\ \nabla_h^T u_h &= 0 \end{aligned} \tag{4.12}$$

in a bounded simply connected domain $\Omega \subset \mathbb{R}^2$. Here $u_h = (u_1, u_2)$ and $\nabla_h = (\partial_1, \partial_2)$ denote the two-dimensional (horizontal) velocity and gradient. Assume that the boundary condition on $\partial\Omega$ is $u_h^T n_h = 0$. Then the two-dimensional Euler equation conserve kinetic energy

$$\frac{d}{dt}\int_\Omega dx \frac{1}{2} u_h^T u_h = 0 \tag{4.13}$$

The boundary terms all vanish. This conservation law suggests an analysis in terms of u_h with the $\mathcal{L}_2$ norm $\| u_h \|^2 = \int_\Omega dx u_h^T u_h$.

Alternatively, one may consider the vorticity

$$\omega_3 := \hat{\nabla}_h^T u_h = -\partial_2 u_1 + \partial_1 u_2 \tag{4.14}$$

where $\hat{\nabla}_h := (-\partial_2, \partial_1)$ is the gradient operator rotated by $\pi/2$. The subscript "3" is to indicate that ω_3 is the vertical component of the vorticity vector. Applying the operator $\hat{\nabla}_h$ to the momentum balance one obtains the vorticity balance

$$\partial_t \omega_3 + (u_h^T \nabla_h)\omega_3 = 0 \tag{4.15}$$

The zero divergence condition $\nabla_h^T u_h = 0$ implies the existence of a stream function ψ such that

$$u = -\hat{\nabla}_h \psi = (\partial_2 \psi, -\partial_1 \psi)^T \tag{4.16}$$

and the Poisson equation

$$\omega_3 = -\hat{\nabla}_h^T \hat{\nabla}_h \psi = -\Delta\psi \tag{4.17}$$

The boundary condition for the stream function becomes $\psi = 0$ on $\partial\Omega$. The vorticity equation conserves enstrophy

$$\frac{d}{dt}\int_\Omega dx \frac{1}{2}\omega_3^2 = 0 \tag{4.18}$$

since, again, the boundary terms vanish. This formulation of the two-dimensional Euler equations suggests an analysis in terms of the vorticity ω_3 with the $\mathcal{L}_2$ norm $\| \omega_3 \|^2 = \int_\Omega dx \omega_3^2$ which is the $\mathcal{L}_2$ norm of $\hat{\nabla}_h^T u_h$. Thus different norms offer themselves for consideration.

3. Linear stability might not imply stability. A counter example is the PDE

$$\partial_t \rho = x\partial_x \rho + \rho^2 \tag{4.19}$$

on the real line. It has the fixed point $\rho_*(x) = 0$. For any initial condition $\rho(x, 0)$ that is constant on an interval I, $\rho(x, 0) = c \neq 0$ for $x \in I$, the solution to the PDE in this interval becomes

$$\rho(x, t) = \frac{c}{1 - ct} \tag{4.20}$$

and develops a singularity in finite time. These initial conditions thus blow up in the $\mathcal{L}_2$ norm. When, however, the linearized problem $\partial_t \rho = x\partial_x \rho$ is multiplied by ρ and integrated over the domain one obtains

$$\frac{d}{dt}\int_{\mathbb{R}} dx \frac{1}{2}\rho^2(x,t) = -\int_{\mathbb{R}} dx \frac{1}{2}\rho^2(x,t) \tag{4.21}$$

if $\rho(x)$ vanishes sufficiently fast as $|x| \to \pm\infty$. The linear problem is thus stable in the $\mathcal{L}_2$ norm. Linear stability does not imply (nonlinear) stability.

In summary, when analyzing the stability of fixed points of PDEs it becomes harder or even impossible to estimate bounds on the growth rate of perturbations, to relate linear to nonlinear stability, and to amalgamate non-equivalent norms. As with all PDEs, the details depend on the type of PDE considered.

The linear stability analysis of a fixed point is a local analysis. It determines the behavior in the infinitesimal neighborhood of the fixed point. A more global analysis is considered next.

4.5 Liapunov Function

The Liapunov function is a global analysis that allows the determination of the stability of a fixed point without explicitly solving the (linearized) equations of motion.

Consider again our standard ODE $\dot{x}(t) = u(x(t))$. Let Ω be an open neighborhood of an isolated fixed point x_*. A function $V : \Omega \mapsto \mathbb{R}$ is called positive definite if $V(x) > 0$ for all $x \neq x_*$ and $V(x_*) = 0$. It is called positive semi-definite if $V(x) \geq 0$ for all $x \neq x_*$ and $V(x_*) = 0$. Negative (semi)-definite functions are defined analogously. In the context of stability analysis positive definite functions V are called Liapunov functions. They have a simple minimum at $x = x_*$. Since

$$\frac{d}{dt}(V \circ x) = \dot{x}^T \nabla V = u^T \nabla V \tag{4.22}$$

$V \circ x$ increases or decreases monotonically with time depending on whether $u^T \nabla V$ is positive or negative. Since the Liapunov function V is positive definite, $u^T \nabla V$ determines whether x increases or decreases in time. This result implies the following theorem:

1. If $u^T \nabla V \leq 0$ in $\Omega \setminus \{x_*\}$ the fixed point is stable.
2. If $u^T \nabla V < 0$ in $\Omega \setminus \{x_*\}$ the fixed point is asymptotically stable.
3. If $u^T \nabla V > 0$ in $\Omega \setminus \{x_*\}$ the fixed point is unstable.

As a simple example consider the ODE $\dot{x} = \sin x$. It has fixed points at $x_*^{(n)} = n\pi$ with $n \in \mathbb{Z}$. Linear stability analysis reveals that the fixed points are linearly stable

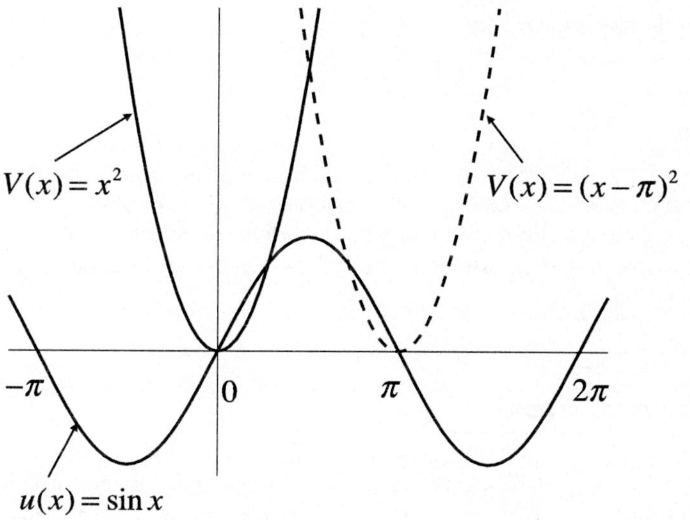

Fig. 4.3 The ODE $\dot{x} = \sin x$ has fixed points at $x_*^{(n)} = n\pi$ for $n \in \mathbb{N}$. The function $V(x) = x^2$ is a Liapunov function for the fixed point $x_* = 0$ with $u^T \nabla V = 2x \sin x > 0$ in the interval $]-\pi, \pi[$. The fixed point is unstable. The function $V(x) = (x-\pi)^2$ is a Liapunov function for the fixed point $x_* = \pi$ with $u^T \nabla V = 2(x-\pi)\sin x < 0$ in the interval $]0, 2\pi[$. This fixed point is asymptotically stable

for odd integers and unstable for even integers. This ODE is also amenable to a Liapunov analysis, as demonstrated in Fig. 4.3.

Note that the Liapunov function provides the stability behavior within the finite neighborhood Ω and does not suffer from any indeterminacy, as was the case for neutrally stable fixed points in the linear analysis. On the other side, it is generally not assured that there exists Liapunov functions and if they exist to find one. Exceptions are gradient and Hamiltonian systems.

A gradient system

$$\dot{x}(t) = -\nabla\Phi(x(t)) \tag{4.23}$$

has fixed points x_* wherever $\nabla\Phi(x_*) = 0$. If the fixed point minimizes the potential Φ then $V(x) = \Phi(x) - \Phi(x_*)$ is a Liapunov function and $-u^T\nabla V = (\nabla V)^T\nabla V \geq 0$ and the fixed point is stable. An additional feature of the gradient system is that the gradient of the potential is perpendicular to the hyper-surfaces $\Phi(x) =$ constant. A particle at a point on the hyper-surface thus moves 'downhill' in a direction of decreasing Φ unless it is at a fixed point. Formally

$$\frac{d}{dt}(\Phi \circ x) = \dot{x}^T\nabla\Phi = -(\nabla\Phi)^T\nabla\Phi \leq 0 \tag{4.24}$$

A particle not at a fixed point can thus never return to its original position. Thus there exist no periodic orbits.

A Hamiltonian system

$$H(q, p) = \frac{p^2}{2m} + U(q) \tag{4.25}$$

with a simple minimum of U at $q = 0$ has a fixed point at $q = p = 0$. The function $V(q, p) = H(q, p) - H(0, 0)$ is positive definite near the fixed point and can serve as a Liapunov function. Since $u^T \nabla H = 0$ the fixed point is stable.

Before we analyze the stability of periodic orbits we introduce an important tool, Poincaré maps, which are useful in the stability analysis of periodic orbits and in their own rights.

4.6 Poincaré Maps

Poincaré maps or *first recurrence maps* are discrete systems constructed from a continuous systems. They are obtained by first identifying a hypersurface in state space that is everywhere transversal to the flow of the system, transversal meaning that the trajectories cross the hyper-surface; they do not lie within it. Such hyper-surfaces are called *Poincaré sections*. Because of their transversality, the trajectories starting on the hyper-surface leave it and one may record where and when they return to the hyper-surface. In particular, one may record where and when they first return to the section. This record of first returns is the Poincaré map or *first recurrence map*. The times between successive crossings might not be equal.

Example: Billiards. Billiards may serve as a visualization. You only record where a billiard ball hits the boundary. In-between the motion is boring, just straight motion at constant speed. As a physics problem, the original state space includes the momentum of the billiard balls but conservation of energy and tangential momentum may be used to reduce the problem to configuration space. Conservation of tangential momentum results in reflection with the incident angle being equal to the reflected angle. Following the balls in configuration space is thus sufficient. As a dynamical systems problem, the boundary becomes the hyper-surface and the recording of the points where the ball hits the boundary becomes the Poincaré map.

Example: Two uncoupled harmonic oscillators. Two uncoupled harmonic oscillators with frequencies ω_1 and ω_2 provide a second example. The multiply periodic motion of the angles θ_1 and θ_2 can be represented in the (θ_1, θ_2)- plane as depicted in Fig. 4.4. Now consider a hyper-surface or line $\theta_2 = \theta_* =$ constant in this plane. Register when the trajectories cut through this line. When a trajectory cuts through this line at time t then it does so again at time $t + T_2$ where $T_2 = 2\pi/\omega_2$. Then $\theta_1(t + T_2) = \theta_1(t) + 2\pi\omega_1/\omega_2$. This construction leads to the Poincaré map

$$P_1(\theta_1) = \theta_1 + \Delta\theta \tag{4.26}$$

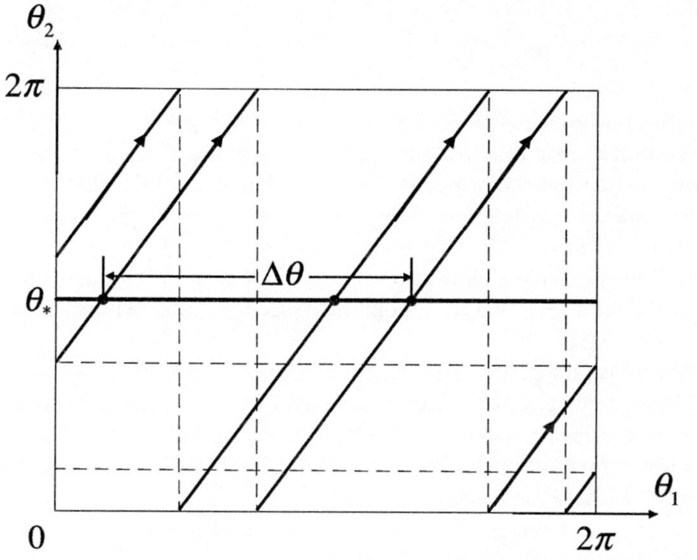

Fig. 4.4 Construction of Poincaré map for two uncoupled harmonic oscillators. The trajectory intersects the line $\theta_2 = \theta_* =$ constant at the points θ_1 which are equally spaced by $\Delta\theta = 2\pi\omega_1/\omega_2$

modulo 2π where $\Delta\theta := 2\pi\omega_1/\omega_2$. At each time step the angle θ_1 where the trajectory crosses the line increases by a constant amount.

If $\omega_1/\omega_2 = p/q$ is rational then the map is periodic with period q. The map will at most produce q different points. Thus every point is a fixed point of the q-times iterated map P_1^q. If the frequency ratio is irrational then the Poincaré map 'fills out' the line; the trajectory of the Poincaré map is dense on the line. Poincaré maps thus allow to easily distinguish between these two cases. The Poincaré map can also be constructed, as we already pointed out, on the torus that is obtained by first joining the top and bottom of the square to form a tube and then joining the two ends of the tube to form a torus. A hypersurface $\theta_2 = \theta_* =$ constant through this torus is then a circle.

Poincaré map. Formally, Poincaré maps are constructed as follows: Let $\{X, (S_t)_{t\in\mathbb{R}}\}$ be a continuous-time dynamical system. An $(N-1)$-dimensional closed hypersurface H is called a Poincaré section or cross section if

(i) for each $x \in X$ there exist times $t_1 > 0$ and $t_2 < 0$ such that $S_{t_1}(x) \in H$ and $S_{t_2}(x) \in H$ and
(ii) every trajectory intersects H transversally.

If such a cross section exists let t_1 be the smallest positive number such that $S_{t_1}(x) \in H$. Then iteration of

$$\begin{aligned} P_1 : H &\mapsto H \\ x &\mapsto S_{t_1}(x) \end{aligned} \tag{4.27}$$

defines a discrete dynamical system, called the Poincaré map or first recurrence map. The Poincaré map is a discrete dynamical system with a state space that is one dimension smaller than the original continuous dynamical system.

Poincaré maps often preserve many properties of the original system and thus allow study of these properties in a lower dimensional space. One important example is that one can relate the stability of a periodic orbit in the original system to the stability of the corresponding fixed point in the first recurrence map, as will be done in the next section.

Poincaré maps are particularly suited and have been used extensively for conservative Hamiltonian systems with 2 degrees of freedom. The state space then has dimension 4 but each trajectory is confined to a three-dimensional hyper-surface because of energy conservation. A Poincaré section is then a two-dimensional surface and can easily be visualized.

Poincaré maps are thus another useful reduced description of a dynamical system, in addition to stroboscopic versions of continuous systems and symbolic dynamics.

4.7 Linear Stability of Periodic Orbits

Periodic orbits are also invariant sets whose stability one might want to investigate. Here we consider a stability analysis based on Poincaré maps.

Continuous periodic orbits. Periodic orbits are those for which there exist times $t_1 \neq t_2$ such that $S_{t_1}(x) = S_{t_2}(x)$ and a time t_3 such that $S_{t_3}(x) \neq x$. The second condition is to exclude fixed points. Periodic orbits are a purely set theoretical concept. When considering such periodic orbits for time-continuous dynamical systems one has, however, to realize that the set of periodic functions includes functions like the Dirichlet function $f : \mathbb{R} \mapsto \mathbb{R}$, $t \mapsto f(t)$ with $f(t) = 1$ if t is irrational and $f(t) = 0$ if t is rational. The Dirichlet function is periodic with all $p \in \mathbb{Q}$ being periods. Such singular cases are, however, excluded when one restricts oneself to functions S_t that are continuous with respect to time. This follows from a theorem (A course in p-adic analysis, Robert, 2000) that states that there are only two kinds of additive subgroups in $\mathbb{R}$, the first ones are the discrete groups $p\mathbb{Z}$ ($p > 0$) and the second ones are non-discrete groups which are dense in $\mathbb{R}$. If we require continuity of S_t with respect to time t then we can exclude the groups that are dense and we are left with the discrete groups $p\mathbb{Z}$. Then there exists a smallest positive period p_0 and all other periods are integral multiples of p_0. If there were an additional period $p' \neq np_0$ for all $n \in \mathbb{N}$ then there exists one n' such that $p'' = p' - n'p_0 < p_0$ and $S_x(t + p'') = S_x(t)$ contrary to the assumption that p_0 is the smallest period. Periodic orbits then represent closed curves that do not intersect and there exists a smallest period p_0 such that $S_{t+p_0}(x) = S_t(x)$ for all times $t \in T$.

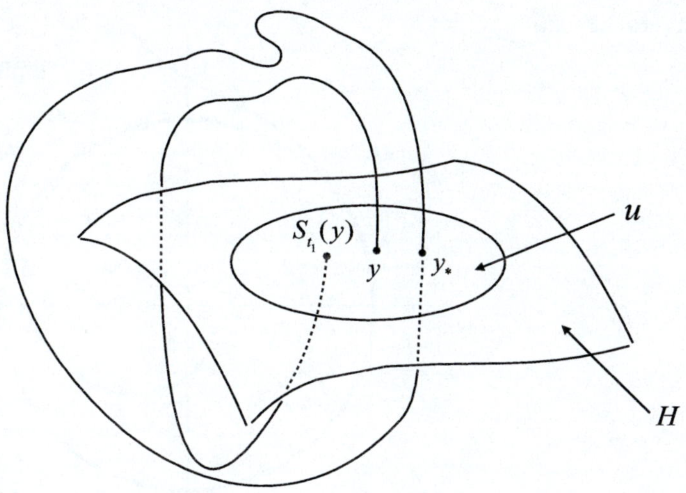

Fig. 4.5 Poincaré map for a periodic orbit

Limit cycles. Periodic orbits that are isolated are called limit cycles. Isolated means that the neighboring trajectories are not periodic. Neighboring points may move either to or away from the limit cycle or may show some neutral behavior but cannot be part of another periodic orbit. Limit cycles only exist for nonlinear systems. As we mentioned before, the periodic orbits of linear systems cannot be isolated since for every periodic orbit $x(t)$ there exist neighboring periodic orbits $cx(t)$ for any $c \neq 0$.

Stability theorem. Assume the dynamical system corresponding to the ODE $\dot{x}(t) = u(x(t))$ with a smooth velocity field and an isolated periodic orbit. The stability of the orbit can be assessed from the stability of the corresponding fixed point of its Poincaré map utilizing the following theorem (See Fig. 4.5).

Assume there exist a cross section H. Let $O(x)$ be a periodic orbit and let y_* be a point where this periodic orbit intersects the cross-section H. Then there exists an open and connected neighborhood U of y_* in H such that for all y in U there exists a t_1 such that $S_{t_1}(y) \in U$ and $S_t(y) \notin U$ for $0 < t < t_1$. One thus has a Poincaré map

$$\begin{aligned} P_1 : U &\mapsto U \\ y &\mapsto S_{t_1}(y) \end{aligned} \tag{4.28}$$

defined on U. The point y_* is a fixed point of this map. The stability of this fixed point determines the stability of the periodic orbit. In particular one can make the following statements. The periodic orbit $O(x)$ of the continuous dynamical system is stable if and only if the corresponding fixed point y_* of the discrete dynamical system P_1 is stable. There is also the additional statement that the periodic orbit $O(x)$ of the continuous dynamical system is asymptotically stable if and only if the

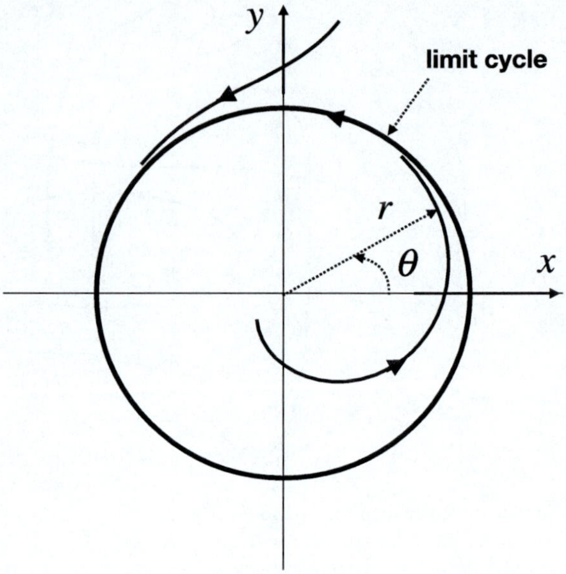

Fig. 4.6 Trajectories of the system 4.29

corresponding fixed point y_* of the discrete dynamical system P_1 is asymptotically stable, though we have not defined yet "asymptotically stable" for invariant sets other than fixed points. In any case, the usually more difficult task of determining the stability of a periodic orbit is thus reduced to the analysis of the stability of a fixed point, albeit of a map that needs first to be constructed. An application of this theorem is given by the following example.

Example. Consider the flow field in the plane $\mathbb{R}^2$ given in polar coordinates by

$$\begin{aligned} \dot{r} &= r(1-r^2) \\ \dot{\theta} &= 1 \end{aligned} \tag{4.29}$$

and shown in Fig. 4.6.

The system has a fixed point at $r = 0$ and an isolated periodic orbit at $r = 1$. All points starting with $r \neq 1, 0$ will eventually circle the periodic orbit at $r = 1$. One thus would like to think of the periodic orbit $r = 1$ as an asymptotically stable limit cycle. This result is indeed borne out by the corresponding Poincaré map. Take as a cross-section the $\theta = 0$ axis. The cross section is just a line and the Poincaré map just a one-dimensional map. Since $\dot{\theta} = 1$ the first return occurs after time $t = 2\pi$. Therefore

$$\int_{r_0}^{r_1} dr \frac{1}{r(1-r^2)} = \int_0^{2\pi} dt = 2\pi \tag{4.30}$$

Evaluation of the integral gives the Poincaré map

$$P_1(r) = [1 + e^{-4\pi}(r^{-2} - 1)]^{-1/2} \tag{4.31}$$

Linearization at $r = 1$ yields

$$\frac{dP_1}{dr}|_{r=1} = e^{-4\pi} \tag{4.32}$$

Since $|e^{-4\pi}| < 1$, the fixed point of the Poincaré map is asymptotically stable.

For discrete-time systems a periodic orbit with period p is stable if the corresponding fixed point of the p-times iterated map is stable.

In addition to the stability of periodic orbits there is also interest in the stability of traveling waves, such as the solitons of the KdV equation.

4.8 Sensitivity to Initial Conditions

A concept closely related to the stability or instability of invariant sets is the concept of sensitivity to initial conditions. It requires the state space to be at least a metric space. One first introduces Liapunov stability.

Liapunov stability. A point $x \in X$, not necessarily a fixed point, is called Liapunov stable if all points sufficiently close to x stay close to x, i.e., for all $\epsilon > 0$ there exists a $\delta > 0$ such that for all points y within the ball $d(x, y) < \delta$ one finds $d(S_t(x), S_t(y)) < \epsilon$ for all $t > 0$.

A point x is called (Liapunov) unstable if there exist a $\epsilon(x) > 0$ such that, for any $\delta > 0$, there exist at least one point y within the ball $d(x, y) < \delta$ and at least one time $t > 0$ such that $d(S_t(x), S_t(y)) > \epsilon$. The trajectories of two arbitrarily close points will eventually separate by a distance $\epsilon(x)$. Note that not all points in any of the neighborhoods of x need to diverge from x but there needs to be at least one.

Sensitivity to initial conditions. A system is called sensitive to initial conditions if every point of the state space (or every point in a invariant subset) is unstable with a common ϵ which is then called the sensitivity constant. Sensitivity to initial conditions is an important characteristic of dynamical systems since it limits predictability. Sensitivity to initial conditions is also called the "butterfly effect", referring to Edward Lorenz (1993) famous question "Does the flap of a butterfly's wings in Brazil set off a tornado in Texas?".

Sensitivity to initial conditions alone does not necessarily imply any irregular long-term or chaotic behavior. It can easily be achieved in an unbounded state space. The system $S_t(x) = xe^t$ exhibits sensitivity to initial conditions in the forward time direction. This system also shows that sensitivity to initial conditions is not an invariant property under a transformation of the state variables. Introduce $y = 1/x$ and $S_t(x) = xe^t$ becomes $\hat{S}_t(y) = ye^{-t}$ which is not sensitive to initial conditions in the forward time direction.

Liapunov exponent. A quantitative characterization of the rate of divergence is given by the Liapunov exponent $\lambda(x)$. In the one-dimensional case, the Liapunov exponent is defined by

$$\lambda(x) := \lim_{|\Delta x|\to 0} \lim_{t\to\infty} \frac{1}{t} \log \frac{\| S_t(x+\Delta x) - S_t(x) \|}{\| \Delta x \|} \tag{4.33}$$

The limit might exist even if the function $S_t(x)$ is not differentiable with respect to x because the time limit is applied first. If the limits exist, the length of an initial infinitesimal displacement changes asymptotically as

$$\| dS_t(x) \| \sim e^{\lambda(x)} \| dx \| \tag{4.34}$$

If $\lambda > 0$ this length grows exponentially. The point is then Liapunov unstable. The Liapunov exponents might be different for different x. If the Liapunov exponent is larger than zero for all x the system is sensitive to initial conditions. The system $S_t(x) = xe^t$ has the Liapunov exponent $\lambda = 1$, independent of x. The Liapunov exponent for Bernoulli shift is $\lambda = \log 2 > 0$ also independent of x. The Bernoulli system thus also exhibits sensitivity to initial conditions.

For higher dimensional systems, the rate of growth might depend on the direction of the initial separation vector dx since $|dS_t(x)|/|dx| = (e^T D_t D_t^T e)^{1/2}$ with unit vector $e = dx/|dx|$ and displacement gradient tensor $D_t = \nabla_x S_t^T$. In this case there is a spectrum of Liapunov exponents. If $\beta(x)$ are the eigenvalues of the matrix

$$\lim_{t\to\infty} (D_t D_t^T)^{1/2t} \tag{4.35}$$

then the Liapunov exponents are defined by

$$\lambda(x) = \ln \beta(x) \tag{4.36}$$

The maximal Liapunov exponent determines the asymptotic behavior. If this exponent is positive for all x then the system exhibits sensitivity to initial conditions.

4.9 Limiting Sets and Attractors

Now we discuss the limiting behavior of non-periodic orbits. Again, the concept of limits requires a topology, which is assumed here to be realized by a norm.

We called a fixed point asymptotically stable if trajectories not only stay in a neighborhood but approach the fixed point. More precisely: a fixed point x_* is called asymptotically stable if it is stable and if there exists a neighborhood Ω_0 of x_* such that all points within Ω_0 approach the fixed point. Formally

$$\lim_{t\to\infty} \| S_t(x) - x^* \| = 0 \tag{4.37}$$

for all $x \in \Omega_0$. The fixed point then acts as an attractor.

Here we generalize this concept from fixed points x_* to arbitrary invariant sets A_*. Trajectories starting within a neighborhood Ω_0 of A_* do not only stay in Ω_0 but approach the invariant set A_*. Such a generalization requires the definition of the concept of approach. Invariant sets may contain more than one point. It may happen that the trajectories still approach the invariant set but that limits of the form (4.37) do not exist because they do not approach a single point of the invariant set. This is the case for the flow (4.29) depicted in Fig. 4.6. The system has a periodic orbit at $r = 1$. All points starting with $r \neq 1, 0$ will eventually circle the periodic orbit at $r = 1$ but will not approach a single point of the periodic orbit. Still one would like to think of the periodic orbit $r = 1$ as asymptotically stable. To cover situations like this one, one defines limiting points and sets.

Limiting points and sets. Positive limiting points x_∞ of a trajectory $S_t(x)$ consist of all points for which there exist an increasing sequence of times t_n with $t_n \to \infty$ such that

$$\lim_{n\to\infty} \| S_{t_n}(x) - x_\infty \| = 0 \tag{4.38}$$

The trajectory comes arbitrarily close to a positive limiting point an infinite number of times.

The set of all positive limiting points is denoted by Ω_∞. It depends on the trajectory. Each point x in state space is thus assigned a positive limiting set $\Omega_\infty(x)$ (which might be empty). If the system is invertible, one can define negative limiting points and sets by replacing $t_n \to \infty$ by $t_n \to -\infty$ in the above definition.

Limiting sets have the following properties:

1. If the orbit is bounded then a limiting set exists, meaning that $\Omega_\infty(x)$ is not empty.
2. The limiting set is a closed set.
3. The limiting set is a positively invariant set. If $y \in \Omega_\infty(x)$ then $S_t(y) \in \Omega_\infty(x)$ for all $t > 0$.
4. If the orbit is bounded its limiting set is connected.
5. The limiting set is transitive, that is if $z \in \Omega_\infty(y)$ and $y \in \Omega_\infty(x)$ then $z \in \Omega_\infty(x)$.

Attractor. An invariant set A_* is called an attractor if there exist an open set Ω_0 that contains A_* such that $\Omega_\infty = A_*$ for all trajectories that begin in Ω_0, or formally: A set A_* is an attractor if there exists an open set Ω_0 such that

$$A_* \subset \Omega_0 \tag{4.39}$$

$$A_* = \Omega_\infty(\Omega_0) \tag{4.40}$$

The largest possible set Ω_0 is called the basin (domain) of attraction

$$B(A_*) := \{x : \Omega_\infty(x) = A_*\} \tag{4.41}$$

To exclude trivial situations, the definition of an attractor often includes the requirements: (i) there exists no smaller attractor within A_* and (ii) the attractor has a basin of attraction larger than the attractor, $A_* \subset B(A_*)$. Attractors are only possible for dissipative systems for which the volume shrinks. Attractors are thus not possible for Hamiltonian systems, the baker transformation, and the doubling map which are all volume conserving.

The simplest attractors are asymptotically stable fixed points and limit cycles. Attractors can, however, have a much more complicated structure, including fractals.

Fractals. Fractals are geometric structures that possess non-trivial structure on all scales. A classical example is the Cantor set which is obtained by dividing the unit interval [0, 1] into three subintervals of equal length and eliminating the middle interval. The remaining intervals are again divided into three subintervals and the middle one again removed, and so forth ad infinitum. The set so obtained is called the Cantor set and consists of all real numbers x in the unit interval which have a representation

$$x = \sum_{n=1}^{\infty} i_n 3^{-n} \tag{4.42}$$

in ternary representation. The Cantor set is nowhere dense in the interval [0, 1].

Box dimension. The box dimension of a set $E \subset \mathbb{R}^d$ is D if the number N of cubes of volume V that one needs to cover the set scales as

$$N(V) \sim V^{-D} \tag{4.43}$$

as $V \to 0$. If one allows coverage by sets other than cubes one arrives at the more general concept of the Hausdorff dimension.

Example: Dissipative baker transformation. The dissipative baker transformation is defined by

$$F(x, y) = \begin{cases} (2x, ay) & \text{for } x \in [0, \tfrac{1}{2}[\\ (2x - 1, ay + 1/2) & \text{for } x \in [\tfrac{1}{2}, 1[\end{cases} \tag{4.44}$$

For $a < 1/2$ this map is dissipative since the 'volume' V, area in this case, shrinks as $V_{n+1} = 2aV_n$. The attractor consists of an infinite set of horizontal lines whose box dimension can be inferred from the self similarity of the attractor in x-direction. Consider 'boxes' of length

$$V_n = a^n \tag{4.45}$$

If one needs $N(V_n)$ boxes of size V_n to cover the attractor then one needs $N(V_{n+1}) = 2N(V_n)$ boxes of size V_{n+1} to cover the attractor at the next iteration or

$$N(V_n) = 2^n N(V_0) = 2^n \tag{4.46}$$

Eliminating n between these two expressions one finds

$$N(V_n) = V_n^{\log 2/\log a} \tag{4.47}$$

and hence

$$D_x = \frac{\log 1/2}{\log a} \tag{4.48}$$

Since the dimension in y-direction is $D_y = 1$ the total dimension of the attractor is $D = D_x + D_y < 2$ for $a < 1/2$. Similarly, one obtains the box dimension $d = \log 2/\log 3$ for the Cantor set.

An attractor of fractal dimension is called a strange attractor.

4.10 State Space Portrait

Dynamical systems theory emphasizes the flow in state space rather than a particular trajectory. Flows are global descriptions and require knowledge of all trajectories. This is a demanding task and one often has to settle for a qualitative or geometric description. Such a description might still be more revealing than an exact trajectory. To illustrate this point consider again the ODE $\dot{x} = \sin x$ which can be 'solved' by separation of variables resulting in an explicit integral (quadrature)

$$\int_0^t dt' = \int_{x_0}^{x(t)} dx' \frac{1}{\sin x'} \tag{4.49}$$

which can be expressed in terms of more elementary functions as

$$t = \log \left| \frac{\csc x_0 + \cot x_0}{\csc x(t) + \cot x(t)} \right| \tag{4.50}$$

where $\csc = 1/\sin$. We may write this equation as $t = g\,(x(t))$, invert g and write the solution as $x(t) = g^{-1}(t)$, possibly multi-valued. Though we have an analytic solutions it is not particular insightful. Simple questions such as how does the solution depend on its initial position are hard to answer. If one plots, however, the velocity field $u(x) = \sin x$ as a function of x, as in Fig. 4.7, one can then easily determine the qualitative behavior. The particle remains at rest at positions $x_*^{(n)} = n\pi$ where the velocity is zero and moves either towards or away from these points. This dynamical systems approach gives a qualitative but insightful description of the system's

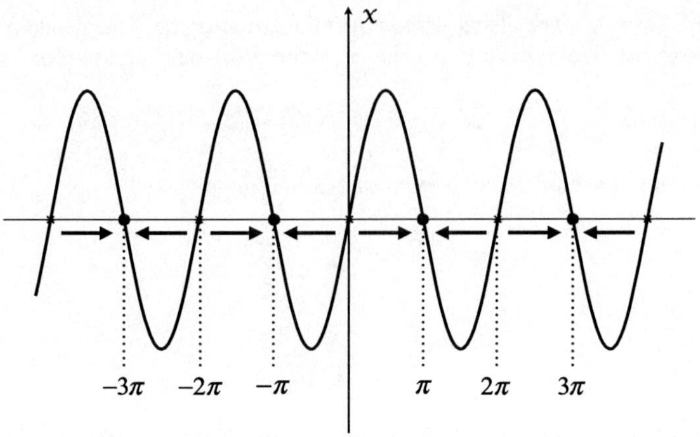

Fig. 4.7 State space portrait of the system $\dot{x} = \sin x$. The closed dots are attractors and the crosses are repellors

behavior. It is often easier to make statements about flows rather than about specific orbits.

As we have already pointed out in the introduction, the emphasis on the flow field is common practice in fluid dynamics. There one is, however, usually satisfied once one has determined the flow field from the governing fluid dynamical equations. Trajectories are only a diagnostic tool for those interested in floats or particle dispersion. Thus the fluid dynamics community spends most of its efforts on calculating the flow field whereas the dynamical systems community spends most on its efforts to determine the properties of a given flow field.

State space portrait. To arrive at a qualitative geometric comprehension of the flow field one tries to build a state space portrait. To this end one identifies the fixed points, the periodic orbits, the attractors of non-periodic orbits and their basins of attraction. There are additional sets that have been found useful in portraying the system such as:

1. If the system has a countable set of attractors A_*^i with $i \in I$ then the separatrix is defined by

$$\partial B := X \setminus \cup_{i \in I} B(A_*^i) \tag{4.51}$$

2. For fixed points, the stable and unstable manifold are defined by

$$W^s(x_*) := \{x \in \Gamma : \lim_{t \to \infty} \parallel S_t(x) - x_* \parallel = 0\} \tag{4.52}$$

$$W^u(x_*) := \{x \in \Gamma : \lim_{t \to -\infty} \parallel S_t(x) - x_* \parallel = 0\} \tag{4.53}$$

These manifolds split the phase space into subspaces for which the fixed point is either an attractor or repellor. If the fixed point is asymptotically stable then W^s is the basin of attraction and W^u is empty.

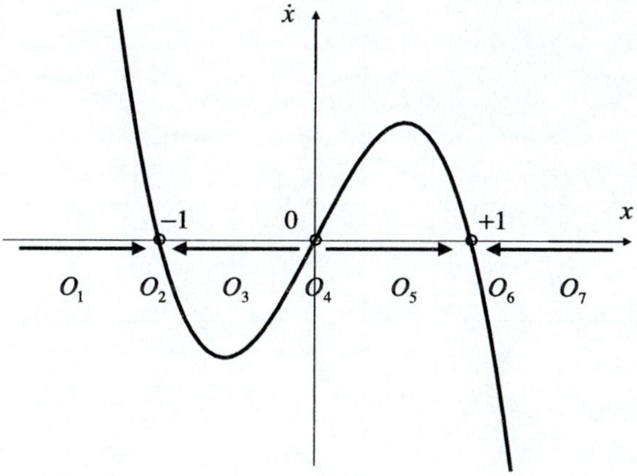

Fig. 4.8 State space portrait of the system $\dot{x} = x(1 - x^2)$. The state space $X = \mathbb{R}$ is made up of the seven orbits: $O_1 =]-\infty, -1[$, $O_2 = \{-1\}$, $O_3 =]-1, 0[$, $O_4 = \{0\}$, $O_5 =]0, 1[$, $O_6 = \{1\}$, and $O_7 =]1, +\infty[$. The points $x = 0, \pm 1$ are fixed points, with $x = \pm 1$ being an attractor and $x = 0$ being a repellor. The basins of attraction for the attractors are $B(\{-1\}) =]-\infty, 0[$ and $B(\{+1\}) =]0, +\infty[$ and their separatrix is $\partial B = \{0\}$. The stable manifolds of the attractors are $W^s(\{-1\}) = [-\infty, 0[$ and $W^s(\{+1\}) =]0, +\infty]$. The unstable manifold of the repellor is $W^u(\{0\}) = (-1, +1)$. The orbits O_3 and O_5 are heteroclinic orbits

3. A connecting arc is an orbit whose limiting sets $\Omega_\infty = x_*^\infty$ and $\Omega_{-\infty} = x_*^{-\infty}$ are fixed points. The arc or orbit is called heteroclinic if $x_*^{-\infty} \neq x_*^\infty$ and homoclinic if $x_*^{-\infty} = x_*^\infty$.

These concepts are illustrated in Fig. 4.8 for the one-dimensional system $\dot{x} = x(1 - x^2)$ and in Fig. 4.9 for the two-dimensional system $\ddot{\theta} = -\sin\theta$ which represents a free pendulum.

Poincaré-Bendixson theorem. There are theorems about what types of orbits are possible in a given state space. One such important theorem is the Poincaré-Bendixson theorem which states:

If the velocity $u(x)$ is smooth and the fixed points are isolated then the possible orbits are limited as follows:

1. Orbits of planar one-dimensional dynamical systems are either (a) fixed points, (b) approach a fixed point or (c) escape to infinity if the state space is unbounded.
2. Orbits of systems on a circle are either (a) fixed points or (b) approach a fixed point, or (c) are periodic but then there are no fixed points.
3. Orbits of two-dimensional systems on a plane, a cylinder or a torus are either (a) fixed points, (b) periodic orbits, or (c) approach a fixed point or a periodic orbit or a connecting arc, or (d) escape to infinity if the state space is unbounded.

The requirement of isolated fixed points make the formulation and proof simpler. There are other formulations of the Poincaré-Bendixson theorem such as: If $u(x)$ is a smooth vector field, the fixed points are isolated, and the forward orbit $S_t(x)$ for

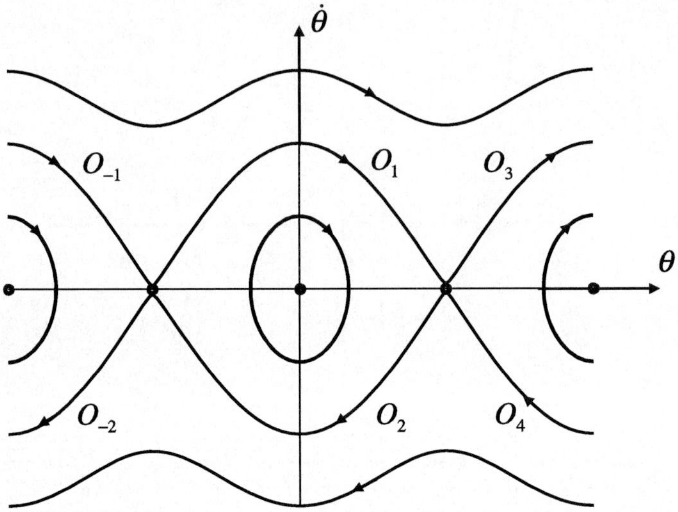

Fig. 4.9 State space portrait of the free pendulum $\ddot{\theta} = -\sin\theta$. The state space is $X = \mathbb{R}^2$ with θ and $\dot{\theta}$ being the two components. The state space consist of an infinite number of orbits. The closed curves represent oscillations and the never ending curves at the top and bottom represent rotations. These regimes are separated by the arcs $\ldots, O_{-2}, O_{-1}, O_1, O_2, \ldots$. There are stable fixed points at $(\theta, \dot{\theta}) = (n\pi, 0)$ with n being an even integer. They are no attractors. There are unstable fixed points at $(\theta, \dot{\theta}) = (n\pi, 0)$ with n being an odd integer. They are saddle points. The stable manifolds of the unstable fixed point $(\pi, 0)$ are the arcs O_1 and O_4. The unstable manifolds are the arcs O_2 and O_3. These arcs are heteroclinic orbits. The stable and unstable manifolds of the stable fixed points are empty

$t \geq 0$ is bounded then the limiting set $\Omega_\infty(x)$ of a 2D planar autonomous flow is either (a) a fixed point, (b) a periodic orbit, or (c) a connecting arc.

The Poincaré-Bendixson theorem makes statements about continuous-time systems. Discrete-time systems (such as the Bernoulli shift) can have all kinds of periodicities.

The important message of the Poincaré-Bendixson theorem is that more complicated orbital behavior can occur only in dimension $N \geq 3$.

4.11 Ingredients of Chaos

The state space portraits that were shown in the previous section are very benign. There exist much more complicated structures, one of which is referred to as chaos though there is no generally accepted definition of chaos. Before we define and discuss chaotic systems we first list some of the ingredients that go into their definition.

Dense orbits. An orbit $O(x)$ is called dense if $\overline{O(x)} = X$. This is just the standard definition of a dense set in topology. A system is said to have dense periodic orbits if there exists a dense set $D \subseteq X$ such that $O(x)$ is periodic for all $x \in D$.

Topological transitivity and mixing. A system is called topologically transitive if for any two non-empty open subsets $A, B \in X$ there exists a time $t > 0$ such that $S_t(A) \cap B \neq \emptyset$. Any open subset A will in finite time overlap with any other open subset B. The system can not be broken up into disjoint parts. The stronger statement that the above condition holds for all $t' > t$ is called topologically mixing.

Topological transitivity is not, as often asserted, equivalent to the existence of a dense orbit. Special conditions must be met for this equivalence to hold.

The over-damped oscillator $S_t(x) = xe^{-\gamma t}$ is a counter-example that the existence of a dense orbit does not imply topological transitivity. For any $x \in \mathbb{R}_+$ the orbit $O(x)$ consists of the complete state space $\mathbb{R}_+$; the orbit is thus trivially dense. Any interval moves to the left towards zero. Thus an interval will never overlap during its evolution with an interval to the right of it.

A restriction of the Bernoulli shift may serve as counter-example that topological transitivity does not imply the existence of a dense orbit. Consider the union Y of all periodic orbits of the Bernoulli shift and consider the Bernoulli shift restricted to this subset Y. The set Y is infinite whereas every periodic orbit is finite. Thus there does not exist a dense orbit. On the other hand, every pair of subintervals shares a periodic orbit. The restricted system is thus topologically transitive.

Topological Mixing does imply sensitivity to initial conditions. One may prove this assertion as follows. Take two arbitrary different points $x_1, x_2 \in X$ and define $\epsilon := d(x_1, x_2)/4$. Topological mixing then implies that

$$\begin{aligned} S_t(B(x,\delta) \cap B(x_1,\epsilon) &\neq \emptyset \\ S_t(B(x,\delta) \cap B(x_2,\epsilon) &\neq \emptyset \end{aligned} \tag{4.54}$$

for all balls $B(x, \delta)$ and all times t larger than some t_*. Thus there exists points $y_1, y_2 \in B(x, \delta)$ with $S_t(y_1) \in B(x_1, \epsilon)$ and $S_t(y_2) \in B(x_2, \epsilon)$. The triangle (polygon) inequality then implies

$$\begin{aligned} d(S_t(y_1), S_t(y_2)) &\geq d(x_1, x_2) - d(x_1, S_t(y_1)) - d(x_2, S_t(y_2)) \\ &\geq 4\epsilon - \epsilon - \epsilon = 2\epsilon \end{aligned} \tag{4.55}$$

Another application of the triangle inequality then implies that either

$$d(S_t(y_1), S_t(x)) > \epsilon \tag{4.56}$$

or

$$d(S_t(y_2), S_t(x)) > \epsilon \tag{4.57}$$

There thus exists a point y in any ball $B(x, \delta)$ such that $d(S_t(y), S_t(x)) > \epsilon$. The system is sensitive to initial conditions.

Expansive maps. A system is called expansive if there exists a constant $\epsilon > 0$ such that for any $x, y \in X$ with $x \neq y$ there exists a time $t > 0$ such that $\| S_t(x) - S_t(y) \| > \epsilon$. The constant ϵ is called the expansion coefficient. Every expanding system is sensitive to initial condition since all nearby points separate by at least ϵ. Sensitivity to initial conditions, on the other hand, only requires that there is at least one nearby point that separates by ϵ. Thus sensitivity to initial condition does not imply an expanding map.

Definitions of chaos. There are many definitions of chaos. One is due to Devaney (2003) and requires that

(1) There exists at least one point with a dense orbit, i.e., there exist at least one $x \in X$ such that $\overline{O(x)} = X$.
(2) There exists a dense set of points with periodic orbits, i.e., there exists a dense set $D \subseteq X$ such that $O(x)$ is a periodic orbit for all $x \in D$.

These two conditions imply that any point is arbitrarily close to both a periodic orbit and a dense orbit. These are very demanding conditions and also hard to visualize if we exclude the trivial case that the state space consists of a single periodic orbit.

The Bernoulli shift meets this definition of a chaotic system. Its orbits $O(x)$ are dense for irrational x and periodic for rational x. Thus there exists more than one dense orbit and a dense set, the rational numbers, with periodic orbits.

Note that any one of the two conditions alone does not imply irregular behavior. To see this consider the linear map on the unit circle on the complex plane given by

$$z_{n+1} = e^{2\pi i \varphi} z_n \tag{4.58}$$

which describes successive rotations by the angle φ, a quite benign behavior. If the number φ is irrational any orbit is dense but there are no periodic orbits. If φ is a rational number the map has periodic orbits and they are dense but there is no single orbit that is dense. The map thus does not meet Devaney's criteria for chaos.

Other definitions for chaos require that the system is:

(1) topologically transitive, and
(2) the periodic orbits are dense.

This definition implies sensitivity unless the system consists of a single periodic orbit. Alternative definitions require the system

(1) to be topologically transitive, and
(2) to be sensitive to initial conditions

or

(1) to be topologically transitive, and
(2) to have positive Liapunov exponents $\lambda(x) > 0$ for all $x \in X$.

The important point that follows from this discussion is:

There is no generally accepted definition of chaos.

There are many theorems which relate the various definitions of chaotic behavior and the various ingredients (denseness of orbits, transitivity, mixing, sensitivity to initial conditions, and Liapunov exponents). These theorems require various assumptions about the underlying topological or metric space and the dynamical system. There are qualifiers in these theorems such that there should exist no isolated points, the state space should not consist of a single periodic orbit. One has to read the fine print.

As we have seen, one-dimensional maps like the Bernoulli shift can be chaotic. Iterative maps can thus be chaotic in all dimensions. For continuous-time systems chaos is only possible for systems with dimension $N \geq 3$. In two dimensions a trajectory can never cross into the interior or exterior of a periodic orbit to 'fill out' the plane. The logistic map $F(x) = rx(1-x)$ shows chaotic behavior for certain parameter values r. The time-continuous logistic map $\dot{x}(t) = rx(t)(1-x(t))$, on the contrary, has the quite benign analytical solution

$$x(t) = \left[1 - \frac{x(0)-1}{x(0)} e^{-rt}\right]^{-1} \tag{4.59}$$

Some systems are chaotic for all $x \in X$ others only for some invariant subsets. These invariant subsets might have a complicated structure. An example is the logistic map $F(x) = rx(1-x)$ for parameter values $r > 4$. For such parameter values most points leave the interval $[0, 1[$ but there are some points that stay within the interval. These points form a Cantor set. It is an invariant set and the map is chaotic on this set.

The notion of chaos has been developed for dynamical systems with a continuous state space. If the state space is discrete and finite and if one chooses the discrete topology where every subset is open (and closed) then one has for every subset $A = \mathring{A} = \overline{A}$. Thus a periodic orbit that visits all points satisfies the above criterion for chaos but does not capture our intuitive notion of chaotic behavior. Dynamical systems on finite state spaces can, however, exhibit, sensitivity to initial conditions.

Chaos is a much celebrated and often overused (see Bricmont Bricmont 1995) concept in dynamical systems theory. It is, however, the sensitivity to initial conditions which is of practical relevance since it limits predictability. Here the important theorems are: The existence of a dense and dense periodic orbits imply the sensitivity to initial conditions for iterative maps, except when the complete state space consists of a single periodic orbit. For time-continuous systems topological mixing (instead of topological transitivity) and dense periodic orbits imply sensitivity to initial conditions. Also, sensitivity to initial conditions does not imply chaos. The system

$S_t(x) = xe^t$ is sensitive to initial conditions but not chaotic in any sense. The important point is, however,:

In applications, sensitivity to initial conditions is more relevant than chaos.

Chaos in a linear system of infinite dimension. Chaos requires nonlinearity, except when the state space has infinite dimension, as the following example shows: Consider the set l^2 consisting of all sequences $x = (x_0, x_1, x_2, \ldots)$ that are square summable, $\| x \|^2 = \sum_{i=0}^{\infty} |x_i|^2 < \infty$. The natural inner product is given by $\langle x, y\rangle := \sum_{i=0}^{\infty} x_i^* y_i$ for $x, y \in l^2$. This inner product makes l^2 a separable infinite dimensional Hilbert space.

Define the shift map T_+ by

$$T_+(x_0, x_1, x_2, \ldots) = (x_1, x_2, \ldots) \tag{4.60}$$

It is a very benign linear operator with $\| B(z) \| \leq \| z \|$. All trajectories converge to the fixed point $x = 0$. Now consider the operator $D := 2T_+$ which is also a linear operator. It is a surprising result, first shown by Rolewicz (1969), that this linear operator is chaotic. We explicitly prove this since it is another example of the fact that properties qualitatively change when the limit $n \to \infty$ is taken.

First of all notice that applying the operator n times yields $D^n(x_0, x_1, \ldots) = 2^n(x_n, x_{n+1}, \ldots)$. Let y be a point of l^2. Then the point

$$x(n) := \left(y_0, \ldots, y_{n-1}, \frac{y_0}{2^n}, \ldots, \frac{y_{n-1}}{2^n}, \frac{y_0}{2^{2n}}, \ldots, \frac{y_{n-1}}{2^{2n}}, \ldots\right) \tag{4.61}$$

is an element of l^2, The orbit $O(x(n))$ is a periodic orbit with period n, and $x(n)$ approaches y as n goes to infinity. Hence these periodic orbits are dense in l^2. In addition, if A and B are two open subsets of l^2 and $x \in A$ and $y \in B$ then we find for the point

$$z(n) := \left(y_0, \ldots, y_{n-1}, \frac{x_0}{2^n}, \ldots, \frac{x_{n-1}}{2^n}, 0, 0, \ldots\right) \tag{4.62}$$

that $z(n) \to y$ and $D^n(z(n)) \to x$ as $n \to \infty$. For large enough n we hence find $z(n) \in B$ and $D^n(z(n)) \in A$. The operator D is transitive, even mixing. Mixing and dense periodic orbits imply chaos.

Quantum chaos. A final remark on chaos concerns the notion of quantum chaos. Quantum systems are linear dynamical systems in an often infinite dimensional Hilbert space but they do not show chaotic behavior. The distance between two states does not change in time because the time evolution is governed by the unitary operator $\exp\{-itH/\hbar\}$, hence

$$\| S_t(x) - S_t(y) \| = |\exp\{-itH/\hbar\}| \, \| x - y \| = \| x - y \| \tag{4.63}$$

since the magnitude of the exponential is one. One of course has to show that this is also true for the exponential of an operator but it is. Unitary operators conserve the scalar product.

What goes under the name of quantum chaos arises from the following problem. Quantum mechanics does not show any chaos but classical systems do. Classical mechanics should correspond to quantum mechanics or should be obtained from QM in the classical limit. Or put another way: the true theory is quantum mechanics. So, how can it happen that classical systems become chaotic. What are the conditions in QM that enable chaos to occur in the classical limit?

4.12 Bifurcations

We found that the trajectories of a dynamical system might depend sensitively on the initial condition. We also found that the trajectories might depend sensitively on parameter values, such as in the case of two uncoupled oscillators where the behavior depends on whether or not the ratio of the two frequencies is rational. Here we consider in more detail the sensitivity to parameter values.

If the dynamical system $(S_t)_{t\in T}$ depends on a parameter p, then the state space portrait will also depend on this parameter. Small smooth changes in this parameter may produce sudden qualitative changes in the portrait and in the behavior of the system. If this is the case the dynamical system is said to have gone through a bifurcation. At the bifurcation the orbits may change their stability, split into new orbits, or merge with other orbits. Bifurcation theory studies these changes, not only for dynamical systems but also for other families of objects that depend on parameters. Bifurcation theory distinguishes between local bifurcations where qualitative changes occur only in the neighborhood of a point, usually a fixed point, and global bifurcations where larger sets get transformed.

Bifurcation analysis of fixed points. Fixed points of ODEs $\dot{x}(t) = u(x(t))$ are defined by $u(x_*) = 0$. Consider a family of velocity fields $u(x; p)$, characterized by a parameter $p \in \mathbb{R}^L$. The equilibrium curve $u(x_*; p) = 0$ then describes the fixed points in $\mathbb{R}^N \times \mathbb{R}^L$ space. A bifurcation or branch point is defined as a point (x_0, p_0) on this curve where the number of fixed points changes in the neighborhood of x_0 when p is smoothly varied in the neighborhood of p_0.

The equilibrium curve $u(x_*; p) = 0$ can only be determined explicitly for special cases. However, one can study and classify bifurcations by a Taylor expansion. This expansion gets of course the more complicated the higher the dimensions N and L are. The simplest case is $N = L = 1$. Without loss of generality assume $(x_0, p_0) = (0, 0)$. Taylor expansion at this point along the equilibrium curve $u(x_*, p) = 0$ gives

$$0 = c_1 x_* + c_2 p + c_3 x_*^2 + c_4 p^2 + c_5 x_* p + \ldots \qquad (4.64)$$

If $c_1, c_2 \neq 0$ one obtains the regular behavior

$$x_*(p) = -\frac{c_2}{c_1} p \tag{4.65}$$

with no bifurcation. If $c_1 = 0$ one has to include the next higher order term and obtains

$$x_*(p) = \pm\sqrt{-p\frac{c_2}{c_3}} \tag{4.66}$$

where the sign of p has to be chosen such that one obtains two real roots. This case describes a bifurcation where the number of fixed points changes from zero to two as the parameter p changes its sign. The equilibrium curve turns at (0, 0) and this bifurcation is therefore called a regular turning point. If $c_3 = 0$ as well one has to include the next higher order term. Using this procedure, the following classification of bifurcation is obtained and shown in Fig. 4.10:

(1) Regular behavior: regular behavior with no bifurcation is obtained when $c_1, c_2 \neq 0$.
Example: $u(x; p) = x - p$
(2) Regular turning point: a regular turning point is a point where dp/dx_* changes sign and where $c_1 = 0$ and $c_2, c_3 \neq 0$.
Example: $u(x, p) = x^2 - p$
(3) Double point: a double point is a singular point through which pass two and only two branches of the equilibrium curve with distinct tangents. It requires $c_5^2 > 4c_3c_4$.
Example: $u(x; p) = x^2 - p^2$
(4) Pitchfork bifurcation: the pitchfork bifurcation is a double point at which dp/dx_* changes sign on one branch of the equilibrium curve.
Example: $u(x; p) = (x^2 - p)x$
(5) Cusp: a cusp is a singular bifurcation where both dp/dx_* and d^2p/dx_*^2 change sign.
Example: $u(x, p) = x^2 - p^3$
(6) Conjugate point: a conjugate point is an isolated fixed point.
Example: $u = x^2 + p^2$

This list exhausts the cases where at least one of the second derivatives is unequal to zero. More cases arise when one needs to include higher order terms in the Taylor expansion. For completeness the regular behavior is included although it does not represent a bifurcation. A bifurcation is called singular if $c_1 = c_2 = 0$, otherwise regular.

The figure also shows the linear stability of the fixed points, determined by the eigenvalue of the linearized flow field at the fixed point. As one observes, the fixed points change their linear stability at the bifurcation point. At the bifurcation the eigenvalue is zero. A change of the parameter thus causes a change of the stability of the system.

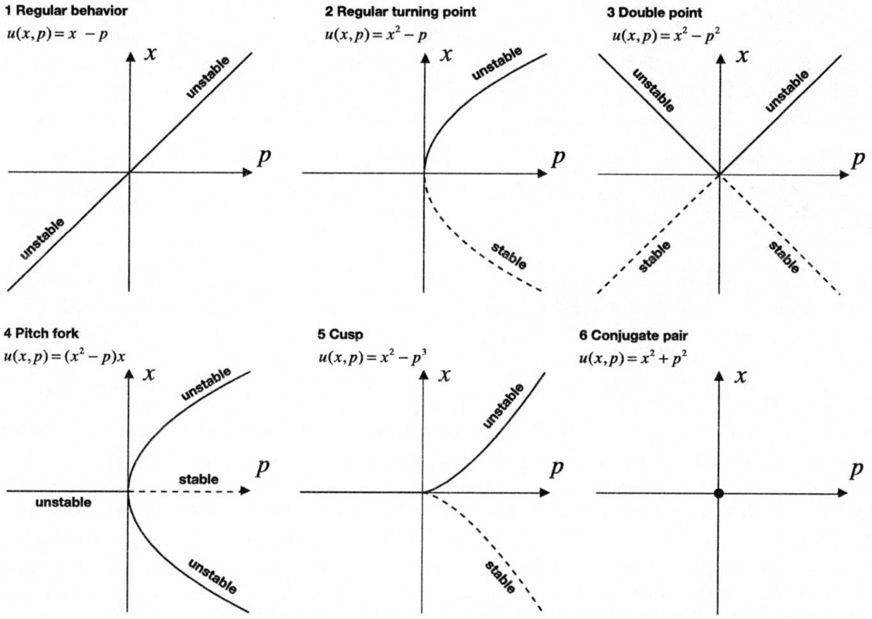

Fig. 4.10 Classification of bifurcations for a one-dimensional flow $u(x, p)$ with an equilibrium curve $u(x_*, p) = 0$ where at least one of the second derivatives is unequal to zero

Similar results hold for higher dimensions $N > 1$. The linear stability is then determined by the eigenvalues of the velocity gradient matrix. At the bifurcation at least one of the eigenvalues lies on the imaginary axis. A change of the parameter then causes a change of the stability of the system. If this eigenvalue is zero, then the bifurcation is often called a steady state bifurcation. If there is a complex pair of eigenvalues at the bifurcation with purely imaginary parts, then the bifurcation is called a Hopf bifurcation. The explicit analysis for such higher dimensional systems gets successively more complicated, as one might imagine, and there exists a plethora of bifurcation, not yet completely categorized and named. Beware that different authors often categorize bifurcations by different names.

Example: The Lorenz system. The bifurcation analysis of the Lorenz system is given in Fig. 4.11.

Iterative maps. The fixed points of iterative maps $x_{n+1} = F(x_n)$ are given by $F(x_*) = x_*$. Bifurcations now have eigenvalues on the unit circle. If the eigenvalue is $+1$ the bifurcation is called a pitchfork bifurcation. If the eigenvalue is -1 the bifurcation is a period-doubling bifurcation and if there is a pair of complex eigenvalues with non-zero imaginary parts it is a Hopf bifurcation. The period doubling bifurcation is a bifurcation in which the system moves to a new behavior with twice the period of the original system. The logistic map is the standard example for this type of bifurcation (see Fig. 4.12). The logistic map shows successive period dou-

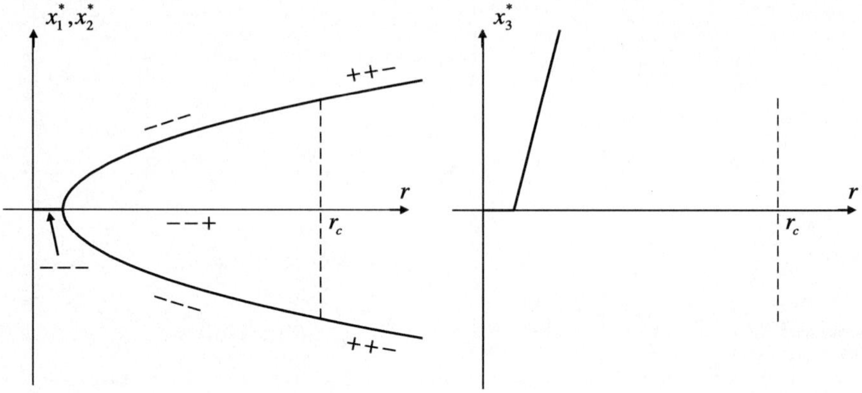

Fig. 4.11 Bifurcation analysis of the Lorenz system. Varying the control parameter r of the Lorenz system corresponds to the case $N = 3$, $L = 1$. For $0 \leq r < 1$ the Lorenz system has one fixed point $x_*^{(1)} = (0, 0, 0)$. This fixed point is stable and corresponds to a state of pure molecular heat conduction without any motion. At $r = 1$ this first fixed point becomes unstable and two additional fixed points appear $x_*^{(2,3)} = (\pm\sqrt{b(r-1)}, \pm\sqrt{b(r-1)}, r-1)$. The new fixed points correspond to a state of convective rolls. They have one real eigenvalue and a pair of complex conjugate eigenvalues. They are stable up to $r = r_c$ where $r_c = \sigma\frac{\sigma+b+3}{\sigma-b-1}$ and are unstable beyond. In the range $r > r_c$ none of the fixed points is stable

bling until the system becomes chaotic. This is a general behavior of many systems and called the Feigenbaum route to chaos. The important point is:

The behavior of a dynamical system can change drastically when a parameter value changes slightly.

Global bifurcations. The successive period doubling of the logistic map is actually an example of a non-local bifurcation. One periodic orbit turns into another periodic orbit of twice the period. The effect of period doubling is not confined to a point but to many points. Generally, global bifurcations occur when such larger invariant sets get transformed. Such larger sets may include periodic orbits but also more complicated sets such as chaotic attractors. Changes in the parameter then cause changes in the state space portrait which cannot be confined to a small neighborhood, as is the case for the bifurcation of fixed points.

Structural stability. One way to study the structural stability of a system with velocity field $u(x; p)$ is to introduces the system $u(x, p, \delta)$ where δ is an additional small parameter and asks what happens to the bifurcations. One calls the system with $\delta = 0$ the perfect system and the perturbed system with $\delta \neq 0$ the imperfect system. One now has to perform a bifurcation analysis with respect to two parameters. Consider especially the system $u(x, p, \delta) = u(x, p) + \delta$ where the perfect system

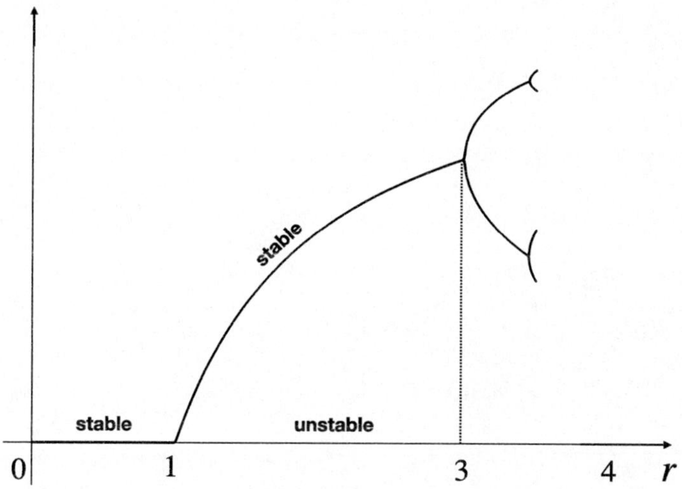

Fig. 4.12 Bifurcation analysis of the logistic map $F(x) = rx(1 - x)$ with the control parameter $r \in [0, 4]$. This map has a fixed point $x_*^{(1)} = 0$. For $0 \le r < 1$ this fixed point is stable. At $r = 1$ the eigenvalue becomes 1 and one encounters a pitchfork bifurcation at which there occurs a second fixed point $x_*^{(2)} = (r - 1)/r$. The first fixed point becomes unstable. The new second fixed point is stable until a bifurcation at $r = 3$ where the eigenvalue becomes -1 and where this second fixed point becomes unstable. A stable orbit of period 2 emerges which is a fixed point of the twice iterated map $F \circ F$. As one further increases the value of r one observes successive period doubling until at the end of this period doubling cascade one reaches a chaotic regime, which is, however, interspersed with periodic regimes. At $r = 4$ the map is chaotic

represents the cases discussed in Fig. 4.10. One then obtains the results shown in Fig. 4.13. The turning point is structurally stable. The imperfection only induces a translation of the bifurcation point. The other cases are structurally unstable. The bifurcation exhibits a qualitative change in character.

It is emphasized that bifurcation analysis compares dynamical systems with different parameter values. Bifurcation analysis does not address the dynamical problem of how a given system responds when a parameter is changed in time even when it is changed slowly. This latter problem leads to a non-autonomous dynamical system with a time-dependent parameter. This is an important point:

Bifurcation analysis does not give the response to a time-varying parameter.

4.13 Summary

This chapter addressed the stability of fixed points and periodic orbits and the long-time behavior of non-periodic orbits. The analysis required the topological concepts

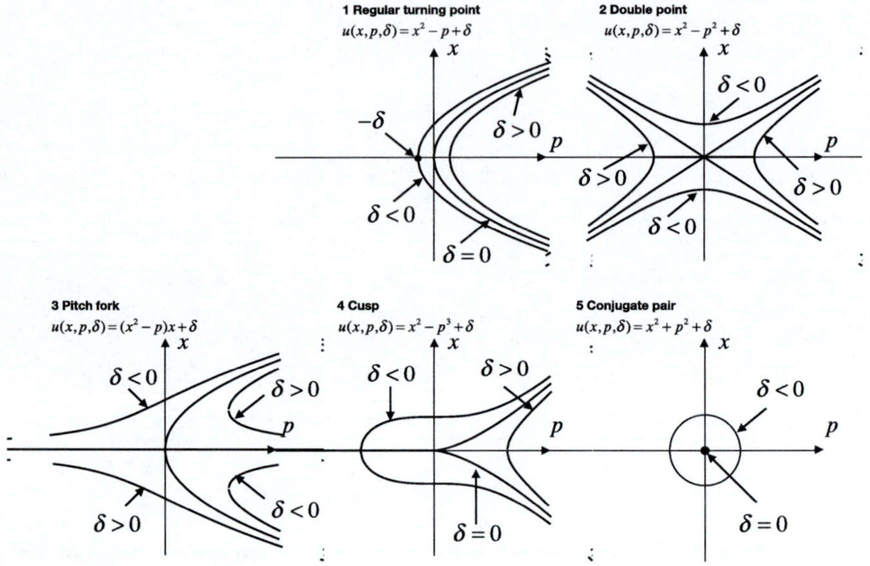

Fig. 4.13 Structural stability of the bifurcations considered in Fig. 4.10

of neighborhood and convergence. In most situations the topology is realized by a metric, norm or scalar product.

The stability of fixed points can be assessed by a local linear analysis which requires the determination of the eigenvalues of the linear tangent map. The linear stability analysis is sufficient to determine the stability of finite-dimensional systems, unless the fixed point turns out to be neutrally stable. Then a higher order analysis is required. For infinite-dimensional systems, such as PDEs, the stability analysis is more complicated and depends on the type of PDE. There might exist different non-equivalent norms and examples where linear stability does not imply (nonlinear) stability. The linear stability analysis of periodic orbits can be reduced to the stability analysis of the corresponding fixed point of the Poincaré map.

A more global view of the stability of fixed points can be obtained with the help of a Liapunov function. This method does not require the calculation of eigenvalues but finding such a Liapunov function requires some serendipity, except for some special cases, among them gradient flows and Hamiltonian systems.

Instability might result in sensitivity to initial conditions, a property of great practical importance since it limits predictability.

To analyze the long-time behavior of non-periodic orbits one introduces limiting sets and attractors. Attractors can have a very complicated structure, including fractal dimension. Fixed points, periodic orbits, attractors and their basins of attraction are the basic elements of the state space portrait of the system, which gives the most comprehensive description of a dynamical system. An important theorem, the Poincaré-Bendixson theorem states that the state space portrait of continuous planar one- and two-dimensional systems consists only of fixed points, periodic orbits, connecting arcs and orbits that approach the fixed points, periodic orbits or connecting arcs or escape to infinity. Discrete-time systems of any dimension and continuous-time systems of dimension three and higher can exhibit much more irregular long-time behavior. One such irregular behavior is chaotic behavior. Chaotic behavior is a fairly abstract topological property. There is no generally accepted definition for chaos. Ingredients of chaos include sensitivity to initial conditions, dense orbits, topological transitivity and mixing. The discovery that fairly simple systems can exhibit chaotic behavior is regarded as a paradigm change and has been studied extensively, mostly for low-dimensional systems.

As parameter values vary smoothly the system might undergo bifurcations, profound changes in its state space portrait.

Overall, nonlinear systems show a much richer behavior than linear systems. Nonlinear systems can have multiple isolated equilibrium points, they may exhibit limit cycles, sensitivity to initial conditions, unpredictability, chaotic behavior and states can reach infinity in finite time.

Further details can be found in Bricmont (1995), Devaney (2003), Drazin (1992), Ford (1992), Lorenz (1993), Schuster (1995), and Strogatz (2000).

References

Bricmont, J.: Science of chaos or chaos in science. Phys. Mag. **17**(3–4), 159–208 (1995)
Devaney, R.I.: An Introduction to Chaotic Dynamical Systems. Westview Press (2003)
Drazin, P.G.: Nonlinear Systems. Cambridge University Press (1992)
Ford, J.: The fermi-pasta-ulam problem: paradox turns discovery. Phys. Rep. **213**(5), 271–310 (1992)
Lorenz, E.: The Essence of Chaos. University of Washington Press, Washington (1993)
Rolewicz, S.: On orbits of elements. Stud. Math. **32**, 17–22 (1969)
Schuster, H.G.: An Introduction: Deterministic Chaos. Wiley-VCH (1995)
Strogatz, S.: Nonlinear Dynamics and Chaos. Perseus Publishing (2000)
Strauss, W.: Tutorial: Notes on stability theory. Stability Workshop, Seattle, June 11-12, (2012)

Chapter 5
Ergodic Theory

5.1 Introduction

The evolution of volumes provides useful information about dynamical systems. Conservative systems that conserve volume cannot have attractors, only dissipative systems that do not conserve volume can. Volume is but a special kind of measure and studying the behavior of more general measures provides profound insights into dynamical systems, allowing to address questions such as whether points return to the set from which they originated, whether they visit other sets, what is the averaged behavior of observables along trajectories, and what may be regarded as typical behavior.

Measures are studied by *measure theory* which is a well-developed mathematical theory. The basis of measure theory is a *measure space*: a triple that consists of a basic set, a σ-algebra, which is the family of its measurable subsets, and a measure that assigns to each measurable subset a non-negative real number, the measure of that subset. We thus assume the state space of the dynamical system to be a measure space. To make the dynamical system compatible with the measure space one assumes that the dynamical map is a measurable function, similarly to assuming that the dynamical map is a continuous function when studying the topological properties of systems. In this chapter we thus consider the quadruple consisting of a state space, a sigma-algebra, a measure, and a family of measurable dynamical maps. Such systems are studied by ergodic theory.

Among the many possible measures we only consider invariant measures. Invariant measures are those that are not changed by the dynamical evolution. The dynamical evolution preserves these measures or is measure preserving. The most succinct results are obtained for finite measures which can be normalized to one. Such normalized measures are probability measures which are considered in detail in part II of this book to characterize uncertainty. It is, however, stressed that the finite invariant measures considered here are a property of the dynamical system. There is nothing

P. Müller, *Handbook of Dynamics and Probability*,
https://doi.org/10.1007/978-3-030-88486-4_5

random or probabilistic about them. The case that a researcher assigns a probability measure to initial conditions to account for incomplete knowledge of the initial condition and then follows the evolution of this probability measure is treated in the Chap. 11, "Time evolution of broadened initial conditions".

A given dynamical system might have no invariant measure, just one invariant measures or many invariant measures. The existence of a finite invariant measure implies two remarkable theorems:

- the recurrence theorem of Poincaré which states that every set of positive measure will be revisited by almost all of its points infinitely often, and
- the existence theorem of Birkhoff and Khinchin which states the existence of the time mean of measurable functions for almost all points of the state space.

An important subclass of invariant normalized measure are ergodic measures. A measure is ergodic if all invariant subsets have either the measure one or zero. Ergodicity implies two important properties: First, the time mean of measurable functions equals the expectation value of the measurable function almost surely. The time mean equals the ensemble mean. Second, ergodic measures are the extreme points of the set of all invariant measures which form a convex set. These two properties may actually serve as alternative definition of ergodic measures. A theorem by Bogoliubov and Krilov states that, for certain classes of systems, there exists at least one invariant normalized measure. A dynamical system together with an ergodic measure is called an ergodic system.

The etymology of the word "ergodic" and the origin of ergodic theory trace back to Boltzmann. He called a stationary distribution on the phase space of a Hamiltonian system a *monode* and an *ergode* a monode characterized by the uniform distribution on the surface of constant energy, a distribution that we now call the *micro-canonical ensemble*. He also asserted that there may exist systems for which a single trajectory 'invades densely' all of the energy surface. This assertion is now known as the ergodic hypothesis. Ergodic theory grew out of studies that aimed at giving a precise meaning to this assertion. It does not mean that a trajectory actually reaches every point in state space which is not possible in higher-dimensional state spaces; a line cannot cover a surface. It also does not mean that a trajectory comes arbitrarily close to every point in state space, in the topological sense. Ergodicity is a measure-theoretical concept. Proper formulations of the ergodic hypothesis are: given two measurable subsets of non-zero measure then there will always be a trajectory starting in one of the subsets that will reach the other subset in finite time, or: a trajectory visits every measurable set of the state space for a duration proportional to the measure of this set.

At its heart ergodicity states that the system is irreducible, it cannot be decomposed into subsystems that can be studied separately. Ergodicity does not imply complexity in any sense. Indeed, we find many ergodic systems which are extremely simple. More complex ergodic systems are obtained when one requires more stringent properties of the ergodic system such as mixing and exactness.

Also be aware that measure theory always assures validity of a statement with the qualifier 'for almost all points', meaning except for a set of measure zero. A set of measure zero might be, as we will find out, quite large by any other standard.

In this chapter, we denote measures by the symbol μ and by P when they are normalized to one. In part II of this book we will interpret such normalized measures as probabilities but in this chapter the intuitive notion of a measure should be one of fraction or volume rather than one of probability.

5.2 The Evolution of Volumes

Volume is a particular measure and we may ask whether or not the volume of a set A changes when transformed by the flow of a dynamical system S_t. From calculus we know that

$$V(A) = \int_A dy = \int_{S_t^{-1}(A)} J_t(x)dx \tag{5.1}$$

where

$$J_t(x) := \frac{\partial(S_t(x))}{\partial(x)} \tag{5.2}$$

is the Jacobian of the map S_t. To be volume preserving the dynamical map must therefore satisfy $J_t(x) = 1$. The infinitesimal form of (5.1) is

$$dy = J_t(x)dx \tag{5.3}$$

If the system is a time-continuous flow with velocity field $u(x)$ time-differentiation of (5.1) yields the Euler formula

$$\frac{1}{J_t}\frac{d}{dt}J_t = \nabla^T u \tag{5.4}$$

which implies that the flow is volume preserving if the velocity is non-divergent.

The deformation of volumes is one of the many characteristics of fluid flows considered in fluid dynamics. Other characteristics include the deformation of surfaces and lines. It should be noted that the description of fluid flows in fluid dynamics is something of an afterthought. The main task of fluid dynamics is the calculation of the fluid velocity field from the basic fluid dynamical equations, such as the Navier Stokes equations. The description of the resulting velocity field is considered mere kinematics.

More importantly, the volumes in fluid dynamics are actual volumes in three-dimensional space, consisting of actual fluid particles. Sets and their volumes in the state space of a dynamical system describe an ensemble of system, possible systems differing in their initial state, a set of parallel universes. This must be kept in mind when using fluid dynamical terms to describe flows in state space.

5.3 Measure Spaces

Measure spaces generalize the concept of volume. Here we list some basic elements of measure theory. More details can be found in Appendix E.13.

Measure spaces. A measure space is the triple $\{X, \mathcal{A}, \mu\}$ consisting of:

(1) A basic set X.
(2) A family of subsets $\mathcal{A}$ of X called a σ-algebra, in which one can perform all countable set operations without leaving the family.
(3) A function $\mu : \mathcal{A} \mapsto [0, \infty]$ that assigns a non-negative number $\mu(A)$, called the measure of A, to each element A of the σ-algebra $\mathcal{A}$.

Borel set. If $X = \mathbb{R}$ then the σ-algebra that contains all finite half open intervals $]a, b]$ is called the Borel set and denoted by $\mathcal{B}$. The Borel set is larger than the set of all countable unions of half open intervals and smaller than the power set of $\mathbb{R}$, though it is hard to explicitly construct non-Borel sets. The analogous construction in $\mathbb{R}^N$ leads to the Borel set $\mathcal{B}^N$.

Lebesgue measure. The measure that assigns to all half open intervals $]a, b]$ on the real line the length $b - a$ can be extended to a measure on the Borel set $\mathcal{B}^1$. This extended measure is called the Lebesgue measure and denoted by μ_L^1. Since the Borel set is smaller than the power set of $\mathbb{R}$ there exists sets which are not measurable. If these sets are subsets of measurable sets of measure zero then one can assign measure zero to these sets as well. This extension of the Lebesgue measure is called the Borel measure. The Lebesgue measure μ_L^N in N dimensions can be defined by an analogous construction. This Lebesgue measure is the proper formulation of what one usually calls the volume.

Measurable functions and their Lebesgue integral. A function $f : X \mapsto \mathbb{R}$ is called measurable if $f^{-1}(B) \in \mathcal{A}$ for all $B \in \mathcal{B}$. An integral can be defined for measurable functions f that generalizes the Riemann integral. It is written as

$$\int_A \mu(dx) f(x) \tag{5.5}$$

and is called the Lebesgue integral. Its exact definition is given in Appendix E.13. The set of measurable functions for which this integral exists is the function space $\mathcal{L}_1$ of all integrable functions.

Finite measures. A measure is called finite if $\mu(X) < \infty$. Such measures can be normalized to 1 and will then by denoted by the special symbol

$$P(A) := \frac{\mu(A)}{\mu(X)} \tag{5.6}$$

These normalized measures constitute the basis of probability theory and are called probability measures in that context. In the context of this chapter, however, measures do not imply anything random or probabilistic but rather denote normalized general volumes. We therefore use the name 'normalized measure' for P in this chapter. When the measure is finite the space of integrable function $\mathcal{L}_1$ consists of all bounded measurable functions $X \mapsto \mathbb{R}$.

For normalized measures the expectation of a measurable function f is denoted by cornered brackets and defined by

$$\langle f \rangle := \int_X P(dx) f(x) \tag{5.7}$$

Densities. Certain normalized measures can be represented by a density $\rho : X \mapsto \mathbb{R}$ with $0 \leq \rho(x) \leq 1$ for all $x \in X$ such that

$$P(A) = \int_A \mu(dx)\rho(x) \;\; \forall \; A \in \mathcal{A} \tag{5.8}$$

where the integral is the Lebesgue integral with respect to another measure μ. Often μ is the Lebesgue measure and then we write $\mu_L^1(x) = dx$ to follow standard notation. A condition for the existence of such a density is given in the Radon-Nikodym theorem discussed in Sect. E.13. The density $\rho(x) = \chi_X(x)/\mu(X)$ represents the uniform distribution. It is constant.

5.4 Invariant Measure

We now endow the state space of the dynamical system with a measure. Our starting point is thus a quadruple $\{X, \mathcal{A}, \mu_0, (S_t)_{t\in T}\}$ consisting of a state space X, a σ-algebra $\mathcal{A}$, a measure μ_0 and a family of dynamical maps $(S_t)_{t\in T}$. For the dynamical map and the measure to be compatible the maps will be assumed to be measurable functions so that $S_t^{-1}(A) \in \mathcal{A}$ for all $A \in \mathcal{A}$.

Induced measure. A dynamical system $(S_t)_{t\in T}$ induces the measures

$$\mu_t(A) := \mu_0(S_t^{-1}(A)) \tag{5.9}$$

or $\mu_t = \mu_0 \circ S_t^{-1}$ for short. The measure μ_t assigns to a set A the measure μ_0 of the pre-image $S_t^{-1}(A)$. The induced measure μ_t acts on the same σ-algebra as μ_0.

To understand the usefulness of this definition recall that a set in the state space of a dynamical system describes an ensemble of systems. The Definition 5.9 insures that no member from this ensemble gets lost from this ensemble during the time evolution.

Expectation. Consider observables $f_0 : X \mapsto \mathbb{R}$ that do not explicitly depend on time. We assume these observables to be measurable functions. The compositions $f_t := f_0 \circ S_t$ are then also measurable functions for which one finds

$$\int_A \mu_t(dy) f_0(y) = \int_{S_t^{-1}(A)} \mu_0(dx) f_t(x) \tag{5.10}$$

which is a straightforward consequence of a change of variables $y = S_t(x)$. If μ_0 is a normalized measure P_0 then one finds for the expectation

$$\begin{aligned} \langle f_t \rangle := & \int_X P_0(dx) f_t(x) \\ = & \int_X P_t(dx) f_0(x) \end{aligned} \tag{5.11}$$

since $S_t^{-1}(X) = X$. This formula shows that the expectation of f_t can be calculated in two different ways: One either keeps the initial normalized measure fixed and lets the observable evolve, or one keeps the initial observable fixed and let the normalized measure evolve. In quantum mechanics, the first line corresponds to the Heisenberg and the second line to the Schrödinger representation.

Invariant measures. In general the two normalized measures μ_0 and μ_t differ. Of particular importance are measures μ_* that do not change in time, for which

$$\mu_*(S_t^{-1}(A)) = \mu_*(A) \tag{5.12}$$

for all $A \in \mathcal{A}$ and all times $t \in T$. Such measures are called invariant (or stationary) measures. Instead of saying that the measure μ_* is invariant under the map S_t one also says that the map S_t preserves the measure μ_* or is measure preserving. Formula (5.10) then takes the form

$$\int_A \mu_*(dx) f_0(x) = \int_{S_t^{-1}(A)} \mu_*(dx) f_t(x) \tag{5.13}$$

and (5.11) implies the stationarity of the expectations.

$$\langle f_0 \rangle = \langle f_t \rangle =: \langle f \rangle \tag{5.14}$$

Next we survey whether such invariant measures exist and whether or not they are unique.

Hamiltonian flows are non-divergent and hence preserve the Lebesgue measure. The over-damped oscillator $\dot{x} = -\gamma x$ is a simple example of a flow that is divergent and hence does not preserve the Lebesgue measure. The Lorenz system is also a flow with a divergent velocity field and hence does not preserve the Lebesgue measure. Among our iterative maps, the doubling map preserves length and the baker transformation preserves area and these maps hence preserve the Lebesgue measure. The logistic map and the Ulam map, on the other hand, do not preserve the Lebesgue measure.

There thus exists dynamical systems that do not preserve the Lebesgue measure. Do they perhaps preserve other invariant measures. The answer is positive. An invariant measure of the over-damped oscillator $\dot{x} = -\gamma x$ is the Dirac measure which concentrates all the measure on the point $x = 0$ and is formally defined by

$$\delta_0(A) := \chi_A(0) \tag{5.15}$$

where χ_A is the indicator function of A. Note that this is true independent of the sign of γ. Also note that this statement requires that the over-damped oscillator is considered on the state space $\mathbb{R}$. If it is considered $\mathbb{R}_+$, the positive real numbers, then there does not exist any invariant measure. The case of the Lorenz system is more complicated. The support of the invariant measure are the attractors and they are hard to specify once the attractors become chaotic. In Sect 11.3 it will also be proven that an invariant measure of the Ulam map is given by the density $\rho_*(x) = (\pi\sqrt{x(1-x)})^{-1}$.

The second question is whether a dynamical system may preserve more than one invariant measure. The answer to this question is positive as well, as shown by the following example.

Example: Hamiltonian systems. Consider Hamiltonian systems with time-independent Hamilton function $H(q, p)$ on the state space Γ. The Lebesgue measure is a non-finite invariant measure. In addition, all measures with densities $\rho_*(q, p) = f(H(q, p))$ that depend solely on the energy are finite invariant measures. A special subset among these are the measures that represent uniform distributions on energy hyper-surfaces $\Gamma_E = \{(q, p)|H(q, p) = E)$. We assume that these measures can be represented by the density

$$\rho_*(q, p) = \frac{\delta(H(q, p) - E)}{\int_\Gamma dqdp\delta(H(q, p) - E)} \tag{5.16}$$

involving a δ function. The denominator serves as a normalization constant. These measures are the backbone of statistical mechanics and are called micro-canonical ensembles there. If the system is the harmonic oscillator $H(q, p) = (q^2 + p^2)/2$ then it is easy to prove that this measure exists and is adequately represented by the δ function. The normalization constant is 2π.

The representation of the micro-canonical ensemble by a δ function also allows an easy explicit proof of its invariance. Without the normalization constant:

$$\begin{aligned} \mu_*(S_t^{-1}(A)) :&= \int_{S_t^{-1}(A)} dqdp\, \delta(H(q, p) - E) \\ &= \int_A dqdp\, \delta(H(S_t(q, p)) - E) \\ &= \int_A dqdp\delta(H(q, p) - E) \\ &= \mu_*(A) \end{aligned} \tag{5.17}$$

The second equality is due to the fact that the Lebesgue measure $dqdp$ is invariant for Hamiltonian flows; the third equality is due to the fact that the Hamilton function is a constant of motion.

These examples show that there are systems with no invariant measures (such as the over-damped oscillator on state space $\mathbb{R}_+$), with just one invariant measure (such as the over-damped oscillator on the state space $\mathbb{R}$), and with more than one invariant measure (such as Hamiltonian systems). Also, the invariant measures can be quite singular, as shown by the Dirac measure and the delta-function densities. The important point is:

Invariant measures are a property of the dynamical system.

Before continuing the analysis of invariant measures we first consider two important consequences of the existence of invariant measures, the Poincaré recurrence theorem and the Birkhoff-Khinchin theorem.

From now on we only consider finite invariant measures normalized to one and denote them by P_*.

5.5 Poincaré Recurrence Theorem

Poincaré recurrence theorem is quite an unexpected implication of the existence of an invariant measure. It holds for a system $\{X, \mathcal{A}, P_*, S_t\}$ with a finite invariant normalized measure $P_*(X) = 1$.

Recurrent points. A point $x \in A$ is called a *recurrent point* in A if for all $t_0 > 0$ there exist a time $t > t_0$ such that $S_t(x) \in A$. Fixed points and points on periodic orbits are trivially recurrent. The qualifier that t needs to be larger than any given $t_0 > 0$ rather than merely $t > 0$ is to stress that the theorem will be a statement about the long term behavior of the dynamical system and to exclude trivial short term behavior. If X were a topological space, A an open subset and S_t a continuous time dynamical system that is continuous in the topological sense then there would exist for every point $x \in A$ an open time interval $t_1 < 0 < t_2$ such that $S_t(x) \in A$ for all $t \in]t_1, t_2[$, and the $t > 0$ criterion would be trivially satisfied.

Poincaré recurrence theorem. We now prove Poincaré's recurrence theorem in three steps following Petersen (1983).

1. If $P_*(A) > 0$ then there exist at least one recurrent point in A. To proof this assertion assume that the contrary is true. Then there are no recurrent points in A and hence $S_t^{-1}(A) \cap A = \emptyset$ for all $t > t_0$ and hence $S_{t'}^{-1}[S_t^{-1}(A) \cap A] = S_{t+t'}^{-1}(A) \cap S_{t'}^{-1}(A)] = \emptyset$ for all $t' > 0$ and all $t > t_0$. Thus the intersections of the preimages are empty if the times are at least t_0 apart. One can thus pick an infinite countable sequence of times $t_1, t_2 \ldots$ such that $S_{t_n}^{-1}(A)$ and $S_{t_m}^{-1}(A)$ are mutually disjoint for $n \neq m$. Then invariance of P_* and σ-additivity give

$$\sum_{n=1}^{\infty} P_*(A) = \sum_{n=1}^{\infty} P_*(S_{t_n}^{-1}(A)) = P_*(\cup_{n=1}^{\infty} S_{t_n}^{-1}(A)) \leq P_*(X) = 1 \tag{5.18}$$

This can only be true if $P_*(A) = 0$, contrary to the assertion. Note that this proof requires $P_*(X) < \infty$, and the invariance of P_*.

2. The set of all non-recurrent points in A has measure zero. This assertion simply follows from the above proof by substituting the set of all non-recurrent points for A.

3. A recurrent point will return to A infinitely often. This assertion follows from repeated application of the above two assertions, making use of the fact that the measure of a countable union of zero measure sets has zero measure.

Together these three assertions constitute Poincaré's recurrence theorem which is usually phrased as follows:

Every set of positive measure in a space of finite measure will be revisited by almost all of its points infinitely often.

The assumption that the state space has a finite measure is important. Consider the flow $\dot{x} = u_0$ with constant velocity u_0 defined on $X = R^N$. The trajectories simply leave any subset and escape to infinity without ever revisiting that subset. The Lebesgue measure is an invariant measure but the recurrence theorem does not apply since the Lebesgue measure of R^N is not finite.

The qualifier 'for almost all points' is also important. Consider the over-damped oscillator $\dot{x} = -\gamma x$ on $X = \mathbb{R}$ with negative γ. All trajectories diverge exponentially. Its invariant measure is given by the Dirac measure δ_0. Consider a finite interval that contains the point $x = 0$. All points in this interval except $x = 0$ will eventually leave the interval and are hence non-recurrent but they carry zero Dirac measure.

Example: Recurrence times of non-harmonic oscillator. Consider a particle in a periodic potential with Hamilton function (Berger 2001)

$$H(q, p) = -\frac{2}{\pi}\left(\cos\frac{\pi q}{2}\right)^2 + \frac{p^2}{2} \tag{5.19}$$

Restrict the analysis to the region X where $-1 \leq q \leq +1$ and $H(q, p) \leq 0$. Since the system is Hamiltonian the Lebesgue measure is an invariant measure. It is also finite. The state space portrait of this region is given in Fig. 5.1.

For $H < 0$ the particle is trapped and describes a periodic motion. For a particle starting at $(q, 0)$ the period is given by

$$T(q) = 2\sqrt{2\pi} \int_0^q dq' \frac{1}{\sqrt{\cos \pi q' - \cos \pi q}} \tag{5.20}$$

Points on these periodic orbits are hence trivially recurrent. Since the integral diverges as

$$T(q) = -\frac{4}{\sqrt{\pi}} \log(1 - q) \tag{5.21}$$

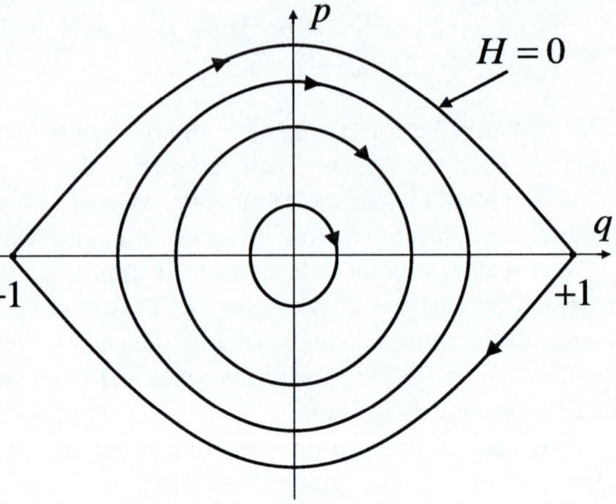

Fig. 5.1 State space portrait of the dynamical system characterized by the Hamilton function (5.19)

as $q \to \pm 1$ the recurrence times can be very long. For $H = 0$ the points ($q = \pm 1$, $p = 0$) are (unstable) fixed points and hence are also trivially recurrent. Points on the arcs connecting the fixed points describe a non-periodic motion, being attracted or repelled by the fixed points. These points on the arc are non-recurrent, but their two-dimensional Lebesgue measure is zero.

This example also shows that recurrent times can be very different for neighboring points. The points (± 1, 0) are fixed points and thus have a recurrence time of zero whereas neighboring points have periods that become the larger the closer these points are to the fixed points.

The recurrence requires that the state space be a measure space. Often the recurrence theorem is formulated for topological spaces in the form: a trajectory will come infinitely often infinitely close to its initial value for almost all initial values. This formulation presupposes that open sets form a σ-algebra, which might not be true. The recurrence theorem is a measure-theoretical statement.

The recurrence theorem has figured prominently in the debate about the foundations of statistical mechanics. Every attempt to derive irreversible macroscopic behavior from microscopic dynamics has to confront the recurrence objection or paradox ("Wiederkehreinwand"). This argument was first raised by Zermelo (1896). Typically it is argued that a gas expanding (irreversibly) from the left half of a container into the whole of the container must eventually, according to the recurrence theorem, return to its original state in the left part of the container, contrary to observations. A possible resolution is that the recurrence times turn out to be extremely long, being estimated to be longer than the age of the universe in the case of a gas in a container. Indeed we will show in Sect. 5.8 below that the mean recurrence time grows exponentially with the number of particles. We come back to these foundational issues in Chap. 19.

5.6 Time Averages

Let P_* be an invariant normalized measure and $f_0 : X \mapsto \mathbb{R}$ be a measurable function. Then the composition $f_t = f_0 \circ S_t$ is also measurable. The existence theorem of Birkhoff-Khinchin states that the time mean or time average

$$\bar{f}(x) := \lim_{T\to\infty} \frac{1}{T} \int_0^T dt f_t(x) \tag{5.22}$$

(or

$$\bar{f}(x) := \lim_{N\to\infty} \frac{1}{N} \sum_{n=0}^{N} f_n(x) \tag{5.23}$$

in the discrete case) exists for nearly all points x or almost everywhere and is a measurable function. The set of points for which it does not exist has P_*-measure zero. Note that the time mean depends on the initial position x. Also note that the invariant measure must be finite. Consider again the flow $\dot{x} = u_0$ with constant velocity u_0 defined on $X = \mathbb{R}^N$ with the invariant Lebesgue measure. The time-average does not exist for the identity $f_0(x) = x$.

As all limits, the time mean depends only on the asymptotic properties and one thus finds

$$\bar{f}_t = \bar{f}_0 =: \bar{f} \tag{5.24}$$

which is the analogue of (5.14). Furthermore, since time and space integration can be interchanged one finds

$$\int_X P_*(dx)\bar{f}(x) = \int_X P_*(dx) f(x) \tag{5.25}$$

using (5.13).

Time averaging provides a way to generate a measure from a single trajectory. One just follows a single trajectory and defines the measure $\bar{P}(A)$ by the relative time that the particle spends in the set A

$$\overline{P_x}(A) := \overline{\chi_A}(x) = \lim_{T\to\infty} \frac{1}{T} \int_0^T dt \chi_A(S_t(x)) \tag{5.26}$$

where χ_A is the indicator function of the set A. This measure is usually presented in the form of a histogram $\overline{P_x}(A_i)$ where $A_i, i = 1, \ldots, N$ is a partition of the state space ($X = \cup_{i=1}^N A_i$ with $A_i \cap A_j = \emptyset$ for $i \neq j$). This measure exists nearly everywhere but may depend on the initial position x. When A is an invariant set then $S_t(A) \subseteq A$ and $\overline{P_x}(A) = 1$ if $x \in A$ and zero otherwise. The questions then arise: what are the conditions that the measure $\overline{P_x}(A)$ is unique and that it is equal to the invariant measure P_*? These questions are addressed next.

5.7 Ergodic Measures

Here we define and discuss the most important subset of finite invariant measures, namely ergodic measures. They are often thought of being associated with highly complex systems, but this is not the case. They only define the measure-theoretical concept of irreducibility. The simpler a system is the more irreducible it might be.

Ergodic measures. There are various equivalent ways to define ergodic measures. Here we define ergodic measures as those invariant measures P_* for which

$$P_*(A) \in \{0, 1\} \tag{5.27}$$

for all invariant sets A of S_t. Recall that a set A is invariant if $S_t^{-1}(A) = A$ for all $t \in T$. The measure of invariant sets is either 0 or 1. This definition captures the notion of irreducibility in the following sense: Assume the measure is not ergodic. Then there exists an invariant set A for which $0 < P_*(A), P_*(X \setminus A) < 1$. Introduce the relative normalized measure measure

$$P(B|A) := \frac{P_*(B \cap A)}{P_*(A)} \tag{5.28}$$

which describe the fraction of B that is in A. This relative measure is also an invariant measure since

$$\begin{aligned} P(S_t^{-1}(B)|A) &= \frac{P_*(S_t^{-1}(B) \cap A)}{P_*(A)} \\ &= \frac{P_*(S_t^{-1}(B) \cap S_t^{-1}(A))}{P_*(A)} \\ &= \frac{P_*(S_t^{-1}(B \cap A))}{P_*(A)} \\ &= \frac{P_*(B \cap A)}{P_*(A)} = P(B|A) \end{aligned} \tag{5.29}$$

Since

$$\begin{aligned} P_*(B) &= P_*(B \cap A) + P_*(B \cap (X \setminus A)) \\ &= P_*(A)P(B|A) + (1 - P_*(A))P(B|X \setminus A) \end{aligned} \tag{5.30}$$

and $0 < P_*(A) < 1$ the measure P_* can be written as a non-trivial convex sum of two measures acting on the two complementary sets A and $X \setminus A$.

Ergodic theorem of Birkhoff. Ergodic measures satisfy an important theorem, the ergodic theorem of Birkhoff, which states that the expectation value or ensemble mean of all measurable functions f with finite mean is equal to the time average of the Birkhoff-Khinchin's existence theorem

$$\langle f \rangle = \bar{f}(x) \tag{5.31}$$

almost everywhere. This property, the equivalence of time and ensemble mean, is important because it shows that ensemble averages (which represent averages over parallel universes) of measurable functions can be inferred from a single trajectory. Especially, one can construct the measure P_* from a single trajectory by choosing $f = \chi_A$ which then implies

$$P_*(A) = \overline{P}_x(A) \tag{5.32}$$

almost everywhere.

To prove the ergodic theorem consider the level sets $A := \{x | a \le \bar{f}(x) < b\}$ for arbitrary $a < b \in \mathbb{R}$. These sets are invariant sets

$$\begin{aligned} S_t^{-1}(A) &= \{x | S_t(x) \in A\} \\ &= \{x | a < \bar{f}(S_t(x)) < b\} \\ &= \{x | a < \bar{f}(x) < b\} = A \end{aligned} \tag{5.33}$$

since $\bar{f}_t = \bar{f}_0$. The measure of these sets is either 0 or 1. This is only possible if $\bar{f}$ is constant almost everywhere. Therefore

$$\int_X P_*(dx)\bar{f}(x) = \bar{f}\int_X P_*(dx) = \bar{f}P_*(X) = \int_X P_*(dx)f(x) \tag{5.34}$$

because of property (5.25). Therefore $\bar{f} = \langle f \rangle$.

The converse is also true. If the ergodic theorem holds then one has for every invariant set A

$$\bar{P}_x(A) = \lim_{T\to\infty} \frac{1}{T}\int_0^T dt \chi_A(S_t(x)) \tag{5.35}$$

which is either one or zero depending on whether x belongs to A or not. The same holds for P_* since $P_*(A) = \bar{P}_x(A)$.

Explicitly, the ergodic theorem may be written as

$$\lim_{N\to\infty} \frac{1}{N}\sum_{n=1}^{N} f_n(x) = \langle f_0 \rangle \tag{5.36}$$

almost everywhere. In probability theory, a sequence $(f_n)_{n\in\mathbb{N}}$ that satisfies this equation is said to satisfy the strong law of large numbers. Equation (5.36) is therefore often called the ergodic form of the law of large numbers. It also includes the continuous-time case where the sum in (5.36) is replaced by an integral. We may thus say that ergodicity states that $(f_t)_{t\in T}$ satisfies the strong law of large numbers.

Ergodic measures are often visualized as those for which a single trajectory 'covers' the whole state space or 'densely invades' it, though this is not the correct language. A countable sequence of points as generated by an iterative map cannot cover an interval on the real line. The trajectory of a time-continuous dynamical system is a one-dimensional curve or line that cannot cover a two or higher-dimensional space. The above theorem also does not imply topological interpretations such as that all trajectories come arbitrarily close to any point. What the above theorem states is: if the system is ergodic then the relative time that the particle spends in a subset is given by the measure of this subset. This property of ergodic measures is also invoked in justifying the use of statistical mechanics, as we will discuss in Sect. 19.6.

Examples. Consider the harmonic oscillator $H(q, p) = (q^2 + p^2)/2$ on $\Gamma = \mathbb{R}^2$. It has already been stated that the Lebesgue measure is an invariant measure albeit a non-finite measure. All measures with densities $\rho(q, p) = f(H(q, p))$ that depend solely on the energy are finite invariant measures. A special subclass of these are the micro-canonical measures $\rho_E(q, p) = \delta(H(q, p) - E)/2\pi$ that distribute the measure uniformly on the energy circle $\Gamma_E := \{(q, p)|H(q, p) = E\}$. These micro-canonical measures are ergodic measures. Indeed $\rho_E(A) = 1$ if $\Gamma_E \subset A$ and $\rho_E(A) = 0$ if $\Gamma_E \not\subset A$.

The multiply periodic system

$$\begin{aligned} \dot{\theta}_1 &= \omega_1 \\ \dot{\theta}_2 &= \omega_2 \end{aligned} \tag{5.37}$$

on a two-dimensional torus represents the evolution of the two angle variables of two uncoupled harmonic oscillators. If ω_1/ω_2 is a rational number, then the trajectories are periodic. The ergodic measures are the uniform measures on these periodic orbits. However, when the ratio of the frequencies is irrational then the trajectories are non-periodic and each trajectory 'covers' the complete torus. The uniform measure on the torus is then the only ergodic measure.

The uniform measure also represents ergodic measures for the doubling map and the baker transformation. The Dirac measure $\delta_{\{0\}}$ is an ergodic measure of the over-damped oscillator $\dot{x} = -\gamma x$. The rotation map $z_{n+1} = e^{i2\pi\varphi} z_n$ on the unit circle in the complex plane has as its ergodic measure the uniform measure on the circle if $\varphi \notin \mathbb{Q}$ and the uniform measures on the periodic orbits if $\varphi \in \mathbb{Q}$.

These examples show that there may exist more than one ergodic measure, as in the case of the single and the two uncoupled harmonic oscillators and that the ergodic measures may depend on subtle differences in the specification of the system, such as in the case of the two uncoupled harmonic oscillators and the rotation map, where the character of the ergodic measure depends on whether a parameter takes on a rational or irrational value. Some more understanding into what makes an invariant measure an ergodic measure is given by the ergodic theorem of Bogoliubov and Krilov.

The ergodic theorem of Bogoliubov and Krilov. Bogoliubov and Krilov consider the set $\mathcal{M}(X, \mathcal{A}, (S_t)_{t\in T})$ of all invariant normalized measure measures for an invertible dynamical system $(S_t)_{t\in T}$ on a compact metric space space X. The theorem of Bogoliubov and Krilov then states:

1. There exists at least one invariant normalized measure, that is, the set $\mathcal{M}(X, \mathcal{A}, S_t)$ is not empty.
2. The set $\mathcal{M}(X, \mathcal{A}, S_t)$ is a convex set. If $P_*^{(1)}, P_*^{(2)} \in \mathcal{M}(X, \mathcal{A}, S_t)$ then $aP_*^{(1)} + (1-a)P_*^{(2)} \in \mathcal{M}(X, \mathcal{A}, S_t)$ for all $a \in [0, 1]$.
3. The extreme points of the convex set $\mathcal{M}(X, \mathcal{A}, S_t)$ are ergodic measures. An invariant measure P_* is thus ergodic if $P_* = aP_*^{(1)} + (1-a)P_*^{(2)}$ implies that a is either zero or one. Of course, if there is only one invariant measure it is ergodic.
4. If $P_*^{(1)}$ and $P_*^{(2)}$ are two ergodic measures then they are mutually singular. There exist disjoint invariant sets A_1 and A_2 such that $X = A_1 \cup A_2$ and $P_*^{(1)}(A_2) = P_*^{(2)}(A_1) = 0$, or more precisely, there exist disjoint sets A_1 and A_2 such that $P_*^{(1)}(A) = 0$ for all measurable subsets $A \subset A_2$ and $P_*^{(2)}2(A) = 0$ for all measurable subsets $A \subset A_1$.

The picture that emerges from these properties, especially property (4), is that the state space can be partitioned into disjunct invariant sets $A_1, \ldots,$ such that each of the ergodic measures $P_*^{(i)}$ has the set A_i as its support, i.e., $P_*^{(i)}(A_j) = 1$ if $i = j$ and $P_*^{(i)}(A_j) = 0$ if $i \neq j$. An ergodic decomposition theorem furthermore implies that every invariant measure can be represented as a weighted sum of mutually singular ergodic measures. This is the consequence of a more general theorem which states that each point of a compact convex set can be represented as a convex combination of its extreme points. Ergodic measures are thus concentrated on invariant subsets that can be studied independently.

To prove that ergodic measures are extreme points assume that $P_*(A)$ is zero or one and that it can be written as $P_* = aP_1 + (1-a)P_2$ with $0 < a < 1$. One can then show that $P_* = P_1 = P_2$ as follows: First, $P_*(A) = 0$ implies $P_{1,2}(A) = 0$ since $a \neq 0, 1$. Hence the Radon-Nikodym theorem assures that there exists densities $\rho_{1,2} \in \mathcal{L}_1$ such that

$$P_{1,2}(A) = \int_A P_*(dx)\rho_{1,2}(x) \tag{5.38}$$

for all $A \in \mathcal{A}$. A change of variables and invariance of the measures $P_{1,2}$ then imply (recall that the dynamical system is invertible)

$$\int_A P_*(dx)\rho_{1,2}(S_{-t}(x)) = \int_{S_t^{-1}(A)} P_*(dx)\rho_{1,2}(x) = \int_A P_*(dx)\rho_{1,2}(x) \tag{5.39}$$

for all $A \in \mathcal{A}$. Hence $\rho_{1,2}(S_t(x)) = \rho_{1,2}(x)$ almost everywhere. Furthermore, $\rho_{1,2}$ must be a constant almost everywhere. Since if it were not then there exist non-empty level sets

$$A_{1,2} := \{x | a < \rho_{1,2}(x) < b\} \tag{5.40}$$

These are invariant sets since

$$
\begin{aligned}
S_t^{-1}(A_{1,2}) &= \{x|S_t(x) \in A_{1,2}\} \\
&= \{x|a < \rho(S_t(x)) < b\} \\
&= \{x|a < \rho(x) < b\} = A_{1,2}
\end{aligned}
\tag{5.41}
$$

which have P_*-measure in $]0, 1[$, contradicting our premise. Furthermore, the constant must be 1 because of the normalization $\int_X P_*(dx)\rho_{1,2} = 1$ and hence $P_{1,2} = P_*$ and P_* is an extreme point.

It is stressed that the application of the ergodic decomposition theorem requires that the state space be a compact metric space. For Euclidean spaces, this requires that the state space is closed and bounded. Generalizations are possible. The finite invariant measures of the harmonic oscillator are the measures that depend only on the energy and one might construct them as a weighted integral of the ergodic measures that are confined to the energy circles. In Chap. 12 we consider in detail the shift map which has a non-countable number of ergodic measures.

5.8 Ergodic System

A dynamical system can have many invariant measures. Here we discuss the systems which have only one.

Ergodic systems. We call a system $\{X, \mathcal{A}, P_*, (S_t)_{t\in T}\}$ ergodic if the measure P_* is the only invariant measure. The measure is then ergodic. An alternative definition sometimes used is that the above system is ergodic if the measure P_* is ergodic. The two definitions subtly differ. Consider the harmonic oscillator. If one chooses as the state space an energy circle then the uniform distribution on this energy circle is the only invariant measure and the system with this measure is hence an ergodic system, according to the first definition. If one chooses as the state space the plane then the uniform distribution on that energy circle is an ergodic measure and the system with this measure an ergodic system, according to the second definition, albeit on a different state space, on the plane rather than on the energy circle. In the second definition one carries around an additional invariant subset which has P_*-measure zero. We stick to the first definition although an enlarged state space might sometimes be implied as when the uniform distribution is characterized by $\rho_E(q, p) = \delta(H(q, p) - E)/2\pi$.

Ergodic systems are the irreducible building blocks of a dynamical system. The ergodic theorem by Bogoliubov and Krilov implies that a invertible continuous-time dynamical system (on a compact metric space) can be decomposed into a set of ergodic subsystems. In the case of the harmonic oscillator, these subsystems are the systems on the energy circles. The evolution on each energy circle can be studied independently from the evolution on other energy circles. Of course, one does not have to study the harmonic oscillator on energy circles. One may also study it on some other (invariant) sets but then the system is not ergodic.

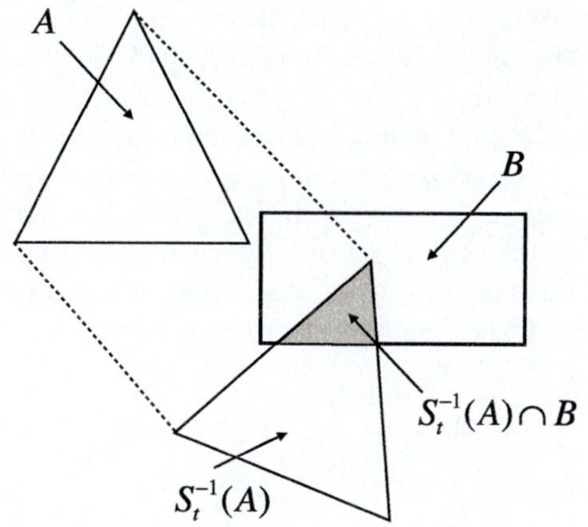

Fig. 5.2 Evolution of ink blobs

Evolution of ink blobs. Within ergodic theory one can now study the evolution of its measurable subsets. Often these subsets are visualized as ink blobs. Given two measurable sets A and B of non-zero measure one may ask what fraction of the set B will be in the set A at time t. This fraction is given by (see Fig. 5.2)

$$P_*(S_t^{-1}(A)|B) := \frac{P_*(S_t^{-1}(A) \cap B)}{P_*(B)} \tag{5.42}$$

If the set B is envisioned as an ink blob then this fraction gives the amount of ink that is in the set A at time t. In Fig. 5.2 we represented the set A and its pre-image by a triangle in order to stress the fact that ergodicity does not require any complicated flows that change the shape of sets.

For ergodic measures we have the following theorem: Let A and B two measurable sets with $P_*(A)$, $P_*(B) \neq 0$ then

$$\lim_{T\to\infty} \frac{1}{T} \int_0^T dt\, P_*(S_t^{-1}(A) \cap B) = P_*(A) P_*(B) \tag{5.43}$$

The time average of the fraction $P_*(S_t^{-1}(A)|B)$ approaches $P_*(A)$ as $t \to \infty$. The time average of the relative fraction becomes equal to 'the relative volume of A in X'. The point is that the time average of the fraction becomes independent of the set B.

The proof goes as follows:

$$\begin{aligned} \lim_{T\to\infty} \frac{1}{T} \int_0^T dt\, P_*(S_t^{-1}(A) \cap B) &= \int_X P_*(dx) \lim_{T\to\infty} \frac{1}{T} \int_0^T dt \chi_{S_t^{-1}(A)\cap B}(x) \\ &= \int_X P_*(dx)\chi_B(x) \lim_{T\to\infty} \frac{1}{T} \int_0^T dt \chi_A(S_t(x)) \\ &= P_*(B) P_*(A) \end{aligned} \tag{5.44}$$

where we have used, in the third line, the fact that the time average equals the ensemble average. The property (5.43) may also be used as a definition of an ergodic measure.

An immediate consequence of (5.43) is there exist a time $t > 0$ such that $S_t^{-1}(A) \cap B \neq \emptyset$, implying that there are points in B that end up in A at time t. The contrary conflicts with (5.43). This formulation is the measure-theoretical equivalent of topological transitivity, where a similar statement was formulated for two open sets rather than two sets of non-zero measure.

If the system is invertible we find

$$\begin{aligned}\lim_{T\to-\infty}\frac{1}{T}\int_0^T dt\,P_*(S_t^{-1}(A)\cap B) &= \lim_{T\to\infty}\frac{1}{T}\int_0^T dt\,P_*(S_{-t}^{-1}(A)\cap B)\\ &= \lim_{t\to\infty}\frac{1}{T}\int_0^T dt\,P_*(A\cap S_t^{-1}(B))\\ &= P_*(A)P_*(B)\end{aligned} \tag{5.45}$$

where the first equality is just a change of variable from t to $-t$, the second equality is a consequence of the invariance of P_*, and the third equality follows from the system being ergodic. Thus ergodicity does not introduce any irreversible behavior into the flow for invertible systems. An invertible dynamical system cannot show time-asymmetric behavior.

Evolution of observables. Observables f_0 evolve according to $f_t = f_0 \circ S_t$. When evolving on a measure space with an invariant normalized measure then, as already pointed out, the expectations do not change in time, $\langle f_t\rangle = \langle f_0\rangle$. If g_0 is another observable then we find additionally

$$\lim_{T\to\infty}\frac{1}{T}\int_0^T dt\langle f_t g_0\rangle = \langle f_0\rangle\langle g_0\rangle \tag{5.46}$$

The proof is again straightforward and requires that one can interchange the time integration and expectation value and the equality of time and ensemble mean. If one chooses $f_0 = \chi_A$ and $g_0 = \chi_B$ one recovers (5.43) above. One can also write the formula (5.46) in the form

$$\lim_{T\to\infty}\frac{1}{T}\int_0^T dt C_{t,0} = 0 \tag{5.47}$$

where

$$C_{t,0} := \langle (f_t - \langle f_t\rangle)(g_0 - \langle g_0\rangle)\rangle \tag{5.48}$$

is the covariance function. The time average of the covariance function between f_t and g_0 converges to zero as time goes to infinity. If the system is invertible the covariance function also converges to zero when time goes to minus infinity.

Recurrence time of N uncoupled oscillators. As an application of the ergodic theorem we estimate the mean recurrence time for a set of N uncoupled oscillators with incommensurable frequencies ω_n. The system is ergodic since the usual Lebesgue measure is the only invariant measure on the finite N-dimensional state space. Specify a volume $\Pi_{n=1}^{N} \Delta\theta_n$ in this state space spanned by the angle variables θ_n. We represent this state space as a N-dimensional periodic cube rather then a N-dimensional torus, for convenience. The fraction of time Δt that a trajectory spends in this volume is then given by

$$\lim_{t\to\infty} \frac{\Delta t}{t} = \Pi_{n=1}^{N} \frac{\Delta\theta_n}{2\pi} \tag{5.49}$$

since we can equate the time mean with the ensemble mean. Introducing the number r of recurrences the mean passage time of an recurrence is defined by

$$\overline{\Delta t} := \lim_{t\to\infty} \frac{\Delta t}{r} \tag{5.50}$$

and the mean recurrence time by

$$\overline{t_r} := \lim_{t\to\infty} \frac{t}{r} \tag{5.51}$$

The mean passage can be shown to be given by

$$\frac{1}{\overline{\Delta t}} = \sum_{n=1}^{N} \frac{1}{t_n} \tag{5.52}$$

where $t_n = \Delta\theta_n / \omega_n$ are just the individual passage times. This assertion can easily be demonstrated for the case $N = 2$. The mean length of a passage $\bar{l}$ is given by the ratio of the volume $\Delta\theta_1 \Delta\theta_2$ divided by its projection d perpendicular to the velocity (see Fig. 5.3). Elementary algebra results in

$$d = \frac{v_1}{|v|}\Delta\theta_2 + \frac{v_2}{|v|}\Delta\theta_1 \tag{5.53}$$

giving

$$\frac{1}{t_1} + \frac{1}{t_2} = \frac{|v|}{\bar{l}} = \frac{1}{\overline{\Delta t}} \tag{5.54}$$

The extension of this proof to higher dimensions can be found in Hemmer et al. (1958).

Using the above formulas the mean recurrence can be expressed as

$$\overline{t_r} = \frac{\Pi_{n=1}^{N} \frac{2\pi}{\Delta\theta_n}}{\sum_{n=1}^{N} \frac{\omega_n}{\Delta\theta_n}} \tag{5.55}$$

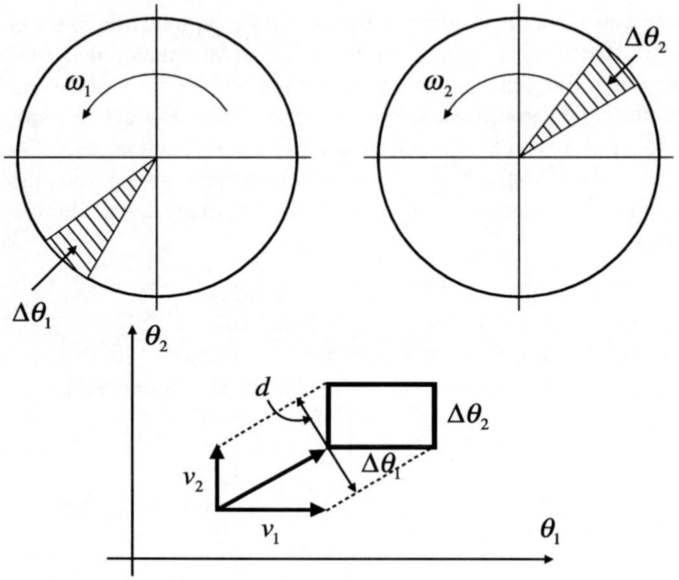

Fig. 5.3 Geometry of recurrence time in two dimensions

This specific example thus results in a mean recurrence time that grows proportional to $\Delta\theta^{-N}$ or, more accurately, proportional to $N^{-1}\Delta\theta^{1-N}$. Take $N = 10$ and $\Delta\theta = 2\pi/100$. Then $\overline{t_r} > 10^{17} T_{min}$ where T_{min} is the minimal period of the oscillators. This recurrence time is much larger than the period observed in the FPU experiment, which represents a quasi-periodicity and not the recurrence time.

As our examples such as the harmonic oscillator and the over-damped oscillator show, extremely simple and benign systems can have ergodic measures. To emphasize this point even more consider the turning of a coin. Its state space is $X = \{H, T\}$ and the measure $P(H) = P(T) = 1/2$ is an ergodic measure. The distinguishing feature of ergodic systems is that it cannot be reduced any further. Thus:

Ergodic systems are not complicated systems but systems that cannot be reduced any further.

5.9 Mixing and Exact Systems

Though ergodicity does not imply any complicated behavior there do exist ergodic systems with more complicated flows.

Mixing systems. An dynamical system S_t with invariant measure P_* is called mixing if it satisfies the stronger property

$$\lim_{t\to\infty} P_*(S_t^{-1}(A) \cap B) = P_*(A)P_*(B) \tag{5.56}$$

for all $A, B \in \mathcal{A}$ instead of (5.43). This property implies that there exists a time t such that $P_*(S_{t'}^{-1}(A)|B) \neq 0$ for all $t' > t$.

Mixing systems are ergodic but not every ergodic system is mixing. To prove that mixing systems are ergodic consider an invariant set $A = S_t^{-1}(A)$ and its complement $B = X \setminus A$. Then $P_*(A \cap B) = P_*(S_t^{-1}(A) \cap B) = 0$ and hence

$$\lim_{t\to\infty} P_*(S_t^{-1}(A) \cap B) = P_*(A)P_*(B) = (1 - P_*(A))P_*(A) = 0 \tag{5.57}$$

and hence $P_*(A) \in 0, 1$. The system is ergodic.

The ergodic measures of the harmonic oscillator, the rotation on the unit circle by an irrational angle, and the flow on the torus with incommensurable frequencies are all not mixing. The doubling map and the baker transformation are mixing. Even the over-damped oscillator is trivially mixing since $\delta_0(S_t^{-1}(A) \cap B) = \delta_0(A)\delta_0(B)$.

Observables in mixing systems have the stronger convergence

$$\lim_{t\to\infty} \langle f_t g_0 \rangle = \langle f_0 \rangle \langle g_0 \rangle \tag{5.58}$$

Mixing is a very suggestive but unfortunate term since it seems to introduce some irreversibility into the flow. As we already pointed out this is not the case. For invertible systems we also find

$$\lim_{t\to-\infty} P_*(S_t^{-1}(A) \cap B) = P_*(A)P_*(B) \tag{5.59}$$

which shows that the system mixes in both time directions. The system first 'unmixes' as it approaches its present state and then mixes as it evolves into the future. An invertible dynamical system cannot show time-asymmetric behavior.

In fluid dynamics one distinguishes between reversible stirring and irreversible mixing. Mixing dynamical systems would be considered stirring flows in fluid dynamics. The fluid dynamical concept of mixing is closer to the concept of exactness in ergodic theory.

Exact systems. The even stronger requirement

$$\lim_{t\to\infty} P_*(S_t(A)) = 1 \tag{5.60}$$

for all A with $P_*(A) > 0$ defines exact systems. Exact systems are necessarily mixing. The proof is difficult and given in Berger (2001) or Lasota et al. (1994).

The requirement (5.60) states that any initial blob of positive measure eventually fills the whole phase space. For invertible systems one has $P_*(S_t(A)) = P_*(A)$. A blob of ink can spread over the whole state space but can never fill out the whole state space. Thus only non-invertible systems can be exact. Thus the harmonic oscillator, Hamiltonian systems, the over-damped oscillator, and the baker transformation cannot be exact. The doubling map with the uniform distribution, on the other hand, is exact. Any initial interval $[0, a[$ is mapped onto the complete state space $[0, 1[$ after a finite number of iterations. The doubling map is dynamically equivalent to the Bernoulli shift whose invariant measures are discussed in the next section in the more general context of shift maps.

5.10 Shift Map

The shift map is a generalization of the Bernoulli shift. It is defined on the state space $X = \mathbb{R}^T$ where $T = \mathbb{N}, \mathbb{Z}$. The one-sided shift map is defined by the iterates of

$$\begin{aligned} T_+ : \mathbb{R}^{\mathbb{N}} &\mapsto \mathbb{R}^{\mathbb{N}} \\ (x_0, x_1, x_2, \ldots) &\mapsto (x_1, x_2, x_3, \ldots) \end{aligned} \tag{5.61}$$

It deletes from any one-sided infinite sequence of real numbers the first number and shifts the remaining numbers to the left. The one-sided shift map is non-invertible. There are two two-sided shift maps: the right shift

$$\begin{aligned} T_r : \mathbb{R}^{\mathbb{Z}} &\mapsto \mathbb{R}^{\mathbb{Z}} \\ (\ldots, x_{-1}, x_0, x_1, \ldots) &\mapsto (\ldots, x_{-2}, x_{-1}, x_0, \ldots) \end{aligned} \tag{5.62}$$

which shifts all numbers of an infinite two-sided sequence to the right by one step. The left shift T_l shifts all numbers to the left. The two-sided shift maps are invertible. Often $X = \{0, 1, \ldots, K-1\}^T$. The one-sided shift on $X = \{0, 1\}^{\mathbb{N}}$ is the Bernoulli shift discussed in the chapter on dynamical systems where we showed that it is dynamically equivalent to the doubling map. There we also showed that the two-sided shift on $X = \{0, 1\}^{\mathbb{Z}}$ is related to the baker transformation. In the following we just consider the one-sided map. Two-sided maps can be treated similarly.

The state space of the shift map can be turned into a measure space. Its σ-algebra is generated by all the cylinder sets

$$C(B_0, B_1, \ldots, B_N) := B_0 \times B_1 \times \ldots \times B_N \times \mathbb{R} \times \mathbb{R} \ldots \tag{5.63}$$

where $B_n \in \mathcal{B}$, $n = 0, 1, \ldots, N$ and $N \in \mathbb{N}$. A measure on this σ-algebra can be generated by assigning a normalized measure to all these cylinder sets. A particular measure is the product measure

$$P(C(B_0, B_1, \ldots, B_N)) = \Pi_{n=0}^{N} P_n(B_n) \tag{5.64}$$

where P_n are measures on the Borel set $\mathcal{B}$. If all these measures are identical, $P_n = P_0$ for all $n \in \mathbb{N}$, then this product measure is called a Bernoulli measure. If the state space is discrete $X = \{0, 1, \ldots, K-1\}^{\mathbb{N}}$ and $P_0(\{k\}) = p_k$ for $k = 0, \ldots, K-1$ then

$$P(C(\{k_0\}, \ldots, \{k_N\})) = \Pi_{n=0}^{N} p_{k_n} \tag{5.65}$$

If $0 < p_k < 1$ only cylinder sets have non-zero normalized measure.

The Bernoulli measure are invariant measures. It is sufficient to show this for cylinder sets. Since $T_+^{-1}(C(B_0, \ldots, B_N)) = \mathbb{R} \times B_0 \times \ldots \times B_N \times \mathbb{R} \times \ldots$ is also a cylinder set we have indeed

$$\begin{aligned} P(T_+^{-1}(C(B_0, B_1, \ldots, B_N)) &= P_0(\mathbb{R}) P_0(B_0) P_0(B_1) \ldots P_0(B_N) \\ &= P_0(B_0) P_0(B_1) \ldots P_0(B_N) \\ &= P(C(B_0, B_1, \ldots, B_N)) \end{aligned}$$

Furthermore, if A and B are two cylinder sets then

$$P(T_+^{-n}(A) \cap B) = P(A)P(B) \tag{5.66}$$

for sufficiently large n. Hence the Bernoulli measures are mixing and hence ergodic. They are even exact since $P_h(T_+^n(C) = 1$ for sufficiently large n. The uniform measure of the doubling map corresponds to Bernoulli shift with measure $p_0 = p_1 = 1/2$. The other Bernoulli measures can not be represented by a density.

5.11 Summary

Endowing the state space with a measure allows to study dynamical systems from yet another angle. The theory requires the state space to be a measure space and the dynamical map to be a measurable function for all times. First, one can ask whether or not a given measure is preserved by the dynamical evolution. If the given measure is the Lebesgue measure the answer divides dynamical systems in conservative systems, for which the Lebesgue measure is conserved and into dissipative systems for which the Lebesgue measure is not conserved. This is an important distinction with implications such that only dissipative systems can have attractors.

In this chapter we considered invariant measures. Their study is particularly useful because these measures are a property of the dynamical system. The set of invariant measures is determined by the dynamics. The existence of a finite invariant measure implies Poincaré's recurrence theorem and the existence of a time mean. The set of finite invariant measures form a convex set. The ergodic measures are the extreme points of this set. The time mean of observables then equals the ensemble mean and the measure of every invariant set is either one or zero. These three properties are equivalent and each can be used to define an ergodic measure. Ergodic measures have their support on invariant subsets that cannot be reduced further. A dynamical system with an ergodic measure is called an ergodic system. An important property of an ergodic system is that the trajectories visit every measurable set of the state space for a duration that is proportional to the measure of this set. Very benign systems are ergodic, such as the harmonic oscillator with the uniform measure on the energy circle and the rotation of the unit circle by an irrational number with the uniform measure on the unit circle. Mixing and exact systems show more complicated behavior.

Measure-theoretical concepts and ergodic theory are often more useful than the topological concepts studied in the previous chapter. Topological concepts are most powerful for low-dimensional systems. A measure-theoretical approach often makes high-dimensional systems amenable to analysis. Invariant measures may correspond to the equilibrium or stationary states of the system. The most prominent example is the uniform distribution on the energy hyper-surface in Hamiltonian dynamics. It is called the micro-canonical ensemble and is the basis for equilibrium statistical mechanics, as we will discuss in detail in the chapter on statistical mechanics.

More details on ergodic theory can be found in Berger (2001), Lasota et al. (1994), Mackey (1989), Petersen (1983), Sinai (1976), and Walters (2000).

References

Berger, A.: Chaos and Chance. Walter de Gruyter (2001)

Hemmer, P.C., Maximon, L.C., Wergeland, H.: Recurrence time of a dynamical system. Phys. Rev. **111**, 689 (1958)

Lasota, A., Mackey, M.C.: Chaos, Fractals, and Noise; Stochastic Aspects of Dynamics. Springer (1994)

Mackey, M.C.: The dynamic origin of increasing entropy. Rev. Mod. Phys. **61**(4) (1989)

Petersen, K.: Ergodic Theory. Cambridge University Press, Cambridge (1983)

Sinai, YA.G.: Introduction to Ergodic Theory. (translated by V. Scheffer) Princeton University Press (1976)

Walters, P.: An Introduction to Ergodic Theory. Springer (2000)

Zermelo, E.: Ann.Phys. (Leipzig) **57**(458) (1896)

Chapter 6
Numerical Algorithms

6.1 Introduction

An important class of discrete dynamical systems arises when continuous dynamical systems are discretized to be solved numerically on a computer. The main concern is to what extent the numerical discrete solution mimics or resembles the solution of the original continuous system. This chapter describes some basic properties of the algorithms that are used to obtain numerical solutions for ODEs and PDEs. It is an abridgment of a chapter in Müller and von Storch (2004).

Numerical algorithms executed on a digital computer have to introduce a finite time step, and a finite grid size in case of PDEs. Numerical algorithms introduce two kind of basic errors: round-off errors and truncation errors. Round-off errors occur since digital computers have only a finite word length, a finite number of bits to represent real numbers. Round-off errors are property of the machine on which the numerical code is executed. Truncation errors occur when the derivative of a function is replaced by a finite difference approximation. Truncation errors are under the control of the modeler. They can be reduced by decreasing the time step or grid spacing.

The most important property of a numerical algorithm is its stability. A numerical solution is stable if the difference between the exact and numerical solution stays bounded. Other desirable properties are consistency and convergence. For PDEs, numerical schemes may also introduce artificial or numerical diffusion and dispersion which are absent in the original problem, and they may not respect the conservation laws of the continuous problem. It requires great ingenuity to design numerical algorithms that avoid or minimize these various problems and can be run efficiently on a computer.

Truncation and round-off errors are mathematical issues arising for all numerical algorithms; numerical diffusion and dispersion and violation of conservation laws are issues that require a specific physical contexts.

P. Müller, *Handbook of Dynamics and Probability*,
https://doi.org/10.1007/978-3-030-88486-4_6

Instead of grids numerical models may also employ *spectral* or *finite element* methods to represent the spatial structure of PDEs.

Numerical algorithms are discrete-time dynamical systems on a finite state space. If invertible, such systems can only have fixed points and periodic orbits. In any practical problem the periods are so large that this fact becomes irrelevant. Such systems can also not exhibit chaos, though they may exhibit sensitivity to initial conditions. Neighboring points may have very different periods.

Numerical algorithms are also important for solving problems other than differential equations. Newton's method for finding the zeros of a continuous function is a prime example. Also, the properties of most iterative maps are obtained by numerical codes. No researcher had the stamina to iterate a map such as the logistic map a thousand times for different parameter values.

6.2 Round-Off Error

A digital computer can only manipulate a finite amount of information. Most real numbers correspond to infinite sequences of a finite set of symbols (digits), so, for computational purposes, they have to be approximated by finite sequences. Computers have a finite word length.

Current computers use 32 or 64 bit words. A real number x is represented by the floating point number

$$\bar{x} = s0.m2^{e} \tag{6.1}$$

where s is the sign, m the mantissa and e the exponent. In a typical 64 bit word-length one bit is used for the sign; 52 bits for the mantissa and 11 bits for the exponent. In decimal representation this covers negative numbers in the range from $\approx -10^{308}$ to $\approx -10^{-324}$ and positive numbers in the range from $\approx 10^{-324}$ to $\approx 10^{308}$ with about 15 significant decimal digits.

The difference between a real number and its finite-sequence representation is called the *round-off error*. In the above example the relative error $|x - \bar{x}|/|x|$ is 10^{-15}. If one subtracts two numbers which are identical in the first k of the n significant digits, these first digits cancel in the subtraction. When the result is normalized, the missing digits are filled in from the right by zeros which is certainly wrong in most cases and the result only has $n - k$ correct digits and hence the relative error increases from 10^{-n} to 10^{n-k}. Beware of subtracting almost equal numbers.

The representation of real numbers in the finite floating point representation represents a coarse graining of the real axis. In the case of a word length of 64 bits it is a coarse graining into $2^{64} + 3$ boxes, the extra 3 boxes being the non-representable large negative and large positive numbers and the non-representative small numbers.

Round-off errors may have grave consequences for the numerical computation of dynamical systems. Consider the Bernoulli shift. If the initial value is represented by a 64 bit binary sequence then all initial values will be mapped onto zero after 64 iterations.

6.3 Truncation Errors

Solutions of differential equations are differentiable functions, which are defined at every point of a continuum, whereas in a numerical algorithm we can only specify such functions at a finite set of points. Therefore derivatives must be replaced by *finite differences.* This introduces truncation errors. We discuss these for ordinary differential equations where time needs to be discretized. The same ideas apply to partial differential equations where also space needs to be discretized.

Consider a smooth (i.e. infinitely differentiable) function $x(t)$ defined on an interval $t \in [0, T]$. Computationally, we can only handle this function at a finite number of times. A finite set of times $0 \leq t_0 < t_1 < \cdots < t_{n-1} < t_n < \cdots < t_{N-1} < t_N \leq T$ is called a *time grid.* See Fig. 6.1.

Often the points are equally spaced, $t_n = n\Delta t$, then Δt is called the time step. The value of the function $x(t)$ at the grid point t_n is denoted by $x_n := x(t_n)$. If the function $x(t)$ satisfies the differential equation

$$\frac{dx}{dt} = u(x) \tag{6.2}$$

then the derivative is obtained from the Taylor expansion

$$x_{n+1} = x_n + \left.\frac{dx}{dt}\right|_n \Delta t + \frac{1}{2!}\left.\frac{d^2x}{dt^2}\right|_n (\Delta t)^2 + \ldots \tag{6.3}$$

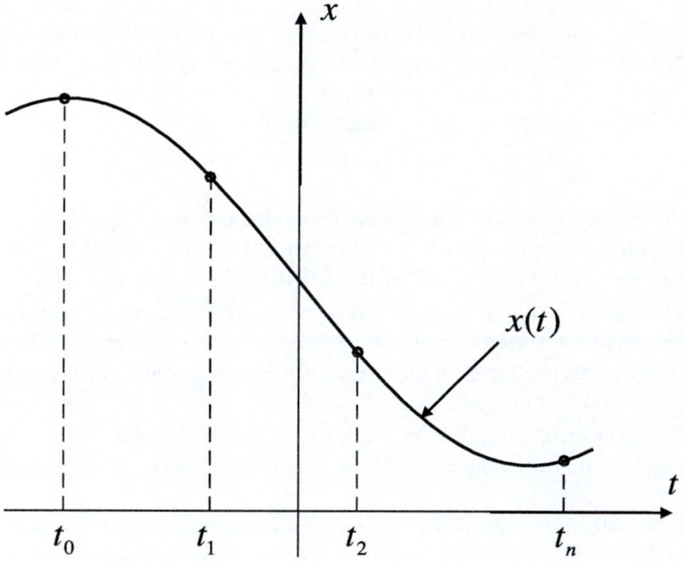

Fig. 6.1 Discretization of continuous function

as

$$\left.\frac{dx}{dt}\right|_n = \frac{x_{n+1} - x_n}{\Delta t} - \frac{1}{2!}\left.\frac{d^2x}{dt^2}\right|_n \Delta t + \ldots \tag{6.4}$$

The first term is the finite difference approximation for the first derivative. The second and all higher terms are the truncation error. For this specific scheme, the truncation error is $O(\Delta t)$. The scheme is said to be of first order accuracy. Generally, one writes $f(x) = O(g(x))$ for $x \to 0$ if there exists a $c > 0$ and a $\delta > 0$ such that $|f(x| < c|g(x)|$ for all $|x| < \delta$.

If one uses the Taylor expansion

$$x_{n-1} = x_n - \left.\frac{dx}{dt}\right|_n \Delta t + \frac{1}{2!}\left.\frac{d^2x}{dt^2}\right|_n (\Delta t)^2 + \ldots \tag{6.5}$$

and subtracts (6.5) from (6.3) the derivative becomes the finite difference approximation

$$\left.\frac{dx}{dt}\right|_n = \frac{x_{n+1} - x_{n-1}}{2\Delta t} + O((\Delta t)^2) \tag{6.6}$$

which is of second order accuracy. The finite difference approximation (6.4) is called forward (or downwind) differencing. The scheme (6.6) is called a centered difference. There is also a backward (or upwind) finite difference

$$\left.\frac{dx}{dt}\right|_n = \frac{x_n - x_{n-1}}{\Delta t} + O(\Delta t) \tag{6.7}$$

which is of first order accuracy. Taylor expansion also provides finite difference approximations for higher order derivatives. Adding (6.3) and (6.5) gives

$$\left.\frac{d^2x}{dt^2}\right|_n = \frac{x_{n+1} - 2x_n + x_{n-1}}{(\Delta t)^2} + O((\Delta t)^2) \tag{6.8}$$

as a finite difference approximation of the second derivative.

The truncation error is under control of the modeler. It can be reduced by decreasing the time step Δt. Of course this will decrease the efficiency as a larger number of time points has to be handled. Also, the larger number of operations increases the round-off errors introduced by the finite machine accuracy. Eventually, the round-off errors tend to become larger than the truncation errors and any further reduction of the time step defeats its purpose.

The truncation error can also be reduced by increasing the order of the finite difference approximation. The trade-off here is that higher order schemes increase the complexity of the code.

The above considerations show that there are various finite difference approximations to the derivative dx/dt. One can also evaluate the velocity field $u(x)$ in different ways. One can evaluate it at x_n or at x_{n+1} or at some weighted average.

Consider the forward scheme (6.4) and evaluate the velocity at x_n. One then obtains the explicit equation

$$x_{n+1} = x_n + \Delta t u(x_n) \tag{6.9}$$

It explicitly gives the $(n+1)$th value in terms of the nth value. One can March through the solution from an initial value x_0. If one evaluates the velocity at x_{n+1} (or equivalently uses the backward scheme) then one obtains the implicit equation

$$x_{n+1} = x_n + \Delta t u(x_{n+1}) \tag{6.10}$$

To obtain x_{n+1} from x_n, one has to invert u. This can be a major computational effort, especially if u represents the velocity in a high dimensional space. One can lessen the effort by using the semi-implicit equation

$$x_{n+1} = x_n + \Delta t \left[u(x_n) + \left.\frac{du}{dx}\right|_n (x_{n+1} - x_n) \right] \tag{6.11}$$

where $u(x_{n+1})$ has been approximated by the first two terms of a Taylor expansion.

6.4 Stability

The most important property of a finite difference approximation is its stability. We discuss stability separately for the numerical solutions of ODEs and PDEs.

Stability of algorithms for ODEs. Let $x_n = x(t_n)$ be the exact solution of the differential equation and $\hat{x}_n$ the solution of the finite difference approximation. The important point here is:

The finite difference solution $\hat{x}_n$ is in general different from the function x_n.

The finite difference solution is stable if the difference $\hat{x}_n - x_n$ stays bounded as n tends to infinity for fixed Δt.

Consider the linear differential equation

$$\frac{dx}{dt} = -\gamma x,$$

with the initial condition $x(t = 0) = x_0$ and $\gamma > 0$. This is our over-damped oscillator. It has the solution $x(t) = x_0 e^{-\gamma t}$, which tends to zero for $t \to \infty$. Using the forward difference or explicit scheme, we get

$$\hat{x}_{n+1} = (1 - \gamma \Delta t)\hat{x}_n.$$

For $|1-\gamma\Delta t| > 1$ or $\Delta t > 2/\gamma$ the value $|\hat{x}_n|$ increases monotonically with n and the solution is unstable. For the backward or implicit scheme, we find

$$\hat{x}_{n+1} = \hat{x}_n - \Delta t \gamma \hat{x}_{n+1},$$

or

$$\hat{x}_{n+1} = \frac{\hat{x}_n}{1+\gamma\Delta t}.$$

In this case $|\hat{x}_n|$ decreases monotonically with n independent of the time step Δt. The implicit scheme is unconditionally stable. The semi-implicit method results in

$$\hat{x}_{n+1} = \frac{1-\gamma\Delta t - \gamma^2\Delta t}{1-\gamma^2\Delta t}\hat{x}_n,$$

which is again unconditionally stable.

As a second example consider the continuous analogue

$$\dot{x}(t) = ax(t)(1-x(t)) \tag{6.12}$$

of the logistic map. It has the quite benign analytical solution

$$x(t; x_0) = \left[1 - \frac{x_0 - 1}{x_0} e^{-at}\right]^{-1} \tag{6.13}$$

Any initial value $0 < x_0 < 1$ monotonically approaches the carrying capacity $x_c = 1$ as $t \to \infty$. The forward difference scheme results in

$$\begin{aligned} x_{n+1} &= x_n + a\Delta t(1-x_n)x_n \\ &= a\Delta t(x_c - x_n)x_n \end{aligned} \tag{6.14}$$

with a modified carrying capacity $x_c := (1+a\Delta t)/(a\Delta t)$. Introducing the scaled variable $\tilde{x}_n := x_n/x_c$ one recovers the logistic map

$$\tilde{x}_{n+1} = r(1-\tilde{x}_n)\tilde{x}_n \tag{6.15}$$

with $r = a\Delta t x_c$. As we pointed out, this discrete logistic map can exhibit quite irregular behavior, such as chaotic behavior, for certain values of the coefficient r. This example is a warning that discretization may not only change the quantitative behavior but also might induce profound changes in the qualitative behavior.

Finite difference approximations should also be consistent and convergent. A finite difference approximation is consistent with the original differential equation if the difference equation converges to the differential equation as the time step Δt goes to zero. For finite difference approximations that are based on Taylor expansions this property is easily checked and satisfied. Consistency generally does not constitute a problem.

The solution of a finite difference approximation is said to converge to the solution of the original differential equation if the difference between the two solutions at a fixed time instant t tends to zero uniformly as $\Delta t \to 0$ and $n \to \infty$. Convergence is much harder to prove. Generally one relies on the Lax equivalence theorem. It states that consistency and stability imply convergence for linear systems.

Diffusion equation. Finite difference approximations become more involved for partial differential equations since both time and space have to be discretized. As an example consider the diffusion equation

$$\partial_t \rho(x, t) = D \partial_x \partial_x \rho(x, t) \tag{6.16}$$

which might describe the diffusion of the density ρ of a property like mass in physical space. Its main characteristic is that any gradients are smeared out in the course of time. Consider a time step Δt such that $t = n\Delta t, n = 1, \ldots, N$ and a grid size Δx such that $x = j\Delta x,\ j = 1, \ldots, J$.

The simplest finite difference scheme then is the *explicit* scheme

$$\rho_j^{n+1} = \rho_j^n + \hat{D}(\rho_{j+1}^n - 2\rho_j^n + \rho_{j-1}^n) + O(\Delta t, (\Delta x)^2) \tag{6.17}$$

where

$$\hat{D} = \frac{\Delta t D}{(\Delta x)^2} \tag{6.18}$$

and where subscripts denote the space grid and superscripts the time grid. One way to asses the (linear) stability is to require that the spatial Fourier modes $\rho_j^n = A_k^n e^{ikj\Delta x}$ do not amplify with time for any wavenumber k. When these Fourier modes are substituted into the explicit finite difference scheme (6.17) one obtains

$$A_k^{n+1} = G_k A_k^n \tag{6.19}$$

with amplification factor $G_k = 1 - 4\hat{D}\sin^2 k\Delta x/2$. The scheme is thus linearly stable if $|G_k| < 1$ for all k which requires $\hat{D} < 1/2$ or a time step $\Delta t < (\Delta x)^2/2D$. The scheme is conditionally stable.

If we violate this criterion the wave with $k\Delta x/2 = \pi/2$ or with wavelength $2\Delta x$, the smallest resolvable wavelength on the grid, is the most unstable, i.e., the fastest growing mode. For diffusion problems in higher dimensional spaces G_k becomes a matrix and the stability criterion is that the spectral radius of the matrix be smaller than one, i.e., all eigenvalues of G_k must have a magnitude smaller than 1.

The constraint of a maximum allowable time step Δt for the explicit scheme (6.17) can be avoided by using the *implicit* scheme

$$\rho_j^{n+1} - \rho_j^n = \hat{D}(\rho_{j+1}^{n+1} - 2\rho_j^{n+1} + \rho_{j-1}^{n+1}) + O(\Delta t, (\Delta x)^2), \tag{6.20}$$

which requires solving the tridiagonal system $-\hat{D}\rho_{j-1}^{n+1} + (1 + 2\hat{D})\rho_j^{n+1} - \hat{D}\rho_{j+1}^{n+1} = \rho_j^n$. In this case $G_k = (1 + 4\hat{D}\sin^2 k\Delta x/2)^{-1}$ which is smaller or equal to one for all k. The scheme is thus unconditionally stable. Arbitrarily large time steps can be taken.

In contrast the *centered* time difference scheme

$$\rho_j^{n+1} - \rho_j^{n-1} = 2\hat{D}(\rho_{j+1}^n - 2\rho_j^n + \rho_{j-1}^n) + O((\Delta t)^2, (\Delta x)^2) \tag{6.21}$$

has $G_k = \frac{1}{2}[a \pm (a^2 + 4)^{1/2}]$, where $a = 8\hat{D}_{)} \sin^2 k\Delta x/2$. Since $|G_k| \geq 1$ for all $\hat{D} > 0$, the scheme is unconditionally unstable. This scheme was, unfortunately, used by L. F. Richardson in his heroic but failed attempt at weather forecasting in the 1920s.

For the two-dimensional diffusion problem, the explicit scheme leads to $\hat{D} \leq \frac{1}{2}\frac{(\Delta y)^2}{(\Delta x)^2+(\Delta y)^2}$. For $\Delta x = \Delta y$ the allowable time step is half that of the the one-dimensional problem. For the three dimensional problem it is one third of the one-dimensional problem.

Advection equation. As a second example consider the advection equation

$$\partial_t \rho + c\partial_x \rho = 0 \tag{6.22}$$

and the scheme

$$\rho_j^{n+1} = \rho_j^{n-1} - C(\rho_{j+1}^n - \rho_{j-1}^n) \tag{6.23}$$

where $C := c\Delta t/\Delta x \leq 1$ is the Courant number. The scheme is stable if $|C| \leq 1$. This is the famous Courant-Friedrich-Levy criterion (CFL). It states that the algorithm may not propagate information faster than the propagation speed c.

Advection-diffusion equation. If we add *diffusion* to the advection equation we arrive at the advection-diffusion equation

$$\partial_t \rho + c\partial_x \rho = D\partial_x\partial_x \rho \tag{6.24}$$

Consider the scheme

$$\rho_j^{n+1} = \rho_j^{n-1} - C(\rho_{j+1}^n - \rho_{j-1}^n) + \hat{D}(\rho_{j+1}^{n-1} - 2\rho_j^{n-1} + \rho_{j-1}^{n-1}) \tag{6.25}$$

for which one finds the stability criterion $C^2 \leq 1 - 2\hat{D}_{)}$, which reduces to the appropriate criteria for the purely advective and diffusive case but implies that the Courant number or time step needs to be reduced when diffusion is added to the advection equation.

6.5 Numerical Diffusion and Dispersion

Discretization can introduce artificial properties into the discretized system that are absent in the original continuous problem. Two such artificial properties are numerical diffusion and numerical dispersion that we demonstrate for the advection equation (6.22). It has the exact solution $\rho(x,t) = \rho_0(x - ct)$ where $\rho_0(x) = \rho(x, t = 0)$ is the initial distribution. The initial distribution just propagates with speed c along the x-axis without any change of form. One may also analyze the advection equation in terms of wave solutions

$$\rho(x,t) = ae^{i(kx-\omega t)} \tag{6.26}$$

with dispersion relation $\omega = ck$, phase speed $c_p := \omega/k = c$ and group velocity $c_g := d\omega/dk = c$.

Numerical diffusion. One might attempt to solve this advection equation with an explicit scheme that is forward in time and *centered* in space

$$\rho_j^{n+1} = \rho_j^n - \frac{1}{2}C(\rho_{j+1}^n - \rho_{j-1}^n) + O(\Delta t, (\Delta x)^2), \tag{6.27}$$

where C is the Courant number. Linear stability analysis gives the amplification factor $|G_k| = (1 + C^2 \sin^2 k\Delta x)^{1/2} \geq 1$. This schemes is unconditionally unstable.

The *upwind* scheme

$$\rho_j^{n+1} = \rho_j^n - C(\rho_j^n - \rho_{j-1}^n) = (1 - C)\rho_j^n + C\rho_{j-1}^n + O(\Delta t, \Delta x) \tag{6.28}$$

is stable if $|C| < 1$. However, it is diffusive. This can be seen by rewriting (6.28) in the form

$$\begin{aligned} \rho_j^{n+1} &= \rho_j^n - C(\rho_j^n - \rho_{j-1}^n) + \frac{1}{2}C(\rho_{j+1}^n - \rho_{j+1}^n) \\ &= \rho_j^n - \frac{1}{2}C(\rho_{j+1}^n - \rho_{j-1}^n) + \frac{1}{2}C(\rho_{j+1}^n - 2\rho_j^n + \rho_{j-1}^n) \end{aligned} \tag{6.29}$$

The first two terms represent the unconditionally unstable advection scheme (6.27). The last term is a second order approximation to a diffusion term $D\partial_x\partial_x\rho$. The upwind scheme actually solves an advection-diffusion equation with a diffusion coefficient $D = \frac{C\Delta x}{2}$. This diffusion is called numerical diffusion. The unstable scheme for the advection term is stabilized by introducing diffusion.

Numerical dispersion. Numerical dispersion occurs when *second order* schemes are introduced for the spatial coordinates. This can be seen by analyzing the second order scheme

$$\frac{d\rho_j}{dt} + c\frac{\rho_{j+1} - \rho_{j-1}}{2\Delta x} = 0 \tag{6.30}$$

It has wave solutions $\rho_j(t) = A_k e^{i(kj\Delta x - \omega t)}$, with dispersion relation

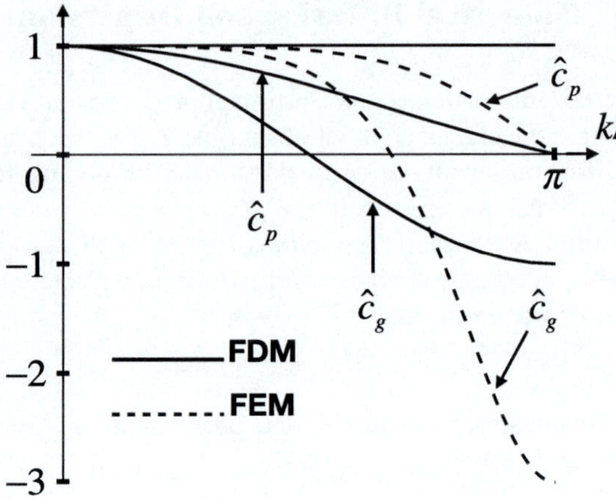

Fig. 6.2 Dispersion relations of the normalized ($c = 1$) advection equation $\partial_t \rho + \partial_x \rho = 0$. The exact phase speeds and group velocities are $c_p = c_g = 1$. The solid lines give the values for the second order finite difference approximation (FDM). The dashed lines give the values for the finite element approximation (FEM), discussed further below

$$\omega(k) = c\frac{\sin k\Delta x}{\Delta x} \tag{6.31}$$

These waves are dispersive. Their phase speed

$$\hat{c}_p(k) = \frac{\omega}{k} = c\frac{\sin k\Delta x}{k\Delta x} \tag{6.32}$$

and their group velocity

$$\hat{c}_g(k) = \frac{d\omega}{dk} = c\cos k\Delta x \tag{6.33}$$

depend on the wavenumber, whereas the original advection equation supports non-dispersive waves with constant phase speed and group velocity c. See Fig. 6.2. Any initial distribution thus disperses instead of staying intact. The problem is especially crucial for small scale features since their phase speed approaches zero as $k \to \pi/\Delta x$. Small scale features do not propagate at all but stay stagnant.

These schemes are all not very satisfactory. The first order schemes introduce numerical diffusion that smears out gradients. The second order schemes introduce numerical dispersion. More elaborative schemes have been designed that combine low order diffusive and high order dispersive schemes in a way to minimize diffusive and dispersive distortions.

6.6 Conservation Laws

Numerical schemes should also respect the conservation laws of the original continuous system. Consider the nonlinear equation

$$\partial_t \rho + \rho\partial_x \rho = 0 \tag{6.34}$$

It can be rewritten as

$$\partial_t \rho + \partial_x \left(\frac{1}{2} \rho^2 \right) = 0 \tag{6.35}$$

In this form the nonlinear equation becomes a conservation equation for a property of density ρ and flux $j = \rho^2/2$. The amount of the property within a 'volume' only changes due to fluxes across the 'surface'

$$\frac{d}{dt} \int_a^b dx \rho = -(j|_{x=b} - j|_{x=a}) \tag{6.36}$$

Any numerical algorithm should be designed to not violate this conservation and to not introduce any artificial sources or sinks. This is most easily done by discretizing the flux form (6.35) rather than the advective form (6.34). In the flux form the finite differences are automatically of the form $j_{j+1} - j_j$ and satisfy (6.36) whereas finite differences of the advective form such as $\rho_j(\rho_{j+1} - \rho_j)$ do not necessarily do so. The fluid dynamical equations represent balance equations for mass, momentum, energy and other properties and should thus be discretized in their flux form.

Moreover, the nonlinear equation (6.34) additionally conserves the variance of ρ or the 'kinetic energy' for periodic boundary conditions

$$\frac{d}{dt} \int_a^b dx \frac{1}{2} \rho^2 = - \int_a^b dx \rho^2 \partial_x \rho = - \int_a^b dx \partial_x \left(\frac{1}{3} \rho^3 \right) = 0 \tag{6.37}$$

The finite difference approximation

$$\frac{d}{dt} \rho_j = -\rho_j \frac{\rho_{j+1} - \rho_{j-1}}{2\Delta x} \tag{6.38}$$

on the other hand gives

$$\begin{aligned}
\frac{d}{dt} \sum_{j=1}^{N} \frac{1}{2} {\rho_j}^2 &= -\frac{1}{2\Delta x} \sum_j (\rho_j^2 \rho_{j+1} - \rho_j^2 \rho_{j-1}) \\
&= -\frac{1}{2\Delta x} \sum_j (\rho_j^2 \rho_{j+1} - \rho_{j+1}^2 \rho_j) \\
&= \frac{1}{2\Delta x} \sum_j \rho_j \rho_{j+1} (\rho_{j+1} - \rho_j)
\end{aligned} \tag{6.39}$$

where the right hand side does not necessarily sum to zero.

The same is true for the flux form

$$\frac{d}{dt} \rho_j = \frac{1}{2} \frac{\rho_{j+1}^2 - \rho_{j-1}^2}{2\Delta x}, \tag{6.40}$$

which leads to

$$\frac{d}{dt}\sum_{j=1}^{N}\frac{1}{2}{\rho_j}^2 = -\frac{1}{2\Delta x}\sum_j \rho_j\rho_{j+1}(\rho_{j+1}-\rho_j) \tag{6.41}$$

and also fails to conserve energy.

A related issue with the discretization of nonlinear equation like (6.34) is that the numerical scheme may become nonlinearly unstable even if it is linearly stable. Most often variance piles up at the smallest resolvable wavelengths $2\Delta x$. The reason for this nonlinear instability is twofold. First, the nonlinearity in the equations couples the Fourier modes. Variance appears in modes that had no variance initially. Second, the smallest resolvable wavelength in a grid is $2\Delta x$. If the nonlinear interactions try to transfer energy to smaller wavelengths it is folded back or aliased into the larger resolvable wavelengths. This leads to build-up of variance at the smallest wavelengths and eventual instability.

The build-up of variance at the smallest wavelengths can be avoided by removing the variance by scale selective filtering or the addition of artificial diffusion or by introducing variance conserving schemes.

6.7 Spectral and Finite Element Models

Instead of representing the spatial structure on a grid, as the finite difference method does, one may also represent it in terms of a truncated set of basis functions. When these basis functions are the eigenfunctions of an associated linear operator then these expansions form the basis of spectral models.

Spectral models. If the function $\rho(x,t)$ is expanded into a generally infinite sum of basis functions $\phi_n(x)$

$$\rho(x,t) = \sum_n \rho_n(t)\phi_n(x) \tag{6.42}$$

then the differential operators act on the coefficients $\rho_n(t)$ and on the basis functions $\phi_n(x)$, rather than on the function $\rho(x,t)$ itself. We used this method in Sect. 3.4 when we determined the dynamical system associated with the diffusion equation on the finite interval $[0,\pi]$. As basis functions we used the eigenfunctions of the differential operator $D\partial_x\partial_x$ which are $\sin nx$, $n = 1, 2, \ldots$. Using the orthogonality of the eigenfunctions we obtained for the expansion coefficients the equation

$$\dot{\rho}_n(t) = -Dn^2\rho_n(t) \tag{6.43}$$

which can of course be solved analytically without any recourse to numerical methods and leads to the solution

$$\rho(x,t) = \sum_{n=1}^{\infty} c_n(0)e^{-Dn^2 t} \sin nx \tag{6.44}$$

Truncation of the expansion at a finite number allows the numerical calculation of the solution, albeit with a truncation error.

Such truncated spectral representations are most efficient if the basis functions diagonalize part of the problem. As an example consider geophysical motions on a sphere. Spectral models then represent the latitudinal (φ) and longitudinal (θ) dependence of a function in terms of spherical harmonics

$$\rho(\varphi, \theta, \ldots) = \sum_n \sum_m u_n^m(\ldots) P_n^m(\mu) e^{im\theta} \tag{6.45}$$

with $\mu = \sin\varphi$ and

$$P_n(\mu) = \frac{1}{2^n n!} \frac{d^n}{d\mu^n}(\mu^2 - 1)^n \tag{6.46}$$

$$P_n^m(\mu) = (1-\mu^2)^{1/2} \frac{d^m}{d\mu^m} P_n(\mu)$$

and $m = \ldots, -2, -1, 0, +1, +2, \ldots$ and $n = 1, 2, 3, \ldots$ with $|m| \leq n$. Here P_n is the Legendre polynomial of order n and P_n^m the associated Legendre polynomial of the first kind of degree n and order m. The basis functions $P_n^m(\mu)e^{im\theta}$ are orthogonal and normalized. The index m is the zonal wavenumber. The index $n - |m|$ is the meridional wavenumber, i.e., the number of zero crossings in the open interval $(-\pi/2, \pi/2)$. The dots in (6.45) are placeholder for other variables such as time and height.

Spectral models offer a couple of advantages. First of all, they represent fields globally. The solution at a point in physical space depends on the solution everywhere else, whereas in a finite difference approximation it only depends on its neighboring points. This global representation often conforms more to the underlying physics where "everything is connected with everything else". Also, spectral models can easily accommodate variance conserving schemes.

The major disadvantage of spectral models is inflexibility. Local increase of resolution in regions of interest is impossible. Also, they are computationally somewhat less efficient than finite difference models. If N is the number of retained basis functions or the number of grid points then the number of operations grows proportional to $N \log N$ for spectral models and proportional to N for finite difference models.

Finite element models. Finite elements are another method where the spatial structure is expanded in terms of a set of basis functions. Unlike the global basis functions of spectral methods, the basis functions of the finite element method are local. They are low-order polynomials that are nonzero only in a limited domain. One can thus increase resolution locally. This is a great advantage when dealing with inhomogeneous flow fields and complex geometries.

The standard basis function for one-dimensional problems is the ‘tent’ function

$$\phi_j(x) = \begin{cases} 0 & \text{for } x < (j-1)\Delta x \\ [x-(j-1)\Delta x]/\Delta x & \text{for } (j-1)\Delta x \leq x \leq j\Delta x \\ [(j+1)\Delta x - x]/\Delta x & \text{for } j\Delta x \leq x \leq (j+1)\Delta x \\ 0 & \text{for } x > (j+1)\Delta x \end{cases} \tag{6.47}$$

Some of the intrinsic advantages and disadvantages of the finite element method can be seen by applying it to the linear advection Eq. (6.22) which results in

$$\frac{1}{6}\left(\frac{d\rho_{j+1}}{dt} + 4\frac{d\rho_j}{dt} + \frac{d\rho_{j-1}}{dt}\right) + c\frac{(\rho_{j+1} - \rho_{j-1})}{2\Delta x} = 0 \tag{6.48}$$

Contrary to the finite difference approximation (6.30) the time derivative is smoothed over adjacent grid points.

A wave dispersion analysis gives the phase speed $\hat{c}_p = \frac{3c}{k\Delta x} \cdot \frac{\sin k\Delta x}{2+\cos k\Delta x}$ and the group velocity $\hat{c}_g = 3c \cdot \frac{1+2\cos k\Delta x}{(2+\cos k\Delta x)^2}$. This phase speed and group velocity is compared in Fig. 6.2 with the phase speed and the group velocity obtained for the second order finite difference method (6.30). The dispersion of the finite element method is less than for the finite difference method. However, the $2\Delta x$ wave is still stagnant and has a group velocity $-3c$, three times faster than for the finite different approximation. Small-scale inaccuracies propagate rapidly through the elements and distort the solution.

6.8 Summary

This chapter discussed some of the basic problems that arise when one attempts to solve differential equations numerically on a computer. Numerical models face two kinds of errors. The first kind of error is the round-off error, caused by the finite word length of digital computers. The second kind of error is the truncation error, caused by the discretization of continuous functions. The overall problem is how well does a numerical solution represent the solution of the original continuous problem. Various concepts were introduced to assess the performance of a numerical algorithm: stability, order of accuracy, numerical diffusion and dispersion, and fidelity to conservation laws. There are additional issues. One is efficiency. How many operations are needed to actually carry out a numerical calculation on a computer. This issue involves the computer architecture and software and has not been covered. There is generally a trade off between accuracy and efficiency. The skill of the numerical modeler is to balance the requirements of the problem with the available computer resources.

Reference

Müller, P., von Storch, H.: Computer Modelling in Atmospheric and Oceanic Sciences Building Knowledge. Springer, Berlin, Heidelberg (2004)

Part II
Probability

When using the word probability one has to distinguish between the concept and the theory of probability.

The *concept* of probability is used in the real world when we try to quantify the uncertainty, our expectation, or our belief that an event will occur under certain circumstances or as the outcome of a certain experiment. The time-honored examples are the probabilities of the outcomes of games of chance such as flipping a coin or rolling a die. The emergence of this concept of probability is traced in Hacking (2007).

The *theory* of probability, on the other hand, is an abstract mathematical theory that was developed by Borel, Kolmogorov, and others. To what extent probability theory is useful in describing phenomena of the real world needs to be established. The relationship between the concept and the theory of probability is thus analogous to the relation between physical space and Euclidean geometry.

Probability theory provides the rules of how to manipulate given probabilities. It introduces probability distributions, determines mean values and fluctuations. The main point is that all these manipulations assume that a basic underlying (or 'a priori') probability distribution is given. This basic probability distribution then determines (in principle) the probability distributions of all quantities of interest. This determination is by mathematical relations and transformations that are provided by probability theory.

The task of statistics is to deduce or infer the basic probability distribution from observations or measurements or samples. In this part, we discuss the basic aspects of probability theory. Statistics will be discussed in the part "Probability in scientific reasoning".

Probability theory can be viewed as part of measure theory. Indeed we will present an axiomatic approach where probability is defined as a finite measure normalized to 1. There is also an approach to probability theory via the generalized Pythagorean theorem which we will discuss in the chapter on quantum mechanics.

A central concept of probability theory is that of a random variable. A random variable is a well-defined mathematical object. It is a measurable function. The adjective 'random' does not signify that anything is random in the colloquial understanding of this word as when it is used for real-world events in phrases like "picking something at random".

Probability theory goes beyond measure theory when it introduces the concepts of conditional probabilities and statistical independence. Again, these concepts are exactly defined in probability theory. How these definitions relate to their real world use in phrases like "picking two things independently" will be discussed in the part "Probability in scientific reasoning".

The first chapter of this part introduces probability spaces, random variables and their distributions, expectation values, bi- and multivariate random variables, and other fundamental notions of probability theory. The chapter stresses the point that calculating probabilities is nothing else than inferring the probability distribution of one random variable from the probability distribution of another functionally related random variable. Finally, this chapter considers the convergence and limits of family of random variables and their limiting distributions. It discusses the weak and strong law of large numbers, the central limit theorem, and extreme value distributions.

Then families of random variables are considered: first infinite sequences called discrete-time stochastic processes. Here, we focus on a subclass called Markov chains. Then, we introduce continuous-time stochastic processes which are families of random variables with a continuous index set, allowing the introduction of calculus tools into their analysis.

This part studies random stochastic processes solely in terms of their probability distributions. These distributions can be used to answer questions about the sample paths or realizations of the stochastic process such as what is the probability that each sample path is bounded, monotonic, continuous, or differentiable; what is the probability that each sample path has a limit; or what is the probability distributions of functionals of the process, such as sums or integrals. Stochastic processes are also generated when dynamical systems act on a probability space. This connection between stochastic processes and dynamics is explored in the part "Probability in dynamical evolution".

How random is a random variable? This important question is addressed by the concepts of information content and information entropy which are discussed in the last chapter of this part.

Chapter 7
Probability Theory

7.1 Introduction

This chapter gives the axiomatic approach to probability theory, due to Kolmogorov. It defines the probability space and then introduces the important notions of conditional probabilities and statistical independence. The central concept of a random variable and its probability distribution and its expectation is introduced next. Moments, characteristic functions, generating functions, and escort distributions are various ways to describe the probability distributions of random variables. Then families of two or a finite number of random variables are considered, so-called bi- and multivariate random variables. Finally, we consider two important theorems which describe the limiting behavior of mean values, the *weak* and *strong law of large numbers*. Limits of sequences of random variables can also be used to *define* probability distributions. One example is the central limit theorem which establishes the importance of the Gaussian distribution. Another example are the generalized extreme value distributions.

7.2 Probability Spaces

Here we present the axiomatic approach to probability theory as put forward by Kolmogorov. This approach has three building blocks: a *sample space*, a *σ-algebra of events* and a *probability measure*.

Sample space. The sample space Ω is the set of all possible outcomes. These outcomes are also called *simple* or *elementary events*.

Sigma-algebras. A family $\mathcal{A}$ of subsets A of the sample space $\Omega \neq \emptyset$ is called a σ-algebra, if

(1)

$$\emptyset \in \mathcal{A} \tag{7.1}$$

P. Müller, *Handbook of Dynamics and Probability*,
https://doi.org/10.1007/978-3-030-88486-4_7

(2)

$$A \in \mathcal{A} \Rightarrow (\Omega \setminus A) \in \mathcal{A} \tag{7.2}$$

(3)

$$A_n \in \mathcal{A}, n \in \mathbb{N} \Rightarrow \bigcup_{n=1}^{\infty} A_n \in \mathcal{A} \tag{7.3}$$

The σ-algebra represents (compound) events or the field of events. The properties (1) and (2) imply that $\Omega \in \mathcal{A}$ and the properties (2) and (3) together with de Morgan's law $A \setminus (B_1 \cup B_2) = (A \setminus B_1) \cap (A \setminus B_2)$ imply that the countable intersection of events is also an event. These properties of a σ-algebra insure that one can perform all countable set operations without leaving the family. On the other hand, one can leave the family by uncountable operations.

Two trivial examples of a σ-algebra are $\mathcal{A} = \{\emptyset, \Omega\}$ which is the smallest σ-algebra of Ω and $\mathcal{A} = \mathcal{P}(\Omega)$, the power set of Ω which is largest σ-algebra of Ω. There are usually additional families of subsets that form a σ-algebra. For any $A \subset X$, the set $\mathcal{A} = \{\emptyset, A, X \setminus A, X\}$ forms a σ-algebra. σ-algebras are thus not given but must be chosen or assigned. As a set, $\mathcal{A}$ may be viewed as a subset of the power set.

There exists an extension theorem for σ-algebras which states: If $\mathcal{E}$ is an arbitrary family of subsets of Ω then there exists a smallest σ-algebra $\mathcal{A}(\mathcal{E})$ of Ω which contains all sets of $\mathcal{E}$. $\mathcal{A}(\mathcal{E})$ is called the σ-algebra generated by $\mathcal{E}$.

The σ-algebra generated by finite half-open intervals of $\mathbb{R}^N$ is called the σ-algebra of Borel sets and denoted by $\mathcal{B}^N$. The Borel set of $\mathbb{R}^N$ is smaller than the power set of $\mathbb{R}^N$. Non-Borel sets are hard to construct. It is somewhat easier to prove that they exist. These Borel sets are very important in what follows.

A σ-algebra of events differs from a topology of open sets. A σ-algebra is closed under countable set operations. The open sets of a topology allow for arbitrary unions but only for finite intersections and it is unclear whether the complement of an open set is open.

Probability measures. A probability measure is a function $P : \mathcal{A} \to [0, 1]$ that satisfies the three Kolmogorov axioms

$$(1) \quad P(A) \geq 0 \tag{7.4}$$

$$(2) \quad P(\Omega) = 1 \tag{7.5}$$

$$(3) \quad P\left(\bigcup_{n=1}^{\infty} A_n\right) = \sum_{n=1}^{\infty} P(A_n) \quad if \quad A_n \cap A_m = \emptyset, \; for \;\; n \neq m \tag{7.6}$$

The third property is called σ-additivity.

These axioms assure that the probability measure is consistent with the σ-algebra. If a probability measure is assigned to sets $A_n \in \mathcal{A}$, $n \in \mathbb{N}$ then a probability measure is also assigned to all sets obtained by countable set operations, such as unions, intersections and complements. Since a probability measure can be assigned to all elements of the σ-algebra these elements are called measurable sets. It is stressed again that one needs countable unions of disjoint subsets in axiom 3.

It especially follows from the axioms that

(1)
$$P(\emptyset) = 0 \tag{7.7}$$

(2)
$$A \subseteq B \longrightarrow P(A) \leq P(B) \tag{7.8}$$

(3)
$$0 \leq P(A) \leq 1 \;\; \forall A \in \mathcal{A} \tag{7.9}$$

(4)
$$P(A \cup B) = P(A) + P(B) - P(A \cap B) \tag{7.10}$$

which is called the sum rule. It reduces to the third axiom (7.6) if the events are disjoint and implies especially

$$P(\Omega \setminus A) = 1 - P(A) \tag{7.11}$$

if $B = \Omega \setminus A$.

The triple $\{\Omega, \mathcal{A}, P\}$ is called a probability space.

$P(A)$ is called the probability of the occurrence of the event A. Specifically, one calls Ω the sure event and $\emptyset$ the impossible event. If $P(A) = 1$ one says that the event occurs 'almost surely'. If $A \subseteq B$ one says that the event A implies the event B, although this does not signify any causal relationship. If $A \cap B = \emptyset$ one says that the events A and B are disjoint. $P(A \cap B)$ and $P(A \cup B)$ are called the probabilities of the events 'A and B' and 'A or B'.

The probability P, as introduced here, is a special case of more general measures μ which satisfy a set of axioms where the axiom $P(\Omega) = 1$ is replaced by $\mu(\emptyset) = 0$. These general measures can take as a value any positive real number and infinity and

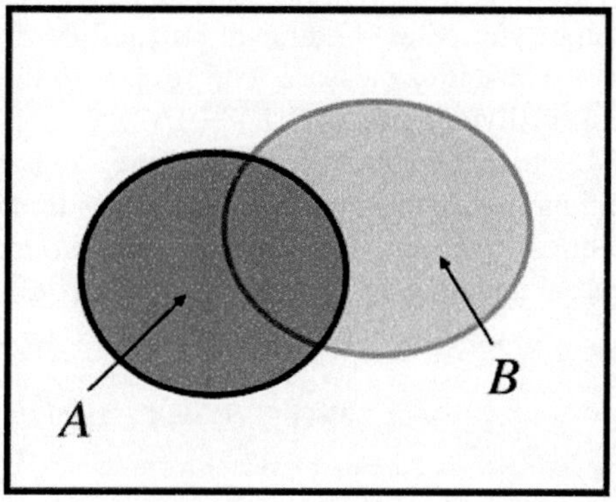

Fig. 7.1 Area representation of probabilities

are not necessarily normalized to one. An important example is the volume measure or Lebesgue measure on $\mathbb{R}^N$. Viewing probability as a normalized measure stresses the fact that there is nothing 'random' about probability at this point. Indeed, we will illustrate probability relations by set (Venn) diagrams in the unit square where the area of sets represents their probability. See Fig. 7.1. Some care must be exercised when using such Venn diagrams since probabilities are only assigned to sets that belong to the σ-algebra and not to all subsets. Throughout the book we denote general measures by μ and probabilities by P.

There is also an extension theorem for probabilities measures: If $\mathcal{E} \in \mathcal{P}$ and $\tilde{P}$ a probability measure on $\mathcal{E}$ then there exists a unique probability measure P on the σ-algebra $\mathcal{A}(\mathcal{E})$ generated by $\mathcal{E}$ if $\mathcal{E}$ meets the conditions spelled out in Sect. E.13.

The probability theory based on Kolmogorov's axioms is a well-established part of mathematics. Within mathematics, probability is a simple concept that cannot be reduced to simpler concepts other than being a normalized measure. Within this axiomatic approach probabilities are assigned only to events. These assignments must satisfy the axioms. These axioms neither tell us what probabilities are nor how to obtain them. What probabilities might be in the real world and how one might determine them is discussed in part IV "Probability in scientific reasoning".

7.3 Conditional Probabilities and Statistical Independence

By introducing conditional probabilities and statistical independence we depart from set and measure theory and enter the realm of probability theory proper.

Conditional probabilities. In our intuitive use of the concept of probability, probabilities often depend on whether or not certain conditions are met. To formalize this concept one introduces the conditional probability measure

$$P(A|B) := \frac{P(A \cap B)}{P(B)} \tag{7.12}$$

for $P(B) \neq 0$. One says that $P(A|B)$ is the probability of event A given event B or knowing that event B occurred. Specifically, $P(A|\Omega) = P(A)$ so all probabilities can be viewed as conditional probabilities. For given B, the conditional probability is a probability measure with respect to the first argument before the vertical bar. Especially one has $P(\Omega|B) = 1$.

Note that conditional probabilities do not imply causal relations or that the conditioning event B must precede the conditioned event A in time. Conditional probabilities do not imply an arrow of time. Conditional probabilities are not transitive as causal relations, in general $P(A|B)P(B|C) \neq P(A|C)$.

The conditional probabilities satisfy the product rule

$$P(A \cap B) = P(A|B)P(B) = P(B|A)P(A) \tag{7.13}$$

The last equality sign implies Bayes formula

$$P(A|B) = \frac{P(B|A)P(A)}{P(B)} \tag{7.14}$$

An expanded version which will turn out to play a fundamental role in (Bayesian) statistical inference is obtained when the sample space is partitioned, $\Omega = \cup_{i=1}^{\infty} A_i$ with $A_i \cap A_j = \emptyset$ for $i \neq j$. Then

$$P(A_i|B) = \frac{P(B|A_i)P(A_i)}{\sum_{j=1}^{\infty} P(A_j)P(B|A_j)} \tag{7.15}$$

since $B = \cup_{j=1}^{\infty}(B \cap A_j)$ and hence $P(B) = \sum_{j=1}^{\infty} P(A_j)P(B|A_j)$.

Statistical independence. Two sets A and B are called statistically (or stochastically) independent if

$$P(A \cap B) = P(A)P(B) \tag{7.16}$$

The conditional probability is then the same as the unconditional probability

$$P(A|B) = P(A), \quad P(B|A) = P(B) \tag{7.17}$$

We use the qualifier "statistical" to distinguish this kind of independence from other kinds of independences as e.g. linear independence in algebra. Also, the concept of independence is unique to probability spaces.

The concept of independence is one of the most central and perhaps the most difficult concept of probability theory. The most intuitive understanding comes perhaps from the statement that conditional and unconditional probabilities are the same. Events A and B are independent if knowing that B occurred does not affect the probability of event A. Independence has nothing to do with the set theoretical concept of two sets being disjoint. Indeed, disjoint sets are generally strongly dependent. Assume $P(A)$, $P(B) \neq 0$ and $A \cap B = \emptyset$ then $P(A \cap B) = 0 \neq P(A)P(B)$.

More generally, events $A_1, \ldots, A_N$ are called pairwise independent if $P(A_n \cap A_m) = P(A_n)P(A_m)$ for all $n \neq m$. They are called (mutually) independent if for any subset $A_{i_1}, \ldots, A_{i_n}$ with $n \leq N$ one has

$$P(A_{i_1} \cap \ldots \cap A_{i_n}) = P(A_{i_1}) \ldots P(A_{i_n}) \tag{7.18}$$

These conditions include the condition $P(A_1 \cap \ldots \cap A_N) = P(A_1) \ldots P(A_N)$ but are more than just this one condition. Mutual independence implies pairwise independence but pairwise independence does not imply mutual independence. Consider three sets A, B and C with the single probabilities $P(A) = P(B) = P(C) = 1/10$ that overlap such that $P(A \cap B \cap C) = 1/100$. Then A, B and C are pairwise independent since $P(A \cap B) = P(A \cap C) = P(B \cap C) = 1/100$ which is the product of the single probabilities. However, the sets are not mutually independent since $P(A \cap B \cap C) = 1/100$ which is not the product of the three single probabilities.

Also, the condition $P(A_1 \cap \ldots \cap A_N) = P(A_1)\ldots P(A_N)$ alone does not imply pairwise independence. If the previous three sets overlap such that $P(A \cap B) = P(A \cap C) = P(B \cap C) = 1/1000$ then these three sets satisfy $P(A \cap B \cap C) = P(A)P(B)P(C)$ but are not pairwise independent since $P(A \cap B) = 1/1000 \neq P(A)P(B)$.

For any three measurable sets $A, B, C \in [0, 1]$ the sets

$$A \times [0, 1] \times [0, 1], \quad [0, 1] \times B \times [0, 1] \quad [0, 1] \times [0, 1] \times C \tag{7.19}$$

in the unit cube are the standard example of three mutually independent and hence pairwise independent sets.

7.4 Random Variables and Their Distributions

Random variables and their distributions are the central concepts of probability theory. The remainder of this part is just an elaboration of these two concepts.

Random variable. Given a probability space $\{\Omega, \mathcal{A}, P\}$, a (univariate) random variable is a function $f : \Omega \mapsto \mathbb{R}$ with the property

$$f^{-1}(B) := \{\omega | f(\omega) \in B\} \in \mathcal{A} \text{ for all } B \in \mathcal{B} \tag{7.20}$$

where $\mathcal{B}$ is the Borel set. In measure theory such a function is called measurable. It is the analogue of a continuous function where $\mathcal{A}$ would be a topology, the family of all open sets.

Distribution of random variable. A measurable function induces a probability measure on $\mathbb{R}$ by the definition

$$P_f(B) := P(f^{-1}(B)) = P(\{\omega \in \Omega : f(\omega) \in B\}) \tag{7.21}$$

for all $B \in \mathcal{B}$. In even shorter notation

$$P_f = P \circ f^{-1} \tag{7.22}$$

This formula is often referred to as the transformation law for probabilities or the conservation law of probability. The probability measure P_f is called the (probability) distribution of f and the probability space $\{\mathbb{R}, \mathcal{B}, P_f\}$ the canonical probability space of f. The value of a random function $f(\omega)$ for a specific ω is often called an outcome or a realization. The set of all possible values of f, the image $f(\Omega)$ is called the ensemble of outcomes. The concept of a random variable formalizes what we might mean by the random outcome of an experiment.

The name random variable for f is very unfortunate. There is nothing random about f. It is just a measurable function. If there is something random at all about a random variable it is the argument and not the function. To avoid such misconception some authors prefer the name stochastic variable. This is an important point:

A random variable is a well-defined measurable function, with nothing 'random' about it.

If $h : \mathbb{R} \mapsto \mathbb{R}$ is a measurable function then $g = h \circ f$ is a random variable as well and one obtains the transformation formula

$$P_g = P \circ g^{-1} = P \circ f^{-1} \circ h^{-1} = P_f \circ h^{-1} \tag{7.23}$$

The distribution of g can thus again be calculated when the distribution of f is given. This formula turns out to be the most important formula of probability theory.

A large part of probability theory is devoted to the rationalization and calculation of the distribution P_f of a random variable f. These calculations rely on the transformation formulas $P_f = P \circ f^{-1}$ and $P_g = P_f \circ h^{-1}$ if $g = h \circ f$. The calculations assume that P is known and then calculate P_f or they assume that P_f is known and calculate P_g. How to assign the original probabilities is not covered by probability theory and will be discussed in the part "Probability in scientific reasoning".

Random variables f and their distribution P_f are the basic elements of probability theory. For all purposes one may disregard the original probability space $\{\Omega, \mathcal{A}, P\}$ and only consider the space $\{\mathbb{R}, \mathcal{B}, P_f\}$ which describes the possible outcomes and their probability distribution.

Probability density function. A random variable is called continuous if its probability measure can not assign a non-zero probability to points on the real line, but possibly to arbitrary small intervals. The probability measure P_f then operates on the Borel sets of the real line, which might be very awkward for practical calculations. Such calculations become much simpler when the probability measure can be represented by a density ρ_f such that

$$\begin{aligned} P_f(B) &= \int_B dx \rho_f(x) \ \forall \ B \in \mathcal{B} \\ \rho_f(x) &\geq 0 \\ \int_{\mathbb{R}} dx \rho_f(x) &= 1 \end{aligned} \tag{7.24}$$

The integrals are Lebesgue integrals (with respect to the Lebesgue measure $dx = \mu_L(dx)$; see Appendix E.13). The density ρ_f is thus an element of the function space $\mathcal{L}_1$, the space of all integrable functions. The Radon–Nikodym theorem states the condition for the existence of such a density (see again Appendix E.13). P needs to be 'larger' than the Lebesgue measure μ_L. All sets with Lebesgue measure zero must have zero probability.

Probability mass function. A random variable f is called discrete if there exists an at most countable set $S = \{x_1, x_2, \ldots\} \subset \mathbb{R}$ such that $P_f(S) = 1$ and

$$p_f(k) := P_f(\{x_k\}) \geq 0 \tag{7.25}$$

The set S is called the support of f or P_f. The weights $p_f(k)$ constitute the probability mass function, satisfy $\sum_{k=1}^{\infty} p_f(n) = 1$, and completely specify the probability distribution. One may disregard points for which $p_f(k) = 0$. If one allows for delta functions, discrete random variables can also be characterized by the probability density function

$$\rho_f(x) = \sum_{k=1}^{\infty} p_f(k)\delta(x - x_k) \tag{7.26}$$

A discrete random variable that is often useful is the indicator function χ_A of a set A. Then $S = \{0, 1\}$ and

$$\begin{aligned} p_{\chi_A}(1) &= P(A) \\ p_{\chi_A}(0) &= 1 - P(A) \end{aligned} \tag{7.27}$$

If the support is of a fractal nature then one enters the complicated theory of multifractals.

Probability distribution function. An alternative characterization of the probability distribution of a random variable is the probability distribution function

$$F_f(y) = P_f[(-\infty, y)] = P_f(\{x | x < y\}) \tag{7.28}$$

which always exists. If F_f is differentiable one has $dF_f/dx = \rho_f$. The probability distribution function allows one to characterize the probability distribution in terms of the median $x_{1/2}$ defined by $F_f(x_{1/2}) = 1/2$, the lower and upper quartiles $x_{1/4}$ and $x_{3/4}$ defined by $F_f(x_{1/4}) = 1/4$ and $F_f(x_{3/4}) = 3/4$ and other percentiles, though these quantities might not always be unique.

We use the expression "probability distribution" if we just want to assert that the probability measure is given regardless of whether it is specified by a probability density function, a probability mass function, or a probability distribution function.

7.5 Expectation Values

Each probability space $\{\Omega, \mathcal{A}, P\}$ induces a Lebesgue integral with an infinitesimal integration element denoted by $P(d\omega)$. Its exact definition can be found in the Appendix E.13. Using this concept one defines the expectation value or average value of a random variable f by

$$\begin{aligned} \langle f \rangle : &= \int_{\Omega} P(d\omega) f(\omega) \\ &= \int_{\mathbb{R}} P_f(dx) x \end{aligned} \tag{7.29}$$

which becomes

$$\langle f \rangle = \int_{\mathbb{R}} dx\, \rho_f(x) x \tag{7.30}$$

or

$$\langle f \rangle = \sum_{k=1}^{\infty} p_f(k) x_k \tag{7.31}$$

when the probability distribution is given by a density or mass function. The expectation value represents the average of all possible outcomes, weighted by their probabilities. The expectation value is a number. It is also denoted by $E[f]$ and also called the ensemble mean or simply the mean. Especially, one finds $\langle \chi_A \rangle = P(A)$. A random variable f is a function from Ω into $\mathbb{R}$. Instead of writing the expectation as $\langle f \rangle$ one also writes it as $\langle x \rangle$.

The concept of the expectation allows one to write the density as

$$\begin{aligned}
\rho_f(x) &= \langle \delta(x - f) \rangle \\
&= \int_{\Omega} P(d\omega) \delta(x - f(\omega)) \\
&= \int_{\mathbb{R}} P_f(dy) \delta(x - y) \\
&= \int_{\mathbb{R}} dy \rho_f(y) \delta(x - y)
\end{aligned} \tag{7.32}$$

If $g = h \circ f$ this formula implies the transformation rule for densities

$$\rho_g(x) = \int_{\mathbb{R}} dx' \rho_f(x') \delta(x - h(x')) \tag{7.33}$$

which is the basis for many of the calculations carried out in probability theory. This point is worth repeating:

Formula (7.33) is the most important formula for determining probability density functions of random variables.

The ensemble average is a linear operation

$$\left\langle \sum_{i=1}^{N} a_i f_i \right\rangle = \sum_{i=1}^{N} a_i \langle f_i \rangle \tag{7.34}$$

$$\langle a \rangle = a \tag{7.35}$$

for $a, a_i \in \mathbb{R}$. Furthermore the ensemble average commutes with differentiation

$$\left\langle \frac{d}{dt} f \right\rangle = \frac{d}{dt} \langle f \rangle \tag{7.36}$$

and has the property

$$\langle\langle f_1\rangle f_2\rangle = \langle f_1\rangle\langle f_2\rangle \tag{7.37}$$

These properties imply that the ensemble average is idempotent

$$\langle\langle f\rangle\rangle = \langle f\rangle \tag{7.38}$$

If we define the deviation from the mean or the fluctuations by

$$\delta f := f - \langle f\rangle \tag{7.39}$$

then these properties also imply

$$\langle\delta f\rangle = 0 \tag{7.40}$$

$$\langle f_1 f_2\rangle = \langle f_1\rangle\langle f_2\rangle + \langle\delta f_1 \delta f_2\rangle \tag{7.41}$$

If $h : X \mapsto X$ then $\langle h \circ f\rangle \neq h(\langle f\rangle)$ unless h is linear in f.

7.6 Moments

There are various ways to characterize probability distributions.

Moments. If $f : \Omega \mapsto \mathbb{R}$ is a random variable then we denote by f^n with $n \in \mathbb{N}$ the function

$$\begin{aligned} f^n : \Omega &\mapsto \mathbb{R} \\ \omega &\mapsto [f(\omega)]^n \end{aligned} \tag{7.42}$$

If f is a random variable then f^n is a random variable as well. Its expectation value $\langle f^n\rangle$ is called the nth moment. The expectation value $\langle(f - \langle f\rangle)^n\rangle$ is called the centered nth moment.

Mean. The mean is the first moment. Often one introduces special symbols

$$\mu_f = E[f] = \langle f\rangle \tag{7.43}$$

Means have the property $\mu_{af+b} = a\mu_f + b$ and $|\langle f\rangle| \leq \langle|f|\rangle$.

Variance. The second centered moment is called the variance and its square root the root mean square or standard deviation. Again one uses special symbols

$$\sigma_f^2 = VAR[f] = \langle(f - \langle f\rangle)^2\rangle \tag{7.44}$$

One finds $\sigma_f^2 = \langle f^2\rangle - \langle f\rangle^2$ and $\sigma_{af+b}^2 = a^2\sigma_f^2$. Invoking Jensen's inequality (see Appendix E.14) one finds $\langle f^2\rangle \geq \langle f\rangle^2$. The variance is larger or equal than zero.

If the variance is zero then the Chebyshev inequality (see Appendix E.14) implies $f(\omega) = \langle f \rangle$ for almost all $\omega \in \Omega$ and $\rho_f(x) = \delta(x - \langle f \rangle)$. The distribution is concentrated on the mean. There are almost surely no fluctuations.

Explicitly, one calculates the moments using either of the following formulas

$$\begin{aligned} \langle f^n \rangle :&= \int_\Omega P(d\omega) f^n(\omega) \\ &= \int_{\mathbb{R}} P_f(dx) x^n \\ &= \int_{\mathbb{R}} dx\, \rho_f(x) x^n \end{aligned} \tag{7.45}$$

We usually assume that probability distributions are uniquely determined by their moments. This is not necessarily true. There are counter examples.

The mean and variance or higher moments of the probability distribution might not exist. An example is the one-sided Cauchy distribution defined for $x \geq x_0$ by the density

$$\rho_C(x; x_0, a) = \frac{2}{\pi} \frac{a}{a^2 + (x - x_0)^2} \tag{7.46}$$

where the parameter x_0 determines the location of the maximum of the distribution and the parameter a the width of the distribution. Neither the mean nor the variance of this distribution (and all higher moments) exist. The same is true for the two-sided Cauchy distribution

$$\rho_C(x; x_0, a) = \frac{1}{\pi} \frac{a}{a^2 + (x - x_0)^2} \tag{7.47}$$

defined on the full real line. Though symmetric about x_0 the mean is the sum of two divergent integrals.

Characteristic function. Given a random variable f then

$$\begin{aligned} \hat{\rho}_f(k) :&= \langle e^{-ikf} \rangle \\ &= \int_\Omega P(d\omega) e^{-ikf(\omega)} \\ &= \int_{\mathbb{R}} P_f(dx) e^{-ikx} \end{aligned} \tag{7.48}$$

defines a mapping $\mathbb{R} \mapsto \mathbb{C}$, $k \mapsto \hat{\rho}_f(k)$ which is called the characteristic function of f. The characteristic function always exists. If f has a density $\rho_f(x)$ then the characteristic function becomes the Fourier transform of the density,

$$\hat{\rho}_f(k) = \int_{\mathbb{R}} dx \rho_f(x) e^{-ikx} \tag{7.49}$$

In this case we also have the inverse formula

$$\rho_f(x) = \frac{1}{2\pi}\int_{\mathbb{R}} dk\hat{\rho}_f(k)e^{ikx} \tag{7.50}$$

The characteristic function has the following properties:

(1) The normalization of the probability measure implies

$$\hat{\rho}_f(0) = 1 \tag{7.51}$$

(2) If the moments exist up to Nth order then the characteristic function is N times differentiable and

$$\langle f^n \rangle = i^n \frac{d^n \hat{\rho}_f(k)}{dk^n}\Big|_{k=0} \tag{7.52}$$

for $n = 1, \ldots, N$.

(3) If $g : \mathbb{R} \mapsto \mathbb{R}$ then

$$\langle g \circ f \rangle\rangle = g\left(i\frac{d}{dk}\hat{\rho}_f(k) \right|_{k=0} \tag{7.53}$$

(4) If $f_1, \ldots, f_N$ are independent random variables then

$$\hat{\rho}_{\sum_{i=1}^N f_n}(k) = \prod_{i=1}^N \hat{\rho}_{f_i}(k) \tag{7.54}$$

(5) The characteristic function of the convolution

$$h(y) := \int_{-\infty}^{\infty} dx f(x)g(y-x) \tag{7.55}$$

is given by the product

$$\hat{h} = \hat{f}\hat{g} \tag{7.56}$$

This property is called the convolution theorem.

These properties are often useful when it comes to actual calculations of probabilities. We will present an example in Sect. 9.4.

Generating function. Consider a discrete random variable f with probabilities mass function $p_f(n)$ for $n = 0, 1, 2, \ldots$. Then the generating function is defined by

$$G_f(z) := \langle z^f \rangle = \sum_{n=0}^{\infty} p_f(n) z^n \tag{7.57}$$

One finds

$$\frac{1}{n!}\frac{d^n G_f(z)}{dz^n}|_{z=0} = p_f(n) \tag{7.58}$$

and

$$\frac{d^n G_f(z)}{dz^n}|_{z=1} = \sum_{m=0}^{\infty} m(m-1)\dots(m-n+1)p_f(m) \tag{7.59}$$

The latter relation is a hierarchy of relations among the moments, the first ones are

$$G'_f(z=1) = \sum_{m=1}^{\infty} m p_f(m) = \langle f \rangle = \mu_f \tag{7.60}$$

$$G''_f(z=1) = \sum_{m=1}^{\infty} m(m-1)p_f(m) = \sigma_f^2 + \mu_f^2 - \mu_f \tag{7.61}$$

Again, these properties turn out to be useful for calculations. An example is again given in Sect. 9.4.

Escort distributions. In dynamical systems theory one often encounters probability distributions on supports which have a complicated often fractal structure. To characterize such distributions one may use escort distributions. Given a probability mass function $p(n) > 0, n = 1, \dots, N$ escort distributions are defined for any $\lambda \in \mathbb{R}$ by

$$p_\lambda(n) := \frac{(p(n))^\lambda}{Z_\lambda} \tag{7.62}$$

with 'partition function'

$$Z_\lambda := \sum_{n=1}^{N} (p(n))^\lambda \tag{7.63}$$

The escort distribution scans the original distribution. The smaller the parameter λ, the more heavily the small probabilities are weighted. For large values of λ the higher values are weighted more heavily. Indeed, for $\lambda \to \infty$ the largest $p(n)$ dominates the sum whereas for $\lambda \to -\infty$ the smallest (non-zero) probability dominates. Of course, $\lambda = 1$ gives the original distribution and $\lambda = 0$ gives the uniform distribution $p_0 = 1/N$.

7.7 Bi- and Multivariate Random Variables

Next we consider pairs and then finite sequences of random variables.

Bivariate random variables. Consider a probability space $\{\Omega, \mathcal{A}, P\}$ and a pair of random variables $f_1, f_2 : \Omega \mapsto \mathbb{R}$. The joint distribution $P_{f_1,f_2} : \mathcal{B}^2 \mapsto [0, 1], A \mapsto P_{f_1,f_2}(A)$ is then defined as the unique extension of the distribution for which

$$P_{f_1,f_2}(A_1 \times A_2) := P(f_1^{-1}(A_1) \cap f_2^{-1}(A_2)) \tag{7.64}$$

for all Borel sets $A_1, A_2 \in \mathcal{B}^1$. The pair (f_1, f_2) is called a bivariate random variable. The probability space $\{\mathbb{R}^2, \mathcal{B}^2, P_{f_1,f_2}\}$ is the canonical probability space of (f_1, f_2). The joint probability density function of the bivariate random variable (f_1, f_2) can, if it exists, obtained from

$$\rho_{f_1,f_2}(x_1, x_2) := \langle \delta(x_1 - f_1)\delta(x_2 - f_2) \rangle = \int_\Omega P(d\omega)\delta(x_1 - f_1(\omega))\delta(x_2 - f_2(\omega)) \tag{7.65}$$

The marginal (or reduced) distributions are the distributions P_{f_1} and P_{f_2} which give the distribution of f_1 or f_2 irrespective of the outcome of the other random variable. In case that the distributions are given by densities the marginal densities are obtained by integrating out or marginalizing the other variable

$$\rho_{f_1}(x_1) = \int_{\mathbb{R}} dx_2 \rho_{f_1,f_2}(x_1, x_2), \quad \rho_{f_2}(x_2) = \int_{\mathbb{R}} dx_1 \rho_{f_1,f_2}(x_1, x_2) \tag{7.66}$$

The *conditional* probability densities are defined by

$$\rho_{f_1|f_2}(x_1|x_2) = \frac{\rho_{f_1,f_2}(x_1, x_2)}{\rho_{f_2}(x_2)} \tag{7.67}$$

The joint probability distribution function is given by

$$F_{f_1,f_2}(x_1, x_2) = P_{f_1,f_2}(\{x_1, x_2'|x_1' < x_1, x_2' < x_2\}) \tag{7.68}$$

Statistical independence. Two random variables f_1 and f_2 are statistically independent when

$$P_{f_1,f_2}(A_1 \times A_2) = P_{f_1}(A_1)P_{f_2}(A_2) = P(f_1^{-1}(A_1))P(f_2^{-1}(A_2)) \tag{7.69}$$

for all $A_1, A_2 \in \mathcal{A}$. Statistical independence of the random variables f_1 and f_2 implies the independence of the sets $f^{-1}(B_1), g^{-1}(B_2) \subset \Omega$. *All* events of f_1 must be independent of *all* events of f_2. The indicator functions χ_{A_1} and χ_{A_2} are independent when the sets A_1 and A_2 are independent.

When the probability distributions are given by densities then the random variables f_1 and f_2 are independent if the joint density function has the separable form

$$\rho_{f_1,f_2}(x_1, x_2) = \rho_{f_1}(x_1)\rho_{f_2}(x_2) \tag{7.70}$$

In this case $\rho_{f_1|f_2}(x_1|x_2) = \rho_{f_1}(x_1)$.

Covariance. The strength of the link between the paired outcomes may be characterized by the covariance of the two variables

$$C_{f,g} := \langle (f - \langle f \rangle)(g - \langle g \rangle) \rangle = \langle fg \rangle - \langle f \rangle \langle g \rangle \tag{7.71}$$

also denoted by $COV[f, g]$. The covariance is a symmetric bilinear form. The covariance is zero if f and g are independent. Additionally the covariance has the following properties:

$$C_{f,f} = \sigma_f^2 \tag{7.72}$$

$$C_{f,g} = \frac{1}{2}(\sigma_{f+g}^2 - \sigma_f^2 - \sigma_g^2) \tag{7.73}$$

$$|C_{f,g}|^2 \leq \sigma_f^2 \sigma_g^2 \tag{7.74}$$

where $0 \leq \sigma_f \sigma_g \leq \infty$ has been assumed. The inequality is the Cauchy–Schwartz inequality since the symmetric bilinear form constitutes an inner product.

Correlation coefficient. If $\sigma_f \sigma_g > 0$ then one can normalize the covariance and obtains the correlation coefficient

$$\kappa_{f,g} := \frac{C_{f,g}}{\sigma_f \sigma_g} \tag{7.75}$$

Covariances vary between $-\infty$ and $+\infty$ and correlations between -1 and $+1$. Correlation coefficients are invariant under translation and scaling,

$$\kappa_{f+a,g+b} = \kappa_{f,g} \tag{7.76}$$

for $a, b \in \mathbb{R}$ and

$$\kappa_{af,bg} = \kappa_{f,g} \tag{7.77}$$

for $a, b > 0$. If the $\kappa_{f,g} = \pm 1$ then there exist $a > 0$ and $b \in \mathbb{R}$ such that $g(\omega) = \pm a f(\omega) + b$ for almost all $\omega \in \Omega$.

If f and g are independent then it follows that $c_{f,g} = 0$ and $\sigma_{f+g}^2 = \sigma_f^2 + \sigma_g^2$. Independence implies zero correlation. However, the converse is not true. Figure 7.2

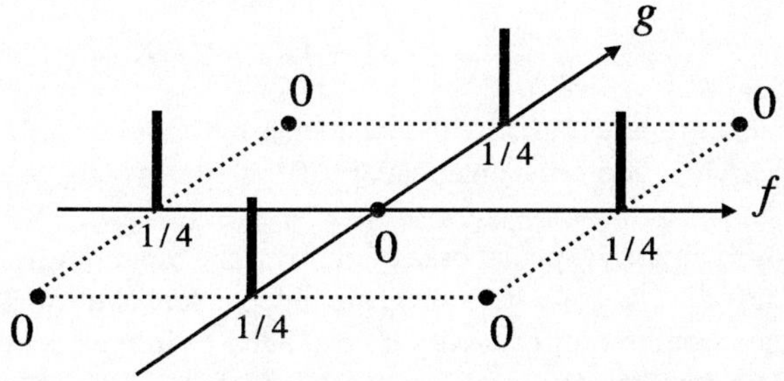

Fig. 7.2 Assume two random variables f and g that can take the values $(-1, 0, 1)$ and assume that the bivariate random variable (f, g) takes the values $(1, 0)$, $(0, 1)$, $(-1, 0)$, $(0, -1)$ with equal probability 1/4 and the other values with zero probability. Then $\langle f \rangle = \langle g \rangle = 0$. One also has $\langle fg \rangle = 0$ and hence $C_{f,g} = 0$. However, the random variables are not independent since, for example, $p_f(x = 0) = p_g(y = 0) = 1/2$ whereas $p_{f,g}(x = 0, y = 0) = 0$

gives a counter example. Being uncorrelated is a weaker statement than being independent. Independence applies to all values of (x, y), correlation only to the average value. This is one important point. The other is:

1. Covariance and correlation are simple concepts that cannot be reduced or derived from something simpler.

The transformation formula (7.33) for densities can be generalized to bivariate random variables. Specifically, the density of the random variable $f + g$ is given by

$$\begin{aligned}\rho_{f+g}(y) &= \langle \delta(y - (f + g)) \rangle \\ &= \int_{\mathbb{R}} \int_{\mathbb{R}} dx dx' \rho_{f.g}(x, x') \delta(y - (x + x')) \\ &= \int_{\mathbb{R}} dx \rho_{f,g}(x, y - x)\end{aligned} \tag{7.78}$$

In the case of independence $\rho_{f,g} = \rho_f \rho_g$ and the integral becomes the convolution of ρ_f and ρ_g.

Multivariate random variables. The definitions and concepts for bivariate random variables can easily be extended to a set of N random variables $(f_1, \ldots, f_N)$, called a multivariate random variable. The canonical probability space is $\{\mathbb{R}^N, \mathcal{B}^N, P_{f_1,\ldots,f_N}\}$ where the joint distribution $P_{f_1,\ldots,f_N}$ and density $\rho_{f_1,\ldots,f_N}$ can be constructed in complete analogy to the construction for bivariate random variables. Marginal and conditional densities are also obtained in a straightforward manner. The covariance becomes the $N \times N$ covariance matrix with elements

$$C_{i,j} := \langle (f_i - \langle f_i \rangle)(f_j - \langle f_j \rangle) \rangle \tag{7.79}$$

which is symmetric in the indexes and non-negative. The random variables $f_1, \ldots, f_N$ are mutually independent if

$$\rho_{f_1,\ldots,f_N}(x_1, \ldots, x_N) = \Pi_{i=1}^N \rho_{f_i}(x_i) \tag{7.80}$$

Mutual independence implies pairwise independence but not vice versa. The indicator functions $\chi_{A_1}, \chi_{A_2}, \chi_{A_3}$ are only independent if the sets A_1, A_2, A_3 are mutually independent.

Finite Bernoulli chain. A special case of multivariate random variables is a finite Bernoulli sequence $f_1, \ldots, f_N$ where all f_n are independent and equally distributed.

These concepts can straightforwardly be extended to infinite sequences of random variables. One important aspect of such sequences is their asymptotic behavior. Before we discuss this asymptotic behavior for specific sequences, such as arithmetic and geometric mean, we first list some common probability distributions.

7.8 Some Probability Distributions

Here we list some of the basic probability distributions and their rationale. Extensive use will be made of the transformation formula for probabilities.

Dirac measure. The simplest case of a discrete distribution is obtained by taking a point $x_0 \in X$ and an arbitrary σ-algebra. Then

$$\delta_{x_0}(A) := \chi_A(x_0) \tag{7.81}$$

with χ_A being the indicator function of A defines a probability measure concentrated on x_0, called the Dirac measure.

Counting measure. Another simple discrete probability measure arises from the counting measure. Let the sample space Ω contain a finite number of elements, the σ-algebra $\mathcal{A}$ be the power set, and $|A|$ be the number of elements in a set A. Then $P(A) = |A|/|\Omega|$ is a probability measure and $\{\Omega, \mathcal{P}(\Omega), P\}$ a probability space called the Laplacian probability space. In analyzing Laplacian probability spaces extensive use is made of combinatorics, which is the art of counting. The basic formulas of combinatorics are listed in the Appendix E.15 "Permutations and combinations".

Consider the Laplacian probability space for the set Ω which consists of all binary sequences $\omega = (\omega_1, \ldots, \omega_N)$ where $\omega_n = 0$ or 1 for $n = 1, \ldots, N$. An element ω of this set is a specific binary sequence. Combinatorics tells us that the number of elements in Ω is 2^N. Each sequence thus has the probability $1/2^N$. Consider the set A_n of all sequences that contain n ones, $n = 0, \ldots, N$. Combinatorics tells us that there are $\binom{N}{n}$ such sequences. The probability distribution is thus

$$P(A_n) = \binom{N}{n} \frac{1}{2^N} \tag{7.82}$$

and is the number of sequences with n ones in it divided by the the total number of sequences.

Another example is the game of poker. A regular deck consists of 52 cards of 4 suites (club, queen, diamond, spade) of 13 distinct face values. A poker hand consists of 5 cards. Because the order does not matter there are $\binom{52}{5} = 2{,}598{,}960$ possible poker hands. A 'full house' is defined as a hand where 3 of the cards are of the same face value (three-of-a-kind) and the two other cards are of the same but different face value (a pair). The number of full houses is given by $\binom{13}{1}\binom{4}{3}\binom{12}{1}\binom{4}{2} = 3{,}744$. The factors can be rationalized as follows. $\binom{13}{1}$ is the number that one pick one face value. $\binom{4}{3}$ is the number that one can pick three cards from the 4 suites. $\binom{12}{1}$ is the number you can pick a different face value and $\binom{4}{2}$ is the number you can pick 2 cards from 4 suites. From these numbers one can again infer the probability of a full house by dividing the number of full houses by the possible number of poker hands.

All our 'Marbles in bowls' examples that we discuss later in the book are also of this kind. There are some intricacies when determining the Laplacian probability space in statistical mechanics. According to quantum mechanics particles are indistinguishable and might satisfy the exclusion principle which results in the Bose–Einstein statistics for indistinguishable particles and the Fermi–Dirac statistics for indistinguishable particles with the exclusion principle.

The strategy of calculating probabilities by counting the favorable outcomes and dividing them by the number of possible outcomes is one of Laplace's prescription to assign probabilities. This is a correct calculation of probabilities if the underlying probability space is indeed a Laplacian probability space, in which all elements have the same probability Of course, one can consider other underlying discrete probability spaces, as we will do next.

Binomial distribution. A slight generalization of the counting measure (7.82) is the binomial distribution. Consider a random variable f that can only take two values x_1 and x_2 with probability

$$\begin{aligned} p(x_1) &= p \\ p(x_2) &= 1 - p \end{aligned} \tag{7.83}$$

and then a sequence of N independent random variables $(f_n), n = 1, \ldots, N$ of this random variable f. This is a finite Bernoulli sequence. The two random variables g_1 and g_2 that count the number of times n_1 and n_2 that x_1 and x_2 occur in each sequence can take any value from 0 to N and is distributed according to the binomial probability distribution given by

$$p_b(n_1, n_2) = \binom{N}{n_1} p^{n_1}(1-p)^{n_2} \tag{7.84}$$

where the combinatorial factor arises because there are $\binom{N}{n_1}$ combinations that contribute to a specific value n_1. Of course $n_2 = N - n_1$ and the binomial distribution reduces to a univariate distribution

$$p_b(n; N, p) = \begin{cases} \binom{N}{n} p^n (1-p)^{N-n} & \text{for } n = 0, \ldots, N \\ 0 & \text{otherwise} \end{cases} \tag{7.85}$$

where we have included the parametric dependence on N and p. This binomial distribution is usually denoted by $B(p, N)$. This binomial distribution has mean $\mu_b = Np$ and the variance $\sigma_b^2 = Np(1-p)$. The binomial distribution is one of the most basic discrete distribution. Its univariate character can also be obtained by choosing $x_1 = 1$ and $x_2 = 0$ and the univariate random variable $g = \sum_{n=1}^{N} f_n$ which is distributed according to (7.85).

Centered binomial distribution. We will also encounter a centered version of the binomial distribution which is obtained by introducing the new variable $m := 2n - N$ and which has the distribution

$$p_{cb}(m; N, p) = \begin{cases} \binom{N}{(N+m)/2} p^m (1-p)^{N-m} & \text{for } m = -N, -N+2, \ldots, N-2, N \\ 0 & \text{otherwise} \end{cases} \tag{7.86}$$

and is called the centered binomial distribution. Using the general formulas $\mu_{af+b} = a\mu_f + b$ and $\sigma^2_{af+b} = a^2\sigma^2_f$ one finds that the centered binomial distribution has mean $\mu_{cb} = 2\mu_b - N = N(2p-1)$ and variance $\sigma^2_{cb} = 4\sigma^2_b = 4Np(1-p)$. The centered binomial distribution will turn up in our analysis of the random walk which considers the problem of a particle moving on a discrete lattice $X = \mathbb{Z} = \{\ldots, -2, -1, 0, 1, 2, \ldots\}$. It is released at time $t = 0$ at the position $k = 0$. At each time step it moves to the left or to the right with equal probability $1/2$. What is the probability that it will be at position m after N time steps? This problem can be approached in many ways. One such way is combinatorial reasoning. It goes as follows: Start the process with $m = 0$. The process can then only be at even (odd) positions m after an even (odd) number N of time steps. To end up at m, $m \geq 0$ the particle must make s moves to the right and $N - s$ moves to the left where $s = (N + m)/2$. There are $\binom{N}{s}$ ways to do this. The probability for being at m is thus given by the centered binomial distribution $p_{cb}(m, N, 1/2)$. This random walk is one of our four random walks which will be described more fully and compared in the next two chapter.

Multinomial distribution. Next consider a random variable f with K possible outcomes $(x_1, \ldots, x_K)$ mapped onto $(1, \ldots, K)$, each with probability $p(k)$, and then an independent sequence $f_1, \ldots, f_N$ of this variable. The outcomes consist of sequences $(k_1, \ldots, k_N)$ where each of the k_n can take any of the values $1, \ldots, K$. Consider the random variable $g = (g_1, \ldots, g_K)$ where each g_k assigns to each sequence the number of times the value k occurs. The probability distribution of g is the multinomial distribution given by

$$p_M(n_1, \ldots, n_K; N, K) = W(n_1, \ldots, n_K; N, K)p(1)^{n_1} \ldots p(K)^{n_K} \tag{7.87}$$

if $\sum_{k=1}^K n_k = N$ and zero otherwise. Here

$$W(n_1, n_2, \ldots, n_K; N, K) = \frac{N!}{n_1!n_2! \ldots, n_k!} \tag{7.88}$$

is the number of ways that one can distribute N different objects into K boxes such that there are n_k objects in box k. The support of this distribution has $\binom{N+K-1}{K-1}$ elements.

We find

$$\langle g_k \rangle = Np_k \tag{7.89}$$

$$C_{k,k} = \sigma_k^2 = Np_k(1 - p_k) \tag{7.90}$$

$$C_{k,l} = -Np_k p_l \quad for \quad l \neq k \tag{7.91}$$

The covariances are negative since an increase in the number n_k requires a decrease in the number n_l.

Poisson distribution. Consider the combinatorial problem that N balls are distributed over K squares. What is the probability that there are n balls in a specific square? This problem maps easily onto the Bernoulli sequence. Pick a specific square and put one ball after another either in the square with probability $p = 1/K$ or into any other square with probability $1 - p$. The result is thus again the binomial distribution. Take the limit $N \to \infty$, $p \to 0 (K \to \infty)$ while keeping the average number of particles per square $\mu := N/K = Np$ constant. The result is the Poisson distribution

$$p_P(n;\mu) := \lim_{\substack{N\to\infty,p\to 0\\ Np=\mu=const.}} p_b(n;N,p) = \frac{\mu^n}{n!}e^{-\mu} \tag{7.92}$$

for $n = 0, 1, \ldots$. The limit is straightforward and requires Euler's formula $\lim_{N\to\infty}\left(1-\frac{\mu}{N}\right)^N = e^{-\mu}$. The Poisson distribution thus describes the distribution of the number of events per unit area or time. The mean and variance of the Poisson distribution are both μ.

The above examples show that the calculation of discrete distributions require the art of counting and often rely on combinatoric arguments. More importantly, they show that one must always prescribe the probability distribution of an underlying probability space (the Laplacian probability space or the Bernoulli sequence).

Uniform distribution. The perhaps simplest among the continuous distributions is the uniform distribution

$$\rho(x) = \frac{\chi_X(x)}{\mu_L(X)} \tag{7.93}$$

where μ_L is the Lebesgue measure of the support of the random variable and χ the indicator function.

The Gaussian or normal distribution. The most distinguished continuous distribution is the Gaussian or normal distribution. Its density and distribution function are

$$\begin{aligned} \rho_G(x) &= \frac{1}{\sqrt{2\pi\sigma^2}}\exp\left[-(x-\mu)^2/2\sigma^2\right] \\ F_G(x) &= \frac{1}{2}\left[1+\operatorname{erf}\left(\frac{x-\mu}{\sqrt{2}\sigma}\right)\right] \end{aligned} \tag{7.94}$$

with mean μ and variance σ^2 and where *erf* denotes the error function. The Gaussian distribution is denoted by $N(\mu, \sigma^2)$. If centered, $\mu = 0$, the odd moments are zero and the nth even moment is given by $\sigma^n(n-1)!!$ where the double factorial $k!!$ is the product of all numbers from 1 to k that have the same parity as k.

The characteristic function of the Gaussian distribution is

$$\hat{\rho}_g(k) = \exp\left[-k^2\sigma^2/2 - ik\mu\right] \tag{7.95}$$

The Gaussian distribution can be obtained from the binomial distribution by a series of approximations and limits. Start from the Gaussian approximation of the centered binomial distribution for $p = 1/2$ which is given by

$$p_{cb}^{Gaussian}(m;N,1/2) \approx \frac{2}{\sqrt{2\pi N}}e^{-\frac{m^2}{2N}} \tag{7.96}$$

Then perform a continuum limit by introducing the normalized variable $x := m\Delta x$. In the limit $N \to \infty$, $m \to \infty$ and $\Delta x \to 0$ such that $\sigma^2 = N(\Delta x)^2$ and x remain finite one obtains the Gaussian distribution $N(0, \sigma^2)$

$$\lim_{\substack{N\to\infty,\Delta x\to 0\\ N(\Delta x)^2=\sigma^2}} \frac{2}{\sqrt{2\pi N}} e^{-\frac{x^2}{2N(\Delta x)^2}} \frac{1}{2\Delta x} = \frac{1}{\sqrt{2\pi\sigma^2}} e^{-\frac{x^2}{2\sigma^2}} \tag{7.97}$$

This derivation is a special case of the Central Limit Theorem which states that the distribution of the average of a series of random variables converges towards a normal distribution under fairly general circumstances. The central limit theorem is discussed further below in Sect. 7.11.

There are other circumstances where the Gaussian distribution distinguishes itself.

1. The sum of two independent Gaussian random variables with distributions $N(\mu_1, \sigma_1^2)$ and $N(\mu_2, \sigma_2^2)$ is a Gaussian random variable with distribution $N(\mu_1 + \mu_2, \sigma_1^2 + \sigma_2^2)$. Since the density of the sum of two independent random variables is given by the convolution of their densities this result may also be stated that a Gaussian density $N(\mu_1, \sigma_1^2)$ convoluted with a Gaussian density $N(\mu_2, \sigma_2^2)$ is the Gaussian density $N(\mu_1 + \mu_2, \sigma_1^2 + \sigma_2^2)$. This is most easily shown by using the product rule for the characteristic function of the sum of independent variables.

2. Consider two independent random variables with joint density $\rho(x, y) = \rho(x)\rho(y)$. Then require that the distribution is invariant under rotation, that is that it depends only on $r^2 = x^2 + y^2$ or $\rho(x)\rho(y) = F(x^2 + y^2)$. This functional relationship is only satisfied for $N(0, \sigma^2)$, requiring continuity. This symmetry argument was originally invoked by Maxwell to justify the Maxwell–Boltzmann distribution for the velocity distribution in gases.

3. The Gaussian distribution is also the distribution that maximizes the entropy, given the mean and variance. This will be proven in Chap. 10.

4. The binomial distribution $p_b(n; N, p)$ is well approximated by a Gaussian distribution for large N. Take the logarithm of the distribution and evaluate it for large values of N, n and $N - n$ using Stirling's formula $\log m! = m \log m - m + \frac{1}{2} \log 2\pi m$ for the factorials. One then obtains

$$\begin{aligned}\log p_b(n; N, p) \approx & -n \log \frac{n}{N} - (N-n) \log\left(1 - \frac{n}{N}\right) \\ & -\frac{1}{2} \log 2\pi n \left(1 - \frac{n}{N}\right) \\ & +n \log p + (N-n) \log(1-p)\end{aligned} \tag{7.98}$$

Expanding this formula around the maximum at $n = Np$ to first order in $|n - Np|/N \ll 1$ one arrives at the Gaussian approximation of the binomial distributions

$$p_b^{Gaussian}(n; N, p) = \frac{1}{\sqrt{2\pi Np(1-p)}} \exp\left\{-\frac{(n-Np)^2}{2Np(1-p)}\right\} \tag{7.99}$$

which is a good one as long as n is not on the tail of the distribution. This result is different though of course related to our previous result that the (centered) binomial distribution approaches the Gaussian distribution in the limit $N \to \infty$. In formula (7.99) n is still discrete and N still finite. The important point is:

The Gaussian distribution is a remarkable distribution indeed.

The log-normal distribution. If the random variable f has the Gaussian distribution $N(\mu, \sigma^2)$ then the random variable e^f has the log-normal distribution

$$\rho_{e^f}(x) = \frac{1}{x}\frac{1}{\sqrt{2\pi\sigma^2}} \exp\left\{-\frac{(\log x - \mu)^2}{2\sigma^2}\right\} \tag{7.100}$$

for $x > 0$. Or conversely, a variable is log-normally distributed if its logarithm is normally distributed. The log-normal distribution has mean and variance

$$\mu_{e^f} = e^{\mu+\sigma^2/2} \tag{7.101}$$

$$\sigma^2_{e^f} = (e^{\sigma^2} - 1)e^{2\mu+\sigma^2} \tag{7.102}$$

Exponential distribution. The exponential distribution arises when decay or waiting times are considered. Assume that the rate at which a radioactive isotope decays is proportional to the number $N(t)$ of radioactive atoms

$$\frac{d}{dt}N(t) = -\lambda N(t) \tag{7.103}$$

with decay constant λ. The solution is $N(t) = N(0)e^{-\lambda t}$. The fraction or probability of atoms that decay in an interval $[t_1, t_2]$ is thus given by $(e^{-\lambda t_1} - e^{-\lambda t_2})$ which can be derived from the probability density

$$\rho_E(t; \lambda) = \lambda \exp(-\lambda t) \tag{7.104}$$

which has mean $\mu = \lambda^{-1}$ and variance $\sigma^2 = \lambda^{-2}$.

Power law distributions. Power law distributions are defined on the positive real line for $n > 1$ by the density

$$\rho(x) = \frac{b}{a^n + x^n} \tag{7.105}$$

where a characterizes the width of the distribution and b is a normalization constant. For $n = 2$ we recover the one-sided Cauchy distribution. Moments exist only for $m < n - 1$. Thus no moments exist for the one-sided Cauchy distribution; only the mean exists for the case $n = 3$ etc.

The chi-squared distribution. Given two independent random variables f_1 and f_2, both with distribution $N(0, 1)$, the χ^2_2-distribution with two degrees of freedom is defined as the distribution of the sum $f_1^2 + f_2^2$. To derive its density we first note that

$$\rho_{f_{1,2}^2}(y) = \begin{cases} \frac{1}{\sqrt{2\pi y}} e^{-\frac{y}{2}} & \text{if } y \geq 0 \\ 0 & \text{otherwise} \end{cases} \tag{7.106}$$

This is the χ_1^2 distribution with one degree of freedom. Therefore we have for $z \geq 0$

$$\begin{aligned} \rho_{f_1^2+f_2^2}(z) &= \int_0^z dx \rho_{f_1^2}(x)\rho_{f_2^2}(z-x) \\ &= e^{-(z/2)} \frac{1}{2\pi} \int_0^z dx \frac{1}{\sqrt{x(z-x)}} \\ &= e^{-(z/2)} \frac{1}{2\pi} \int_0^1 du \frac{1}{\sqrt{u(1-u)}} \end{aligned} \tag{7.107}$$

where we have substituted $x = zu$ in going from the second to the last line. We do not have to evaluate the integral since we know that the density must integrate to one. The result thus is

$$\rho_{f_1^2+f_2^2}(z) = \begin{cases} \frac{1}{2} e^{-z/2} & \text{if } z \geq 0 \\ 0 & \text{otherwise} \end{cases} \tag{7.108}$$

which is the χ_2^2 distribution of two degrees of freedom.

The distribution of random variable $g = \sqrt{f_1^2 + f_2^2}$ is given by $\rho_g(r) = r \exp(-r^2/2)$ and called the Rayleigh distribution.

Normal versus power law distributions. The Gaussian distribution and its relatives are very benign distributions. Their mean and variance exist and extreme events are extremely (exponentially) rare. Using asymptotic formulas for the error function one finds

$$P(|x| > k\sigma) \approx \frac{2}{k} \frac{1}{\sqrt{2\pi}} e^{-k^2/2} \tag{7.109}$$

which becomes $2 \cdot 10^{-6.54..}$ for $k = 5$. The probability of values larger than five standard deviations away from the mean is less than one in a million. Extreme events are extremely rare.

Contrast this result with the power law distributions. For large x one finds

$$\frac{P(2x_0 \leq x < \infty)}{P(x_0 \leq x < \infty)} = \frac{1}{2^{n-1}} \tag{7.110}$$

which is independent of x_0. No matter how far you go out you only reduce the probability by a constant fraction, one half in the case of the one-sided Cauchy distribution $n = 2$.

The Gaussian distribution and its variants apply to situation where a value fluctuates about the mean because of measurement errors or other random influences which tend to cancel each other out. The power law distributions apply to phenomena where the 'winner takes all', such as wealth distributions, book and record sales, and financial products. Of course, the Gaussian distribution and the power law distributions represent extremes and applications may fall somewhere in the middle.

7.9 Convergence of Sequences of Random Variables

Here we consider infinite sequences $(f_n)_{n\in\mathbb{N}}$ of random variables and ask whether or not these sequences converge to some random variable f in the limit $n \to \infty$. The Kolmogorov axioms allow taking such limits of infinite sequences but the usual care must be taken when reordering such sequences, interchanging the order of sums and limits, and other manipulations. Here we only consider specific infinite sequences, those arising from taking the arithmetic or geometric mean or the block maximum of an ever increasing number of random variables. Various laws of large numbers apply to these sequences. These infinite sequences and laws are still considered probability theory proper. More general sequences that are supposed to represent the random evolution of a system are called stochastic processes and are discussed in the next chapters.

First we have define convergence. Recall that a sequence of points in a normed space $(x_n)_{n\in\mathbb{N}}$ converges to a point x, written $x_n \to x$ or $\lim_{n\to\infty} x_n = x$, if for every $\epsilon > 0$ there exists a n_0 such that $\| x_n - x \| < \epsilon$ for all $n > n_0$. We must generalize this concept first to functions and then to functions $f_n : \Omega \mapsto \mathbb{R}$ where Ω has a probability structure. One must distinguish different kinds of convergences, the random functions themselves may converge or their distribution may converge. One distinguishes the following cases:

Sure convergence. A sequence f_n is said to converge surely or everywhere or point-wise to f if it converges to f for all $\omega \in \Omega$, that is

$$\lim_{n\mapsto\infty} f_n(\omega) = f(\omega) \quad , \quad \forall \omega \in \Omega \tag{7.111}$$

This is the strongest form of convergence and does not invoke any probabilistic concepts.

Almost sure convergence. A sequence f_n is said to converge almost surely or almost certainly to f if it converges to f for all $\omega \in \Omega$ except for a set of probability zero, that is

$$P(|\lim_{n\to\infty} f_n - f| = 0) = 1 \tag{7.112}$$

The sequence f_n converges to f for almost all $\omega \in \Omega$. The probability of convergence is one.

Convergence in the mean. A sequence f_n is said to converge to f in the rth mean ($r \geq 1$) if the expectation value converges,

$$\lim_{n\to\infty} \langle |f_n - f|^r \rangle = 0 \tag{7.113}$$

which implies $\lim_{n\to\infty}\langle |f_n|^r\rangle = \langle |f|^r\rangle$. If $r = 1$ one simply says f_n convergence in the first mean, or simply in the mean. If $r = 2$ one says f_n converges in the second mean, or in the mean square. Convergence in the r-th mean implies convergence in the s-th mean for $1 \leq s < r$.

Convergence in probability. A weaker type of convergence than almost sure convergence is obtained if one requires that the probability of the deviation of f_n from f approaches zero as $n \to \infty$. This convergence is called convergence in probability or stochastic convergence. For every $\epsilon > 0$ one must have

$$\lim_{n\to\infty} P(|f_n - f| < \epsilon) = 1 \tag{7.114}$$

The limit of the probability is zero. This limit is also called the stochastic limit.

Convergence in distribution. If the probability distribution P_n is characterized by the probability distribution function $F_n(x) := P_n[(-\infty, x)]$ then f_n is said to converge to f in distribution if

$$\lim_{n\to\infty} F_n(x) = F(x) \tag{7.115}$$

for all values $x \in \mathbb{R}$ for which F is continuous.

One can prove the following relations between these various types of convergences (see Fig. 7.3):

- Convergence in the rth mean implies convergence in probability.
- Almost sure convergence implies convergence in probability.
- Convergence in probability implies convergence in distribution.

Textbooks on probability theory, of course, list examples of sequences that converge in one sense but not in another.

In the next sections we consider the convergence of the specific sequences of random variables that lead to the laws of large numbers.

7.10 Weak and Strong Law of Large Numbers

There are various laws of large numbers. Here we discuss the weak and the strong law. Given a sequence $(f_n)_{n\in\mathbb{N}}$ these laws consider the arithmetic mean

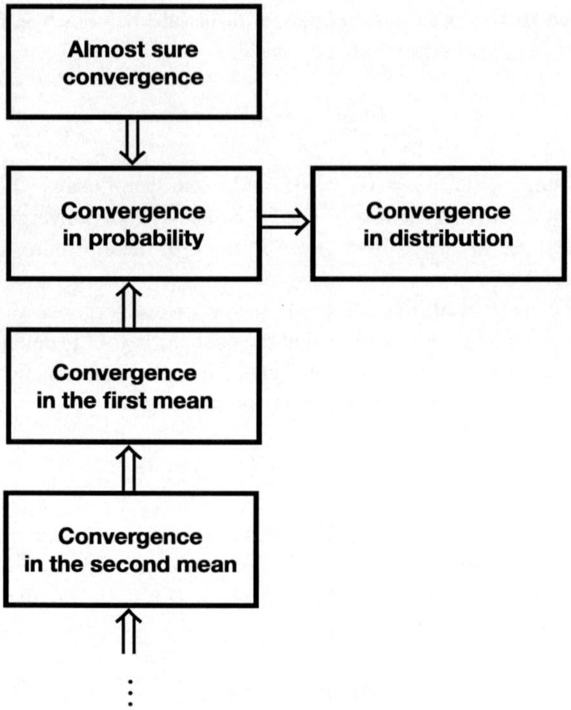

Fig. 7.3 Relation between the various types of convergence. The two boxes "Convergence in the mean" and the dots are to indicate that there is a whole hierarchy of implications, namely that convergence in the rth mean implies convergence in the sth mean for $1 \leq s < r$

$$\bar{f}_N := \frac{1}{N} \sum_{n=1}^{N} f_n \tag{7.116}$$

and ask under what circumstances and in what sense this arithmetic mean converges to its mean or, equivalently, under what circumstances and in what sense the fluctuations

$$\delta \bar{f}_N := \frac{1}{N} \sum_{n=1}^{N} (f_n - \langle f_n \rangle) \tag{7.117}$$

which have zero mean converge to zero.

Weak law of large numbers. If the fluctuations $\delta \bar{f}_N$ satisfy

$$\lim_{N \to \infty} P(|\delta \bar{f}_N|) < \epsilon) = 1 \tag{7.118}$$

for all $\epsilon > 0$ then the sequence $(f_n)_{n \in \mathbb{N}}$ is said to satisfy the weak law of large numbers. The fluctuations converge in probability to zero. Sufficient conditions are

(1) the mean values exist, that is $|\langle f_n \rangle| \leq \langle |f_n| \rangle < \infty$,
(2) the variances can only grow such that

$$\lim_{N\to\infty} \frac{1}{N^2} \sum_{n=1}^{N} \sigma_{f_n}^2 = 0 \tag{7.119}$$

which implies $\sigma_{f_n}^2 < \infty$,
(3) the random variables are pairwise uncorrelated, that is $C_{f_n,f_m} = \sigma_{f_n}^2 \delta_{nm}$.

The proof first considers the variance of $\delta \bar{f}_N$ which is given by

$$\sigma_{\delta \bar{f}_N}^2 = \frac{1}{N^2} \sum_{n=1}^{N} \sigma_{f_n}^2 \tag{7.120}$$

because the f_n are pairwise uncorrelated, according to assumption (3). Furthermore, because of assumption (2), this variance converges to zero as N approaches infinity. From Chebyshev's inequality (see Appendix E.14) it then follows that for arbitrary $\epsilon > 0$

$$P(|\delta \bar{f}_N| \geq \epsilon) \leq \frac{\sigma_{\bar{f}_N}^2}{\epsilon^2} \tag{7.121}$$

converges to zero for $N \to \infty$, which implies the weak law. A consequence of the weak law is the asymptotic equipartition theorem discussed in Sect. 10.5. If the probability distributions can be characterized by densities then the weak law implies

$$\lim_{N\to\infty} \rho_{\delta \bar{f}_N}(x) = \delta(x) \tag{7.122}$$

Strong law of large numbers. If a sequence $(f_n)_{n\in\mathbb{N}}$ of random variables satisfies

$$P(\lim_{N\to\infty} |\hat{f}_N - \langle f_1 \rangle| = 0) = 1 \tag{7.123}$$

it is said to satisfy the strong law of large number. Sufficient conditions are that the random variables

(1) have a finite mean,
(2) are identically distributed, and
(3) are pairwise independent.

The strong law states that the probability of convergence is one. The arithmetic mean

$$\lim_{N\to\infty} \hat{f}_N = \langle f_1 \rangle \tag{7.124}$$

converges to $\langle f_1 \rangle$ almost surely, for all $\omega \in \Omega$ except for a set of probability zero. The strong law implies the weak law since almost sure convergence implies convergence in probability. The weak law states that the limit of the probability is one, whereas the strong law states that the probability of the limit is one.

The strong law for sequences of random variables that satisfy the above three conditions is often called the probabilistic law of large numbers to distinguish it from the ergodic law of large numbers which we considered in Sect. 5.7. The ergodic law has the same form as (7.124) but f_n is an observable of a dynamical system and the expectation is taken with respect to an ergodic measure. In Sect. 12.2 we prove that the probabilistic law of large numbers is as a special case of the ergodic law of large numbers.

The law of large number is of paramount importance. At its heart is the fact that Eq. 7.124 holds for almost all realizations. Thus, a single realization of the arithmetic mean provides information about the probability distribution of the random variable, it determines its mean. Actually, the complete probability distribution of the random variable can be obtained from a single realization. Consider a sequence $(g_n)_{n\in\mathbb{N}}$ of mutually independent, identically distributed random variables of finite mean. For every measurable set $A \subset \mathbb{R}$ the random variables

$$f_n(\omega) := \chi_A(g_n(\omega)) = \begin{cases} 1 \text{ if } \; g_n(\omega) \in A \\ \\ 0 \text{ otherwise} \end{cases} \tag{7.125}$$

have the expectation

$$\langle f_n \rangle = P_{g_n}(A) \tag{7.126}$$

The arithmetic mean $\bar{f}_N$ gives the relative frequency with which a realization $(g_1(\omega), \ldots, g_N(\omega))$ takes values within A. The sequence $(f_n)_{n\in\mathbb{N}}$ satisfies all the assumptions of the strong law which then states that

$$\lim_{N\to\infty} \bar{f}_N = P_{g_1}(A) \tag{7.127}$$

for almost all realizations. The relative frequency of the occurrence of the event A converges in the limit $N \to \infty$ to $P_{g_1}(A)$ almost surely. This form of the strong law is invoked in statistics to justify the estimation of probabilities from relative frequencies. In Sect. 14.3 we discuss the validity of this assertion.

On a more philosophical level the strong law of large numbers may be viewed as bridging the gap between the real world where all what we have is one realization and the fictitious world of probability with all its possible outcomes. Also, the law of large numbers states that probability of a certain event is one. This is as close as one can get to deterministic statements about random events.

7.11 Central Limit Theorem

The laws of large number state the conditions under which the arithmetic mean converges to the ensemble mean. Under somewhat stronger conditions one can also determine the asymptotic distribution of the arithmetic mean and related quantities. These asymptotic distribution are obtained from the central limit theorem. The theorem assumes that the sequence $(f_n)_{n\in\mathbb{N}}$

(1) is identically distributed,
(2) is mutually or jointly independent, and
(3) has non-zero and finite variance $0 < \sigma_1^2 < \infty$.

Arithmetic mean. In case of the arithmetic mean the central limit theorem states that the distribution of the random variable

$$\sqrt{N}\delta\bar{f}_N = \frac{1}{\sqrt{N}}\sum_{n=1}^{N}(f_n - \langle f_1\rangle) \tag{7.128}$$

converges, as $N \to \infty$, to the centered Gauss distribution $N(0, \sigma_1^2)$ in distribution

$$\lim_{N\to\infty} P(\sqrt{N}\delta\bar{f}_N \in [a,b]) = \frac{1}{\sqrt{2\pi\sigma_1^2}}\int_a^b dx e^{-x^2/2\sigma_1^2} \tag{7.129}$$

for all $a, b \in \mathbb{R}$ with $-\infty \leq a, b \leq +\infty$. This theorem again demonstrates the importance of the Gaussian or normal distribution. The theorem only states the convergence in distribution. It hence also applies to discrete random variables. It is remarkable that only the variance σ_1^2 survives the limiting process. Under the additional assumption

$$\gamma_1 := \langle |f_1 - \langle f_1\rangle|^3\rangle < \infty \tag{7.130}$$

one even has an error estimate (Berry–Esseen theorem)

$$|P(\sqrt{N}\delta\bar{f}_N \in [a,b]) - \frac{1}{\sqrt{2\pi\sigma_1^2}}\int_a^b dx e^{-x^2/2\sigma_1^2}| \leq 0.8\frac{\gamma_1}{\sigma^3\sqrt{N}} \tag{7.131}$$

for all $N \in \mathbb{N}$.

The central limit theorem provides the following useful approximations to the density functions for $N \gg 1$:

$$\begin{aligned} \rho_{\sum_{n=1}^N f_n}(x) &\approx \frac{1}{\sqrt{2\pi N\sigma_1^2}}\exp\left\{-\frac{(x-N\langle f_1\rangle)^2}{2N\sigma_1^2}\right\} \\ \rho_{\frac{1}{\sqrt{N}}\sum_{n=1}^N f_n}(x) &\approx \frac{1}{\sqrt{2\pi\sigma_1^2}}\exp\left\{-\frac{(x-\sqrt{N}\langle f_1\rangle)^2}{2\sigma_1^2}\right\} \\ \rho_{\frac{1}{N}\sum_{n=1}^N f_n}(x) &\approx \frac{1}{\sqrt{2\pi\sigma_1^2/N}}\exp\left\{-\frac{(x-\langle f_1\rangle)^2}{2\sigma_1^2/N}\right\} \end{aligned} \tag{7.132}$$

The last line converges to $\delta(x - \langle f_1 \rangle)$ in the limit $N \to \infty$.

We now see that our derivation of the normal distribution from the centered binomial distribution in Sect. 7.8 is just a special case of the central limit theorem. If f_n, $n = 1, \ldots, N$ is a Bernoulli sequence with probabilities $p(-1) = p(1) = 1/2$ then the random variable $\sum_{n=1}^{N} f_n$ has a centered binomial distribution. The central limit theorem then implies that the limiting distribution of the random variable $\frac{1}{\sqrt{N}} \sum_{n=1}^{N} f_n$ is the Gaussian distribution $N(0, 1)$, since $\sigma_{f_1}^2 = 1$ in this case.

Geometric mean. The central limit theorem also provides the asymptotic behavior of products of random variables

$$f_N := \Pi_{n=1}^{N} f_n \tag{7.133}$$

or of the geometric mean $(f_N)^{1/N}$ since the logarithm of f_N is given by

$$\log f_N = \log\left(\Pi_{n=1}^{N} f_n\right) = \sum_{n=1}^{N} \log f_n \tag{7.134}$$

If the $\log f_n$ satisfy the assumption of the central limit theorem then $\log f_N$ has the Gaussian

$$\rho_{\log f_N}(y) \approx \frac{1}{\sqrt{2\pi N \sigma_{\log f_1}^2}} \exp\left\{-\frac{(y - N\langle f_1 \rangle)^2}{2N\sigma_{\log f_1}^2}\right\} \tag{7.135}$$

as its limiting distribution and f_N hence the log-normal distribution

$$\rho_{f_N}(x) \approx \frac{1}{x\sqrt{2\pi N \sigma_{\log f_1}^2}} \exp\left\{-\frac{(\log x - N\langle f_1 \rangle)^2}{2N\sigma_{\log f_1}^2}\right\} \tag{7.136}$$

with expectation

$$\langle f_N \rangle \approx \exp\left\{N\left(\langle f_1 \rangle + \frac{\sigma_{\log f_1}^2}{2}\right)\right\} \tag{7.137}$$

The weak and strong law of large numbers and the central limit theorem are cornerstones of the theory of probability. They apply, however, only to distributions which satisfy the given assumptions, which is often not the case for some power law distributions.

7.12 Extreme Value Theory

The arithmetic mean (times $\sqrt{N}$) converges to the Gaussian distribution. A different set of distributions is obtained when one considers extreme values.

Block maximum. Start with a sequence of iid random variables (f_n), $n = 1, \ldots, N$. The theory is most easily developed when the probability distribution is characterized by the probability distribution function $F_f(x) := P(f < x)$. Consider the maximum

$$g_N := max\{f_1, \ldots, f_N\} \tag{7.138}$$

called the block maximum. The probability distribution function of the block maximum g_N is then given by

$$\begin{aligned} F_{g_N}(y) = P(g_N < y) &= P(x_1 < y, \ldots, x_N < y) \\ &= P(x_1 < y) \cdot \ldots \cdot P(x_N < y) \\ &= F_f^N(y) \end{aligned} \tag{7.139}$$

because of the assumption of independence. As $N \to \infty$, the probability distribution function $F_{g_N}(y)$ tends to the degenerate form which is zero for all y with $F(y) < 1$ and one for all y with $F(y) = 1$. A way out is to introduce the normalized and scaled variable

$$g_N^* := \frac{g_N - b_N}{a_N} \tag{7.140}$$

where $a_N > 0$ and b_N are sequences of numbers. Then the following theorem holds:

If there exist a sequence of numbers a_N and b_N such that the probability distribution function $F_{g_N^*}$ tends to a non-degenerate probability function F_{g^*} as $N \to \infty$ then F_{g^*} belongs to one of the three distributions:

(1)

$$F_{g^*}(y) = \exp\left[-\exp\left(-\frac{y-b}{a}\right)\right] \tag{7.141}$$

(2)

$$F_{g^*}(y) = \begin{cases} 0 & \text{if } y \leq b \\ \exp\left[-\exp\left(-(\frac{y-b}{a})^{-\alpha}\right)\right] & \text{if } y > b \end{cases} \tag{7.142}$$

(3)

$$F_{g^*}(y) = \begin{cases} \exp\left[-\exp\left(-(\frac{y-b}{a})^{\alpha}\right)\right] & \text{if } y \leq b \\ 1 & \text{if } y > b \end{cases} \tag{7.143}$$

where $a > 0, -\infty < b < \infty$ and $\alpha > 0$. Proof can be found in Coles (2004). The first distribution is called the Gumbel distribution, the second one the Fréchet distribution and the third one the Weibull distribution.

Example: Exponential distribution. As an example consider a sequence which is distributed according to the exponential distribution

$$F_f(x) = 1 - \exp(-\lambda x) \tag{7.144}$$

for $x > 0$ and zero otherwise. Then

$$F_{g_N}(y) = \left[1 - \exp[-\lambda y]\right]^N \tag{7.145}$$

and

$$F_{g_N^*}(y) = \{[1 - \exp[-\lambda(a_N y + b_N)]\}^N \tag{7.146}$$

If one chooses

$$a_N = \lambda^{-1}, \quad b_N = \lambda^{-1} \log N \tag{7.147}$$

then the right hand side of (7.146) tends in the limit $N \to \infty$ to the simple form $\exp[-\exp(-y)]$, which is the Gumbel distribution.

Generalized extreme value distributions. The Gumbel, Fréchet and Weibull distributions are also called the generalized extreme value (GEV) distributions and can be written in compact form as

$$F_{GEV}(y; \mu, \sigma, \xi) = \exp\left\{-\left[1 + \xi\left(\frac{y-\mu}{\sigma}\right)\right]^{-1/\xi}\right\} \tag{7.148}$$

This distribution is defined on the set $\{y : 1 + \xi(y - \mu)/\sigma > 0\}$. It contains three parameters: a location parameter $\mu \in \mathbb{R}$, a scale parameter $\sigma \in \mathbb{R}_+$ and a shape parameter ξ. If the y value is such that the content of the square bracket is negative then the understanding is that the square bracket is replaced by zero, giving $F_{GEV}(y) = 1$ for these y values. The limiting case with $\xi \to 0$ is the Gumbel distribution. The GEV distributions with $\xi = 1$ and -1 are the Fréchet and Weibull distributions, respectively.

The normalization and scaling of the threshold by the sequences a_N and b_N looks very contrived but represents no problem when estimating the parameters of the GEV distribution in practice. If one asserts that the distribution function $F_{g_N^*}$ of the normalized and scaled finite block maximum is approximated by a GEV distribution F_{GEV}

$$F_{g_N^*} \approx F_{GEV} \tag{7.149}$$

then the distribution function of the original block maximum g_N

$$F_{g_N}(y) \approx F_{GEV}\left(\frac{y - b_n}{a_n}\right) \tag{7.150}$$

is also a GEV distribution.

The quantiles y_P are defined as the values that are exceeded with probability P. Since the probability of the block maximum exceeding y is just $1 - F_{GEV}(y)$ we find for these quantiles

$$y_P = \begin{cases} \mu - \frac{\sigma}{\xi}[1 - \{-\log(1-P)\}^{-\xi}] & \text{for } \xi \neq 0 \\ \mu - \sigma \log\{-\log(1-P)\} & \text{for } \xi = 0 \end{cases} \tag{7.151}$$

The common terminology is that y_p is exceeded once every $1/P$ blocks.

Pareto distribution. A well-known drawback in using the block maximum approach described above is that it is wasteful as it disregards all the 'runners-up' in the blocks, even though they may also be large. However, as described by Coles (2004), the distribution of the *individual* values of a variable above a high threshold can also be obtained. Assume that $P(g_N < y)$ can be approximated by a GEV distribution. Then we find for the conditional probability

$$\lim_{u\to\infty} P(f > u + y | f > u) = \lim_{u\to\infty} \frac{1 - F_{GEV}(u+y)}{1 - F_{GEV}(u)} = 1 - \left(1 + \frac{\xi y}{\tilde{\sigma}}\right)^{-1/\xi} \tag{7.152}$$

where $\tilde{\sigma} := \sigma + \xi(u - \mu)$. This limiting distribution is known as the generalized Pareto distribution. As an example consider the Gumbel distribution $F_{GEV}(y) = \exp[-\exp(-y)]$. Then the Pareto distribution becomes

$$\lim_{u\to\infty} \frac{e^{-(u+y)}}{e^{-u}} = e^{-y} \tag{7.153}$$

and represent the case $\xi = 0$ and $\tilde{\sigma} = 1$, using Euler's formula.

Overall, extreme value theory studies the probabilistic aspects of extreme events, including rates of convergence to the GEV distributions, their domains of validity, extension of the theory to dependent and non-stationary random variables, and models for thresholds.

7.13 Summary

In this chapter we have introduced probability theory as an abstract axiomatic theory. A probability space is a triple $\{\Omega.\mathcal{A}, P\}$ consisting of a sample space Ω, a σ-algebra $\mathcal{A}$ of events, and a probability measure P. Probability is a normalized measure. Probability theory goes beyond measure theory by introducing the additional concepts of conditional probability and statistical independence. The conditional probability $P(A|B)$ tells us how the probability of an event A changes when another event B occurs. If no change occurs then the events are statistically independent. The central concept of probability theory is that of a random variables f, which is a real-valued function from Ω into $\mathbb{R}$ with an associated probability space $\{\mathbb{R}, \mathcal{B}, P_f\}$. The random variable formalizes what we might mean by the random outcome of an experiment. Probability theory mostly considers the situation where the probability distribution of a random variable is given by its probability density, mass, or distribution function. A further important concept is the expectation value of a random variable, which allows one to characterize a probability distributions by its moments. If complete specification of the probability distribution is neither attainable nor required one often is satisfied with the specification of a few lower order moments. The generalization to bi- and multivariate random variables is straightforward.

We also introduced different types of limits and convergence of random variables. First one must distinguish whether the functions themselves converge (almost surely) or whether their distributions converge. The distributions can converge in different manners and one distinguishes convergence in the rth mean, convergence in probability, and convergence in distribution.

There are two important theorems: the weak and strong law of large numbers state under which conditions and in what sense the arithmetic mean of random variables converges to its expectation value or fluctuations about the mean converge to zero. The central limit theorem additionally states that the fluctuations appropriately scaled approach a Gaussian distribution. Though of paramount importance the Gaussian distribution is not the only relevant limiting distribution. As a different example we considered the generalized extreme value distributions.

Reference

Coles, S.: An Introduction to Statistical Modeling of Extreme Values. Springer, Berlin (2004)

Chapter 8
Discrete-Time Stochastic Processes

8.1 Introduction

Here we start to extend the concept of random variables to families of random variables $(f_t)_{t \in T}$ where the index set is either $T = \mathbb{Z}$ ($\mathbb{N}$) or $\mathbb{R}$ ($\mathbb{R}_+$). The first case represents *discrete-time stochastic processes*, the second case represents *continuous-time stochastic processes*. We thus progress up through the general classification of families of random variables

$$\begin{array}{ll} T = \{1\} & \Longleftrightarrow \text{univariate random variable} \\ T = \{1, 2\} & \Longleftrightarrow \text{bivariate random variable} \\ T = \{1, \ldots, N\} & \Longleftrightarrow \text{multivariate random variable} \\ T = \mathbb{Z}(\mathbb{N}) & \Longleftrightarrow \text{discrete-time stochastic process} \\ T = \mathbb{R}(\mathbb{R}_+) & \Longleftrightarrow \text{continuous-time stochastic process} \\ T = \mathbb{R}^d & \Longleftrightarrow d\text{-dimensional stochastic field} \end{array}$$

This chapter considers discrete-time stochastic processes, the next one continuous-time processes. We do not consider stochastic fields.

We already encountered infinite sequences of random variables in the previous chapter where we considered the limiting behavior of the arithmetic mean and of the block maximum of random variables. There the sequences of random variables were typically thought of as being generated by repeating an experiment. The order was not particular important. Here we interpret the index set T as time since stochastic processes are used to model the non-deterministic time evolution of systems. Realizations of stochastic processes then become time series, discrete or continuous. Use will be made of the fact that time can be ordered.

As for dynamical systems we shorten the adjectives "discrete-time" and "continuous-time" to just "discrete" and "continuous". Thus a discrete stochastic process is one for which the time T is discrete. A continuous stochastic process is one for which time T is continuous. If the support of the random variables is discrete or continuous we speak of a stochastic process on a discrete or continuous state

P. Müller, *Handbook of Dynamics and Probability*,
https://doi.org/10.1007/978-3-030-88486-4_8

space. As in dynamical system theory one must thus distinguish the four cases where either the time index is discrete or continuous or the support is discrete or continuous. Indeed, we will consider the random walk for all these four cases.

Given infinite families of random variables, whether discrete or continuous, the first order of business is to define the *canonical probability spaces* for such families. Here we make use of the extension theorem by Daniell and Kolmogorov which states that the joint distribution of such infinite families of random variables is determined by the set of all finite marginal distributions. The explicit formulation of this theorem is the same for discrete and continuous processes and given here in its generality though there are some non-trivial issues for continuous processes which will be discussed in the next chapter. Though the existence of a probability distribution for stochastic processes is assured by the extension theorem the construction of a consistent set of finite marginal distributions is a challenging task and we only discuss in detail discrete independent processes which are completely characterized by a set of one-point probabilities and discrete Markov processes which are completely characterized by a set of one-step transition probabilities.

Independent processes are the simplest processes. The outcome at a certain time does not depend on the outcome at any previous time. Markov processes are next in line where the outcome at a certain time only depends on the outcome at the previous time. The future distributions are fully determined by the present distribution, similar to dynamical systems where the time evolution is determined by the present state and does not depend on the history of how this present state is reached. The transition probabilities of Markov processes evolve according to a linear recursion formula with a linear operator called the Markov operator. The evolution of these probabilities thus constitutes a linear dynamical system, exhibiting all the nice properties of such systems. Fixed points of this dynamical systems are stationary processes. Whether or not arbitrary processes converge towards these stationary processes can be inferred from the eigenvalues of the Markov operator.

Random walks are the simplest examples of such Markov processes and are discussed in some detail to show the richness of results that can be obtained from such a simple process.

8.2 The Canonical Probability Space

Stochastic processes $(f_t)_{t\in T}$ are families of random variables, where the index set is $T = \mathbb{Z}$ ($\mathbb{N}$) for discrete processes and $T = \mathbb{R}$ ($\mathbb{R}_+$) for continuous processes. For each fixed $t \in T$, the mapping $\omega \mapsto f_t(\omega)$ is a random variable. For each fixed $\omega \in \Omega$ the mapping $t \mapsto f_t(\omega)$ is a *realization* or sample path of the stochastic process. The probability space for both discrete and continuous stochastic processes can be constructed as follows:

Sample space. Recall that the Cartesian product of $N + 1$ sets $B_0, \ldots, B_N$ is defined by

$$\Pi_{n=0}^{N} B_n := \{(x_0, \ldots, x_N) | x_n \in B_n; n = 0, \ldots, N\}, \tag{8.1}$$

and consists of all $(N+1)$-tuples $(x_0, \ldots, x_N)$ with $x_n \in B_n$. This Cartesian product can also be written

$$\Pi_{n=0}^{N} B_n = \{x : n \mapsto x_n | n \in \{0, \ldots, N\}; x_n \in B_n; \} \quad (8.2)$$

and consists of all functions x with arguments $\{0, \ldots, N\}$ and values $x_n \in B_n$. In this form the definition of the Cartesian product can easily be generalized to arbitrary index sets T to give

$$\Pi_{t \in T} B_t := \{x : t \mapsto x(t) | t \in T; x(t) \in B_t\}. \quad (8.3)$$

Using this generalized Cartesian product the sample space Ω of a stochastic process $(f_t)_{t \in T}$ is

$$\mathbb{R}^T := \{x : t \mapsto x(t) | x(t) \in \mathbb{R}, t \in T\} \quad (8.4)$$

which is the set of all functions from T into $\mathbb{R}$. If $T = \mathbb{Z}$ ($\mathbb{N}$) we call these functions sequences and write $x(n) = x_n$.

The Borel set. The Borel set on $\mathbb{R}^T$ is constructed in the following way. For each finite subset $\{t_0, \ldots, t_N\} \subset T$ and each set of sets $B_{t_0}, \ldots, B_{t_N} \in \mathcal{B}$ (where $\mathcal{B}$ is the Borel set of $\mathbb{R}$) one can define the sets

$$C(B_{t_0}, \ldots, B_{t_N}) := B_{t_0} \times \cdots \times B_{t_N} \times \Pi_{t \notin (t_0, \ldots, t_N)} \mathbb{R} \subset \mathbb{R}^T \quad (8.5)$$

which consist of all functions with values in B_{t_n} for $t_0, \ldots, t_N$ and arbitrary values in $\mathbb{R}$ for all other times. These sets are called cylinder sets and are subsets of the sample space. The σ-algebra generated by all these cylinder sets is the Borel set $\mathcal{B}^T$.

Probability measure. Similarly, each finite subset $(f_{t_0}, \ldots, f_{t_N})$ constitutes a multivariate random variable whose joint probability measure $P_{t_0, \ldots, t_N}$ on $\mathcal{B}^N$ is given by the unique extension of the probability measures for which

$$P_{t_0, \ldots, t_N}(B_{t_0} \times \cdots \times B_{t_N}) = P(f_{t_0}^{-1}(B_{t_0}) \cap \cdots \cap f_{t_N}^{-1}(B_{t_N})) \quad (8.6)$$

where P is the probability of the underlying probability space. Given the set of all finite joint distributions $P_{t_0, \ldots, t_N}$, the Daniell Kolmogorov extension theorem states that there exist a unique joint probability measure P_h operating on $\mathcal{B}^T$ that has these joint distributions as its marginal distributions

$$P_h(C(B_{t_0}, \ldots, B_{t_N})) = P_{t_0, \ldots, t_N}(B_{t_0} \times \cdots \times B_{t_N}) \quad (8.7)$$

This joint probability measure is regarded as the probability measure of the stochastic process. The subscript h stands for the random variable $h : \Omega \mapsto \mathbb{R}^T$ that assigns to each $\omega \in \Omega$ its sample path or realization in $\mathbb{R}^T$. The canonical probability space of a stochastic process $(f_t)_{t \in T}$ is thus $\{\mathbb{R}^T, \mathcal{B}^T, P_h\}$. The difference between dynamical systems and stochastic processes is that a (invertible) dynamical system prescribes a unique trajectory through every point of the state space whereas a stochastic process considers all possible trajectories and assigns a probability to them.

Though stochastic processes constitute an infinite family of random variables the Daniell Kolmogorov extension theorem keeps their description in the finite realm. It is sufficient to just consider finite marginal distributions. The extension theorem of measure theory (described in Appendix E.13) further simplifies their description in that it is sufficient to just consider the rectangular sets $B_0 \times \cdots \times B_N$ of the Borel set $\mathcal{B}^{N+1}$. There still remain some problems in the case that the index set is continuous which will be discussed in the next chapter.

Symmetry and compatibility conditions. The Daniell Kolmogorov extension theorem states the existence and uniqueness of the probability measure P_h but does not provide any means to actually construct P_h explicitly. One difficulty is that the set of all the marginal distributions $P_{t_0,\ldots,t_N}$ that define P_h must satisfy symmetry and compatibility conditions. First, the symmetry relations require

$$P_{t_0,\ldots,t_N}(B_{t_0} \times \cdots \times B_{t_N}) = P_{t_{i_0},\ldots,t_{i_N}}(B_{t_{i_0}} \times \cdots \times B_{t_{i_N}}) \tag{8.8}$$

where $(i_0, \ldots, i_N)$ is a permutation of $(0, \ldots, N)$. Because of the symmetry relations, the order of the indexes in the joint densities does not matter and we will therefore assume that $t_N > \cdots > t_0$ and we will order the time indexes such that the most recent time is the first index. This will turn out to be convenient.

The joint probabilities must furthermore satisfy a hierarchy of compatibility (or consistency) relations that assure that the joint probability of any subset g containing m of the $N+1$ random variables $(f_{t_N}, \ldots, f_{t_0})$ is indeed obtained by marginalizing the complementary set h of $N+1-m$ random variables. These compatibility relations form a hierarchy whose two lowest order members are

$$P_{t_1}(B_{t_1}) = \int_{B_{t_1}} \int_{\mathbb{R}} P_{t_1,t_0}(dx_1 dx_0) \tag{8.9}$$

$$P_{t_0}(B_{t_0}) = \int_{\mathbb{R}} \int_{B_{t_0}} P_{t_1,t_0}(dx_1 dx_0) \tag{8.10}$$

and

$$P_{t_1,t_0}(B_{t_1} \times B_{t_0}) = \int_{\mathbb{R}} \int_{B_{t_1}} \int_{B_{t_0}} P_{t_2,t_1,t_0}(dx_2 dx_1 dx_0) \tag{8.11}$$

$$P_{t_2,t_0}(B_{t_2} \times B_{t_0}) = \int_{B_{t_2}} \int_{\mathbb{R}} \int_{B_{t_0}} P_{t_2,t_1,t_0}(dx_2 dx_1 dx_0) \tag{8.12}$$

$$P_{t_2,t_1}(B_{t_2} \times B_{t_1}) = \int_{B_{t_2}} \int_{B_{t_1}} \int_{\mathbb{R}} P_{t_2,t_1,t_0}(dx_2 dx_1 dx_0) \tag{8.13}$$

The relation (8.12) is called the Chapman–Kolmogorov condition and will play a special role.

8.3 Discrete Stochastic Processes

We now consider discrete stochastic processes $(f_n)_{n\in\mathbb{N}(\mathbb{Z})}$ where the index n is a time index such that $t_n = n\Delta t$. Conforming with dynamical systems theory, we call the support of the random variables the state space and denote it by X. If $T = \mathbb{N}$ we call the process one-sided and realizations are then one-sided infinite sequences of real numbers, $(x_0, x_1, \ldots)$. Here we use the convention of starting the counting at $n = 0$ since the index set denotes time rather than starting at $n = 1$ as we did in the previous chapter where we had the repetition of experiments in mind. If $T = \mathbb{Z}$ we call the process two-sided and realizations are two-sided infinite sequences of integers, $(\ldots, x_{-1}, x_0, x_1, \ldots)$. One-sided processes correspond to non-invertible dynamical systems, two-sided processes to invertible ones.

We formulate the theory for a discrete state space $\{\ldots, x_{-1}, x_0, x_1, \ldots\}$ which can be mapped onto the set of integers $(\ldots, -1, 0, 1, \ldots)$. Most of our examples and most applications are of this kind. Discrete stochastic processes on a discrete state space are called chains. We thus will formulate everything in terms of probability mass functions p. For a continuous state space one would need to use probability density functions ρ, replace sums by integrals, and Kronecker Deltas by delta functions. We also formulate the theory for two-sided processes, $T = \mathbb{Z}$. The one-sided case, $T = \mathbb{N}$, can be treated the same way. The probability distribution P_h is then characterized by all the joint probability mass functions

$$p_{N,N-1,\ldots,M+1,M}(k_N, k_{N-1}, \ldots, k_{M+1}, k_M) \tag{8.14}$$

where $N, M \in \mathbb{Z}$ and $N > M$. These probabilities describe the marginal probabilities that the process takes a value k_N at time N, a value k_{N-1} at time $N-1$, etc., without reference to the values of the random variables outside the finite segment. If we need to be specific we call a marginal probability with N arguments a N-point joint probability. The time indexes have been ordered with the future times being to the left. As we stated, these finite marginal probability mass functions cannot arbitrarily be prescribed but must satisfy the compatibility conditions such as (8.9) through (8.13). The relation (8.9) takes the form

$$p_N(k_N) = \sum_{k_{N-1}=-\infty}^{\infty} p_{N,N-1}(k_N, k_{N-1}) \tag{8.15}$$

Many of the notions developed for bi- and multivariate random variables can easily be generalized to discrete stochastic processes but there are also a few new concepts.

Moments. Moments are defined by

$$\langle f_N \ldots f_M \rangle := \sum_{k_N=-\infty}^{\infty} \cdots \sum_{k_M=-\infty}^{\infty} p_{N,\ldots,M}(k_N, \ldots, k_M) k_N \ldots k_M \tag{8.16}$$

Especially important are the mean

$$\mu_n := \langle f_n \rangle = \sum_{k=-\infty}^{\infty} p_n(k)k \tag{8.17}$$

which is now a function of the time index n, and the covariance function

$$C_{n,n'} := \langle (f_n - \mu_n)(f_{n'} - \mu_{n'}) \rangle = \sum_{k=-\infty}^{\infty} \sum_{k'=-\infty}^{\infty} p_{n,n'}(k,k')(k - \mu_n)(k' - \mu_{n'}) \tag{8.18}$$

which is now a function of the two time indexes n, n'. The variance is given by $\sigma_n^2 = C_{n,n}$.

Conditional probabilities. The bivariate formulas generalize to formulas like

$$p_{N,N-1,\ldots,L|L-1,\ldots,M}(k_N, k_{N-1}, \ldots, k_L | k_{L-1}, \ldots, k_M) := \frac{p_{N,\ldots,M}(k_N, \ldots, k_M)}{p_{L-1,\ldots,M}(k_{L-1}, \ldots, k_M)} \tag{8.19}$$

which describes the probability of the outcome $(k_N, \ldots, k_L)$ at times $(N, \ldots, L)$ given the conditions $(k_{L-1}, \ldots, k_M)$ at previous times $(L-1, \ldots, M)$. In the above example the conditions precede the outcomes as one expect for causal processes but causality is not implied by the concept of conditional probabilities. One might as well consider the reverse case $p_{L-1,\ldots,M|N,\ldots,L}(k_{L-}, \ldots, k_M | k_N, \ldots, k_L)$ which describes the probability of the outcome $(k_{L-1}, \ldots, k_M)$ given the conditions $(k_N, \ldots, k_L)$ at future times.

Stationary processes. A two-sided stochastic process is called stationary if it is invariant under time translation. All marginal probabilities must satisfy

$$p_{N+K,N-1+K,\ldots,M+K} = p_{N,N-1,\ldots,M} \tag{8.20}$$

for all $N, M, K \in \mathbb{Z}$ an d $N > M$. As you move back and forth in time nothing changes. Wherever you are in time the future looks the same and the past looks the same. The one-point probabilities p_n thus do not depend on time and the two-point probabilities $p_{n,n'}$ depend only on the time difference or lag $n - n'$. Consequently, the mean and variance do not depend on time, $\mu_n = \mu_0 = \mu_*$, $\sigma_n^2 = \sigma_0^2 = \sigma_*^2$, and the covariance function is a function of the time lag $n - n'$, $C_{n,n'} = C_{n'-n,0} = C^*_{n'-n}$. When the covariance function tends to zero sufficiently fast as the lag increases to infinity (such as an exponential decay), then the process is said to have a finite memory. A possible finite memory is one of the features that distinguishes a stochastic process from a deterministic process.

A one-sided stochastic process is called stationary if

$$p_{N+K,N-1+K,\ldots,M+K} = p_{N,N-1,\ldots,M} \tag{8.21}$$

for all $N, M, K \in \mathbb{N}$ and $N > M$. Such a one-sided stationary process can be extended to the two-sided process $\tilde{p}$ defined by

$$\tilde{p}_{N,\dots,M} := p_{N-M,\dots,0} \tag{8.22}$$

for $N, M \in \mathbb{Z}$ which is also stationary and identical to p for $N < M \in \mathbb{N}$.

Reversible processes. Given a two-sided discrete stochastic process $(f_n)_{n\in\mathbb{Z}}$ one can define for each $K \in \mathbb{Z}$ a time-reversed process $\tilde{f}_{n;K}$ by requiring

$$\tilde{p}_{N,N-1,\dots,M+1,M} := p_{K-N,K-N+1,\dots,K-M-1,K-M} \tag{8.23}$$

to hold for all $N, M \in \mathbb{Z}$. The time-reversed process is a time mirror of the original process, mirrored at the time K. It is the original process evolving backwards. It is the movie played backwards. If the original process is stationary then the time-reversed process is stationary. It is stressed that time is not reversed but that the time-reversed process is a different process evolving forward in time as the original process evolves backward in time. We could have also introduced the same concept for dynamical systems by defining a 'time-reversed' dynamical system by $\tilde{S}_N := S_{K-N}$.

A reversible stochastic process is one for which all time reversed processes are equal to the original process. Thus

$$p_{N,N-1,\dots,M+1,M} = p_{K-N,K-N+1,\dots,K-M-1,K-M} \tag{8.24}$$

holds for all $N, M, K \in \mathbb{Z}$. The special case $K = N + M$ results in

$$p_{N,N-1,\dots,M+1,M} = p_{M,M+1,\dots,N-1,N} \tag{8.25}$$

A reversible process is stationary since

$$p_{N,\dots,M} = p_{K_1-N,\dots,K_1-M} = p_{N+K_2-K_1,\dots,M+K_2-K_1} \tag{8.26}$$

Two successive reversions at K_1 and K_2 result in a translation by $K_2 - K_1$. A reversible process is a stationary process where the future looks the same as the past from wherever you are in time. The dynamical systems analogue is the trivial identity mapping since $S_N = S_{K-N}$ for all $K, N \in \mathbb{Z}$ implies $S_K = S_0$.

Ergodic processes. There are various definitions of an ergodic stochastic process. They are all intended to assure that various moments of the process can be inferred from single realizations. At the very least one requires that

$$\lim_{N\to\infty} \frac{1}{N} \sum_{n=0}^{N-1} f_n(\omega) = \langle f_0 \rangle \tag{8.27}$$

holds for almost all realizations $f_n(\omega)$. The time mean of the process equals the ensemble mean of the process almost everywhere. An ergodic process defined this way satisfies the strong law of large numbers. There are also definitions which imply that higher order moments satisfy

$$\lim_{N\to\infty}\frac{1}{N}\sum_{n=0}^{N-1} f_n(\omega)f_{n+\Delta n_1}\cdots f_{n+\Delta n_K} = \langle f_0 f_{\Delta n_1}\cdots f_{\Delta n_K}\rangle \tag{8.28}$$

almost surely for all $K \in \mathbb{N}$ and all possible lags $\Delta n_1, \ldots, \Delta n_K$. We give a definition of an ergodic stochastic process in Chap. 12 which implies all the above relations and conforms to the definition of an ergodic measure.

Ergodic processes are important since they allow to determine expectation values from a single realization and are assumed when estimating moments from observed time series, though that estimation is complicated by the fact that one only has finite time series. The issues surrounding estimation are treated in detail in the chapter on "Probability in scientific reasoning".

A useful classification of stochastic processes is obtained by asking: on which previous outcomes does the outcome at a certain time depend? One obtains a sequence of ever more complicated processes. The simplest case are independent processes where the outcome at a certain time does not depend on any of the previous outcomes. Next in line are Markov processes where the outcome at a certain time depends only on the outcome at the previous time. We will only discuss these two cases. Of course, more complicated dependencies reaching further into the past can be considered or might be called for.

8.4 Independent Chains

Independent chains are intended to model the succession of independent events. The prime example is the repeated tossing of a coin.

The independence condition. Formally, independent chains are characterized by the independence condition which requires that the forward conditional probabilities satisfy

$$p_{N+1\,N,\ldots,M}(k_{N+1}|k_N,\ldots,k_M) = p_{N+1}(k_{N+1}) \tag{8.29}$$

for every choice of $N > M \in \mathbb{Z}$ and every choice of integers $(k_{N+1},\ldots,k_M)$. If the time N represents the present then this conditional probability states that the future probability depends neither on the present nor on the past. The independence condition is equivalent to

$$p_{N,\ldots,M}(k_N,\ldots,k_M) = \prod_{n=M}^{N} p_n(k_n) \tag{8.30}$$

The random variables at different times are statistically independent. The covariance matrix of an independent process is given by the diagonal matrix

$$C_{n,m} := \langle (f_n - \mu_n)(f_m - \mu_m) \rangle = \sigma_n^2 \delta_{nm} \tag{8.31}$$

Independent identically distributed chains. If in addition to being independent the random variables have the same distribution, $p_n = p_*$ for all $n \in \mathbb{Z}(\mathbb{N})$, then the chain is called *independent identically distributed* (or iid for short). Iid chains are stationary. If they have a finite mean then the strong law of large numbers holds for them.

The stationary Bernoulli chain. The simplest non-trivial iid chain is the Bernoulli chain. It has as its support two real numbers, $\{x_0, x_1\}$, with $x_0 < x_1$ and the probability distribution

$$p_*(x_0) = p \ , \quad p_*(x_1) = 1 - p \tag{8.32}$$

and is abbreviated as $B(x_0, x_1; p)$. This Bernoulli chain is stationary, reversible, and ergodic. It forms the basis of many applications. Its realizations are Bernoulli sequences, sequences of the numbers x_0 and x_1. If $x_0 = 0$ and $x_1 = 1$ and $p_* = 1/2$ it assigns to all possible sequences of zeros and ones the same probability. The relation between Bernoulli chains and the Bernoulli shift, which is a dynamical system, is discussed in Chap. 12.

8.5 Markov Chains

Markov chains are used to model systems where the stochastic evolution depends only on the present state, similar to the evolution of dynamical systems. An example are two gamblers who pay each other one dollar depending on the outcome of the tossing of a fair coin. The expected amount in their purse after the next toss depends only on the amount they presently have, not on the way they got to that amount. A non-Markov chain would be obtained if the two players start out with a certain number of nickels, dimes, and quarters in their purse and exchange coins picked randomly from their purse. Then the expected amount in their purse after the next exchange does not only depend on the present amount but also on the history since the history determines how many nickels, dimes and quarters are in the purse and hence the probability of picking any of these.

Markov condition. Formally, Markov chains satisfy the Markov condition which requires that the forward conditional probabilities satisfy

$$p_{N+1\,N,\ldots,M}(k_{N+1}|k_N, \ldots, k_M) = p_{N+1\,N}(k_{N+1}|k_N) \tag{8.33}$$

for every choice of $N > M \in \mathbb{Z}$ and every choice of integers $(k_{N+1}, \ldots, k_M)$. The probabilities at time $N + 1$ only depend on the outcome at the previous time N.

The Markov property implies

$$p_{N+L,\dots,N+1\,N,\dots,M}(k_{N+L},\dots,k_{N+1}|k_N,\dots,k_M) = p_{N+L,\dots,N+1\,N}(k_{N+L},\dots,k_{N+1}|k_N) \tag{8.34}$$

for all $L \geq 2$. If the time N denotes the present then this formula states that the future is fully determined by the present and independent of the past. The Markov condition also implies

$$p_{N+L,\dots,N+1,N-1,\dots,M|N} = p_{N+L,\dots,N+1\,N}\,p_{N-1,\dots,M|N} \tag{8.35}$$

which states that the future and the past are statistically independent when conditioned on the present.

Transition probabilities. The conditional probabilities $p_{N+1|N}$ are called one-step transition probabilities. These transition probabilities are the key ingredient of Markov chains. Indeed, any given set of one-step transition probabilities $p_{N+1|N}$ with $N \in \mathbb{N}$ together with an initial one-point probability p_0 defines a one-sided Markov chain with joint probabilities

$$p_{N,\dots,0}(k_N,\dots,k_0) = p_{N|N-1}(k_N|k_{N-1})p_{N-1\,N-2}(k_{N-1}|k_{N-2})\dots p_{1|0}(k_1|k_0)p_0(k_0) \tag{8.36}$$

A set of one-step transition probabilities defines a family of Markov chains, a family of one-sided chains which differ in their initial one-point probability. It is stressed that the transition probability $p_{N+1|N}(k_{N+1}|k_N)$ is a forward probability. It asserts a direction of time. It gives the probability of the outcome k_{N+1} at the future time $N+1$ given the condition k_N at the present time N. From these forward probabilities one can construct only one-sided Markov chains in the forward direction of time. The backward transition probability, the probability of the outcome k_N at the previous time N given the condition k_{N+1} at present time $N+1$, can also be calculated and is given by

$$p_{N|N+1}(k_N|k_{N+1}) = \frac{p_{N+1\,N}(k_{N+1}|k_N)p_N(k_N)}{p_{N+1}(k_{N+1})} = \frac{p_{N+1\,N}(k_{N+1}|k_N)p_N(k_N)}{\sum_{l_N=-\infty}^{\infty} p_{N+1\,N}(k_{N+1}|l_N)p_N(l_N)} \tag{8.37}$$

It can, however, not be used to construct a chain backwards in time, from the present time $N+1$ to the past time N, since it depends on the to-be-constructed psst probability p_N. A set of forward transition probabilities thus implies a family of one-sided chains in the forward time direction. Of course, if the resulting one-sided Markov chain is stationary it can be extended to a stationary two-sided Markov chain.

Independent processes are of course Markov processes where the probability of reaching k_{N+1} is the same, independent of where you are at the previous time step.

Homogeneous Markov chains. Homogeneous chains are an important subset of Markov chains:. Homogeneous Markov chains are Markov chains for which the 1-point transition probabilities are independent of time such that

$$p_{n+1|n} =: \pi \tag{8.38}$$

for all $n \in \mathbb{N}$. The construction of homogeneous Markov chains only requires the specification of one transition probability π and an initial one-point probability p_0. Homogeneity does not imply that the chain is stationary. Only the transition probability is invariant under time translation, not the joint probabilities. On the other hand, stationary Markov chains are homogeneous. Thus stationary chains are a strict subset of homogeneous chains. A sufficient condition for a homogeneous Markov chain to be stationary is that the one-point probability is stationary.

Reversible Markov chains. A subset of stationary Markov chains are reversible Markov chains which satisfy

$$p_{N+1,N} = p_{N,N+1} \tag{8.39}$$

for all $N \in \mathbb{Z}$. This condition can also be written as

$$\pi(k|l)p_N(l) = \pi(l|k)p_N(k) \tag{8.40}$$

and is then called detailed balance. This condition is often inferred from a time symmetry of an underlying problem. Summation over l implies $p_{N+1}(k) = p_N(k)$. The one-point probability and hence the induced chain are stationary. A special case is $\pi(k|l) = \pi(l|k)$. The stationary one-point probability is then the uniform distribution, if the state space is finite. Reversible Markov chains are a strict subset of stationary Markov chains.

Example: Random walk. The most prominent example is the random walk on the discrete lattice $\mathbb{Z}$ with time-independent transition matrix

$$\pi(k|l) = \begin{cases} p \text{ for } k = l + 1 \\ 1 - p \text{ for } k = l - 1 \\ 0 \text{ otherwise} \end{cases} \tag{8.41}$$

If $p_0(k) = \delta_{k0}$ it may be interpreted as a particle released at time $t = 0$ at the position $k = 0$. At each time step it moves to the right with probability p or to the left with probability $1 - p$. The random walk is not stationary for any initial probability. The special case $p = 1/2$ will be discussed in detail below.

Example: The faulty clock. The faulty clock is a family of homogeneous Markov chains on $\mathbb{Z}_N := \mathbb{Z} \bmod N$ defined by the transition probabilities

$$\pi(k|l) = \begin{cases} 1 - \epsilon \text{ if } k = l + 1 \\ \epsilon \text{ if } k = -l \\ 0 \text{ otherwise} \end{cases} \tag{8.42}$$

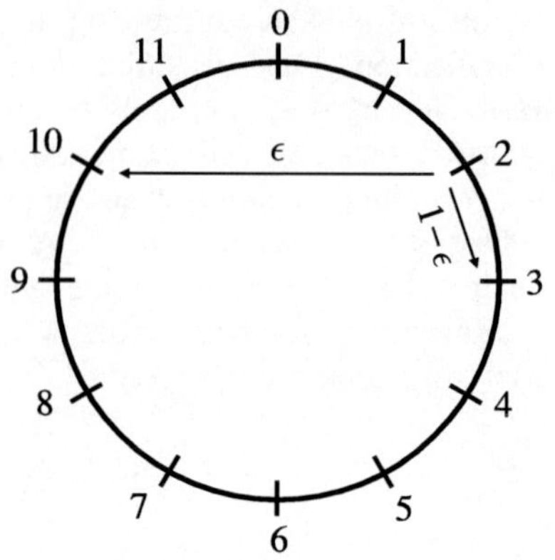

Fig. 8.1 The faulty clock

as shown in Fig. 8.1. The faulty clock has one stationary 1-point probability, namely the uniform probability $p_*(k) = 1/N$ which thus induces a stationary Markov chain. The stationary faulty clock chain is not in detailed balance and hence not reversible. One cannot reverse time, not even with a faulty clock.

Example: The golden mean chain. The golden mean chain has the two possible outcomes 0 and 1 and the time-independent transition probability

$$\begin{pmatrix} \pi(0|0) & \pi(0|1) \\ \pi(1|0) & \pi(1|1) \end{pmatrix} = \begin{pmatrix} \frac{1}{\alpha} & 1 \\ \frac{\alpha-1}{\alpha} & 0 \end{pmatrix} \tag{8.43}$$

where $\alpha \geq 1$. The specific value $\alpha = (1 + \sqrt{5})/2$ would satisfy the golden mean ratio $(\alpha - 1)/1 = 1/\alpha$. There are no consecutive ones in each realization. The golden mean chain also has one stationary 1-point probability, namely

$$p_* = \left(\frac{\alpha}{2\alpha - 1}, \frac{\alpha - 1}{2\alpha - 1} \right)^T \tag{8.44}$$

which also induces a stationary Markov chain. This stationary golden mean chain is in detailed balance and hence reversible.

These different kinds of Markov chains are summarized in Fig. 8.2.

8.6 The Markov Operator

Here we discuss the time evolution of two important probabilities, the one-point probabilities and the higher-step transition probabilities. Use will be made of the fact that the one-step transition probability is a matrix and hence a linear operator on an

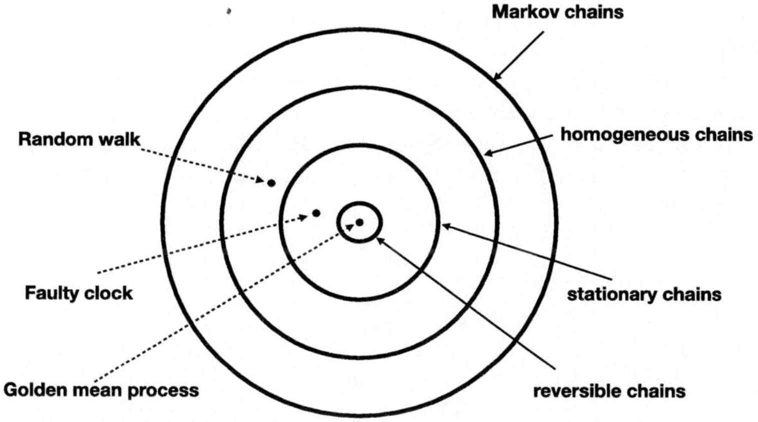

Fig. 8.2 The different kinds of Markov chains. Reversible chains (or chains in detailed balance) are a strict subset of stationary chains which are a strict subset of homogeneous chains. The random walk is homogeneous but not stationary; the faulty clock is stationary but not reversible; the golden mean chain is reversible

appropriate vector space. For simplicity we consider homogeneous Markov chains on a finite discrete state space with elements $\{1, \ldots, K\}$.

Recursion formula for one-point probabilities. The one-point probabilities evolve according to the recursion formula

$$p_{n+1}(k_{n+1}) = \sum_{k_n=1}^{K} \pi(k_{n+1}|k_n) p_n(k_n) \tag{8.45}$$

If one introduces the K-dimensional probability vector

$$p_n := (p_n(1), \ldots, p_n(K))^T \tag{8.46}$$

and the $K \times K$ transition matrix M with elements

$$M_{kl} = \pi(k|l) \tag{8.47}$$

then this recursion formula can be written

$$p_{n+1} = M p_n \tag{8.48}$$

and hence $p_n = M^n p_0$. The matrix M is column stochastic, meaning that its elements are from the interval [0, 1] and that the values in each column must add up to one. If M is a stochastic matrix then M^n for $n \geq 1$ is a stochastic matrix as well. The components of the probability vector are also from the interval [0, 1] and add up to one. They are stochastic vectors. The matrix M thus maps stochastic vectors onto stochastic vectors. Such a stochastic matrix is called a Markov operator. It is a linear operator.

Independent processes are also covered by these formulas with transition matrices where all the columns are identical.

Recursion formula for higher-step transition probabilities. Similarly the higher-step transition probabilities $p_{n|0}$ can be calculated for all $n \geq 0$ by the forward recursion formula

$$\begin{aligned} p_{n+1|0}(k_{n+1}|k_0) :&= \frac{p_{n+1,0}(k_{n+1}, k_0)}{p_0(k_0)} \\ &= \frac{\sum_{l_n=1}^{K} p_{n+1,n,0}(k_{n+1}, l_n, k_0)}{p_0(k_0)} \\ &= \frac{\sum_{l_n=1}^{K} p_{n+1|n,0}(k_{n+1}|l_n, k_0) p_{n,0}(l_n, k_0)}{p_0(k_0)} \\ &= \sum_{l_n=1}^{K} \pi(k_{n+1}|l_n) p_{n|0}(l_n|k_0) \end{aligned} \tag{8.49}$$

when starting from $p_{0|0}(l_0|k_0) = \delta_{k_0 l_0}$. The first equality is the definition of the conditional probability. The second equality is a consistency condition, namely the Chapman–Kolmogorov condition (8.12). The third equality again incorporates the definition of the conditional probability. The last equality uses the Markov condition. This recursion formula is thus a consequence of the Chapman–Kolmogorov condition.

The higher-step transition probabilities are probabilities with respect to the argument before the bar. If one introduces for fixed starting time $n = 0$ and fixed value k_0 the K-dimensional vector q with the K components

$$q_n = (p_{n|0}(1|k_0), \ldots, p_{n|0}(K|k_0))^T \tag{8.50}$$

then the forward recursion formula (8.49) can be written as

$$q_{n+1}(k_{n+1}) = \sum_{l_n=1}^{K} \pi(k_{n+1}|l_n) q_n(l_n) \tag{8.51}$$

or in matrix notation

$$q_{n+1} = M q_n \tag{8.52}$$

The notation is a little awkward. The components of the vector are given by the argument in parentheses. The subscripts denote the times.

There also exists a backward recursion formula. For fixed terminal time N and fixed k_N one finds

$$p_{N|n-1}(k_N|k_{n-1}) = \sum_{l_n=1}^{K} p_{N|n}(k_N|l_n)\pi(l_n|k_{n-1}) \tag{8.53}$$

which allows one to calculate all higher step transition probabilities $p_{N|n}$ for all n with $N \geq n \geq 0$ when starting from $p_{N|N}(k_N|l_N) = \delta_{k_N l_N}$. If one introduces for a fixed terminal time N and a fixed terminal state k_N the vector

$$\tilde{q}_n := (p_{N|n}(k_N|1), \ldots, p_{N|n}(k_N|K))^T \tag{8.54}$$

then the backward recursion formula (8.53) becomes

$$\tilde{q}_{n-1}(k_{n-1}) = \sum_{l_n=1}^{K} \tilde{q}_n(l_M)\pi(l_n|k_{n-1}) \tag{8.55}$$

or in matrix notation $\tilde{q}_{n-1}^T = \tilde{q}_n^T M$ or

$$\tilde{q}_{n-1} = M^T \tilde{q}_n \tag{8.56}$$

The conditional probabilities $\tilde{q}_n$ satisfy the adjoint problem.

Though one has forward and backward recursion formulas it is stressed that the conditional probabilities $p_{n|m}(k_n|k_m)$ that appear in these formulas are all forward conditional probability. They are the probability of k_n occurring at time n when k_m occurred at an earlier time $m < n$. The two recursion formulas are equivalent. They differ in what variables are held constant, either the ones before or after the bar. One has not tinkered with the direction of time. The forward recursion formula is useful when asking the question: starting at a fixed initial state k_0 at time zero which terminal state k_N has the highest probability at a later time $N > 0$. For the answer one calculates $p_{N|0}(k_N|k_0)$ for fixed k_0 from the forward recursion formula. The backward recursion formula is useful when asking the question: given a fixed terminal state k_N at time N (where you get a reward or punishment) which initial state k_0 yields the highest probability to get this reward (or lowest probability to avoid punishment). For the answer you calculate $p_{N|0}(k_N|k_0)$ for fixed k_N from the backward recursion formula.

Both, the evolution of the one-point probabilities and the higher-step transition probabilities are governed by the linear Markov operator $M_n := M^n$ and represents a linear dynamical system on the state space of probability vectors. One usually calculates the higher-step transition probabilities because one gets more information for the same amount of effort since the one-point probabilities can be inferred from the higher-step transition probabilities by

$$p_n(k_n) = \sum_{k_0=1}^{K} p_{n|0}(k_n|k_0)p_0(k_0) \tag{8.57}$$

Note that these formulas can in general not be inverted. The matrix M might not be invertible or if the inverse matrix exists it might not be a stochastic matrix. The inverse of a stochastic matrix M is a stochastic matrix only if it is double stochastic, the rows also add up to one, and orthogonal, $M^T M = MM^T = I$, since then the inverse M^{-1} is the transpose M^T. For $K = 2$ only the trivial evolutions

$$M = \begin{pmatrix} 1 & 0 \\ 0 & 1 \end{pmatrix} , \begin{pmatrix} 0 & 1 \\ 1 & 0 \end{pmatrix} \tag{8.58}$$

can be inverted. Generally, the Markov operator represents a linear non-invertible dynamical system.

Had we assumed a discrete state space of countably infinite dimension then the probability vectors would represent an infinite sequence of non-negative numbers that add up to one. They would be elements of the space l_1 of all summable sequences. The Markov operator then becomes an infinite matrix acting on l_1. Had we assumed a Markov process with continuous support then the probability vectors would have become probability densities which are elements of the space $\mathcal{L}_1$ of integrable functions. The summations would have been replaced by integrations and the Markov operator M would have become an integral operator acting on $\mathcal{L}_1$. Had we assumed time-dependent transition probabilities then we would have arrived at non-autonomous dynamical systems. In all these cases one can use results from the theory of linear dynamical systems to analyze such Markov chains. The important point is:

The evolution of higher-step transition probabilities is governed by a linear operator and constitutes a linear dynamical system.

8.7 Asymptotic Behavior

Here we address the asymptotic behavior of the time evolution of probabilities, by asking questions such as: Are there stationary chains and if so how many? If there is only one stationary probability p_* do Markov chains that start at some other probability $p_0 \neq p_*$ converge towards this stationary chain, and if so in what sense? How do realizations behave? Does a state eventually reach any other state? Does a state eventually return to itself? The most succinct answers are obtained for homogeneous Markov chains on a finite state space using an eigen-decomposition of the transition matrix M.

Existence of stationary chains. The stationarity condition $Mp_* = p_*$ states that p_* is a fixed point of the matrix M or that p_* is a (right) eigenvector with eigenvalue 1. The existence of an eigenvalue 1 can be inferred from the fact that the matrix M is column stochastic. It has at least one eigenvalue of 1 since $v = (1, \dots, 1)$ is an eigenvector with eigenvalue 1 of the transposed problem, $M^T v = v$, and M has the same eigenvalues as M^T. A fixed point theorem applied to the convex set of all stationary distributions assures that the corresponding eigenvector can also be normalized to be a stochastic vector. Note that this argument only works for finite matrices or chains since in the infinite case the vector $v = (1, 1, \ldots)$ cannot be normalized. The above arguments thus assure that there exist at least one stationary 1-point probability p_* and hence one stationary chain. If the eigenvalue 1 has algebraic multiplicity 1 then there exists only one stationary chain.

We exemplify these results by considering a few simple transition matrices. The first one is

$$M_B := \begin{pmatrix} 1/2 & 1/2 \\ 1/2 & 1/2 \end{pmatrix} \tag{8.59}$$

Since it has identical columns it represents an independent chain. The matrix has eigenvalues $\lambda_B^{(1)} = 1$ and $\lambda_B^{(2)} = 0$ and has thus one and only one stationary one-point probability, namely

$$p_* = (1/2, 1/2)^T \tag{8.60}$$

which is the probability of the stationary Bernoulli chain introduced in Sect. 8.4. For this reason the chains generated by the matrix M_B are sometimes called non-stationary Bernoulli chains.

The chains defined by the transition matrix

$$M_1 = \begin{pmatrix} 1/2 & 1/2 & 0 \\ 1/2 & 1/2 & 1/2 \\ 0 & 0 & 1/2 \end{pmatrix} \tag{8.61}$$

have eigenvalues $\lambda^{(1)} = 1$, $\lambda^{(2)} = 1/2$ and $\lambda^{(1)} = 0$ and therefore also have one unique stationary state, namely

$$p_* = (1/2, 1/2, 0)^T \tag{8.62}$$

On the other hand the chains defined by the transition matrix

$$M_2 = \begin{pmatrix} 1/2 & 1/2 & 0 \\ 1/2 & 1/2 & 0 \\ 0 & 0 & 1 \end{pmatrix} \tag{8.63}$$

have eigenvalues $\lambda^{(1)} = 1$, $\lambda^{(2)} = 1$ and $\lambda^{(1)} = 0$. The eigenvalue 1 has algebraic and geometric multiplicity 2. There exist a whole suite of stationary one-point probabilities since linear systems cannot have more than one isolated fixed point. They can all be written as convex combinations

$$p_* = \alpha p_*^{(1)} + (1 - \alpha) p_*^{(2)} \tag{8.64}$$

where $0 \leq \alpha \leq 1$ and

$$p_*^{(1)} = (1/2, 1/2, 0)^T \quad p_*^{(2)} = (0, 0, 1)^T \tag{8.65}$$

The set of stationary one-point probabilities forms a convex set and the vectors $p_*^{(1)}$ and $p_*^{(2)}$ are the extreme points of this set.

Convergence towards stationary chains. Assume that there exists only one stationary chain with stationary one-point probability p_* or equivalently that the eigenvalue 1 has algebraic multiplicity of 1. When do arbitrary initial probability vectors p_0 converge towards the stationary probability vector p_* as

$$\lim_{n\to\infty} p_n = \lim_{n\to\infty} M^n p_0 = p_* \tag{8.66}$$

To address this question note that the additional eigenvalues have an absolute value smaller or equal to 1. The proof makes use of the fact that an eigenvector e with eigenvalue λ satisfies $M^n e = \lambda^n e$. If $|\lambda|$ were larger than 1 the right hand side would increase without bound as $n \to \infty$. This requires that at least one element of M^n is larger than 1, contradicting the fact that M^n is a stochastic matrix. If the matrix is non-defective the eigenvectors provide a complete basis. Arbitrary initial distributions then converge to the stationary distribution if all other eigenvalues have a magnitude strictly less than one. To see this perform an eigen-analysis of the matrix problem. The stationary chain is the eigenvector with eigenvalue 1. All the other eigenvectors have eigenvalues with magnitudes smaller than 1 and their amplitudes approach zero upon successive iterations.

Since the matrix M does not only govern the evolution of the 1-point probability p_n but also the evolution of the n-step transition probability q_n one also finds

$$\lim_{n\to\infty} p_{n|0}(k_n|l_0) = p_*(k_n) \tag{8.67}$$

independent of l_0, if one eigenvalue is one and all the other eigenvalues have a magnitude strictly less than one.

If the matrix is defective, and there are stochastic matrices that are defective such as

$$M = \frac{1}{5}\begin{pmatrix} 2 & 2 & 1 \\ 1 & 2 & 1 \\ 2 & 1 & 3 \end{pmatrix} \tag{8.68}$$

with eigenvalues $\lambda_1 = 1$ and $\lambda_2 = 1/5$ then a generalized eigenvalue analysis has to be performed.

Convergence is thus assured for Bernoulli chains and for the chains M_1 but not for the 'coin turning' chain

$$M_{CT} = \begin{pmatrix} 0 & 1 \\ 1 & 0 \end{pmatrix} \tag{8.69}$$

which has eigenvalues $\lambda_{CT}^{(1)} = 1$ and $\lambda_{CT}^{(2)} = -1$ and arbitrary initial one-point probabilities do not approach its stationary distribution

$$p_* = (1/2, 1/2)^T \tag{8.70}$$

The analysis of the asymptotic behavior of Markov chains by an eigen-analysis of the Markov operator is straightforward but not very intuitive and instructive. The eigenvectors are not necessarily probability vectors. What does it mean that an eigenvalue equals $-1/2$?. For this reason mathematicians have developed a number of additional more intuitive concepts.

Irreducibility and aperiodicity. A chain is called irreducible if for any pair of states k, l there exist a time $n_{k,l} \geq 0$ such that $(M^{n_{k,l}})_{kl} > 0$. Any state k can be reached in finite time with finite probability from any state l. The Bernoulli chain and the coin turning chain are irreducible with $n_{1,2} = n_{2,1} = 1$ in both cases. The process M_1 is not irreducible since the state "3" can not be reached from the states "1" and "2'. For such non-irreducible chains the state space can be reduced or decomposed into irreducible subsets. For the chain M_1 these irreducible subsets consist of the sets $\{1, 2\}$ and $\{3\}$. The subset $\{1, 2\}$ consists of 'recurrent' states. The chain cannot leave this subset with non-zero probability. The subset $\{3\}$ on the other hand is 'transient'. The chain leaves this subset with finite probability. The chain M_2 is also not irreducible. It has the two irreducible subsets $\{1, 2\}$ and $\{3\}$, both with recurrent states.

A state k is called periodic with period

$$T = gcd\{n > 0 | (M^n)_{kk} > 0\} \tag{8.71}$$

if $T > 1$ where gcd is the greatest common divisor. If $T = 1$ the state is called aperiodic. It can then be shown that there exists a time n_k such that $(M^{n_k})_{kk} > 0$ for all $n > n_k$. Each state k will be revisited for all times $n > n_k$ with finite probability. If all states are aperiodic the chain is called aperiodic. The Bernoulli chain is aperiodic. All elements of M^n are positive, thus $n_1 = n_2 = 1$. The chains M_1 are also aperiodic. The coin turning chain, on the other hand, is not aperiodic since $(M^n)_{kk} = 0$ for odd n. This follows from the property that $M_{CT}^n = M_{CT}$ for odd n.

There are various theorems that relate these kinds of properties to the existence of stationary chains and convergence towards such chains. One such theorem states that irreducible and aperiodic chains on a finite state space have a single stationary chain and that other chains converge toward this unique stationary chain. Being irreducible and aperiodic is a sufficient condition, not a necessary condition as our chain M_1 demonstrates.

After all this theory let us discuss the most important example of a Markov chain, random walks.

8.8 Random Walks

The perhaps most famous example of a Markov chain is the random walk on a discrete state space. We reiterate some of the basic characteristics to demonstrate that such a simple model gives rise and answers to quite sophisticated questions. One distinguishes between infinite and finite state spaces. The random walk on an infinite discrete state space is our random walk I.

Random walk I. This first random walk is a homogeneous Markov chain on a discrete lattice $X = \mathbb{Z}$ with transition matrix

$$\pi(k|l) = \begin{cases} 1/2 \text{ for } k = l \pm 1 \\ 0 \text{ otherwise} \end{cases} \tag{8.72}$$

It is usually interpreted as a particle moving on a discrete grid. At each time step it moves to the left or to the right with equal probability. Of course other choices for these probabilities are also possible. The one-point probabilities evolve according to

$$p_{n+1}(k) = \frac{1}{2} p_n(k+1) + \frac{1}{2} p_n(k-1) \tag{8.73}$$

which is solved by the centered binomial distribution (7.86) when starting at $p_0(k) = \delta_{k0}$. The random walk has mean $\mu_n = 0$ and variance $\sigma_n^2 = n$. The variance grows linearly in n. The random walk process does not reach a stationary state. There are no stationary states. For large n and $|k| \ll n$ the centered binomial distribution is well approximated by the Gaussian distribution $2e^{-k^2/2n}/\sqrt{2\pi n}$.

In Sect. 7.8 we have deduced the binomial distribution for the random walk by combinatoric arguments. The advantage of describing the random walk by a transition probability is that one can not only determine the evolution of the 1-point distributions but one can also easily calculate the evolution of higher-order joint distributions $p_{N,\dots,0}$ by using formula (8.36).

Finite state space. One can define random walks on a bounded domain $-L \le k \le L$ by specifying the transition probabilities at the boundaries $k = \pm L$, mimicking absorption or reflection or some other mechanism. One such choice is

$$\pi(k| \pm L) = \begin{cases} 1/2 \text{ for } k = \pm L, k = L-1, k = -L+1 \\ 0 \text{ otherwise} \end{cases} \tag{8.74}$$

Contrary to the random walk on an infinite domain the random walk on a bounded domain allows for a stationary Markov chain. The transition probability is in detailed balance and the stationary one-point distribution is hence the uniform distribution.

Random walk II. Random walks can also be defined on continuous state spaces. Indeed, all concepts that we developed in this chapter can easily be applied to continuous state spaces as well. One only needs to replace the probabilities by probability density functions and sums by integrals. The prime example is the Gaussian random walk on $\mathbb{R}$ which is defined by the transition density

$$\pi(x_{n+1}|x_n) = \frac{1}{\sqrt{2\pi\sigma_1^2}} \exp\left\{-\frac{(x_{n+1}-x_n)^2}{2\sigma_1^2}\right\} \tag{8.75}$$

For the initial condition $\rho_0(x) = \delta(x)$ the 1-point density at time step n is a series of convolutions of Gaussian distributions and hence evolves as

$$\rho_n(x) = \frac{1}{\sqrt{2\pi n\sigma_1^2}} \exp\left\{-\frac{x^2}{2n\sigma_1^2}\right\} \tag{8.76}$$

At time step n it has mean zero and the variance $\sigma_n^2 = n\sigma_1^2$ increases proportional to n. This is our random walk II.

The main difference of this random walk to the one on a discrete lattice is not that the support is continuous rather than discrete but that the continuous random walk can move in a single time step to arbitrarily large values with finite probability. Nevertheless, for large values of n and small values of $k \ll n$ the Gaussian approximation of the centered binomial distribution implies $p_{cb}(k; n, 1/2) \approx 2\sigma_1\rho_n(k\sigma_1)$ and assures that these two processes are similar for small excursions.

An important generalization is the process with transition probability

$$\pi(x_{n+1}|x_n) = \frac{1}{\sqrt{2\pi\sigma_1^2}} \exp\left\{-\frac{(x_{n+1}-ax_n)^2}{2\sigma_1^2}\right\} \tag{8.77}$$

for $a \in \mathbb{R}$. For $|a| < 1$ this process has as a stationary solution which is a Gaussian distribution with mean zero and variance $\sigma_1^2/(1-a^2)$. This process appears in the discussion of autoregressive processes in Sect. 13.2.

Generalizations of the random walks to higher dimensional lattices and state spaces are straightforward.

The random walk on a discrete state space, infinite or finite, can be used as the basic model for many games of chances. Although very simple it allows to address some intricate and important questions relevant to gambling and to other processes that can be modeled by random walks. Here we just look at two such questions, the first passage and the gambler's ruin problem.

First passage problem. If the random walk starts at $k_0 = 0$. What is the probability that the random walk returns to zero after $N = 2M$ time steps. To answer this question one introduces the random variable h_0 defined by

$$h_0 := \min\{n | f_n = 0\} \tag{8.78}$$

where f_n denotes the random walk process. For each realization $\omega \in \Omega$ it registers the smallest time step n for which $f_n(\omega)$ becomes zero. If $f_n \neq 0$ for all $n \in \mathbb{N}$ then $h_0 = \infty$. The probability to return to zero after two ($M = 1$) steps is one half, $P(h_0 = 2) = 1/2$. The probability for $M > 1$ can be inferred by the following reasoning. Assume that the first step is positive, $k_1 = 1$. To be able to reach zero for the first time in $2M$ steps the walk must be again at $k = 1$ at the step $2M - 1$, $k_{2M-1} = 1$. In between it must have performed a path that never crosses zero. The number of these paths is equal to the number that one can arrange $(M - 1)$ open and closed brackets in a mathematically consistent manner. This is a combinatoric problem. The answer is given by the Catalan number

$$c_{M-1} = \frac{1}{M}\binom{2M-2}{M-1} \tag{8.79}$$

For $M > 1$ the probability is thus given by

$$P(h_0 = 2M) = \frac{2}{M}\binom{2M-2}{M-1}\frac{1}{2^{2M}} \tag{8.80}$$

The extra factor 2 arises from the fact that we could as well have started with a negative first step. Since

$$\sum_{M=1}^{\infty} P(h_0 = 2M) = 1 \tag{8.81}$$

the path returns to zero in finite time with probability one. The path actually returns to zero infinitely often with probability one since we can apply the above considerations again, once the path has reached zero.

A related question is: If the walk is at zero at time step $2M$ what is the probability that it is the first return. The number of ways to get to zero is obtained from the centered binomial distribution (7.86) as $\binom{2M}{M}$. The number of ways to return to zero for the first time is $\frac{2}{M}\binom{2M-2}{M-1}$. The ratio is $\frac{1}{2M-1}$.

The gambler's ruin. Have the random walk start at $k_0 > 0$. Ask whether it will reach $k_* > k_0$ or zero first, whether a person with initial capital k_0 will win $k_* - k_0$ or loose the initial capital. For this consider the random variable

$$h_{0,k_*} := \min\{n, m | f_n = 0, f_m = k_*\} \tag{8.82}$$

For each realization it registers the time steps where f hits $k = k_*$ or $k = 0$ and then takes the minimum. Again

$$\sum_{N=1}^{\infty} P(h_{0,k_*} = N) = 1 \tag{8.83}$$

One will reach either $k = k_*$ or $k = 0$ in finite time with probability one. One reaches $k = k_*$ first with probability

$$P(f_{h_{0,k*}} = k_*) = \frac{k_0}{k_*} \tag{8.84}$$

and one reaches $k = 0$ first with probability

$$P(f_{h_{0,k*}} = 0) = 1 - \frac{k_0}{k_*} \tag{8.85}$$

These results can be obtained by considering the probability

$$w(k_0) := P(f_{h_{0,k*}} = k_*) \tag{8.86}$$

as a function of k_0 which must satisfy difference equation

$$w(k_0) = \frac{1}{2}w(k_0 + 1) + \frac{1}{2}w(k_0 - 1) \tag{8.87}$$

as can be seen by evolving the random walk by one time step. Solving this difference equation with boundary condition $w(k_*) = 1$ and $w(0) = 0$ results in (8.84).

The above results are relevant for evaluating gambling strategies. There are many games where the odds are stacked against the gambler according to the above results, therefore the name 'gambler's ruin'. One such doomed game is the game where two players start out with k_1 and k_2 dollars each. Flipping a fair coin, a dollar is transferred from the looser to the winner. Taking $k_* = k_1 + k_2$ in the above formula the probability of player 1 going broke first is

$$P_1 = 1 - \frac{k_1}{k_1 + k_2} = \frac{k_2}{k_1 + k_2} \tag{8.88}$$

The probability of player 2 going broke first is the probability of player 1 reaching k_* first and is hence given by

$$P_2 = \frac{k_1}{k_1 + k_2} \tag{8.89}$$

So the player with the smaller initial capital has a higher chance of going broke first. Casinos usually have more initial capital than the players and will hence come out ahead. Additionally, casinos skew the odds in their favor. Do not gamble, buy casino stocks!

The perhaps less dramatic implications of the above results are: If k_* goes to infinity then $P(f_{h_{0,k_*}} = 0) = 1$. The path will reach zero in finite time with probability one. Path starting at zero will reach any k value eventually with probability one. With probability one every state will be reached infinitely often.

In the above discussion we have sneaked in the expression "odds" which is not quite the same as probability. The gambling community calls the ratio $P(A)/(1 - P(A))$ the odds in favor of A and $1 - P(A)/P(A)$ the odds against A. In the above example, the odds that the casino goes broke first are $P_{casino}/P_{player} = k_{player}/k_{casino}$, extremely small even for high rollers.

8.9 Summary

Time-discrete stochastic processes are a straightforward extension of multivariate random variables. Indeed, a discrete stochastic process is fully determined by the infinite set of all finite-point joint probabilities. Most applications consider a discrete state space as well. The concepts of moments, conditional probabilities, and statistical independence are easily generalized to stochastic processes. New concepts are stationary processes which are invariant under a time-translation, reversible processes which are stationary processes for which the future and past look alike, and ergodic processes for which the ensemble mean equals the time mean. Discrete-time processes on a discrete state space are called chains.

A special class of chains are Markov chains where the conditional probabilities depend only on the previous outcome. The future depends only on the present and is independent of the past. An important subset are homogeneous Markov chains which are characterized by a single time-independent transition probability. The evolution of one-point probabilities and higher-step transition probabilities then constitutes a linear dynamical system governed by an operator called the Markov operator. The analysis is particular simple for homogeneous chains on a discrete finite state space. The Markov operator is then a stochastic matrix. An eigen-analysis of this stochastic matrix demonstrates that there always exists at least one stationary chain and gives conditions under which there exists only one stationary chain and arbitrary chains converge to this one stationary chain.

The random walk in its various forms is the prime example a homogeneous Markov chain. Though random walks are simple prescription they can answer many intricate questions and form the basis for many applications in physics, gambling and finance.

Chapter 9
Continuous-Time Stochastic Processes

9.1 Introduction

In this chapter we consider families of random variables $(f_t)_{t \in T}$ where time T is continuous, either $T = \mathbb{R}$ or $T = \mathbb{R}_+$. We call these families continuous stochastic processes.

As in the case of discrete stochastic processes the Daniell–Kolmogorov extension theorem asserts that the probability distribution of a continuous stochastic process is specified by the set of all finite-point joint distributions. These finite-point joint distributions must satisfy compatibility relations and construction of a compatible set of finite-point joint distributions remains in general a formidable task.

There arise issues in the case of continuous processes that have to do with the fact that probability theory only provides means for the manipulation of countable sets of events, not for uncountable sets as occur for continuous processes. This issue can be resolved by considering separable modifications of processes. Separable processes are stochastic processes whose properties are almost surely determined by properties on a dense countable subset of T.

Again we concentrate on Markov processes which are again defined as processes where the conditional probabilities only depend on the most recent outcome. These Markov processes are characterized by transition rates, rather than transition probabilities. The rates must satisfy a compatibility relation called the Chapman–Kolmogorov condition. One must distinguish between processes with discontinuous sample paths and with continuous sample paths. The processes with discontinuous sample paths are called jump processes. Processes on a discrete state space are trivially jump processes.

For jump processes the Chapman–Kolmogorov condition can be turned into an integro-differential equation, called the master equation, which becomes the basis for their analysis. We discuss two jump processes, a random walk on a discrete lattice, dubbed random walk III, and the Poisson process used to describe the decay of atoms or the arrival of customers.

P. Müller, *Handbook of Dynamics and Probability*,
https://doi.org/10.1007/978-3-030-88486-4_9

For processes with continuous sample paths the Chapman–Kolmogorov condition can be turned into a partial differential equation, called the Fokker–Planck equation. It describes a combination of drift and diffusion processes. Pure diffusion processes are governed by a diffusion equation. The Wiener process is the most important example of a pure diffusion process and constitutes our random walk IV. Pure drift processes are governed by the Liouville equation and describe a degenerate kind of stochastic process where the evolution is deterministic. The Liouville equation is thus not very relevant in the context of stochastic processes but plays an important role in the analysis of dynamical systems with random initial conditions, discussed in Chap. 11. As an example of a combined diffusion and drift process we discuss the Ornstein–Uhlenbeck process.

White noise is an important continuous stochastic process. It is an iid process which may be considered the time derivative of the Wiener process. White noise has some peculiar properties, most importantly it has an infinite variance.

9.2 Separable Modifications

Given a continuous stochastic process $(f_t)_{t\in T}$ where $T = \mathbb{R}$, $(\mathbb{R}_+)$ or some interval of the real line we have first to face some issues arising from the fact that time is now continuous. The sample paths or realizations $t \mapsto f_t(\omega)$ are functions that are amenable to the concepts of calculus. Most of the concepts used in calculus such as boundedness, monotonicity, continuity and differentiability involve continuous (or non-countable) operations. Let us exemplify this fact with the concept of boundedness by asking the question what is the probability that the sample paths or realizations fall into a closed interval $B \in \mathcal{B}$. For fixed t all realizations in $f_t^{-1}(B) := \{\omega | f_t(\omega) \in B\}$ will do. Since f_t is a measurable function $f_t^{-1}(B)$ is an element of the σ-algebra $\mathcal{A}$ and one can hence assign a probability to $f_t^{-1}(B)$. The realizations that fall into B for all times within an open interval I of T are the ones in the intersection

$$\bigcap_{t\in I} f_t^{-1}(B) := \{\omega | f_t(\omega) \in B \ \ \forall t \in I\} \tag{9.1}$$

We cannot assign a probability to this set since probability theory only provides the means to deal with countable operations. Probability theory is silent about uncountable operations.

Separable processes. To circumvent this problem one considers processes such that the behavior of their sample paths is essentially determined by their behavior on a dense countable subset. To be more precise, let Q be a dense countable subset of T. Then a process is called separable if for all I and B the set

$$\bigcap_{t\in I\cap Q} f_t^{-1}(B) := \{\omega | f_t(\omega) \in B \ \ \forall t \in I \cap Q\} \tag{9.2}$$

differs from the set (9.1), if at all, on a set which is the subset of a set with probability zero. This definition has to be a little convoluted since one can assign a probability to the set (9.2) (it is a member of the σ-algebra $\mathcal{B}^I$) whereas one might not be able to do so for the set (9.1). If one completes the probability measure such that the subset of every set of measure zero has the measure zero (which one can always do) then the definition will simplify to that the two sets differ only on a set of probability zero. The name "separability" is taken from topology. There a set is called separable if it contains a countable dense subset.

The sample paths of a separable stochastic process are determined almost surely by their values on a countable dense subset. This property implies that important sets and functionals connected with a separable stochastic process are measurable.

Equivalent processes. One can even go further. To do so we need to introduce equivalent processes. Random variables f and g are called equivalent if they only differ on a set with zero probability or if $P(\{\omega | f(\omega) = g(\omega)\}) = 1$. Equivalent random variables have the same distribution and hence all probabilistic statements about the two variables are the same. Similarly, we call two stochastic processes f_t and g_t equivalent if $P(\{\omega | f_t(\omega) = g_t(\omega)\}) = 1$ for all $t \in T$. They are also called two different versions of the same process. Clearly, equivalent versions of a stochastic process have the same finite dimensional marginal distributions. Equivalent processes can, however, have very different sample paths. The standard example assumes $T = [0, 1]$ and the uniform distribution on $\Omega = [0, 1]$. Then the process $f_t(\omega) = 0$ for all $t \in T$ and $\omega \in \Omega$ and the process

$$g_t(\omega) = \begin{cases} 0 \text{ if } \omega \neq t \\ \\ 1 \text{ if } \omega = t \end{cases} \tag{9.3}$$

are equivalent. However, every sample path of f_t is continuous on T whereas none of the sample paths of g_t is continuous on T. See Fig. 9.1. The process $(f_t)_{t \in T}$ is separable whereas the process $(g_t)_{t \in T}$ is not.

One might assert an even stronger version of similarity between stochastic processes. Two processes are called indistinguishable if $P(\{\omega : | f_t(\omega) = g_t(\omega), \forall t \in T\}) = 1$. Now, the qualifier $\forall t \in T$ is in the argument of the probability. When this probability is written as $P(\bigcap_{t \in T} (f_t - g_t)^{-1}(\{0\}))$ one again encounters the need to calculate the probability of the intersection of an uncountable number of sets in Ω.

There is an important theorem (Doob 1953) that states: to any process $(f_t)_{t \in T}$ there exist an equivalent separable stochastic process $(\tilde{f}_t)_{t \in T}$, called a separable modification. This is a somewhat surprising result. It states that though we are considering continuous stochastic process we are not really leaving the realm of countable operations, when considering probability statements. Because of this theorem, we can restrict ourselves to such separable modifications in the following. Everything we will be doing is then well within the confines of probability theory. Let us demonstrate this for the concept of continuity.

Continuity of sample paths. According to one definition in calculus, a sample path $f(t)$ is continuous at t_0 if for every $\epsilon > 0$ one can find a $\delta > 0$ such that $| f(t) -$

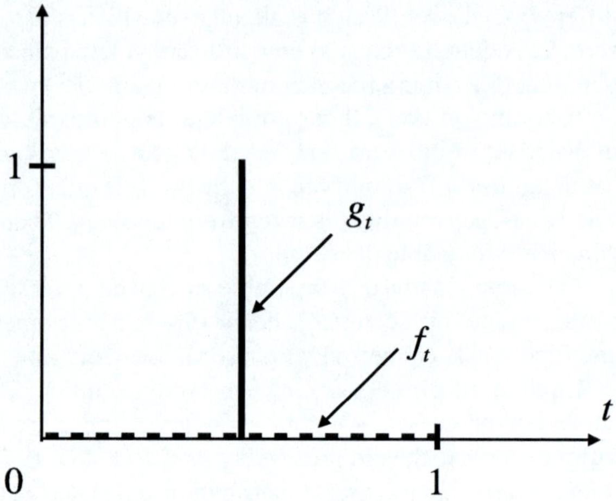

Fig. 9.1 Sample paths of the processes $(f_t)_{t \in T}$ and $(g_t)_{t \in T}$

$f t_0| < \epsilon$ for all $|t - t_0| < \delta$. If one considers all sample paths one has to consider the set

$$A_1 = \{\omega || f_t(\omega) - f_{t_0}(\omega)| \leq \epsilon, \forall |t - t_0| < \delta\} \tag{9.4}$$

which is exactly of the form (9.1) used in the definition of a separable process and can be replaced by

$$A_2 = \{\omega || f_{t_n}(\omega) - f_{t_0}(\omega)| \leq \epsilon, \forall |t_n - t_0| < \delta \ , \ n \in \mathbb{N}\} \tag{9.5}$$

which has a probability. If $P(A_2) = 1$ one says the process has almost surely continuous sample paths at t_0. Indeed, when one rewrites this condition using limits

$$P(| \lim_{t_n \to t_0} f_{t_n} - f_{t_0}| = 0) = 1 \tag{9.6}$$

we recover the almost sure convergence of sequences. One also says that the sample paths are continuous at t_0 for almost all $\omega \in \Omega$. A process with almost surely continuous sample paths is thus one where only continuous functions have non-zero probability.

There exist a famous criterion by Kolmogorov that a stochastic process has almost surely continuous sample paths if there exist positive real numbers a, b and c such that

$$\langle |f_t - f_s|^a \rangle \leq c|t - s|^{1+b} \tag{9.7}$$

This criterion only restricts the two-time marginal distribution.

In Sect. 7.9 we introduced other kinds of limits of random variables. In analogy one says a process f_t is mean square continuous if

$$\lim_{t \to t_0} \langle |f_t - f_{t_0}|^2 \rangle = 0 \tag{9.8}$$

or continuous in probability at t_0 if

$$\lim_{t \to t_0} P(|f_t - f_{t_0}| > \epsilon) = 0 \tag{9.9}$$

for all $\epsilon > 0$. Here we do not need to invoke separability of the process because the limit is outside the probability measure. The probability measure is applied to sets which are measurable.

9.3 Markov Processes

In the previous chapter on discrete stochastic processes we first introduced independent chains and then Markov chains, in order of increasing complexity. It is actually difficult to define independent continuous processes outright. They are most naturally introduced as derivatives of certain Markov processes, with which we will hence start.

Markov condition. Markov processes are also defined in complete analogy to the discrete case. A stochastic process is called a Markov process if the forward conditional probability densities satisfy the Markov condition

$$\rho_{t_{N+1}|t_N,\ldots,t_0}(x_{N+1}|x_N,\ldots,x_0) = \rho_{t_{N+1}|t_N}(x_{N+1}|x_N) \tag{9.10}$$

for every choice of N, $t_{N+1} > \cdots > t_0$ and $x_{N+1},\ldots,x_0$. The distribution at time t_{N+1} only depends on the outcome at the previous time t_N. The conditional densities $\rho_{t_{N+1}|t_N}$ are called the transition densities. Again, stationary Markov processes are those for which all joint densities are invariant under a time translation. Homogeneous Markov processes are those for which the transition densities $\rho_{t|t'}$ only depend on the time difference $t - t'$. Sample paths are continuous iff the transition densities satisfy the Lindeberg condition

$$\lim_{\Delta t \to 0} \frac{1}{\Delta t} \int_{|x-y|>\epsilon} dx\, \rho_{t+\Delta t|t}(x|y) = 0 \tag{9.11}$$

for any $\epsilon > 0$. The probability for the final position x to be finitely different from the initial position y goes to zero faster than Δt goes to zero.

Chapman–Kolmogorov condition. A set of transition densities again specifies a class of Markov processes. The task of specifying a Markov process is, however, not trivial since the transition densities must satisfy the compatibility condition

$$\rho_{t_2|t_0}(x_2|x_0) = \int_{\mathbb{R}} dx_1\, \rho_{t_2|t_1}(x_2|x_1)\rho_{t_1|t_0}(x_1|x_0) \tag{9.12}$$

which is called the Chapman–Kolmogorov condition and constitutes a functional relation between the transition densities. It must hold for all $t_2 > t_0$ and all $x_2, x_0 \in \mathbb{R}$ and for all t_1 with $t_2 > t_1 > t_0$. In the limit $t_2 = t_0$ the compatibility condition becomes

$$\rho_{t|t}(x|y) = \delta(x - y) \tag{9.13}$$

If one finds transition densities that satisfy the compatibility or Chapman–Kolmogorov condition one has a tool to construct Markov processes. There are, of course, a few well-known solutions to the Chapman–Kolmogorov equation. Two of them are the Wiener and the Cauchy process.

Wiener process. The Wiener process is defined by the conditional density

$$\rho_{t+\Delta t|t}(x|y) = (4\pi D\Delta t)^{-1/2} \exp\{-(x - y)^2/4D\Delta t\} \tag{9.14}$$

This conditional density satisfies the Chapman–Kolmogorov condition and approaches the $\delta(x - y)$ for $\Delta t = 0$, as required. (The formula is actually one of the definitions of the delta function). The Lindeberg condition (9.11) is satisfied, so the sample paths are continuous. The Wiener process will explicitly be treated in Sect. 9.6.

Cauchy process. The Cauchy process is defined by the conditional density

$$\rho_{t+\Delta t|t}(x|y) = \frac{\Delta t}{\pi[(x - y)^2 + \Delta t^2]} \tag{9.15}$$

This conditional density also satisfies the Chapman–Kolmogorov condition and approaches $\delta(x - y)$ for $\Delta t = 0$. The resulting process is called the Cauchy process. The Lindeberg condition is not satisfied. The sample paths are discontinuous.

In general, it is, however, not trivial to construct a consistent set of transition densities, even for homogeneous Markov processes where the transition densities are invariant under a time translation. Construction is, however, simplified by the fact that the Chapman–Kolmogorov relation can be turned into an integro-differential equation, the master equation, for processes with discontinuous sample paths and into a partial differential equation, the Fokker–Planck equation, for processes with continuous sample paths, as we demonstrate next. Our derivation and justification of both the master and Fokker–Planck equations is fairly heuristic. The proper mathematics, which is quite involved, can be found in Gardiner (1990).

9.4 Master Equation

The following derivation is analogous to the discrete case. At its base is a transition rate instead of a transition probability. Instead of a forward and backward recursion formulas we will obtain forward and backward integro-differential equations.

Transition rate. Assume that there exists a transition rate (transition probability per unit time)

$$W_t(x|z) := \lim_{\Delta t \to 0} \frac{1}{\Delta t} \rho_{t+\Delta t|t}(x|z) \tag{9.16}$$

This transition rate $W_t(x|z)$ is non-negative. It is not a probability with respect to the first argument x. In its definition, integration over x does not commute with the time limit. Thus, there is no restriction on W_t other than it has to be non-negative.

Using this transition rate and $\rho_{t|t}(x|z) = \delta(x - z)$ as an initial condition the conditional probability $\rho_{t+\Delta t|t}(x|z)$ can be expanded to first order in Δt as

$$\rho_{t+\Delta t|t}(x|z) = \left[1 - \int dy W_t(y|x) \Delta t\right] \delta(x - z) + W_t(x|z)\Delta t \tag{9.17}$$

assuring that the conditional probability is properly normalized, the integral over x is one. Interpreting the state space as particle positions, this formula shows that a particle at position z stays with a finite probability, given by the coefficient in front of $\delta(x - z)$, at its original position z and moves to positions x with probability, given by the last term.

Forward master equation. Setting $t_2 = t_1 + \Delta t$ in the Chapman–Kolmogorov condition one obtains

$$\rho_{t_1+\Delta t|t_0}(x|y) = \int_{\mathbb{R}} dz \rho_{t_1+\Delta t|t_1}(x|z) \rho_{t_1|t_0}(z|y) \tag{9.18}$$

and upon substitution of the above expansion

$$\begin{aligned} \rho_{t_1+\Delta t|t_0}(x|y) &= \int_{\mathbb{R}} dz \delta(x - z) \left[1 - \int_{\mathbb{R}} dy' W_t(y'|x) \Delta t\right] \rho_{t_1|t_0}(z|y) \\ &+ \int_{\mathbb{R}} dz W_t(x|z) \Delta t \rho_{t_1|t_0}(z|y) \end{aligned} \tag{9.19}$$

which turns in the limit $\Delta t \to 0$ into the integro-differential equation

$$\partial_t \rho_{t|t'}(x|y) = \int_{\mathbb{R}} dz [W_t(x|z)\rho_{t|t'}(z|y) - W_t(z|x)\rho_{t|t'}(x|y)] \tag{9.20}$$

called the forward master equation. If the conditional probability does not satisfy the Lindeberg condition (9.11) then one has to stop here. The sample paths are discontinuous and consist of jumps at irregular times.

If the state space is discrete the evolution equation for the transition probability takes the form

$$\frac{d}{dt} p_{t|t'}(i|j) = \sum_{k=-\infty}^{\infty} \left(W_t(i|k) p_{t|t'}(k|j) - W_t(k|i) p_{t|t'}(i|j)\right) \tag{9.21}$$

and the process naturally consists of jumps. There are, however, also jump processes on continuous state spaces, such as the Cauchy process (9.15).

The master equation must be solved subject to the initial condition

$$\rho_{t|t}(x|y) = \delta(x - y) \tag{9.22}$$

and suitable boundary conditions. Specifying suitable boundary conditions is often challenging. Here we assume that the state space is $\mathbb{R}$ and invoke the boundary condition $\rho_{t|t'}(x|y) \to 0$ as $|x| \to \infty$.

The evolution of the 1-point density

$$\rho_t(x) = \int_{\mathbb{R}} dy \rho_{t|t'}(x|y)\rho_t(y) \tag{9.23}$$

is governed by

$$\begin{aligned}\partial_t \rho_t(x) &= \int dy \partial_t \rho_{t|t'}(x|y)\rho_{t'}(y) \\ &= \int dz[W_t(x|z)\rho_t(z) - W_t(z|x)\rho_t(x)]\end{aligned} \tag{9.24}$$

and has the same form as the equation for the transition probabilities. Often one settles for solving just the equation for the 1-point densities. For the same amount of effort one can, however, solve for the transition probabilities and obtains a more complete description of the process.

All these formulas are easily generalized to the case that the state space is $\mathbb{R}^N$ with $N > 1$.

Backward master equation. One can also derive a backward master equation

$$\partial_{t'} \rho_{t|t'}(x|y) = \int dz W_{t'}(z|y)[\rho_{t|t'}(x|y) - \rho_{t|t'}(x|z)] \tag{9.25}$$

which must be solved with (9.22) as the final condition. The backward equation is equivalent to the forward equation. The two forms differ in the variables that are held constant. In the forward equation, we hold y and t' constant and look for solutions at $t \geq t'$ with (9.22) as initial condition. In the backward equation, we hold x and t constant and look for solutions for $t' \leq t$ that have (9.22) as their final condition.

If one writes the forward equation

$$\partial_t \rho_{t|t'}(x|y) = L[\partial_t \rho_{t|t'}(x|y)] \tag{9.26}$$

with the linear operator

$$L[\rho_{t|t'}(x|y)] = \int dz[W_t(x|z)\rho_{t|t'}(z|y) - W_t(z|x)\rho_{t|t'}(x|y)] \tag{9.27}$$

then the backward equation can be written as

$$\partial_{t'} \rho_{t|t'}(x|y) = -L^{\dagger}[\partial_{t'} \rho_{t|t'}(x|y)] \tag{9.28}$$

where

$$L^{\dagger}[\partial_{t'} \rho_{t|t'}(x|y)] = -\int dz W_{t'}(z|y)[\rho_{t|t'}(x|y) - \rho_{t|t'}(x|z)] \tag{9.29}$$

is the adjoint operator.

All these results are generalizations of the results for discrete-time Markov processes. These are characterized by a set of 1-step transition densities $\rho_{N+1|N}$. Higher-step transition densities $\rho_{N|0}$ can be constructed from forward and backward recursion formulas that are adjoint problems. The evolution is governed by a discrete-time Markov operator acting on the space of all integrable functions or summable sequences. Continuous-time Markov processes are characterized by the transition rates $W_t(x|z)$. Higher-step transition densities $\rho_{t|0}$ can be obtained from the forward and backward form of the master equation. The solution to the master equation implies a continuous-time Markov operator acting on the space of all integrable functions or summable sequences. In the discrete case one has a basic fixed step size Δt and iterative formulas. In the continuous case this step size becomes the infinitesimal time increment dt and one has differential formulas.

We consider two examples of jump processes on a discrete lattice, which is our random walk III and the Poisson process, both of which are homogeneous.

Random walk III. Our random walk III is a jump process where a particle jumps on a discrete one-dimensional lattice at a rate d to the left or right. The transition rate is given by

$$W_*(k|l) = \begin{cases} d \text{ if } k = l \pm 1 \\ 0 \text{ otherwise} \end{cases} \tag{9.30}$$

independent of t and leads to the master equation

$$\frac{d}{dt} p_{t|t'}(k|l) = d[p_{t|t'}(k-1|l) + p_{t|t'}(k+1|l) - 2p_{t|t'}(k|l)] \tag{9.31}$$

which must be solved subject to the initial condition $p_{t|t}(k|l) = \delta_{kl}$.

A solution of the master equation (9.31) can be obtained by the use of the characteristic function (Fourier transformation)

$$\hat{p}_{t|t'}(\kappa|l) := \sum_{k=-\infty}^{\infty} p_{t|t'}(k|l) e^{-i\kappa k} \tag{9.32}$$

The characteristic function is a periodic function of the wavenumber κ with period 2π. The evolution of the characteristic function is governed by

$$\frac{d}{dt}\hat{p}_{t|t'}(\kappa|l) = d(e^{i\kappa} + e^{-i\kappa} - 2)\hat{p}_{t|t'}(\kappa|l) \tag{9.33}$$

which is solved by

$$\hat{p}_{t|t'}(\kappa|l) = \exp\{(e^{i\kappa} + e^{-i\kappa} - 2)dt\}\hat{p}_{0|t'}(\kappa|l) \tag{9.34}$$

Consider now the case that the process starts at time $t' = 0$ at $l = 0$. Then $\hat{p}_{0|0}(\kappa|0) = 1$ and

$$\begin{aligned}\hat{p}_{t|0}(\kappa|0) &= e^{-2dt}e^{2dt\cos\kappa} \\ &= e^{-2dt}\sum_{k=-\infty}^{\infty} I_k(2dt)e^{-i\kappa k}\end{aligned} \tag{9.35}$$

where we have used the series expansion of $e^a \cos\kappa$ in terms if the modified Bessel function I_k of order k (see e.g., Abramowitz and Stegun 1964). Converting back we obtain

$$p_t(k) = p_{t|0}(k|0) = e^{-2dt}I_k(2dt) \tag{9.36}$$

The moments of this distribution can be determined by introducing the generating function

$$\begin{aligned}G(z) &= \sum_{k=-\infty}^{\infty} z^k p_t(k) = e^{-2dt}\sum_{k=-\infty}^{\infty} z^k I_k(2dt) \\ &= e^{-2dt}e^{dt(z+1/z)}\end{aligned} \tag{9.37}$$

again consulting Abramowitz and Stegun. From the generating function one infers the moments as follows

$$\frac{dG}{dz}|_{z=1} = \sum_{k=-\infty}^{\infty} kp_t(k) = \mu_t = e^{-2dt}e^{dt(z-1/z)}dt(1-1/z^2)|_{z=1} = 0 \tag{9.38}$$

$$\frac{d^2G}{dz^2}|_{z=1} = \sum_{k=-\infty}^{\infty} k(k-1)p_t(k) = \sigma_t^2 = 2dt \tag{9.39}$$

The mean is zero. The variance increases linear in time. The zero mean could of course also be inferred from the symmetry of the distribution since $I_{-k} = I_k$.

The master equation (9.31) can be obtained from the Markov chain with transition probability

$$p_T(k|l) = \begin{cases} a & \text{if } k = l \pm 1 \\ 1 - 2a & \text{if } k = l \\ 0 & \text{otherwise} \end{cases} \tag{9.40}$$

which implies the relation

$$p_{(n+1)\Delta t|n\Delta t}(k|l) - p_{n\Delta t|n\Delta t}(k|l) = a[p_{n\Delta t|n\Delta t}(k+1|l) + p_{n\Delta t|n\Delta t}(k-1|l) - 2p_{n\Delta t|n\Delta t}(k|l)] \tag{9.41}$$

between the transition probabilities. Dividing by Δt and taking the limits $n \to \infty$, $\Delta t \to 0$ and $a \to 0$ such that $t = n\Delta t$ and $d = a/\Delta t$ remain constant we obtain the master equation (9.31). The continuous-time random walk on a discrete lattice can be viewed as the continuous-time limit of the discrete-time random walk on a discrete lattice.

Poisson process. The Poisson process is a process on the discrete state space $\mathbb{N}$ governed by

$$W_T(n|m) = \begin{cases} d \text{ if } n = m + 1 \\ 0 \text{ otherwise} \end{cases} \tag{9.42}$$

where d is the probability per unit time that another event takes place (such as an atom decaying, a customer arriving, etc.). The master equation thus takes the form

$$\frac{d}{dt} p_{t|t'}(n|n') = d[p_{t|t'}(n-1|n') - p_{t|t'}(n|n')] \tag{9.43}$$

The solution for the case that there have not been any events at time $t = 0$ is the Poisson distribution

$$p_{t|0}(n|0) = \frac{(td)^n}{n!} e^{-td} \tag{9.44}$$

again best proven by use of the characteristic function. The Poisson process has mean $\mu_{f_t} = td$ and variance $\sigma^2_{f_t} = td$. The Poisson process also does not become stationary.

9.5 Fokker-Planck Equation

The master equation is a linear integro-differential equation. If the Lindeberg condition (9.11) is satisfied the sample paths are continuous and the master equation can be turned into a partial differential equation by the Kramers–Moayal expansion.

Kramers–Moyal expansion. The expansion is facilitated by expressing the transition rate in terms of the size $x - y$ of the jump

$$\tilde{W}_t(y, x-y) := W_t(x|y) \tag{9.45}$$

Then the master equation results in

$$\begin{aligned}\partial_t \rho_{t|t'}(x|y) &= \int_{\mathbb{R}} dz[\tilde{W}_t(z, x-z)\rho_{t|t'}(z|y) - \tilde{W}_t(x.z-x)\rho_{t|t'}(x|y)] \\ &= \int_{\mathbb{R}} ds[\tilde{W}_t(x-s, s)\rho_{t|t'}(x-s|y) - \tilde{W}_t(x|s)\rho_{t|t'}(x|y)] \end{aligned}\tag{9.46}$$

where the integration over z has been changed to an integration over the size of the jump $s = x - z$ in the first term and to an integration over the size of the jump $s = z - x$ in the second term. If the size s of the jumps is small and $\tilde{W}_t$ well-behaved one Taylor expands the integrand in the first term around x to obtain

$$\partial_t \rho_{t|t'}(x|y) = \sum n = 1^{\infty} \frac{(-1)^n}{m!} \frac{\partial^m}{\partial x^m} \left\{ \left[\int_{\mathbb{R}} dss^n \tilde{W}_t(x, s) \right] \rho_{t'|t}(x|y) \right\} \tag{9.47}$$

The zeroth order term in the expansion is canceled by the second term in (9.46). Introduce the jump moments

$$\begin{aligned} a_t^{(n)}(x) &= \int_{\mathbb{R}} dss^n \tilde{W}_t(x, s) \\ &= \int_{\mathbb{R}} dy'(y'-x)^n W_t(y'|x) \\ &= \lim_{\Delta t \to 0} \frac{1}{\Delta t} \int_{\mathbb{R}} dy'(y'-x)^n \rho_{t+\Delta t|t}(y'|x) \\ &= \lim_{\Delta t \to 0} \frac{1}{\Delta t} < (f_{t+\Delta t} - f_t)^n > |_{f_t = x} \end{aligned}\tag{9.48}$$

where f is the process that starts with sharp initial condition $f_t = x$. Then the expansion results in

$$\partial_t \rho_{t|t'}(x|y) = \sum_{n=1}^{\infty} \frac{(-1)^n}{m!} \frac{\partial^m}{\partial x^m} [a_t^{(n)}(x) \rho_{t'|t}(x|y)] \tag{9.49}$$

There is not much gained by this expansion. There are, however, various cases where it can be shown that the moments higher than $n = 2$ are zero or negligible. One such case is treated in Sect. 13.5 where the stochastic process results from a stochastic differential equation driven by Gaussian white noise.

The Fokker Planck equation. When moments $n > 2$ are neglected one obtains the Fokker–Planck equation. The Fokker–Planck equation is usually written in the form (Risken 1996)

$$\partial_t \rho_{t|t'}(x|y) = -\frac{\partial}{\partial x}[u_t(x)\rho_{t|t'}(x|y)] + \frac{\partial^2}{\partial x \partial x}[D_t(x)\rho_{t|t'}(x|y)] \tag{9.50}$$

and describes a combined drift and diffusion process with u_t being the drift velocity and D_t being the diffusion coefficient. Given the initial condition $\rho_{t|t}(x|y) = \delta(x - y)$ we find to first order in Δt

$$\rho_{t+\Delta t|t}(x|y) = (2\pi 2D_t(y)\Delta t)^{-1/2} \exp\{-(x - y - u_t(y)\Delta t)^2/4D_t(y)\Delta t\} \tag{9.51}$$

which is a Gaussian distribution with mean $y + u_t(y)\Delta t$ and variance $2D_t(y)\Delta t$. A particle released at y at time t moves with a systematic drift velocity u_t and fluctuates around the path with variance $2D_t(y)\Delta t$. The sample paths are almost surely continuous.

The Ornstein–Uhlenbeck process. The Ornstein–Uhlenbeck process is a solution of the Fokker–Planck equation

$$\partial_t \rho_{t|0}(x|x_0) = \partial_x[\gamma x \rho_{t|0}(x|x_0)] + D_0 \partial_x^2 \rho_{t|0}(x|x_0) \tag{9.52}$$

with a time-independent drift velocity $u_t(x) = -\gamma x$ and a constant diffusion coefficient D_0. For initial condition $\rho_{0|0}(x|x_0) = \delta(x - x_0)$ the solution is a Gaussian

$$\rho_{t|0}(x|x_0) = \frac{1}{\sqrt{2\pi\sigma_t^2}} \exp\left[\frac{(x - x_0 - \mu_t)^2}{2\sigma_t^2}\right] \tag{9.53}$$

with mean

$$\mu_t = x_0 e^{-\gamma t} \tag{9.54}$$

and variance

$$\sigma_t^2 = \frac{D_0}{\gamma}[1 - e^{-2\gamma t}] \tag{9.55}$$

The proof can be found in Gardiner (1990).

The covariance function is given by

$$\langle f_s f_t \rangle - \langle f_s \rangle \langle f_t \rangle = \frac{D}{\gamma}(e^{-\gamma|s-t|} - e^{-\gamma(s+t)}) \tag{9.56}$$

In the limit $t, s \to \infty$ a stationary process is obtained with mean x_0 and covariance function $C_{s,t} = D_) e^{-\gamma|t-s|}/\gamma$. The spectrum, the Fourier transform of the covariance function becomes $S(\omega) = 2D/(\gamma^2 + \omega^2)$. Most contributions to the variance come from low frequencies, the stationary Ornstein–Uhlenbeck process is thus also called red noise.

Liouville equation. If only the drift term is present the Fokker–Planck equation reduces to the Liouville equation

$$\partial_t \rho_{t|t'}(x|y) = -\frac{\partial}{\partial x}[u_t(x)\rho_{t|t'}(x|y)] \tag{9.57}$$

and the expansion (9.51) reduces in the limit $D_t(y) \to 0$ to

$$\rho_{t+\Delta t|t}(x|y) = \delta(x - y - u_t(y)\Delta t) \tag{9.58}$$

The Liouville equation thus describes a degenerated stochastic process where a particle moves deterministically (with probability one) from position y to position $x = y + u_t(y)\Delta t$ during the infinitesimal time interval Δt. For this reason the Liouville equation is not very relevant in the context of stochastic processes but plays a major role in the analysis of dynamical systems with random initial conditions, discussed in Sect. 11.

If only the diffusion term is present one obtains the diffusion equation, discussed next.

9.6 The Wiener Process

The diffusion equation

$$\partial_t \rho_{t|t'}(x|y) = \frac{\partial^2}{\partial x \partial x}\left[D_t(x)\rho_{t|t'}(x|y)\right] \tag{9.59}$$

with initial condition $\rho_{t|t}(x|y) = \delta(x - y)$ is solved to first order in Δt by

$$\rho_{t+\Delta t|t}(x|y) = (2\pi 2D_t(y)\Delta t)^{-1/2} \exp\{-(x-y)^2/4D_t(y)\Delta t\} \tag{9.60}$$

which is a Gaussian with mean y and variance $2D_t(y)\Delta t$. A particle released at time t at position y fluctuates around y with linearly increasing variance $2D_t(y)\Delta t$. The sample paths are almost surely continuous.

Wiener process. The Wiener process is the special case that the diffusion coefficient is independent of space and time, $D_t(x) = D_0 =$constant. With initial condition $\rho_{t|t}(x|y) = \delta(x - y)$ the diffusion equation is then solved by

$$\rho_{t_2|t_1}(x_2|x_1) = (4\pi D_0(t_2 - t_1))^{-1/2} \exp\left\{-(x_2 - x_1)^2/4D_0(t_2 - t_1)\right\} \tag{9.61}$$

where $D_0 > 0$ and $t_2 > t_1$ are assumed. It is the most basic example of a continuous Markov process on a continuous state space. It goes back to Norbert Wiener (1894–1964) and is called a Wiener process. The various densities of the Wiener process are generally denoted by w rather than by ρ, a convention that we follow.

The one-point density

$$w_t(x) = \int_{\mathbb{R}} dy w_{t|0}(x|y) w_0(y) \tag{9.62}$$

is given by the Gaussian

$$w_t(x) = (4\pi D_0 t)^{-1/2} \exp\left\{-x^2/4D_0 t\right\} \tag{9.63}$$

if the initial condition is $w_0(x) = \delta(x)$. It also satisfies the diffusion equation. The transition density $w_{t|0}$ acts as a kernel. The higher-point joint densities also turn out to be jointly Gaussian. The first two moments of w_t are

$$\langle w_t \rangle = \int_{\mathbb{R}} w_t(x) x = 0 \tag{9.64}$$

and

$$\begin{aligned} \langle w_t w_{t'} \rangle &= \begin{cases} \int_{\mathbb{R}} \times \mathbb{R} \, dx dy \rho_{t'|t}(y|x) \rho_t(x) xy \text{ if } t' > t \\ \int_{\mathbb{R}} \times \mathbb{R} \, dx dy \rho_{t|t'}(y|x) \rho_s(x) xy \text{ if } t' < t \end{cases} \\ &= 2D_0 \min(t', t) \end{aligned} \tag{9.65}$$

The variance is thus given by

$$\sigma_{w_t}^2 = \langle w_t w_t \rangle = 2D_0 t \tag{9.66}$$

The Wiener process does not become stationary.

A normalized version of the Wiener process is obtained by setting $D_0 = 1/2$.

The paths of the Wiener process are almost surely continuous but nowhere differentiable. The non-differentiability may be inferred by considering the process $h := (w_{t+\Delta} - w_t)/(\Delta t)$. The probability that the magnitude of h is larger than a prescribed value a is given for small Δt by

$$P(|h| > a) = 1 - 2\int_0^{a\Delta t} dx \frac{1}{\sqrt{4\pi D_0 \Delta t}} \exp\left\{\frac{x^2}{4D_0 \Delta t}\right\} \approx 1 - \frac{2a\Delta t}{\sqrt{4\pi D_0 \Delta t}} \tag{9.67}$$

which approaches one as Δt approaches zero. The probability that the derivative is larger than any given value a is hence one. This may be interpreted as the derivative being infinite at each time.

All these results can easily be generalized to Wiener processes in higher dimensional state spaces $\mathbb{R}^N$ with $N > 1$.

Random walk IV. The Wiener process is a continuous-time random walk on a continuous state space. It is the fourth of the possible random walks.

Fig. 9.2 Relations between the various random walks

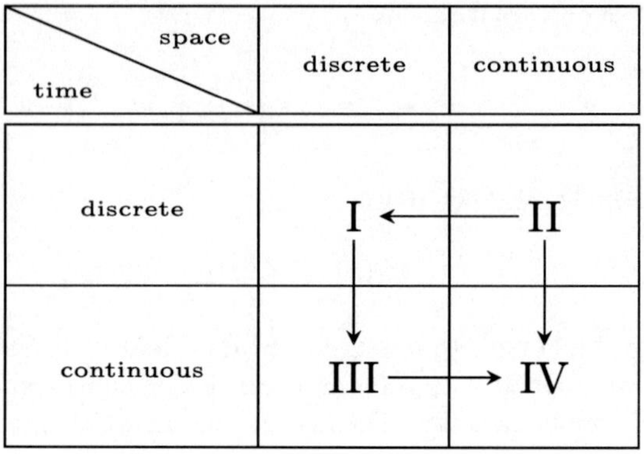

I: $p_0(k) = \delta_{k0}$, $p_N(k) = \binom{N}{\frac{N+k}{2}} \left(\frac{1}{2}\right)^N$, $\sigma_N^2 = N$

II: $\rho_0(x) = \delta(x)$, $\rho_N(x) = \frac{1}{\sqrt{2\pi N\sigma_1^2}} e^{-\frac{x^2}{2N\sigma_1^2}}$, $\sigma_N^2 = N\sigma_1^2$

III: $\rho_0(k) = \delta(k0)$, $\rho_t(k) = e^{-2dt}$, $\sigma_t^2 = 2dt$

IV: $\rho_0(x) = \delta(x)$, $\rho_N(x) = \frac{1}{\sqrt{2\pi 2Dt}} e^{-\frac{x^2}{4Dt}}$, $\sigma_t^2 = 2Dt$

II → I	$N \gg 1$ $\frac{x}{\sigma_1} = k \ll N$	
I → III	$N \to \infty$ $\Delta t \to 0$ $a \to 0$	$t = N\Delta t$ finite $d = a/\Delta t$ finite
II → IV	$N \to \infty$ $\Delta t \to 0$ $\sigma_1^2 \to 0$	$t = N\Delta t$ finite $2D = \sigma_1/\Delta t$ finite
III → IV	$\Delta x \to 0$ $k \to \infty$ $d \to \infty$	$x = k\Delta x$ finite $D = d\,(\Delta x)^2$ finite

- Random walk I: time discrete, space discrete
- Random walk II: time discrete, space continuous
- Random walk III: time continuous, space discrete
- Random walk IV: time continuous, space continuous.

These four random walks are all related by appropriate limits, as depicted in Fig. 9.2. We already showed how to transition from II to I and from I to III. The Wiener process can also be obtained as the continuous limit in time and space.

II → IV: The Wiener process is obtained from the random walk II

$$\rho_n(x) = \frac{1}{\sqrt{2\pi n\sigma_1^2}} \exp\left\{-\frac{x^2}{2n\sigma_1^2}\right\} \tag{9.68}$$

by taking the limits $n \to \infty$, $\Delta t \to 0$ and $\sigma_1^2 \to 0$ such that $t = n\Delta t$ and $2D_0 := \sigma_1^2/\Delta t$ remain constant.

III $\to$ IV: The Wiener process is most easily obtained from the random walk III by considering the characteristic function of the random walk III which is given by

$$\hat{p}_{t|0}(\kappa|0) = \exp\{(\cos\kappa - 1)2dt\} \tag{9.69}$$

In physical space we want to take the limit $k \to \infty$, $\Delta x \to 0$ with $x = k\Delta x$ remaining constant. In wavenumber space this limit corresponds to the limit $\kappa \to 0$, $\Delta x \to 0$ with $\tilde{\kappa} = \kappa/\Delta x$ remaining constant. Substituting $\kappa = \tilde{\kappa}\Delta x$ into the characteristic function, expanding the cosine, and taking the limits $\Delta x \to 0$ and $d \to \infty$ with $\tilde{\kappa}$ and $D_0 = d(\Delta x)^2$ remaining constant we obtain

$$\lim_{\Delta x \to 0, d \to \infty} \hat{p}_{t|0}(\kappa|0) = \exp\{-\tilde{\kappa}^2 D_0 t\} \tag{9.70}$$

which is characteristic function of the Wiener process $w_{t|0}(x|0)$.

Brownian bridge. The Wiener process w_t is the mother of many other processes. One example is the process

$$b_t := w_t - t w_1 \tag{9.71}$$

on the time interval [0, 1] which is called a Brownian bridge. It is fixed at both ends. $b_0 = w_0 = 0$ and $b_1 = 0$ and has mean $\mu_b(t) = 0$ and variance $\sigma_b^2 = 2Dt(1-t)$. Generalizations of this process are often used to interpolate 'stochastically' between two values.

The exponential process. A second important process derived from the Wiener process is e^{w_t}. Since the Wiener process w_t is a process with normal distribution $N(0, 2Dt)$ the transformation formula for densities gives as the distribution for e^{w_t} the log-normal distribution

$$\rho_{e^{w_t}}(x) = \frac{1}{x}\frac{1}{\sqrt{2\pi 2Dt}} \exp\left\{-\frac{[\log x]^2}{4Dt}\right\} \tag{9.72}$$

with mean

$$\mu_{e^{w_t}} = e^{Dt} \tag{9.73}$$

and variance

$$\sigma_{e^{w_t}}^2 = (e^{2Dt} - 1)e^{2Dt} \tag{9.74}$$

9.7 Gaussian White Noise

An important process in the theory of stochastic processes is a process $\dot{w}_t$ called white noise. It is not a real stochastic process as can be seen by its definition as the derivative of the Gaussian process

$$\dot{w}_t := \frac{d}{dt} w_t \tag{9.75}$$

This definition is problematic since we just demonstrated that the Wiener process is nowhere differentiable. If we, however, only require that time differentiation and expectation commute then the mean becomes

$$\langle \dot{w}_t \rangle = \left\langle \frac{d}{dt} w_t \right\rangle = \frac{d}{dt} \langle w_t \rangle = 0 \tag{9.76}$$

and the covariance function

$$\begin{aligned}
\langle \dot{w}_t \dot{w}_{t'} \rangle &= \frac{\partial^2}{\partial t \partial t'} \langle w_t w_{t'} \rangle \\
&= \frac{\partial^2}{\partial t \partial t'} (2Dmin\{t, t'\} \\
&= 2D \frac{\partial^2}{\partial t \partial t'} (t'\theta(t - t') + t\theta(t' - t)) \\
&= 2D \frac{\partial}{\partial t} (\theta(t - t') - t'\delta(t - t') + t\delta(t' - t) \\
&= 2D\delta(t - t')
\end{aligned} \tag{9.77}$$

where we have used $t\delta(t' - t) = t'\delta(t - t')$ in the last step. White noise may be viewed as a stationary 'process' with zero mean and infinite variance that is pair-wise uncorrelated in time. It is called white noise because its spectrum, the Fourier transform of the covariance function, is a constant, hence all frequencies contribute equally to the variance. Often one reserves the name white noise for the normalized case $D = 1/2$, for which the spectrum is one.

The integral form of (9.75) allows to explicitly construct realizations of w_t from realizations of $\dot{w}_t$ by the

$$w_t = w_0 + \int_0^t dt' \dot{w}_{t'} \tag{9.78}$$

Gaussian white noise is ergodic in the sense that

$$\lim_{T \to \infty} \frac{1}{T} \int_0^T dt \dot{w}_t = \lim_{T \to \infty} \frac{w_T - w_0}{T} = 0 = \langle \dot{w}_t \rangle \tag{9.79}$$

almost surely since $(w_T - w_0)/T$ has a $N(0, 2D/T)$ distribution and therefore almost surely converges to zero in the limit $T \to \infty$.

White noise is the continuous-time analog of the Bernoulli process. It will be the essential ingredient of stochastic differential equations considered in Sect. 13.5.

9.8 Summary

This chapter considered continuous stochastic processes which are families of random variables where time is continuous. According to the Daniell–Kolmogorov theorem such processes are still determined by the set of all finite dimensional marginal distributions. Problems arise when one considers realizations of such continuous stochastic processes. These realizations are functions on the real line and concepts of calculus such as boundedness, continuity and differentiability may be applied to them. Unfortunately, these concepts involve sets which are obtained by non-countable number of operations and their probability cannot be assessed. There is another important theorem due to Doob which states that to every continuous stochastic process there exist an equivalent separable modification. A separable process is one whose behavior is determined by the behavior on a dense countable subset of time. If one restricts oneself to such separable processes then no problems arise in applying probability concepts to these processes. The theorems by Daniell–Kolmogorov and Doob assure that all the processes we considered in this chapter can be put on mathematically solid grounds.

The basic aim of this chapter was to establish the continuous-time stochastic processes which are analogous to the discrete-time processes studied in the previous chapter. The most important class of continuous stochastic processes are again Markov processes where conditional probabilities depend only on the most recent outcome. They are characterized by transition rates (instead of transition probabilities). These transition rates have to satisfy the Chapman–Kolmogorov condition from which both the master equation and Fokker–Planck equation can be derived, under appropriate assumptions. These equations describe three distinct processes

- jump processes: they are governed by the master equation, their sample paths are discontinuous.
- drift processes: they are governed by the Liouville equation, they describe a deterministic evolution.
- diffusion processes: they are governed by the diffusion equation, their sample paths are continuous.

We discussed two jump processes, a continuous-time random walk on a discrete lattice, dubbed random walk III, and the Poisson process which is also a jump process on a discrete lattice. The most important diffusion process is the Wiener process which is a continuous-time random walk on a continuous state space, dubbed random walk IV. As its discrete counterpart the Wiener process does not become stationary. It's variance grows proportional to time. Its paths are almost surely continuous but also almost nowhere differentiable. An important process that combines diffusion with drift is the Ornstein–Uhlenbeck process.

White noise is an independent process and was introduced as the (generalized) time derivative of the Wiener process.

References

Doob J.I.: Stochastic Processes. Wiley, New York (1953), 654pp

Gardiner, C.W.: Handbook of Stochastic Methods in Physics, Chemistry, and the Natural Sciences. Springer, Berlin (1990)

Risken, H.: The Fokker-Planck Equation. Springer, Berlin (1996)

Chapter 10
Information Entropy

10.1 Introduction

The concepts of information content and entropy grew out of attempts to quantify randomness. Early ideas go back to Boltzmann, but the definitive step was taken by Claude Shannon (1948) in his treatise "A Mathematical Theory of Communication" which lay the groundwork for what is known today as information theory. More recent references are Mackey (2003) and Cover and Thomas (2005). Information theory studies sequences of letters or symbols from a finite alphabet. It presumes that a source produces these letters with a given probability distribution and then studies how information is degraded when storing, processing, or transmitting data. The initial probability distribution encodes the statistical properties of the information source. Important theorems are the source and channel coding theorems.

Here, we define and discuss only those aspects of information entropy that are relevant to dynamics. Formally, the information entropy of a discrete random variable is a concave, bounded, and non-negative functional of the probability mass function. It is zero if the probability mass function is concentrated on just one point and takes its maximal value for a uniform distribution. This information entropy is often 'interpreted' as the amount of randomness or the amount of missing information in a random variable. The larger the entropy, the more random is the random variable or the more information is missing. These and other 'interpretations' are not particularly revealing since no independent definitions of randomness or missing information exist. The discussion about whether or not information entropy is the right measure of randomness is futile, similar to the discussion of whether or not temperature is the right measure of the hotness (or coolness) of a body. One may define randomness by other measures. The variance also characterizes the broadness of a distribution. Contrary to the variance, the entropy functional has, however, a number of desirable properties such that it is additive for independent random variables and that it is independent of the way one determines it, all at once or sequentially. Indeed, if one requires a set of sensible axioms, the Khinchin axioms, then the entropy turns out

P. Müller, *Handbook of Dynamics and Probability*,
https://doi.org/10.1007/978-3-030-88486-4_10

to be the only functional satisfying them. Information entropy has turned out to be a very useful quantity, otherwise, it would have been eliminated by the Darwinian selection process in science.

The concept of entropy can also be adapted to comparing two or more random variables by introducing joint, conditional, and other entropies. They satisfy important inequalities, which all derive from Gibbs' inequality which is a special case of Jensen's inequality for convex functions. When infinite sequences of random variables are studied, the concept of information entropy allows defining typical sequences of letters. These typical sequences have some surprising properties, as stated in the asymptotic equipartition theorem, with far-reaching consequences for coding.

When the entropy functional is generalized to continuous random variables, it becomes a concave functional of the probability density function which still not only satisfies the Gibbs inequality and some other desirable properties but also has some very undesirable properties such as not necessarily being bounded, neither from below nor from above. Ensuing problems can be circumvented by working with the Kullback–Leibler divergence (also called the relative entropy) which measures how similar two random variables are with respect to their randomness. If the probability space has a finite Lebesgue measure, then the entropy is bounded from above by the entropy of the uniform distribution and one can get away with still working with the entropy.

Consistent with our intuitive notion of randomness, we demonstrate that coarse-graining and smoothing a probability distribution will never decrease the entropy and neither will the diffusion of a probability density. A mere drift of the probability density, however, does not change the entropy.

Finally, we introduce the principle of maximum entropy. At its heart is the probability distribution that maximizes the entropy functional subject to some constraints. This maximizing probability distribution can be obtained by a straightforward application of the calculus of variation. The constraints are incorporated by Lagrange multipliers. When the maximizing probability distribution is substituted into the entropy functional, one obtains the entropy function, which is a function of the constraints. The various quantities obtained by these manipulations are related by Legendre transformation and are the blueprint for equilibrium thermodynamics. The principle of maximum entropy is the most important principle. It allows one to infer a probability distribution from constraints, which will be measurements or data in applications. If one equates entropy with randomness, then the inferred probability distribution is the most random one consistent with the constraints. This principle provides a rationalization of both equilibrium and non-equilibrium thermodynamics and is used in statistics and other forms of scientific reasoning.

10.2 Information Content

Let us first look at how information content and entropy are motivated and defined in information theory. Information theory was developed to quantify the information content of sequences of symbols transmitted from a source to a receiver. Formally, it considers a finite 'alphabet' $X = \{x_1, \ldots, x_K\}$ of K 'letters' x_k each occurring with a probability $P(\{x_k\}) = p(k) > 0$. The English alphabet consisting of 26 letters $a, b, c, \ldots, x, y, z$ plus a blank (to delineate words) constitutes an example. The probabilities can be estimated from the relative frequency with which these letters occur in a *reference library* (i.e., a collection of transmitted texts). We stress that the concept of information used in information theory is distinct from the colloquial use of the word 'information'. This colloquial use is concerned with sentences, sequences of specific letters. These sentences can be meaningful or not meaningful, true or false, comprehensible or incomprehensible, significant or insignificant, and more or less informative. The sentence "All cars in my street are red" is more informative than the sentence "One car in my street is red". It is, however, not this kind of information that information theory is concerned with. Rather, sentences are regarded as realizations of sequences of random variables, and the information content of these realizations measures how probable or improbable these realizations are, regardless of whether they are meaningful to a human being. Information theory is concerned with the design of communication systems. It asks how much information one gains if one receives letters from a source, irrespective of what these letters might signify to a user of such communication devices. It then studies how this information changes when transmitting, storing, and processing such information.

The information gain is the larger the less probable the letter is. Observe something unexpected and you gain information. Information theory thus defines the information content $h(k)$ of the letter x_k or self-information by

$$h(k) := k_B \log \frac{1}{p(k)} = -k_B \log p(k) \tag{10.1}$$

If $p(k) = 1$, then one will receive the letter x_k with certainty. There is no surprise. The outcome is predictable. The information content $h(k)$ is zero. If $p(k) = 1/K$ for $k = 1, \ldots, K$, then $h(k) = k_B \log K$ for each k.

The definition contains a constant k_B. It determines the units in which the information content is measured. A particular choice is that it is measured in bits. If a letter with probability $p(k) = 1/2$ is assigned the information content of one bit, then $1 = -k_B \log\{1/2\}$ and $k_B = 1/\log\{2\}$ and $h(k) = -\log_2\{p(k)\}$. If $k_B = 1$ is chosen, the unit is called 'nat'. The subscript 'B' stands for Boltzmann because in statistical dynamics k_B will be the Boltzmann constant. In this chapter, we set $k_B = 1$ for economy of notation.

Of course, any monotonically increasing function of $1/p(k)$ satisfies the requirement that the information content of a letter be the larger the less probable the letter is. The above definition, however, has two very desirable properties. First, it assigns the information content zero to a letter x_k that occurs with probability $p(k) = 1$, since we know for sure that we will receive it. No information is gained. Second, the information content is additive in the following sense. Consider the information content of the word $x_{k,l} := (x_k, x_l)$, which consists of the letter x_k followed by the letter x_l. Its information content is given by

$$h(k,l) = -\log p(k,l) \tag{10.2}$$

If the letters x_k and x_l are independent, i.e., if $p(k,l) = p(k)p(l)$, then

$$h(k,l) = -\log\{p(k)p(l)\} = h(k) + h(l) \tag{10.3}$$

Conversely, if one requires information content to be a continuous additive functional of the probabilities, then this functional relation implies the logarithm.

10.3 Entropy of a Discrete Random Variable

An alphabet $X = (x_1, \ldots, x_K)$ with probabilities $P(\{x_k\}) = p(k) > 0$ is just a finite discrete random variable f that takes the values x_k with probability $p(k)$. Probability theory defines the information entropy of a discrete random variable f with probability mass function p as the averaged information of all outcomes

$$H[p] := -\sum_{k=1}^{K} p(k) \log p(k) \tag{10.4}$$

In the case of a finite alphabet, information entropy is thus a measure of the averaged amount of information one obtains upon receiving a single letter from the source or a measure of the averaged a priori uncertainty as to which letter will emanate from a source. The entropy $H[p]$ depends on the probability distribution p and on nothing else. This dependence might also be explicated by $H[p(1), \ldots, p(K)]$ or $H[P_f]$ or $H[f]$. The inclusion of outcomes with zero probability requires the limit $\lim_{\epsilon\to 0} \epsilon \log 1/\epsilon = 0$.

If the discrete random variable has only two outcomes x_1 and x_2, then

$$H[p] = -[p(1)\log p(1) + (1 - p(1))\log(1 - p(1))] \tag{10.5}$$

This entropy is shown as a function of $p(1)$ in Fig. 10.1. It is a concave function of $p(1)$. Basically, everything discussed in this chapter follows from the fact that the entropy is a concave function of the probabilities.

Entropy is often 'interpreted' as the amount of randomness in a random variable. This interpretation is not a logical proposition since there is no independent definition

Fig. 10.1 The entropy $H[p]$ as a function of p

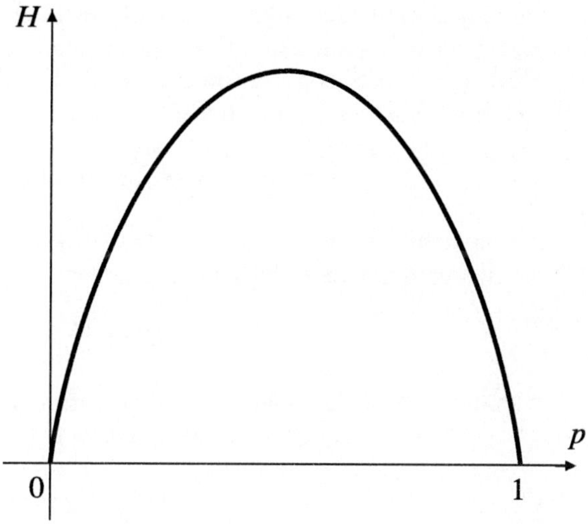

of randomness. One just calls entropy the amount of randomness. The same is true for other 'interpretations' of entropy such as the following:

- Entropy measures the flatness of the probability distribution: The larger the entropy, the flatter the distribution.
- Entropy measures the uncertainty of the outcome of the random variable. The larger the entropy, the more uncertain is the outcome.
- Entropy measures the amount of missing information. The larger the entropy, the more information is missing.
- Entropy measures the surprise of the observer when a specific outcome is realized. The larger the entropy, the more surprised is the observer.

All these phrases just give other names to entropy since independent definitions of flatness, uncertainty, missing information, and surprise are lacking.

In thermodynamics, entropy is often interpreted as a measure of disorder. The concept of order or disorder is hard to define, especially for probability distributions. What is an orderly distribution? Where order can meaningfully be defined, it is not necessarily related to entropy.

The value of the entropy does correspond to the smallest number of binary questions that need to be asked on average to find out which value a random variable has taken. This is a theorem. For a uniformly distributed random variable with $K = 8$, one needs $H = -\log_2(1/8) = 3$ questions to find the value. One needs to ask three times whether the value is in the first half or second half of the (remaining) possible values.

Let us recapitulate and add some more properties of the information entropy functional:

(1) Entropy is a functional of the probability distribution p only. It does not depend on any other properties of the random variable f.
(2) Entropy is a continuous functional of p.
(3) Entropy is a symmetric functional,

$$H[p(1), \ldots, p(K)] = H[p(i_1), \ldots, p(i_K)] \tag{10.6}$$

for any permutation $\{i_1, \ldots, i_K\}$ of $\{1, \ldots, K\}$.
(4) The entropy is bounded by zero from below

$$H[p] \geq 0 \tag{10.7}$$

with the equality sign holding if and only if $p(k) = 1$ for one k.
(5) The entropy is bounded by $\log K$ from above

$$H[p] \leq \log K \tag{10.8}$$

with the equality sign holding if and only if p is the uniform distribution $p(k) = 1/N$ for all k.
(6) The entropy does not change when an additional outcome with zero probability is added:

$$H[p(1), ..., p(K)] = H[p(1), ..., p(K), p(K+1) = 0] \tag{10.9}$$

requiring the limit $\lim_{\epsilon \to 0} \epsilon \log 1/\epsilon = 0$.
(7) Let $p_{1,2}$ be the joint probability distribution of two random variables. If these two random variables are independent $(p_{1,2}(k, l) = p_1(k)p_2(l)$, then the entropies are additive:

$$H[p_{1,2}] = H[p_1] + H[p_2] \tag{10.10}$$

(8) If the two random variables are not independent, $p_{1,2}(k, l) = p_{1|2}(k|l)p_2(l)$, then

$$H[p_{1,2}] = H[p_2] + \sum_{l=1}^{K} p_2(l) H_l[p_{1|2}] \tag{10.11}$$

where

$$H_l[p_{1|2}] := -\sum_{k=1}^{K} p_{1|2}(k|l) \log p_{1|2}(k|l) \tag{10.12}$$

is the entropy of the conditional probability $p_{1|2}$ which is a probability in its first argument k for each condition l. This result states that entropy is independent of the way it is calculated, either from the compound system or successively from the subsystems.

(9) The entropy is a concave functional of p. If we have two distributions p_1 and p_2 and define $p_3 := \lambda p_1 + (1-\lambda)p_2$ for $0 \leq \lambda \leq 1$, then

$$H[p_3] \geq \lambda H[p_1] + (1-\lambda)H[p_2] \tag{10.13}$$

This concavity has important consequences as we will see.

We have defined information entropy by (10.4) and it has the above properties. Conversely, if one requires the properties (1), (2), (5), (6), and (8), called the Khinchin axioms, then the entropy functional (10.4) is the only functional that satisfies these axioms.

The entropy functional has been used as a characterization of the amount of randomness in a random variable. Of course, entropy is just one possible measure that one might use for such a characterization. Entropy distinguishes itself by the many useful and intuitively appealing properties listed above and by its importance in statistical mechanics, but there are other measures that one might choose as well. Indeed, if one makes a seemingly innocuous change in the Khinchin axioms and replaces the Khinchin axiom (8) by (7), i.e., if one only requires that the entropy be additive for independent random variables, then all the information entropies

$$H_\lambda[p] = \frac{1}{1-\lambda}\log Z_\lambda \tag{10.14}$$

with $\lambda \geq 0$ and 'partition function'

$$Z_\lambda := \sum_{k=1}^{K}(p(k))^\lambda \tag{10.15}$$

become consistent with these modified Khinchin axioms. They are called Rényi entropies. One finds $\lim_{\lambda\to 1} H_\lambda[p] = H[p]$ and $H_0 = \log K$. As λ approaches infinity, the Rényi entropy becomes more and more determined by the highest probabilities, and one might have perhaps reason to emphasize these higher probabilities. In any case, the major argument for using the entropy H is that its use has produced many useful results.

If the discrete random variable has infinite but countable support, one still uses the definition (10.4) for the entropy which might then not be bounded anymore from above.

10.4 Entropy of Bi- and Multivariate Random Variables

The previous section already considered more than one random variable such as in the statement that entropy is additive for independent variables. In this section, we extend more systematically the concept of entropy to more than one variable. The statistical dependence of random variables manifests itself by inequalities that can all be derived from Gibbs' inequality, which we use as starting point.

Gibbs' inequality. Two probability distributions satisfy the most important inequality, namely Gibbs' inequality. Gibbs' inequality is a special case of Jensen's inequality (see Appendix E.14) which states that any convex function ϕ satisfies $\langle\phi(f)\rangle \geq \phi(\langle f\rangle)$. Specifying $\phi(y) = -\log y$ and $f = p_2/p_1$ where p_1 and p_2 are two probability mass functions, one deduces

$$-\sum_{k=1}^{K} p_1(k) \log \frac{p_2(k)}{p_1(k)} \geq -\log\left\{\sum_{k=1}^{K} p_1(k) \frac{p_2(k)}{p_1(k)}\right\} = 0 \tag{10.16}$$

or

$$-\sum_{k=1}^{K} p_1(k) \log p_2(k) \geq -\sum_{k=1}^{K} p_1(k) \log p_1(k) \tag{10.17}$$

The equality signs hold if and only if $p_1 = p_2$.

The boundedness of the entropy results from a simple application of Gibbs' inequality. Simply set $p_2(k) = 1/K$ for all $k = 1, \ldots, K$ in Gibbs' inequality and obtain $H[p] \leq \log K$. This is but one example of the important point:

Gibbs' inequality is the most important inequality. All inequalities involving entropies derive from it.

We next define point, conditional, mutual, and relative entropies in a way that is consistent with our intuitive notion of these concepts.

Joint entropy. Consider a bivariate random variable (f, g) with the joint probability $p_{f,g}$. The two random variables do not necessarily have to have the same support. Its entropy is given by

$$H[f, g] := -\sum_{k=1}^{K}\sum_{l=1}^{L} p_{f,g}(k, l) \log p_{f,g}(k, l) \tag{10.18}$$

This joint entropy is sub-additive:

$$H[f, g] \leq H[f] + H[g] \tag{10.19}$$

As already noted, the equality holds if and only if the two random variables are independent, i.e., if and only if $p_{f,g} = p_f p_g$. The generalization to N random variables is

$$H[f_1, \ldots, f_N] \leq H[f_1] + \ldots + H[f_n] \tag{10.20}$$

Conditional entropy. The conditional entropy $H[f|g]$ is defined by

$$H[f|g] := \sum_{l=1}^{K} p_g(l) H_l(f|g) = H[f, g] - H[g] \tag{10.21}$$

where $H_l(f|g)$ is given by (10.12). The conditional entropy is to be interpreted as the amount of randomness that remains about f when g is known. The word 'missing information' may be substituted for the 'amount of randomness' in this sentence. The conditional entropy is not symmetric in its arguments, i.e., $H[f|g] \neq H[g|f]$ in general. Also, we find $H[f|f] = 0$ and $0 \leq H[f|g] \leq H[f]$. If the two random variables are independent, then the conditional entropy is $H[f|g] = H[f]$. The fact that $H[f|g] \leq H[f]$ tells us that additional information never increases the entropy. The generalization of this statement to N random variables is

$$H[f_1, \ldots, f_L | f_{L+1}, \ldots, f_K] := H[f_1, \ldots, f_K] - H[f_{L+1}, \ldots, f_K] \leq H[f_1, \ldots, f_L] \tag{10.22}$$

and

$$H[f_1, \ldots, f_L | f_{L+1}, \ldots, f_K] \leq H[f_1, \ldots, f_L | f_{L+1}, \ldots, f_{K-1}] \tag{10.23}$$

When the definition (10.21) is written in the form

$$\begin{aligned} H[f, g] &= H[f|g] + H[g] \\ &= H[g|f] + H[f] \end{aligned} \tag{10.24}$$

it is called the chain rule for the entropy. The entropy is independent of the way it is gathered, whether from the compound system or successively from its subsystems. The generalization of the chain rule to N random variables is

$$H[f_1, ..., f_N] = \sum_{i=1}^{N} H[f_i | f_{i-1}, ..., f_1] \tag{10.25}$$

Mutual information. Mutual information is defined by

$$I[f, g] := H[f] - H[f|g] \tag{10.26}$$

and can be interpreted as the reduction of the amount of randomness in f from knowing g. One finds $I[f, g] \geq 0$, $I[f, f] = H[f]$, $I[f, g] = 0$ if f and g are independent, and $I[f, g] = H[f] + H[g] - H[f, g]$. The mutual information is the difference between the single entropies and the joint entropy. Indeed, it can be shown that the mutual information is the expectation of the logarithm of the correlation function. The mutual information is symmetric in its arguments f and g and can hence also be interpreted as the reduction of the amount of randomness in g resulting from knowing f. The generalization of (10.26) to N random variables is

$$I[f_1, \ldots, f_N] = H[f_1] + \ldots + H[f_N] - H[f_1, \ldots, f_n] \tag{10.27}$$

Kullback–Leibler divergence. The Kullback–Leibler divergence is defined for any two probability distributions p_f and p_g on the same support by

$$D[p_f||p_g] := k \sum_{k=1}^{K} p_f(k) \log \left\{ \frac{p_f(k)}{p_g(k)} \right\} \tag{10.28}$$

It measures the similarity between p_f and p_g with respect to their entropies. It is also called the Kullback–Leibler distance or Kullback–Leibler entropy or Kullback–Leibler information or relative entropy. It is not really a distance since it is not symmetric, $D[p_f||p_g] \neq D[p_g||_f]$ in general. From Gibbs' inequality, it follows that $D[p_f||p_g] \geq 0$ and $D[p_f||p_g] = 0$ if and only if $p_f = p_g$. Since

$$I[p_f; p_g] = D[p_{f,g}||p_f p_g] \tag{10.29}$$

the mutual information is a special case of the relative entropy.

Let us summarize the inequalities for random variables with the same support:

$$\begin{aligned} 0 &\leq H[f] \leq \log K \\ 0 &\leq H[f|g] \leq H[f] \\ 0 &\leq D(f||g) \\ 0 &\leq I[f, g] \end{aligned} \tag{10.30}$$

10.5 Asymptotic Equipartition Theorem

As before, we may progress from multivariate random variables to infinite sequences $(f_n)_{n \in \mathbb{N}}$ of random variables. Then the asymptotic behavior becomes important. This asymptotic behavior includes the existence and uniqueness of stationary distributions and the conditions under which arbitrary initial distributions approach these stationary distributions. Section 8.7 states conditions for the approach toward a unique stationary distribution for Markov chains. Here, we consider the evolution of the entropy for infinite sequences of random variables that represent a finite alphabet. The main result is the asymptotic equipartition theorem. The evolution of the entropy and the Kullback–Leibler divergence is also of great importance when dynamics and probability are combined and is discussed in Part III.

Entropy rate. Let $(f_n)_{n \in \mathbb{N}}$ be a stationary stochastic process on a finite discrete support. Then

$$\begin{aligned} H[f_n|f_{n-1},\ldots,f_1] &\leq H[f_n|f_{n-1},\ldots,f_2] \\ &= H[f_{n-1}|f_{n-2},\ldots,f_1] \end{aligned} \tag{10.31}$$

The first inequality expresses the fact that less conditioning increases the entropy. The second equality follows from the stationarity of the process. Thus, $H[f_n|f_{n-1},\ldots,f_1]$ is a non-increasing positive function bounded from below which therefore has a limit

$$H_\infty[f] := \lim_{n\to\infty} H[f_n|f_{n-1},\ldots,f_1] \tag{10.32}$$

Furthermore, the chain rule (10.25) implies

$$\begin{aligned} \lim_{n\to\infty} \frac{1}{n} H[f_n, f_{n-1},\ldots,f_1] &= \lim_{n\to\infty} \frac{1}{n} \sum_{m=1}^{n} H[f_m|f_{m-1},\ldots,f_1] \\ &= \lim_{n\to\infty} H[f_n|f_{n-1},\ldots,f_1] = H_\infty[f] \end{aligned} \tag{10.33}$$

where use has been made of

$$\lim_{n\to\infty} \frac{1}{n} \sum_{m=1}^{n} a_m = a \tag{10.34}$$

if $\lim_{m\to\infty} a_m = a$. The joint entropy thus grows (asymptotically) linearly at a rate $H_\infty[f]$ which is called the entropy rate.

If $(f_n)_{n\in\mathbb{N}}$ is a stationary homogeneous Markov process on a discrete state space with transition probability $\pi(k|l)$ and stationary one-point probability p_*, then

$$H_\infty[f] = H[f_2|f_1] = -\sum_{k,l=1}^{K} p_*(k)\pi(k|l)\log\pi(k|l) \tag{10.35}$$

If $(f_n)_{n\in\mathbb{N}}$ is an iid process, then

$$H_\infty[f] = H[f_1] \tag{10.36}$$

Asymptotic equipartition theorem. The major result is obtained when $(f_n)_{n\in\mathbb{N}}$ is an iid process. Then

$$h_n := -\log(p\circ f_n) = -\log p_n \tag{10.37}$$

is also an iid process with expectation

$$\langle h_n\rangle = -\sum_{k=1}^{K} p_n(k)\log p_n(k) = H[f_n] =: H_\infty[f] \tag{10.38}$$

The entropy rate $H_\infty[f]$ can thus be written as an expectation of the random variable h_n. This is a very powerful representation since one now has all the properties of expectations at one's disposal.

Consider the arithmetic mean

$$\begin{aligned}\bar{h}_N &:= \frac{1}{N}\sum_{n=1}^{N} h_n = -\frac{1}{N}\sum_{n=1}^{N}\log p_n \\ &= -\frac{1}{N}\log\left\{\prod_{n=1}^{N} p_n\right\} \\ &= -\frac{1}{N}\log p_{1,\ldots,N} \end{aligned} \tag{10.39}$$

where the last line is simply the definition of the joint probability $p_{1,\ldots,N}$ for independent random variables. The weak law of large numbers tells us that this arithmetic mean converges to $\langle h_n\rangle = H_\infty[f]$ in probability as $N \to \infty$. The weak law of large numbers applied to $(h_n)_{n\in\mathbb{N}}$ is called the asymptotic equipartition theorem. The weak law states explicitly

$$\lim_{N\to\infty} P\left(|-\frac{1}{N}\log p_{1,\ldots,N} - H_\infty[f]| < \epsilon\right) = 1 \tag{10.40}$$

for all $\epsilon > 0$.

What is the use of this theorem or what does it tell us? To answer these questions, consider realizations $(k_1, \ldots, k_N)$. Such a realization is an element of the set A_ϵ^N of typical sequences if

$$|-\frac{1}{N}\log p_{1,\ldots,N}((k_1, \ldots, k_N) - H_\infty[f]| < \epsilon \tag{10.41}$$

The asymptotic equipartition theorem implies that the typical set has probability one in the limit $N \to \infty$. How do typical sequences look like? Consider the Bernoulli chain with the binomial distribution $B(0, 1; p)$. The typical sequences are those for which the relative frequency of zeros is p. In the same limit, we can make the additional statements:

(1) The probability of any sequence $x \in A_\epsilon^N$ is tightly bounded from above and below by the entropy as follows:

$$e^{-N(H_\infty[f]+\epsilon)} < p_{1,\ldots,N}(k_1, \ldots, k_N) < e^{-N(H_\infty[f]-\epsilon)} \tag{10.42}$$

The probabilities of all typical sequences are hence the same. Hence, the name equipartition theorem.

(2) The typical set has approximately $e^{NH_\infty[f]}$ elements since each element has the same probability and the probability of all elements is one.

(3) Most sequences are not typical since

$$\frac{|A_\epsilon^N|}{|X|^N} = \frac{e^{NH_\infty[f]}}{e^{N\log K}} = e^{-N(\log K - H_\infty[f])} \tag{10.43}$$

which goes to zero as $N \to \infty$ unless the f_n have a uniform distribution for which $H_\infty[f] = \log K$.

We thus arrive at the conclusion that in the limit $N \to \infty$, the relative number of typical sequences goes to zero while the probability of all typical sequences remains at one. Thus, $NH_\infty[f]$ bits suffice for specifying realizations of N iid random variables. This result has, of course, implications for coding. The condition that f_n is an iid sequence can be relaxed as long as the weak law of large numbers remains applicable.

10.6 Entropy of Continuous Random Variables

Let us now consider continuous random variables whose probability distribution is given by a density ρ_f. The entropy may then be defined by

$$H[f] := -\int_X dx \rho_f(x) \log \rho_f(x) \tag{10.44}$$

which is also written as $H[\rho_f]$. The entropy of the Gaussian distribution $\rho_G = N(\mu, \sigma^2)$ becomes according to this definition

$$H[\rho_G] = \frac{1}{2}(1 + \log 2\pi\sigma^2) \tag{10.45}$$

independent of the mean μ.

The entropy for continuous random variables shares with the discrete version several important properties:

(1) The 'continuous' entropy is a concave functional

$$H[\lambda\rho_1 + (1-\lambda)\rho_2] \geq \lambda H[\rho_1] + (1-\lambda)h[\rho_2] \tag{10.46}$$

(2) The 'continuous' entropy is sub-additive

$$H[f, g] \leq H[f] + H[g] \tag{10.47}$$

with the equality holding if and only if f and g are statistically independent.

(3) The 'continuous' entropy is invariant under a translation $x' = x + c$. To see this, consider the random variable $g = f + c$ which has the density $\rho_g(x) = \rho_f(x - c)$. Then

$$\int_{\mathbb{R}} dx \rho_f(x) \log \rho_f(x) = \int_{\mathbb{R}} dy \rho_g(y) \log \rho_g(y) \tag{10.48}$$

This translational invariance exhibits itself in that the entropy of the Gaussian distribution $N(\mu, \sigma)$ is independent of the mean μ.

(4) The 'continuous' entropy satisfies Gibbs' inequality

$$-\int_X dx \rho_1(x) \log \frac{\rho_2(x)}{\rho_1(x)} \geq -\log \int_X dx \rho_1(x) \frac{\rho_2(x)}{\rho_1(x)} = 0 \tag{10.49}$$

or

$$-\int_X dx \rho_1(x) \log \rho_2(x) \geq -\int_X dx \rho_1(x) \log \rho_1(x) \tag{10.50}$$

with the equality signs holding if and only if $\rho_1 = \rho_2$. If the state space (or the support of the density) has a finite Lebesgue measure $\mu_L(X)$, then the uniform distribution $\rho_{uni}(x) = 1/\mu_L(X)$ has the entropy

$$H[\rho_{max}] = \log \mu_L(X) \tag{10.51}$$

and Gibbs' inequality again implies that it is larger than the entropy of any other density on the same support.

The entropy for continuous random variables also has some undesirable properties.

(1) The 'continuous' entropy is not the continuous limit of the 'discrete' entropy. Define

$$p(k) := \rho(x_k)\Delta x = \int_{k\Delta x}^{(i+1)\Delta x} dx \rho(x) \tag{10.52}$$

Then

$$\begin{aligned}\lim_{\Delta x \to 0} (-\sum_k p(k) \log p(k)) &= \lim_{\Delta x \to 0} (-\sum_k \rho(x_k)\Delta x \log (\rho(x_k)\Delta x)) \\ &= -\int dx \rho(x) \log \rho(x) - \lim_{\Delta \to 0} \log \Delta x \end{aligned} \tag{10.53}$$

This expression diverges since $\lim_{\Delta x \to 0} \log \Delta x = -\infty$.

(2) If the sure event $\{x_0\}$ is defined by $\rho(x) = \delta(x - x_0)$, then the entropy does not exist.

(3) The 'continuous' entropy is also not invariant under a scaling transformation $x' = ax$. To see this, consider the random variable $g = af$ with density $\rho_g(x) = a^{-1}\rho_f(a^{-1}x)$. Then

$$\int_{\mathbb{R}} dx \rho_f(x) \log \rho_f(x) = \int_{\mathbb{R}} dy \rho_g(y) \log \rho_g(y) + \log a \tag{10.54}$$

(4) The 'continuous' entropy is neither bounded from below nor from above which can be seen by the entropy (10.45) of the Gaussian which approaches $+\infty$ for $\sigma \to \infty$ and $-\infty$ for $\sigma \to 0$.
(5) Also, $\rho(x)$ might have a dimension and one cannot take the logarithm of a dimensional number.

Kullback–Leibler divergence. The mathematical way out of these undesirables is to introduce the Kullback–Leibler divergence (or relative entropy)

$$D[\rho_1||\rho_2] := \int_{\mathbb{R}} dx\, \rho_1(x) \log \left\{ \frac{\rho_1(x)}{\rho_2(x)} \right\} \tag{10.55}$$

This divergence is invariant under coordinate transformations. If the state space has a finite Lebesgue measure $\mu_L(X) < \infty$, then the uniform distribution $\rho = 1/\mu_L(X)$ has entropy $H[1/\mu_L(X)] = \log \mu_l(X)$ which is larger than the entropy of any other density. Substituting this uniform distribution for ρ_2, one finds

$$D[\rho_1 \parallel 1/\mu_L(X)] = -H[\rho_1] + H[1/\mu_L(X)] \tag{10.56}$$

The divergence between any density ρ_1 and the uniform density is the difference between their entropies, a result which will be generalized later. The important point is:

The Kullback–Leibler divergence is the proper concept for random variables with continuous support.

10.7 Coarse-Graining and Smoothing

We now formalize the intuitive notions that coarse-graining and smoothing cannot reduce entropy.

Coarse-graining of discrete random variables. Let us start with a random variable with a discrete state space X and a probability mass function p. The cardinality of the state space may be countably infinite. Partition the state space X into N disjoint bins B_n (or groups or cells or boxes), $X = \cup_{n=1}^{N} B_n$, where the number of boxes N may also be infinite. We require a non-trivial partition $0 < |B_n| < \infty$. One can then define three additional probability distributions:

(1) The probability that the outcomes fall into bin B_n which is given by

$$\tilde{p}(n) := \sum_{k \in B_n} p(k) \quad n = 1, \ldots, N \tag{10.57}$$

The cardinality of the support of this probability distribution is the number of boxes N.

(2) The probability distributions within each bin n which is given by

$$\hat{p}_n(k) = p(k)/\tilde{p}(n) \tag{10.58}$$

where $k \in B_n$. These are N different probability distributions, and the cardinality of their support is given by $|B_n|$, the number of elements in B_n.

(3) A coarse-grained distribution

$$p^{cg}(k) := \sum_{n=1}^{N} \frac{\tilde{p}(n)}{|B_n|} \chi_{B_n}(k) \tag{10.59}$$

This coarse-grained distribution replaces the original distribution with the uniform distribution within each box. Its support is X.

How are all these probability distributions p, $\tilde{p}$, $\hat{p}$, and p^{cg} related? First, one can easily demonstrate that

$$H[p] = H[\tilde{p}] + \sum_{n=1}^{N} \tilde{p}(n) H[\hat{p}_n] \tag{10.60}$$

The entropy of p does not depend on the way it is calculated, either all at once or successively from the bins. Binning is just a different representation of the probability distribution. One describes p by $\tilde{p}$ and $\hat{p}_n$. The original distribution can be recovered by $p(k) = \sum_{n=1}^{N} \tilde{p}(n)\hat{p}_n(k)$. No information is lost. Note that the relation (10.60) implies the inequality $H[p] \geq H[\tilde{p}]$ since all terms in (10.60) are positive. The entropy of the probability $\tilde{p}$ of the bins cannot exceed the entropy of the original probability p. This result might look counter-intuitive. However, the support of $\tilde{p}$ is smaller than that of p. There are fewer choices. Information is, however, lost when one replaces p by the coarse-grained distribution p^{cg}. Both have the same support. Indeed, one can show

$$H[p^{cg}] \geq H[p] \tag{10.61}$$

Coarse-graining never reduces the entropy. The proof of this assertion will be given for continuous random variables, considered next.

Coarse-graining of continuous random variables. Assume the state space to be measure space $\{X, \mathcal{A}, \mu\}$ and the probability distribution to be given by a probability density ρ. Usually, one considers $\{\mathbb{R}^K, \mathcal{B}^K, \mu_L\}$. Again partition the state space into disjoint subsets, $X = \cup_{n=1}^{N} B_n$ where again the number N of subsets can be infinite. Assume non-trivial partitions for which $0 < \mu(B_l) < \infty$ for all $n = 1, \ldots, N$. One can then again define the probabilities:

(1) The probability of the subsets

$$\tilde{p}(n) = \int_{B_n} \mu(dx)\rho(x) \quad n = 1, \ldots, N \tag{10.62}$$

(2) The probability density within each of the N subsets

$$\hat{\rho}_n(x) = \frac{\rho(x)}{\tilde{p}(n)} \chi_{B_n}(x) \tag{10.63}$$

(3) The coarse-grained density

$$\rho^{cg}(x) := \sum_{l=1}^{L} \frac{\tilde{p}(l)}{\mu(B_l)} \chi_{B_l}(x) \tag{10.64}$$

These probability distributions satisfy the binning relation

$$H[\rho] = H[\tilde{p}] + \sum_{n=1}^{N} \tilde{p}(n) H[\hat{\rho}_n] \tag{10.65}$$

from which we cannot infer $H[\rho] \geq H[\tilde{p}]$ any more since the $H[\hat{\rho}_n]$ are not necessarily positive. Using, however, the Kullback–Leibler divergence one finds

$$D[\rho_1 \parallel \rho_2] = D[\tilde{p}_1 \parallel \tilde{p}_2] + \sum_{n=1}^{N} \tilde{p}(n) D[\hat{\rho}_{1,n} \parallel \hat{\rho}_{2,n}] \tag{10.66}$$

where now all three terms are positive, and we can conclude $D[\rho_1 \parallel \rho_2] \geq D[\tilde{p}_1 \parallel \tilde{p}_2]$. We reiterate that using $\tilde{p}$ and $\hat{\rho}_n$ is just a different representation of ρ which can be recovered by $\rho(x) = \sum_{n=1}^{N} \tilde{p}(n)\hat{\rho}_n(x)$.

Applying the property (10.65) to the coarse-grained density, one finds

$$H[\rho^{cg}] = H[\tilde{p}] + \sum_{n=1}^{N} \tilde{p}(n) \log \mu(B_n) \tag{10.67}$$

since $\tilde{p}^{cg} = \tilde{p}$ and $\hat{\rho}_n^{cg} = 1/\mu(B_n)$. Again, we cannot infer $H[\rho^{cg}] \geq H[\tilde{p}]$ since the $\log \mu(B_n)$ are not necessarily positive. For the Kullback-Leibler divergence we find, however, $D[\rho_1^{cg} \parallel \rho_2^{cg}] = D[\tilde{p}_1 \parallel \tilde{p}_2]$. The distance between ρ_1^{cg} and ρ_2^{cg} is the same as the distance between $\tilde{p}_1$ and $\tilde{p}_2$. Combining this result with the result obtained from (10.66), we obtain

$$D[\rho_1^{cg} \parallel \rho_2^{cg}] = D[\tilde{p}_1 \parallel \tilde{p}_2] \leq D[\rho_1 \parallel \rho_2] \tag{10.68}$$

One thus easily infers that coarse-graining reduces the 'divergence' between the two distributions. The coarse-grained distributions are more similar than the original distributions.

Another particular application of the formula (10.67) is the heuristic justification of the quantum-mechanical correction of classical statistical mechanics. There one consider the special situation that the state space is partitioned into cells of equal volume ΔV, with $\Delta V = (2\pi\hbar)^{3N}$ for an N-particle system. Then one obtains the correct quantum-mechanical entropy $H[\tilde{p}]$ from the classical entropy $H[\rho^{cg}]$ by subtracting $\log \Delta V$

$$H[\tilde{p}] = H[\rho^{cg}] - \log \Delta V \tag{10.69}$$

Of course, this is only a heuristic argument since the state space in quantum mechanics is not the classical state space.

It is much harder to prove that the entropy satisfies

$$H[\rho^{cg}] \geq H[\rho] \tag{10.70}$$

Here is an explicit proof of this important theorem:

$$\begin{aligned}
H[\rho^{cg}] &= -\int_X \mu(dx) \sum_{l=1}^{L} \frac{\tilde{p}(l)}{\mu(B_l)} \chi_{B_l} \log \left\{ \sum_{k=1}^{L} \frac{\tilde{p}(k)}{\mu(B_k)} \chi_{B_k} \right\} \\
&= -\sum_{l=1}^{L} \frac{\tilde{p}(l)}{\mu(B_l)} \int_{B_l} \mu(dx) \log \left\{ \sum_{k=1}^{L} \frac{\tilde{p}(k)}{\mu(B_k)} \chi_{B_k} \right\} \\
&= -\sum_{l=1}^{L} \frac{\tilde{p}(l)}{\mu(B_l)} \log \left\{ \frac{\tilde{p}(l)}{\mu(B_l)} \right\} \int_{B_l} \mu(dx) \\
&= -\sum_{l=1}^{L} \frac{\tilde{p}(l)}{\mu(B_l)} \log \left\{ \frac{\tilde{p}(l)}{\mu(B_l)} \right\} \mu(B_l) \\
&= -\sum_{l=1}^{L} \int_{B_l} \mu(dx) \rho(x) \log \left\{ \frac{\tilde{p}(l)}{\mu(B_l)} \right\} \\
&\geq -\sum_{l=1}^{L} \int_{B_l} \mu(dx) \log \rho(x) \\
&= -\int \mu(dx) \rho(x) \log \rho(x) \\
&= H[\rho]
\end{aligned} \tag{10.71}$$

The first line inserts ρ^{cg} into the definition of the entropy functional. The second line takes the sum out of the integral and makes use of the definition of the indicator function. Since the integration is only over B_l, only the term $\tilde{p}(l)/\mu(B_l)$ survives in the sum under the logarithm (third line). The fourth line implements the definition of $\mu(B_l)$ and the fifth line the definition of $\tilde{p}(l)$. The sixth line is then the crucial application of Gibbs' inequality to the density $\rho_1 = \rho$ and $\rho_2 = \tilde{p}(l)/\mu(B_l)$ which are both normalized to $\tilde{p}(l)$ in the box B_l.

To reiterate, coarse-graining never decreases the entropy and never increases the Kullback–Leibler divergence.

Smoothing. To quantify the concept of smoothing, we first define a smoothing function $\varphi : X \mapsto \mathbb{R}$ which is a function that is larger or equal to zero and is normalized to one

$$\varphi(x) \geq 0$$
$$\int_X dx \varphi(x) = 1 \tag{10.72}$$

The smoothing function has the same properties as a density. We will also need that φ is sufficiently often differentiable and decays sufficiently fast to zero for $|x| \to \infty$. The smoothing function is thus a smooth function. Given a function $f : X \mapsto Y$, the smoothed version $\overline{f} : X \mapsto Y$ of f is then defined by

$$\overline{f}(x) := \int_X dx' f(x')\varphi(x - x') \tag{10.73}$$

It is the convolution between the function f and the smoothing function φ. Note that $\overline{f}$ is also a function from X to Y.

Smoothing is a linear operation (if the spaces X and Y are vector spaces)

$$\overline{a_1 f_1 + a_2 f_2} = a_1 \overline{f_1} + a_2 \overline{f_2} \tag{10.74}$$

for $a_1, a_2 \in \mathbb{C}(\mathbb{R})$ and commutes with differentiation since

$$\begin{aligned} \nabla_x \overline{f} &= \int_X dx' \; f(x')\nabla_x \varphi(x - x') \\ &= -\int_X dx' f(x')\nabla_{x'} \varphi(x - x') \\ &= \int_X dx' \nabla_{x'} f(x') \; \varphi(x - x') \\ &= \overline{\nabla f} \end{aligned} \tag{10.75}$$

since the smoothing function vanishes at $|x| = \infty$. Note that smoothing does not have some of the other perhaps desirable properties. It is in general not an idempotent operation, $\overline{\overline{f}} \neq \overline{f}$, which implies $\overline{f - \overline{f}} \neq 0$ in general. Also if $g : Y \mapsto Z$, then $\overline{g \circ f} \neq \overline{g} \circ \overline{f}$ in general. Smoothing and composing do not commute. Composing two functions first and then smoothing is in general different from first smoothing the functions and then composing them. The qualifier 'in general' is necessary to take account of the trivial cases where $\varphi(x) = \delta(x)$ and f or g are the identity mappings.

If one takes for f a probability density ρ, then the entropy functional satisfies

$$
\begin{aligned}
H[\overline{\rho}] &= -\int_X dx \overline{\rho}(x) \log \overline{\rho}(x) \\
&= -\int_X dx \int_X dx' \rho(x') \varphi(x - x') \log \overline{\rho}(x) \\
&= -\int_X dx \int_X dx' \varphi(x') \rho(x - x') \log \overline{\rho}(x) \\
&= -\int_X dx' \varphi(x') \int_X dx \rho(x - x') \log \overline{\rho}(x) \\
&\geq -\int_X dx' \varphi(x') \int_X dx \rho(x - x') \log \rho(x - x') \\
&= H[\rho]
\end{aligned}
\qquad (10.76)
$$

where the inequality sign is again Gibbs' inequality and where we made use of the fact that the entropy is invariant under translation. Smoothing never decreases the entropy. Similarly, one finds $H[\overline{\rho}] \geq H[\varphi]$. The important point is:

Coarse-graining and smoothing never reduce the entropy.

10.8 The Principle of Maximum Entropy

We now come to a very important principle. Probability theory studies the properties of given probability distributions and their transformation under various operations. Probability theory is silent about how to determine these probability distributions for real-world applications. Statistics, which will be covered in Part IV, provides methods to determine probability distributions from samples. One principle that turns data into probabilities is the principle of maximum entropy. It considers the situation that the probability distribution itself is not known but some data, treated as constraints. Many probability distributions are consistent with these constraints. The principle of maximum entropy asserts that among these distributions, one should choose that distribution that maximizes the information entropy or missing information. The rationale is that such a choice provides the most random distribution consistent with the constraints. Technically, the maximization of the entropy functional under constraints is an exercise in the calculus of variations, with the constraints being handled by Lagrange multipliers (see Appendix E.16).

The principle of maximum entropy. The task is thus to find the probability density ρ_{max} that maximizes the entropy functional

$$
H[\rho] = -\int_X dx \rho(x) \log \rho(x) \qquad (10.77)
$$

subject to constraints. One constraint that must always be incorporated is the normalization of the density

$$\int_X dx \rho(x) = 1 \tag{10.78}$$

Additional constraints are usually associated with observables $f : X \mapsto \mathbb{R}$. The theory takes a particular elegant form if the expectation values of K observables

$$\langle f_k \rangle := \int_X dx \rho(x) f_k(x) \quad k = 1, \ldots, K \tag{10.79}$$

are given. Here, $\langle f_1 \rangle, \ldots, \langle f_K \rangle$ are K given numbers. The calculus of variations then states that one has to find the maximum of the functional

$$F[\rho] = -\int dx \rho(x) \log \rho(x) - \lambda_0 [\int dx \rho(x) - 1] - \sum_{k=1}^{K} \lambda_k [\int dx \rho(x) f_k(x) - \langle f_k \rangle] \tag{10.80}$$

with constant Lagrange multipliers $\lambda_0, \ldots, \lambda_K$. The maximum (or more generally an extreme) is obtained when the variation

$$\delta F = -\int dx [\log \rho(x) + 1 + \lambda_0 + \sum_{k=1}^{K} \lambda_k f_k(x)] \delta\rho(x) \tag{10.81}$$

vanishes for arbitrary small variations $\delta\rho(x)$. This is only possible if the term in square brackets vanishes which gives

$$\rho_{max}(x; \lambda_0, \lambda_1, \ldots, \lambda_K) = \exp\{-(\lambda_0 + 1) - \sum_{k=1}^{K} \lambda_k f_k(x)\} \tag{10.82}$$

This condition can also be formulated in terms of the functional derivative (see Appendix E.16). The density ρ_{max} is a function of the Lagrange multipliers (λ_0, λ_1, ..., λ_K) which must be chosen such that the distribution ρ_{max} satisfies the normalization and reproduces the given values $\langle f_1 \rangle, \ldots, \langle f_K \rangle$. This can be done explicitly for λ_0. Upon substitution of ρ_{max}, the normalization condition becomes

$$\exp\{-(\lambda_0 + 1)\} \int_X dx \exp\left\{-\sum_{k=1}^{K} \lambda_k f_k(x)\right\} = 1 \tag{10.83}$$

and allows to eliminate λ_0 from (10.82), giving

$$\rho_{max}(x; \lambda_1, \ldots, \lambda_K) = \frac{1}{Z} \exp\left\{-\sum_{k=1}^{K} \lambda_k f_k(x)\right\} \tag{10.84}$$

where

$$Z(\lambda_1, \ldots, \lambda_K) := \int_X dx \exp\left\{-\sum_{k=1}^{K} \lambda_k f_k(x)\right\} \tag{10.85}$$

is called the partition function. The distribution in the form (10.84) is called the canonical distribution.

The remaining Lagrange multipliers $\lambda_1, \ldots, \lambda_K$ have to be determined from the K relations

$$\begin{aligned}\langle f_i\rangle &= \frac{1}{Z(\lambda_1, \ldots, \lambda_K)} \int_X dx\, f_i(x) \exp\left\{-\sum_{k=1}^{K} \lambda_k f_k(x)\right\} \\ &= \frac{1}{Z(\lambda_1, \ldots, \lambda_K)} (-\frac{\partial}{\partial \lambda_i} \int dx \exp\left\{-\sum_{k=1}^{K} \lambda_k f_k(x)\right\}) \\ &= -\frac{\partial \log Z(\lambda_1, \ldots, \lambda_K)}{\partial \lambda_i} \end{aligned} \tag{10.86}$$

Once ρ_{max} and $Z(\lambda_1, \ldots, \lambda_K)$ have been determined, the expectation values of all other observables are determined as well. One finds for the covariance

$$\begin{aligned}\langle f_i f_j\rangle - \langle f_i\rangle\langle f_j\rangle = -\frac{\partial\langle f_i\rangle}{\partial \lambda_j} &= -\frac{\partial\langle f_j\rangle}{\partial \lambda_i} \\ &= \frac{\partial^2 \log Z(\lambda_1, \ldots, \lambda_K)}{\partial \lambda_i \partial \lambda_j}\end{aligned} \tag{10.87}$$

and similarly for higher moments. The important message is that the higher moments are determined and cannot be chosen. If higher moments are also given, one has to redo the maximization with these additional constraints.

This is the standard and straightforward exposition of the principle of maximum entropy. The derivation does not demonstrate that the solution exists, that it is indeed a maximum (and not a minimum), and that it is unique. Indeed, we will point out below, when we discuss the Gaussian distribution, a few problems where the partition function does not exist or where the set (10.86) does not have solutions. The principle of maximum entropy is very unforgiving, and one cannot sneak ill-posed problems by it. However, if ρ_{max} does exist, Gibbs' inequality implies

$$\begin{aligned}H[\rho] &\leq -\int_X dx \rho(x) \log \rho_{max}(x) \\ &= -\int_X dx \rho(x)[-\log Z(\lambda_1, \ldots, \lambda_K) - \sum_{k=1}^{K} \lambda_k f_k(x)] \\ &= \log Z(\lambda_1, \ldots, \lambda_K) + \sum_{k=1}^{K} \lambda_k \langle f_k\rangle \\ &= H[\rho_{max}]\end{aligned} \tag{10.88}$$

with the equality sign holding if and only if $\rho = \rho_{max}$. Thus, $H[\rho_{max}]$ is an absolute maximum. Furthermore, the Lagrange multipliers have unique values. Otherwise, we have a contradiction to the result that $H[\rho_{max}]$ is an absolute maximum.

The Kullback–Leibler divergence between $\rho_{max}(K)$ and any other density ρ that yield the same constraints is given by

$$\begin{aligned}
D[\rho \parallel \rho^K_{max}] &= -H[\rho] - \int dx \rho(x) \log \rho_{max}(x; K) \\
&= -H[\rho] + \int dx \rho(x) \log Z(K) + \sum_{k=1}^{K} \lambda_k \int dx \rho(x) f_k(x) \\
&= -H[\rho] + \log Z(K) + \sum_{k=1}^{K} \lambda_k \langle f_k \rangle \\
&= -H[\rho] + \int dx \rho_{max}(x; K) \log Z(K) + \sum_{k=1}^{K} \lambda_k \int dx \rho_{max}(x; K) f_k(x) \\
&= -H[\rho] + H[\rho^K_{max}]
\end{aligned} \tag{10.89}$$

where the crucial step is from the third to the fourth line where the expectation with respect to ρ could be replaced by the expectation with respect to $\rho_{max}(K)$.

The uniform distribution. The principle of maximum entropy also works when the constraints are the actual values of observables. In this case, the principle of maximum entropy gives the uniform distribution on that part of the state space that is compatible with these constraints. A special case is when there are no constraints at all. Then the principle of maximum entropy gives the uniform density $\rho_{max}(x) = \chi_X(x)/\mu_L(X)$ which only exists when the Lebesgue measure of the state space is finite. In this case, formula (10.89) also recovers formula (10.56).

The Gaussian distribution. One standard problem for $X = \mathbb{R}$ is to find the maximum entropy distribution when both the mean μ and the variance σ^2 are given. The result is the Gaussian distribution $\rho_{max} = \rho_G = N(\mu, \sigma^2)$ since for any other density ρ

$$\begin{aligned}
0 \leq D(\rho || \rho_G) &= \int dx \rho(x) \log \rho(x) - \int dx \rho(x) \log \rho_G(x) \\
&= \int dx \rho(x) \log \rho(x) - \int dx \rho(x) \log \left\{ \frac{1}{\sqrt{2\pi\sigma^2}} \right\} + \int dx \rho(x) \frac{(x-\mu)^2}{2\sigma^2} \\
&= \int dx \rho(x) \log \rho(x) - \int \rho_G(x) \log \left\{ \frac{1}{\sqrt{2\pi\sigma^2}} \right\} + \int dx \rho_G(x) \frac{(x-\mu)^2}{2\sigma^2} \\
&= -H[\rho] + H[\rho_G]
\end{aligned} \tag{10.90}$$

from which it follows that $H[\rho_G] \geq H[\rho]$ for any ρ. The crucial third line in the proof is valid because ρ and ρ_G have the same mean and variance. This result again shows the special role of the Gaussian distribution.

A few variations on this problem are illuminating. Assume that only the mean μ is prescribed. The principle of maximum entropy then leads to the distribution $e^{-\lambda x}$ which is non-normalizable if $X = \mathbb{R}$. Indeed, the prescription of the mean is not sufficient to single out a particular distribution. The entropy of the Gaussian distribution increases without bound with σ^2 for any μ. When we state the problem for $X = \mathbb{R}_+$, however, then the principle of maximum entropy gives the exponential distribution $\rho_{max} = e^{-x/\mu}/\mu$. Prescribing just the variance is not possible since the definition of the variance requires knowing the mean, since $\sigma^2 = \langle (f - \mu)^2 \rangle$. When prescribing $\langle f^2 \rangle$, the principle of maximum entropy gives the Gaussian distribution $N(0, \langle f^2 \rangle)$. This distribution has zero mean. One might wonder how this zero mean squares with the translational invariance of the entropy. To resolve this apparent problem, we have to note that $\langle f^2 \rangle = \sigma^2 + \mu^2$. If one increases μ one has to decrease the variance σ^2 and hence the entropy. The maximum of the entropy for fixed $\langle f^2 \rangle$ is indeed obtained for $\mu = 0$.

These examples again demonstrate the central role of the Gaussian distribution and the fact that not every constraint leads to a ρ_{max}.

Dependence on number of constraints. If one increases the number of constraints to $L \geq K$, one expects

$$H[\rho_{max}(L)] \leq H[\rho_{max}(K)] \tag{10.91}$$

since more constraints have to be satisfied. This is indeed true since Gibbs' inequality implies

$$\begin{aligned} H[\rho_{max}(L)] &\leq -\int_X dx\, \rho_{max}(x; L) \log \rho_{max}(x; K) \\ &= -\int_X dx\, \rho_{max}(x; L) \left[\log Z(K) - \sum_{k=1}^{K} \lambda_k f_k(x) \right] \\ &= -\int_X dx\, \rho_{max}(x, K)[\log Z(K) - \sum_{k=1}^{K} \lambda_k f_k(x)] \\ &= H[\rho_{max}(K)] \end{aligned} \tag{10.92}$$

where the replacement of $\rho_{max}(L)$ by $\rho_{max}(K)$ is again justified since the two densities yield the same expectations. For the Kullback–Leibler divergence, one finds

$$D[\rho || \rho_{max}(L)] \leq D[\rho || \rho_{max}(K)] \tag{10.93}$$

For illustration, consider $X = [0, 1[$. When there are no constraints, then the uniform distribution over this interval is the maximum entropy distribution and $H[\rho_{max}(0)] = 0$. Consider next the observables f_i that assign to each real number in the interval the ith digit of its binary representation. When the first digit f_1 is prescribed, the set of compatible states is reduced to an interval of length $1/2$ and the principle of maximum entropy implies a uniform distribution over this interval and hence $H[\rho_{max}(1)] = -\log 2$. When successive digits $f_1, \ldots, f_L$ are prescribed, the compatible states become confined to a set of intervals with total length $1/2^L$, and the principle of maximum entropy predicts a uniform distribution over this set of intervals with $H[\rho_{max}(L)] = -L \log 2$. As more and more digits are prescribed, the entropy changes from zero at the beginning to $-L \log 2$ after L digits have been revealed. The entropy becomes smaller and smaller. It is reduced by one bit as one more digit is revealed. This decrease reflects the fact that more and more information about the initial condition becomes available.

Entropy function. Substituting ρ_{max} into the expression for the entropy functional H, one obtains the entropy function

$$\begin{aligned} S := H[\rho_{max}] &= -\int dx \rho_{max}(x) \log \left\{ \frac{1}{Z} \exp \left\{ -\sum_{k=1}^{K} \lambda_k f_k(x) \right\} \right\} \\ &= \int dx f \rho_{max}(x)(\log Z(\lambda_1, \ldots, \lambda_K) + \sum_{k=1}^{K} \lambda_k g_k(x)) \\ &= \log Z(\lambda_1, \ldots, \lambda_K) + \sum_{k=1}^{K} \lambda_k \langle f_k \rangle \end{aligned} \tag{10.94}$$

The entropy function S depends on the constraints and the Lagrange multipliers. However, we find

$$\frac{\partial S}{\partial \langle f_i \rangle} = \lambda_i \tag{10.95}$$

and

$$\begin{aligned} \frac{\partial S}{\partial \lambda_i} &= \langle f_i \rangle + \frac{\partial \log Z}{\partial \lambda_i} \\ &= \langle f_i \rangle + \frac{1}{Z} \int dx (-f_i) \exp \left\{ -\sum_{k}^{K} \lambda_k g_k \right\} \\ &= \langle f_i \rangle - \langle f_i \rangle = 0 \end{aligned} \tag{10.96}$$

Thus, $S = S(\langle f_1 \rangle, \ldots, \langle f_K \rangle)$ and S becomes the Legendre transform of $-\log Z$.

Formula (10.95) is the analogue of formula (10.86). The analogue of formula (10.87) is

$$\frac{\partial^2 S}{\partial \langle f_i \rangle \partial \langle f_j \rangle} = \frac{\partial \lambda_i}{\partial \langle f_j \rangle} = \frac{\partial \lambda_j}{\partial \langle f_i \rangle} \tag{10.97}$$

It is important to distinguish between the entropy functional H and the entropy function S. H is a function(al) of ρ. It describes the 'broadness' or 'randomness' of any distribution ρ. S on the other hand is a function of the constraints $\langle f_i \rangle$. It is obtained by substituting ρ_{max} into $H[\rho]$ and describes the 'randomness' of the maximum entropy solution as a function of the constraints.

We stop the discussion of the principle of maximum entropy here. It is a powerful principle. As we demonstrate in Part V, it implies the structure both of equilibrium and non-equilibrium statistical mechanics. It is also the principle of choice to determine the a priori probability distribution in Bayesian inference, as we argue in Part IV.

10.9 Summary

Information entropy is a central concept of this book. It is a quantitative measure of what one might refer to as the amount of randomness or uncertainty or missing information. The information entropy is a property solely of the probability distribution of the random variable. It does not depend on anything else. The information entropy is independent of the way it is inferred, either all at once or sequentially. The entropy is a concave functional of the probability mass or density function and satisfies Gibbs' inequality, which is needed in the proofs that coarse-graining and smoothing cannot reduce the entropy. One can extend the concept of entropy to bi- or multivariate random variables by defining joint and conditional entropies, in analogy to joint and conditional probabilities. When applying the weak law of large numbers, one obtains the asymptotic equipartition theorem which states that the relative number of typical sequences goes to zero while their probability remains at one, a result with important implications for coding.

When the state space is a finite discrete space, as the alphabet of information theory, then the entropy is bounded from below by zero for a probability mass function which is concentrated on one point and bounded from above by the logarithm of the cardinality of the state space for a uniform distribution.

When extending the concept of entropy to continuous random variables, one encounters a few undesirable properties. The entropy functional is not invariant under scaling transformations. It might neither be bounded from below nor from above, though it is bounded from above by the entropy of the uniform distribution if it exists, as it does when the state space has finite Lebesgue measure. Mathematical statements become more exact and more rigorous if one introduces the Kullback–Leibler divergence (or relative entropy) which compares the entropy gain from one distribution to another. One may still argue with entropies, if one is aware of its shortcomings and some inherent subtleties.

In this chapter, we also introduced for the first time a principle that determines probability distributions. Probability theory just tells us how to manipulate given probabilities. The principle that determines probabilities is the principle of maximum entropy. It assumes that data or measurements act as constraints on the probability distribution and then determines that distribution which maximizes the entropy. The probability distribution so determined is the most 'random' or most 'unbiased' distribution consistent with the data.

References

Cover, T.M., Thomas, J.A.: Elements of Information Theory. Wiley Interscience (2005)
MacKay, D.J.C.: Information Theory, Inference, and Learning Algorithms. Cambridge University Press (2003)
Shannon, C.E.: A mathematical theory of communication. Bell Syst. Tech. J. **27**, 379-423, 623-656 (1948)

Part III
Probability in Dynamical Evolution

Dynamical systems theory and probability theory are combined for various purposes. In the chapter on ergodic theory we used normalized measures to characterize dynamical systems. Though these normalized measures are formally probability measures, there were devoid of any notion of 'randomness' and were interpreted as 'volumes'. In this part, we add probability distributions to a dynamical system in order to account for or generate randomness. Dynamics quantifies the laws of nature. Probability quantifies uncertainty.

When one applies dynamical systems theory to actual systems one generally finds that the theory requires information that one does not have. One situation is that one does not know precisely the initial conditions required for integrating the dynamical equations. The prime example is a gas described by a N-particle Hamilton function. We do not know or cannot control the initial conditions of all the gas molecules. We only know or control some macroscopic variables such as the total mass, volume, and pressure. A second situation is that some process in the dynamical law is not exactly known. The prime example here is the Brownian motion of ink particles where again one cannot exactly follow the molecules that push the ink particles around. To quantify these and other situations one introduces concepts from probability and information theory. Specifically, we use

1. probability distributions to account for incomplete knowledge of the initial conditions of a dynamical system,
2. random initial conditions to generate stochastic processes from observables of dynamical systems,
3. stochastic processes to represent unresolved processes in the dynamical law.

If the initial conditions are not exactly known the system can be in many states. Our knowledge thus does not represent a point but a broadened distribution in state space. Such broadened initial distributions can be described by a probability distributions. Here, we consider the case that this distribution is given by a probability density function. The evolution of such densities can be inferred from the dynamics and one arrives at a dynamical system not for points in the state space but for densities on the state space. The concept of state is thus generalized from a point to a probability distribution. There are interesting consequences that result from this transition such that densities can do things that points cannot do. They can have an attractor even if the underlying dynamics does not have one. The transition from points to densities, however, does not break any time symmetry. The entropy does not change for invertible systems.

Dynamical systems and probability theory can also be combined to generate stochastic processes, indeed a very common method. Introducing random initial conditions makes the dynamical system a stochastic process, albeit a fairly trivial one. The different initial conditions evolve independently, carrying their initial probability with them. Formally, the evolution can be viewed as a deterministic Markov process, transitioning from one state to another state with probability one. The situation changes when observables are considered which are many-to-one functions. The evolution of the initial state is determined by the initial state of the system and not by the initial value of the observable. Hence different future values of the observable may be obtained for different initial states compatible with the same initial value of the observable. The value of the observable can transition to many different values, with a certain probability. The evolution of the observable becomes non-trivial stochastic processes. Our main examples are coarse-graining and the elimination of variables, operations that can be formulated in terms of observables.

Stochastic processes are introduced into a dynamical law when some process cannot be fully resolved. The dynamical system then becomes a stochastic dynamical system. Our main examples are that a forcing term or a growt rate needs to be represented by a stochastic process. In the discrete-time case, this leads to autoregressive and multiplicative processes. In the continuoustime case, it leads to stochastic differential equations where a coefficient or a forcing term is represented by a stochastic process. The Langevin equation used to describe Brownian motion is a prime example. The proper treatment of stochastic differential equations , however, requires a new kind of calculus, the Ito calculus, since the irregularity of stochastic processes such as white noise does not mix well with differentiability.

Chapter 11
Time Evolution of Broadened Initial States

11.1 Introduction

In this chapter, we study the time evolution of broadened initial conditions under the action of a dynamical map. The broadened initial conditions are represented by probability density functions, in contrast to pure initial conditions which are represented by points. The evolution of densities is a special case of the evolution of measures which is studied in the chapter on ergodic theory. The emphasis is on invariant measures which are solely a property of the dynamical system. The study of invariant measures turned out to provide useful insights into the character of dynamical systems, and so will the study of the evolution of densities. Broadened initial conditions and their associated densities are also studied here for an additional reason, to quantify our ignorance. One might be unable, for practical or conceptual reasons, to specify the precise state of a system which corresponds to a point in state space. One is only able to ascertain or control that the initial states are distributed according to a certain probability density function.

Given an initial density function, we address the questions:

- How does the initial density evolve in time?
- Under which circumstances do invariant or stationary densities exist?
- Do arbitrary initial densities converge toward a stationary density, and if so in what sense?
- How does the entropy evolve.

The evolution of densities is governed by an operator, called the Frobenius–Perron operator. The Frobenius–Perron operator is a linear operator, even if the dynamical system is highly nonlinear, a situation similar to a Markov process being governed by the linear Markov operator. In the case of a continuous-time invertible system, the Frobenius–Perron operator is determined by the solution of a Liouville equation. Since the Frobenius–Perron operator is linear, the analysis of the evolution of densities is amenable to the powerful tools of linear algebra. Stationary densities are fixed points of the Frobenius–Perron operator. If there exists a unique stationary density,

P. Müller, *Handbook of Dynamics and Probability*,
https://doi.org/10.1007/978-3-030-88486-4_11

then the system is ergodic and arbitrary initial densities converge to this stationary density in the Cesàro sense. Weak and point-wise convergences are obtained for mixing and exact systems.

11.2 Broadened Versus Pure Initial States

Here we enlarge the concept of state.

Broadened state. Given a dynamical system $\{X, (S_t)_{t\in T}\}$, broadened initial conditions are defined by a probability measure P_0. One thus endows the state space with a probability $\{X, \mathcal{A}, P_0\}$. We only consider the case that the initial probability measure P_0 can be represented by an initial density function ρ_0 in the form

$$P_0(A) = \int_A \mu(dx)\rho_0(x) \tag{11.1}$$

where μ is a σ-finite measure. The initial densities are non-negative members of the function space $\mathcal{L}_1$ which contains all functions $\rho : X \mapsto \mathbb{R}$ with $\int_X \mu(dx)|\rho(x)| < \infty$. Note that two functions that differ on a set of measure zero are considered to be equivalent.

We stress that we only consider broadened initial conditions that can be represented by a density being an element of $\mathcal{L}_1$. The major case that P_0 cannot be represented by such a density is the case that P_0 is given by a probability mass function $p_0(k)$, $k \in \mathbb{Z}$. For some purposes, we have found it useful to represent such probability mass functions as a weighted sum of delta functions, but this case is excluded here since the delta function is not a proper function and not a member of the space $\mathcal{L}_1$. This chapter thus considers only dynamical systems on a continuous state space.

Pure state. States $x \in X$ are called pure states in the context of this chapter. They may formally be characterized by the delta function $\rho_0(y) = \delta(x - y)$ or the Dirac measure $\delta_x(A) = \chi_A(x)$ but are excluded from our analysis since these are no densities. Indeed, we will see that pure states and broadened states can show very different behavior.

11.3 Frobenius–Perron Operator

Broadened initial conditions are represented by a probability measure P_0. This probability measure P_0 may be viewed as the induced probability measure of a random variable

$$\begin{aligned} f_0 : \Omega &\mapsto X \\ \omega &\mapsto f_0(\omega) \end{aligned} \tag{11.2}$$

such that $P_0 = P_{f_0}$.

Induced density. The composition $S_t \circ f_0$ is a random variable as well with an induced probability measure

$$P_t(A) = P_0(S_t^{-1}(A)) \tag{11.3}$$

which assures the conservation of probability. In the case that P_0 can be represented by a density, this induced probability measure P_t is given by

$$P_t(A) == \int_{S_t^{-1}(A)} \mu(dx)\, \rho_0(x) \tag{11.4}$$

for all $t \in T$ and all $A \in \mathcal{A}$.

We now prove that P_t has a density as well if the map S_t is non-singular, i.e., whenever $\mu(A) = 0$ then $\mu(S_t^{-1}(A)) = 0$. The proof uses the Radon–Nikodym theorem which states: A measure P_t has a density with respect to a σ-finite measure μ if $\mu(A) = 0$ implies $P_t(A) = 0$. This is indeed the case. First, $\mu(A) = 0$ implies $\mu(S_t^{-1}(A)) = 0$ since the transformation S_t is assumed to be non-singular. Second, $\mu(S_t^{-1}(A)) = 0$ implies $P_0(S_t^{-1}(A)) = P_t(A) = 0$, according to (11.4). Thus, there exists a density ρ_t such that

$$P_t(A) = \int_A \mu(dx)\, \rho_t(x) \tag{11.5}$$

for all $A \in \mathcal{A}$ and $t \in T$. Using the transformation rule for densities, one finds

$$\rho_t(x) = \langle \delta(x - S_t \circ f_0) \rangle = \int_X \mu(dx')\, \rho_0(x')\delta(x - S_t(x')) \tag{11.6}$$

If $X = \mathbb{R}^N$ and S_t is an invertible transformation, then a change of variables from x' to $y = S_{-t}(x)$ transforms this equation into the explicit solution

$$\rho_t(x) = J_{-t}(x)\rho_0(S_{-t}(x)) \tag{11.7}$$

where the Jacobian

$$J_t(x) := \frac{\partial(S_t(x))}{\partial(x)} \tag{11.8}$$

accounts for the volume changes. The change in density compensates for the change in volume such that probability is conserved. This result may also be obtained from the conservation of probability in the form $\rho(y)dy = \rho(x)dx$ where $y = S_{-t}(x)$ and noting that infinitesimal volumes transform as $dy = \frac{\partial(y)}{\partial(x)}dx$. For transformations that conserve volume, $J_t(x) = 1$, the density is constant along trajectories. Note that ρ is not an observable. It transforms as above and not as $f_t(x) = f_0(S_t(x))$ as (time-independent) observables do. This is an important point:

Probability densities transform differently from observables.

When the map $S_t(x)$ and the Jacobian $J_{-t}(x)$ are given, then formula (11.7) provides an easy explicit solution, as is the case for the following two examples. Of particular importance are stationary or invariant densities ρ_* for which $\rho_t = \rho_*$ for all $t \in T$.

Example: The over-damped oscillator. The dynamical system representing the over-damped oscillator $\dot{x} = -\gamma x$ is $S_t(x) = x\exp[-\gamma t]$ with $J_t(x) = \exp[-\gamma t]$ and hence

$$\rho_t(x) = e^{\gamma t}\rho_0(xe^{\gamma t}) \tag{11.9}$$

There is no stationary density.

Example: Baker transformation. The forward and backward evolution of the baker transformation is given by

$$\rho_{n+1}(x,y) = \begin{cases} \rho_n(x/2, 2y) & \text{for } 0 \le y < \frac{1}{2} \\ \rho_n(x/2 + 1/2, 2y - 1) & \text{for } \frac{1}{2} \le y < 1 \end{cases} \tag{11.10}$$

$$\rho_{n-1}(x,y) = \begin{cases} \rho_n(2x, y/2) & \text{for } 0 \le x < \frac{1}{2} \\ \rho_n(2x - 1, y/2 + 1/2) & \text{for } \frac{1}{2} \le x < 1 \end{cases} \tag{11.11}$$

The uniform density is the stationary density.

For non-invertible systems, one has to use formula (11.6). In the special case that $X \subseteq \mathbb{R}$ and μ is the Lebesgue measure μ_L, then one can integrate this relation and obtains

$$\int_{-\infty}^{x} dx'\rho_t(x') = \int_{S_t^{-1}((-\infty,x))} dx'\rho_0(x') \tag{11.12}$$

Differentiating with respect to x, one finds

$$\rho_t(x) := \frac{d}{dx}\int_{S_t^{-1}((-\infty,x))} dx'\rho_0(x') \tag{11.13}$$

This approach yields the non-invertible evolution for the doubling map and the Ulam map.

Example: Doubling map. The evolution of the doubling map $x_{n+1} = 2x_n mod 1$ is given by

$$\rho_{n+1}(x) = \frac{1}{2}[\rho_n(x/2) + \rho_n(1/2 + x/2)] \tag{11.14}$$

This can be proven by differentiating the relation

$$\int_0^x dx' \rho_{n+1}(x') = \int_0^{x/2} dx' \rho_n(x') + \int_{1/2}^{1/2+x/2} \rho_n(x') \tag{11.15}$$

which is obtained by looking at the pre-images. The uniform density $\rho_*(x) = 1$ is a stationary density.

Example: Ulam map. The evolution of the Ulam map $x_{n+1} = 4(1 - x_n)x_n$ is given by

$$\rho_{n+1}(x) = \frac{1}{4\sqrt{1-x}}\left[\rho_n\left(\frac{1-\sqrt{1-x}}{2}\right) + \rho_n\left(\frac{1+\sqrt{1-x}}{2}\right)\right] \tag{11.16}$$

and has the stationary density

$$\rho_*(x) = \frac{1}{\pi}\frac{1}{\sqrt{1-(2x-1)^2}}\frac{d}{dx}(2x-1) = \frac{1}{\pi}\frac{1}{\sqrt{x(1-x)}} \tag{11.17}$$

One important lesson from the last example is that stationary densities do not need to be the uniform density.

Frobenius–Perron operator. Given the result (11.6), there exists an operator $\tilde{F}_t$

$$\begin{aligned} \tilde{F}_t : \tilde{\mathcal{L}}_1 &\mapsto \tilde{\mathcal{L}}_1 \\ \rho_0 &\mapsto \rho_t \end{aligned} \tag{11.18}$$

where $\tilde{\mathcal{L}}_1$ is the space of all probability densities, i.e., of all non-negative normalized integrable functions. Formally,

$$\tilde{F}_t[\rho_0] = \int_X \mu(dx')\, \delta(x - S_t(x'))\rho_0(x') \tag{11.19}$$

This integral operator can be extended to $\mathcal{L}_1$ and is then called the Frobenius–Perron operator and denoted by F_t. The Frobenius–Perron operator is a linear operator. It satisfies $F_t \circ F_{t'} = F_{t+t'}$. The pair $\{\mathcal{L}_1, (F_t)_{t\in T}\}$ thus constitutes a deterministic linear dynamical system. By introducing broadened initial conditions, we have converted the generally nonlinear problem for the evolution of pure states into the linear problem for the evolution of densities, however, in a function space of infinite dimensions. Nevertheless, there exist powerful tools for the analysis of linear systems such as an analysis in terms of the eigenfunctions and eigenvectors of the Frobenius–Perron operator. If the dynamical system is invertible, the Frobenius–Perron operator is invertible as well. One can integrate it forward and backward in time. Considering broadened states instead of pure states does not change this characteristic property of a dynamical system. This is also an important point:

If the dynamical system is invertible, then the Frobenius–Perron operator is invertible.

Observables. Expectations of (time-independent) observables $A : X \to \mathbb{R}$ are given by

$$\begin{aligned}\langle f_t^A \rangle := &\int_X dx \rho_0(x) f_t^A(x) \\ = &\int_X dx \rho_t(x) f_0^A(x)\end{aligned} \tag{11.20}$$

where $f_t^A := A \circ S_t$. There are thus two ways to calculate expectations, either fix the density and let the observable evolve or fix the observable and let the density evolve. Observables are generally many-to-one functions, and their evolution faces the closure problem. The evolution of observables generally does not constitute a dynamical system. The same is true for the expectation of observables.

11.4 Liouville Equation

Consider continuous-time invertible systems whose flows are given by a set of ODEs $\dot{x}(t) = u(x(t), t)$. In this case, the Jacobian can be calculated from

$$J_t(x) = \exp\left\{\int_0^t dt' (\nabla_x^T u)(S_{t'}(x), t')\right\} \tag{11.21}$$

which together with $S_t(x)$ can be used in formula (11.7) to arrive at a global solution. This approach again requires that S_t is known. Instead, one often works with a local differential version of the conservation of probability, called the Liouville equation.

Liouville equation. One way to derive the Liouville equation is to implement conservation of probability in the form

$$\frac{d}{dt}\int_{V(t)} dx \rho(x, t) = 0 \tag{11.22}$$

where $V(t)$ is the volume that moves with velocity u. Reynolds transport and Gauss theorems from analysis then imply

$$\begin{aligned}\frac{d}{dt}\int_{V(t)} dx \rho(x, t) = &\int_{V(t)} dx \partial_t \rho(x, t) + \int_{\partial V(t)} ds n^T(s) u(s, t) \rho(s, t) \\ = &\int_{V(t)} dx [\partial_t \rho(x, t) + \nabla_x^T [u(x, t) \rho(x, t)]\end{aligned} \tag{11.23}$$

where $\partial V(t)$ is the surface enclosing the volume $V(t)$ with outward normal vector n. The integral on the right hand is zero for all volumes that move with velocity u. This can only be the case if the density satisfies the Liouville equation

$$\partial_t \rho_t(x) + \nabla_x^T (u(x,t)\rho_t(x)) = 0 \tag{11.24}$$

where the time dependence of the density is denoted by a subscript.

A more direct derivation of the Liouville equation is obtained by taking the time derivative of the global solution $\rho_t(x) = J_{-t}(x)\rho_0(S_{-t}(x))$ which results in

$$\frac{D}{Dt}\rho_t(x) = \rho_0(S_{-t}(x))\frac{D}{Dt}J_{-t}(x) \tag{11.25}$$

where

$$\frac{D}{Dt}\rho_t(x) := (\partial_t + u^T\nabla_x)\rho_t(x) \tag{11.26}$$

is the advective time derivative, the time derivative at constant initial position $y = S_{-t}(x)$. The time derivative of volume is given by

$$\frac{d}{dt}J_t(y) = J_t(y)(\nabla^T u)(y) \tag{11.27}$$

which together with $J_t J_{-t} = 1$ leads to

$$\frac{D}{Dt}\rho_t(x) = -\rho_t(x)(\nabla^T u)(x) \tag{11.28}$$

which is the Liouville equation.

If the divergence of the velocity is zero, as for Hamiltonian systems, then

$$\frac{D}{Dt}\rho_t(x) = 0 \tag{11.29}$$

and the solution becomes $\rho_t(x) = \rho_0(S_{-t}(x))$. The density is constant along trajectories.

The Liouville equation is the differential or local formulation of the conservation of probability whereas Eq. (11.7) is the integral or global formulation. Again, note that the Liouville equation is linear in ρ_t although the velocity $u(x)$ may be a highly nonlinear function of x. Stationary densities require $\nabla_x^T(u(x)\rho_t(x)) = 0$. The Liouville equation also governs Markov processes where only the drift term is non-zero.

Example. The over-damped oscillator. Reconsider the over-damped oscillator $\dot{x} = -\gamma x$. Its Liouville equation is given by

$$\partial_t \rho_t(x) - \gamma\rho_t(x) - \gamma x \partial_x \rho_t(x) = 0 \tag{11.30}$$

Stationary densities would have to satisfy

$$x\frac{d\rho_*}{dx} = -\rho_* \tag{11.31}$$

One set of solutions are of the form $\rho_*(x) = a/x$ for $a \in \mathbf{R}$, but these solutions are not densities since they are partly negative and cannot be normalized. The other 'solution' is the delta function, $\rho_*(x) = \delta(x)$, which is also not a density. The over-damped oscillator does not have any stationary densities.

Time evolution of moments. Moments of the densities

$$m_n(t) := \int_{\mathbb{R}} dx x^n \rho_t(x) \tag{11.32}$$

are observables. There thus does not, in general, exist a closed equation for any of these moments. We demonstrate this explicitly for the one-dimensional case (to avoid complicating tensorial notation) and a time-independent velocity field $u(x)$. Time differentiation of (11.32) yields the evolution equation

$$\begin{aligned}\frac{d}{dt}m_n(t) &= \int_{\mathbb{R}} dx x^n \partial \rho_t(x) \\ &= -\int_{\mathbb{R}} dx x^n \partial_x (u(x)\rho_t(x)) \\ &= -\int_{\mathbb{R}} dx \partial_x (x^n u(x)\rho_t(x)) + \int_{\mathbb{R}} dx n x^{n-1} u(x)\rho_t(x)\end{aligned} \tag{11.33}$$

For the sake of argument, disregard the first term which can be converted into a boundary term and assume that the velocity field can be expanded as

$$u(x) = a + bx + cx^2 + \ldots \tag{11.34}$$

Then the evolution of the nth moment is given by

$$\frac{d}{dt}m_n(t) = nam_{n-1}(t) + nbm_n(t) + ncm_{n+1}(t) + \ldots \tag{11.35}$$

and depends on moments other than m_n unless the velocity field is linear. One thus obtains an infinite hierarchy of coupled equations for the moments. To arrive at a finite closed system of equations, one has to truncate this hierarchy by a *closure hypothesis*. Closure at nth order means that all moments of order $n+1$ and higher are expressed in terms of moments of order n and smaller. A first-order closure thus expresses $m_n,\ n \geq 2$ in terms of m_1.

A similar infinite hierarchy of coupled equations is obtained when one marginalizes the density, as is demonstrated in Sect. D.5 for the BBGKY hierarchy in statistical mechanics.

11.5 Stationary Densities

Here, we elaborate on the existence, uniqueness, and properties of stationary densities.

Stationary densities. Stationary densities ρ_* are those for which $\rho_t = \rho_*$ for all times. They do not change under the action of the Frobenius–Perron operator,

$$F_t[\rho_*] = \rho_* \tag{11.36}$$

The stationary densities are fixed points of the Frobenius–Perron operator. Since it is a linear operator, the stationary densities are eigensolutions with eigenvalue one. If a stationary density exists, it implies an invariant probability measure

$$P_*(A) = \int_A \mu(dx)\rho_*(x) \tag{11.37}$$

Conversely, if P_* is an invariant probability measure that has a density, then this density is stationary. This invariant probability measures P_* must be distinguished from the basic (σ-finite) measure μ that allows us to introduce densities. The measure μ may or may not be invariant and may or may not be finite; it only needs to be σ-finite. In all our examples, this measure μ was the Lebesgue measure μ_L. We can, however, make the following statement: if the measure μ is finite and invariant then the uniform density

$$\rho_*(x) = \frac{\chi_X(x)}{\mu(X)} \tag{11.38}$$

is stationary, and conversely, if the uniform density ρ_* is stationary then the measure μ is finite and invariant, and $P_* = \mu$ in both cases.

Ergodic theory studies the properties of such invariant (probability) measures. The invariant measures generally form a convex set with the ergodic measures being the extreme points of this set. A dynamical system is called ergodic if it has only one invariant measure. If one imposes further restrictions on this invariant measure, then one arrives at mixing and exact systems. When applying these results from ergodic theory to the questions of whether or not stationary densities exist and whether or not they are unique, we can only make the statements:

1. There are dynamical systems for which stationary densities do not exist, even when invariant measures exist since they may not have a density. Our over-damped oscillator is an example.

2. An ergodic system can at most have one stationary density since ergodic systems have only one invariant measure, but this again might not have a density. Thus, the uniform density is the only stationary density for the doubling map and baker transformation since they are ergodic systems.

Example: The harmonic oscillator. When the harmonic oscillator is considered on the plane, then there exist many stationary densities. All densities that depend just on the energy are stationary densities. The uniform density is not among them since the Lebesgue measure of the plane is not finite. When one considers the harmonic oscillator on an energy circle, i.e., for just a specific energy value E, then the uniform density on the energy circle is the only stationary density.

11.6 Convergence Toward Stationary Densities

Let us now assume that there exists a unique stationary density ρ_*. Do initial densities converge toward this stationary density and in what sense? Densities are elements of the function space $\mathcal{L}_1$, the space of all integrable functions. The convergence of densities will thus be a convergence in this function space. As is pointed out in the chapter on the stability of dynamical systems, one needs to endow any space with a topology to study such convergence. Here, we introduce a topology in $\mathcal{L}_1$ by defining a scalar product and a norm. The scalar product $\langle g, f \rangle$ is defined between a function $g \in \mathcal{L}_1$ and a function $f \in \mathcal{L}^\infty$ (the space of all bounded functions) by

$$\langle g, f \rangle := \int_X \mu(dx) g(x) f(x) \tag{11.39}$$

and the norm $\| g \|$ of any function $g \in \mathcal{L}_1$ by

$$\| g \| := \int_X \mu(dx) |g(x)| \tag{11.40}$$

If the measure μ is finite, then bounded functions are integrable.

Convergence of integrable functions. Having a scalar product and a norm, one distinguishes three kinds of convergence:

(1) **Cesàro convergence**. A sequence $(g_n)_{n\in\mathbb{N}}$ of integrable functions converges toward g_* in the Cesàro sense if

$$\lim_{N\to\infty} \frac{1}{N} \sum_{n=1}^{N} \langle g_n, f \rangle = \langle g_*, f \rangle \tag{11.41}$$

for all $f \in \mathcal{L}^\infty$.

(2) **Weak convergence**. A sequence $(g_n)_{n\in\mathbb{N}}$ converges weakly if

$$\lim_{n\to\infty} \langle g_n, f \rangle = \langle g_*, f \rangle \tag{11.42}$$

for all $f \in \mathcal{L}^\infty$.

(3) **Point-wise convergence**. A sequence $(g_n)_{n\in\mathbb{N}}$ converges point-wise if

$$\lim_{n\to\infty} \| g_n - g_* \| = 0 \tag{11.43}$$

This is the usual convergence criterion in normed spaces. It is called point-wise convergence because it implies $\lim_{n\to\infty} g_n = g_*$ for almost all x.

Convergence of densities. With these definitions, we can then make the following three statements (for proofs and details, see Mackey 1989 and Lasota and Mackey 1994):

(1) Iff the system is ergodic, then the sequence of densities ρ_n converges toward the uniform density ρ_* in the Cesàro sense.
(2) Iff the system is mixing, then the densities ρ_n converge weakly toward the uniform density ρ_*.
(3) Iff the system is exact, then the densities ρ_n converge point-wise toward the density ρ_*.

Example: Harmonic oscillator. The harmonic oscillator on the energy circle is ergodic. The only invariant measure is given by the uniform density on the energy circle, and it is easily proven that any initial density converges to the uniform density in the Cesàro sense. The harmonic oscillator is neither mixing nor exact.

Example: The baker transformation. The baker transformation is mixing. The only invariant measure is given by the uniform density on its state space. The densities thus converge weakly to this stationary density. The baker transformation is, however, not exact and does not exhibit point-wise convergence. Indeed, one finds $\|\rho_{n+1} - \rho_*\| = \|\rho_n - \rho_*\|$. The distance between any density and the invariant density does not change.

The weak convergence in the case of a mixing system is especially important. Identify g_n with ρ_n and f with an observable, then the scalar product

$$\langle \rho_n, f \rangle = \int_X \mu(dx)\rho_n(x) f(x) = \langle f_n \rangle \tag{11.44}$$

becomes the expectation of the observable at time t and weak convergence implies

$$\lim_{n\to\infty} \langle f_n \rangle = \langle f_* \rangle \tag{11.45}$$

The expectation of an observable converges toward its stationary value. Thus, densities and observables can do things that pure states cannot do, namely converge toward a stationary state even if the system does not have a fixed point. It is tempting to identify this convergence toward the stationary state as the irreversible approach toward equilibrium in statistical mechanics. It only requires the dynamical system to be mixing. A caveat is that arbitrary initial states converge to the stationary state forward and backward in time if the dynamical system is invertible. The time symmetry is not broken. These two points are worth repeating:

Densities can do things that points cannot do. They might converge (weakly) to a stationary state though the dynamical system might not have a fixed point.

If the dynamical system is invertible, then arbitrary initial densities converge toward a stationary density both in the forward and backward time directions.

11.7 Behavior of the Entropy

The convergence of broadened initial states toward stationary densities has been suggested to describe the irreversible approach of non-equilibrium states toward thermal equilibrium. In thermodynamics, such an irreversible approach is accompanied by an increase in entropy. Here, we discuss how the information entropy evolves.

Entropy. Let us recall the properties of the information entropy that are relevant for this discussion. The information entropy is defined by

$$H[\rho] := -\int_X dx \rho(x) \log \rho(x) \tag{11.46}$$

One has to distinguish between densities ρ_* that are stationary and densities ρ_{max} that maximize the entropy functional. If we consider densities on the real line, then there are no densities ρ_{max} that maximize the entropy functional. The entropy is not bounded, neither from below nor from above as exemplified by the Gaussian distribution whose entropy approaches plus infinity for the variance approaching infinity and minus infinity for the variance approaching zero. If we consider densities on an interval of finite length L, then the maximum entropy density is the uniform density $\rho_{max} = 1/L$, or more generally $\rho_{max}(x) = \chi_X(x)/\mu_L(X)$ for state spaces of finite Lebesgue measure. Stationary densities ρ_*, on the other hand, might or might not be maximum entropy densities. The stationary density of the doubling map is the uniform density; the stationary density of the Ulam map is not.

Kullback–Leibler divergence. The entropy of continuous random variables has some non-desirable properties that can be avoided by using the Kullback–Leibler divergence or relative entropy

$$D[\rho_1 \parallel \rho_2] := \int_X dx \rho_1(x) \log \left\{ \frac{\rho_1(x)}{\rho_2(x)} \right\} \tag{11.47}$$

Only this Kullback–Leibler divergence is always well defined and invariant under coordinate transformations. From Gibbs' inequality, it follows that $D[\rho_1 \parallel \rho_2] \geq 0$ and equal to zero only if $\rho_1 = \rho_2$. Note that functions that differ on a set of measure zero represent the same element of $\mathcal{L}_1$. If ρ_2 is the uniform density, then $D[\rho_1 \parallel \rho_2] = H[\rho_2] - H[\rho_1]$ with $H[\rho_2] = -\log \mu_L(X)$ and the Kullback–Leibler divergence is the negative of the entropy, except for a constant..

Time evolution. For the Kullback–Leibler divergence, one can now prove (Voigt 1981):

$$D[F_t[\rho_1] \parallel F_t[\rho_2]] \leq D[\rho_1|\rho_2] \tag{11.48}$$

where F_t is the Frobenius–Perron operator. The relative entropy never increases. Specifically, if $\rho_2 = \rho_*$ one finds that $D[F_t[\rho] \parallel \rho_*]$ is a non-increasing function bounded from below by $D[\rho_* \parallel \rho_*] = 0$.

The entropy does not show such a definite behavior. If the stationary density is the uniform density $\rho_*(x) = 1/\mu_L(X)$, then one finds that $H[F_t(\rho)] \geq H[\rho]$. The entropy does not decrease. The stationary density of the Ulam map is not the uniform density. There one finds that the entropy increases or decreases, depending on whether one starts from a density with a lower entropy or higher entropy than that of the stationary density.

However, and this is a big 'however': If the Frobenius–Perron operator is invertible, then one also has

$$\begin{aligned} D[F_{t+t'}[\rho_1] \parallel F_{t+t'}[\rho_2]] &= D[F_t \circ F_{t'}[\rho_1] \parallel F_t \circ F_{t'}[\rho_2]] \\ &\leq D[F_t[\rho_1] \parallel F_t[\rho_2]] \\ &\leq D[\rho_1 \parallel \rho_2] \end{aligned} \tag{11.49}$$

for all t, t'. Choose $t' = -t$ and one obtains

$$D[\rho_1 \parallel \rho_2] \leq D[F_t[\rho_1] \parallel F_t[\rho_2]] \leq D[\rho_1 | \rho_2] \tag{11.50}$$

for all t. The Kullback–Leibler divergence does not change under the action of an invertible Frobenius–Perron operator.

This important result can also explicitly be proven for an invertible dynamical system that is given by an ODE $\dot{x} = u(x, t)$. The evolution of the density is then given by the Liouville equation. Let ρ_2 be the stationary density ρ_*. Write

$$D[\rho_t \parallel \rho_*] = \int_X dx \rho_t(x) \left\{ \log \left[\frac{\rho_t(x)}{\rho_*(x)} \right] + \frac{\rho_*(x)}{\rho_t(x)} - 1 \right\} \tag{11.51}$$

where the two added terms cancel upon integration. Then one has

$$\begin{aligned} \frac{d}{dt} D[\rho_t \parallel \rho_*] &= \int_X dx \partial_t \rho_t \log \frac{\rho_t}{\rho_*} \\ &= -\int_X dx \nabla^T (u \rho_t) \log \frac{\rho_t}{\rho_*} \\ &= \int_X dx \frac{\rho_*}{\rho_t} \nabla^T (\rho_t / \rho_*)(u \rho_t) \\ &= -\int dx \nabla^T (u \rho_*) \frac{\rho_t}{\rho_*} \\ &= 0 \end{aligned} \tag{11.52}$$

upon two integrations by parts and the fact that ρ_* is stationary.

The situation for the entropy is more complicated. If the flow is non-divergent and invertible, then $\rho_t(x) = \rho_0(S_{-t}(x))$ and one obtains the stronger result

$$\begin{aligned} H[\rho_t] &= -\int_X dx \rho_t(x) \log \rho_t(x) \\ &= -\int_X dx \rho_0(S_{-t}(x)) \log \rho_0(S_{-t}(x)) \\ &= H[\rho_0] \end{aligned} \tag{11.53}$$

upon variable transformation and $J_t(x) = 1$ for non-divergent systems. However, when the flow is divergent, then the entropy changes. One, for example, finds $H[\rho_t] = H[\rho_0] + \gamma t$ for the over-damped oscillator.

Thus, we have the basic result that the Kullback–Leibler divergence $D[\rho_t \parallel \rho_*]$ remains constant for invertible ergodic and mixing systems though the densities converge in the Cesàro or weak sense to the stationary density. If the flow is additionally non-divergent, then we have the stronger result that $H[\rho_t]$ remains constant. These results cause problems for the foundations of non-equilibrium statistical mechanics as discussed in Chap. 19. This is an important point:

Invertible dynamical systems do not change their entropy. Random initial conditions do not break time symmetry.

Evolution for exact systems. The Kullback–Leibler divergence can only decrease for non-invertible dynamical systems. For exact systems, ρ_t converges point-wise to ρ_*. Therefore,

$$\lim_{t\to\infty} D[F_t[\rho] \| \rho_*] = 0 \tag{11.54}$$

The converse is also true. If (11.54) holds the system is exact. In the special situation where ρ_* is the uniform density,

$$\lim_{t\to\infty} H(F_t[\rho]) = H[\rho_*] \tag{11.55}$$

Discrete state space. Although this chapter exclusively dealt with probability measures that are given by densities which are elements of $\mathcal{L}_1$, the above results for the entropy also carry over to systems where the state space is discrete with a finite number K of elements. Then the entropy is given by

$$H[p] = -\sum_{k=1}^{K} p(k) \log p(k) \tag{11.56}$$

and bounded from below by zero and from above by $\log K$. If S_t is an invertible system, then it represents a permutation of the state space and the entropy does not change, $H[p_t] = H[p_0]$. If the system is non-invertible, then at least two different points must map onto a single point. Collecting all the points that map into the same point into a bin represents a partition of the state space and the binning property implies $H[p_t] \geq H[p_0]$. The entropy cannot decrease.

11.8 Summary

Broadened or random initial conditions introduce probability concepts into the dynamics but in a fairly trivial manner. Each initial condition evolves according to the dynamical law without being distorted by some random influences. Only the starting point, the initial condition, is random. Once a random initial condition is realized, the evolution is deterministic. Formally, broadened initial conditions are defined by probability density functions. These densities evolve in time according to a linear operator, called the Frobenius–Perron operator. If the dynamical system is governed by an ODE, the operator is the solution of the Liouville equation. Stationary densities are fixed points of the Frobenius–Perron operator. In the case that there exists a unique stationary density, we gave necessary and sufficient conditions for the convergence of arbitrary initial densities toward this stationary density. This convergence toward stationary densities is a distinct feature of broadened initial states. While pure states might endlessly wander around in state space, broadened states might converge. Though the convergence of broadened initial conditions toward a stationary density has been linked to the irreversible approach of non-equilibrium states toward equilibrium in statistical dynamics, one has to deal with the fact that broadened initial conditions converge toward the stationary density both forward and backward in time for invertible dynamical systems. For such invertible systems, the Kullback–Leibler divergence also does not change in time. Broadened initial conditions cannot break the time symmetry of a dynamical system. Only when the system is exact, the Kullback–Leibler divergence converges to zero.

References

Lasota, A., Mackey, M.C.: Chaos, Fractals, and Noise; Stochastic Aspects of Dynamics. Springer (1994)

Mackey, M.C.: The dynamic origin of increasing entropy. Rev. Mod. Phys. **61**(4) (1989)

Voigt, J.: Commun. Math. Phys. **81**, 31 (1981)

Chapter 12
Stochastic Processes Generated by Observables

12.1 Introduction

Broadened or random initial conditions inject randomness into dynamical systems. When only the time evolution of the dynamical system is considered, this randomness is of a very limited kind. The time evolution of random initial conditions remains deterministic and is given by the linear Frobenius–Perron operator. This situation changes when the evolution of observables is considered which may be many-to-one functions. An initial value of the observable might then not determine uniquely the future values of this observable. The time evolution becomes non-deterministic. When endowed with random initial conditions, this non-deterministic evolution of the observable results in a stochastic process. Observables generate stochastic processes. This is indeed a common way to generate stochastic processes in practice. The converse is also true. To each stochastic process, there exists a dynamical system such that an appropriate observable and an appropriate random initial condition reproduce the stochastic process. We first explicate these assertions formally and then consider two main examples: (1) coarse-graining where some emphasis is put on the evolution of the entropy and (2) the elimination of variables, especially the elimination of fast-responding variables, resulting in equations of the Langevin type.

12.2 Relation Between Stochastic Processes and Dynamical Systems

Here, we look in a formal manner first at the degenerated stochastic process associated with a dynamical system, then at the genuine stochastic processes generated by observables, and finally at the dynamical system associated with a stochastic process.

Degenerated stochastic process generated by a dynamical system. When random initial conditions are introduced, the state space of the dynamical system becomes a probability space $\{X, \mathcal{A}, P_0\}$. The initial measure P_0 of these random

P. Müller, *Handbook of Dynamics and Probability*,
https://doi.org/10.1007/978-3-030-88486-4_12

initial conditions may be thought of as the induced measure of a random variable f_0. Then $(S_t \circ f_0)$ becomes a random variable as well and the family $(S_t \circ f_0)_{t\in T}$ a stochastic process which needs to be characterized by all its N-point joint probabilities, not just by its one-point probability whose evolution is governed by the Frobenius–Perron operator, in the case that the probability is given by a density. These N-point joint probabilities can, however, be constructed in a rather trivial manner. Again in terms of densities, one finds for the two-point joint density

$$\begin{aligned}\rho_{t_2,t_1}(x_2, x_1) &= \langle \delta(x_2 - S_{t_2} \circ f_0)\delta(x_1 - S_{t_1} \circ f_0)\rangle \\ &= \langle \delta(x_2 - S_{t_2-t_1} \circ S_{t_1} \circ f_0)\delta(x_1 - S_{t_1} \circ f_0)\rangle \\ &= \delta(x_2 - S_{t_2-t_1}(x_1))\langle \delta(x_1 - S_{t_1} \circ f_0)\rangle \\ &= \delta(x_2 - S_{t_2-t_1}(x_1))\rho_{t_1}(x_1)\end{aligned} \tag{12.1}$$

The joint probability density of finding a particle at time t_2 at position x_2 when it was at time t_1 at position x_1 is infinite when x_1 and x_2 are on the same trajectory and zero otherwise. This result reflects the fact that the time evolution remains deterministic. Corresponding results are true for the higher order joint densities. The conditional density

$$\rho_{t_2|t_1}(x_2|x_1) = \delta(x_2 - S_{t_2-t_1}(x_1)) \tag{12.2}$$

is thus a 'deterministic' transition density. The evolution of broadened initial conditions is a Markov process albeit with this 'deterministic' transition probability. The Frobenius–Perron operator is the Markov operator of this 'deterministic' Markov process.

Stochastic processes generated by observables. The situation becomes more interesting when the evolution of observables is considered. Any (measurable) observable $A : X \mapsto Y$ induces

$$(f_t^A)_{t\in T} := (A \circ S_t)_{t\in T} \tag{12.3}$$

which is also a stochastic process but generally less trivial than the stochastic process $(S_t)_{t\in T}$.

To see this, consider, as in Sect. 2.6 on observables, the 'function'

$$R_t^A(y) := f_t^A \circ A^{-1}(y) \tag{12.4}$$

for $y \in Y$. If A were a bijection, then R_t^A would represent a dynamical system equivalent to S_t. If an initial value y_0 of the observable were realized, its value y_t at a later time would be determined. The situation becomes different when the observable is a many-to-one function. Then there are many points in the pre-image $A^{-1}(y_0)$. They all move according to the dynamics and might end up at time t at points where the values of the observable are different (see Fig. 2.8). If they do, then the value of the observable at time t is not determined by the value at time zero. The function R_t^A becomes a multi-valued function. Multi-valued functions are not proper

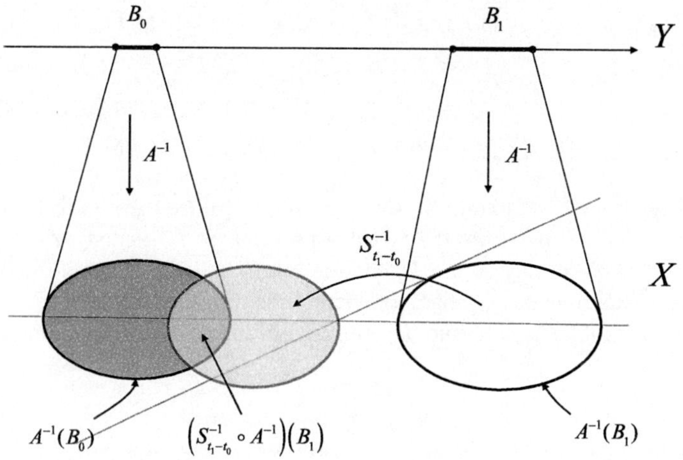

Fig. 12.1 Construction of the argument of the joint probability $P_0((f^A_{t_1})^{-1}(B_1) \cap A^{-1}(B_0))$

functions. Properly, R^A_t is a function from the power set $\mathcal{P}(Y)$ into the power set $\mathcal{P}(Y)$. For such multi-valued functions, one can, however, determine the probability that the observable has a value $y_1 \in B_1$ at time t_1 when it had the value in $y_0 \in B_0$ at time $t_0 = 0$, which is given by

$$P_{t_1,t_0}(B_1 \times B_0) = P_0((S^{-1}_{t_1-t_0} \circ A^{-1})(B_1) \cap A^{-1}(B_0)) \tag{12.5}$$

and shown in Fig. 12.1. The conditional probability $P_{t_1|t_0}(B_1|B_0)$ may thus be smaller than $P_0(A^{-1}(B_0))$ since not all points in B_0 map to points in B_1. The family $(f^A_t)_{t\in T}$ becomes a non-trivial stochastic process.

The canonical probability space of the stochastic process $(f^A_t)_{t\in T}$ is thus $\{\mathbb{R}^T, \mathcal{B}^T, P_h\}$ where P_h is the joint probability defined by all marginal distributions

$$P_{t_N,\ldots,t_0}(B_{t_N} \times \cdots \times B_{t_0}) = P_0((f^A_{t_N})^{-1}(B_{t_N}) \cap \cdots \cap f^{-1}_{t_0}(B_{t_0})) \tag{12.6}$$

Details will depend on the dynamical system, the observable, and the initial probability distribution. Here and in the following, we make use of the fact that it is sufficient to consider the Cartesian product of Borel sets since these generate the complete σ-algebra.

In Sect. 2.6, we called the fact that R^A_t might be multi-valued the closure problem. The function R^A_t does not represent a deterministic dynamical system. By introducing random initial conditions, we convert this indeterminate evolution into a stochastic evolution, a stochastic process. This conversion is often called a stochastic closure.

If the initial probability measure P_0 is an invariant measure, then the stochastic process is stationary since

$$
\begin{aligned}
P_{t_N,\dots,t_0}(B_N \times \cdots \times B_0) &= P_0((f_{t_N}^A)^{-1}(B_N) \cap \cdots \cap (f_{t_0}^A)^{-1}(B_0)) \\
&= P_0(S_{\Delta t}^{-1}[(f_{t_N}^A)^{-1}(B_N) \cap \cdots \cap (f_{t_0}^A)^{-1}(B_0)]) \\
&= P_0((f_{t_N+\Delta t}^A)^{-1}(B_N) \cap \cdots \cap (f_{t_0+\Delta t}^A)^{-1}(B_0)) \\
&= P_{t_N+\Delta t,\dots,t_0+\Delta t}(B_N \times \cdots \times B_0) \qquad (12.7)
\end{aligned}
$$

The first equality is the definition, the second one follows from the invariance of P_0, the third one follows from the fact that pre-images f^{-1} satisfy $f^{-1}(A \cap B) = f^{-1}(A) \cap f^{-1}(B)$ for any A and B, and the fourth equality is again the definition. The joint probabilities do not change under a time translation.

If the initial probability measure P_0 is an ergodic measure of the dynamical system, then

$$
\lim_{N\to\infty} \frac{1}{N} \sum_{n=0}^{N-1} f_n^A = \langle A \rangle \qquad (12.8)
$$

almost surely. The stochastic process $(f_n^A)_{n\in\mathbb{N}}$ satisfies the strong law of large numbers and Eq. (12.8) is called the ergodic form of the law of large numbers.

Dynamical system associated with stochastic process. Every dynamical system and an observable generate a stochastic process. The converse is also true. To every stochastic process $(f_n)_{n\in\mathbb{N}}$, there exists a dynamical system $(S_n)_{n\in\mathbb{N}}$, an observable A, and an initial probability distribution P_0 such that $(f_n)_{n\in\mathbb{N}} = (f_n^A)_{n\in\mathbb{N}}$ where f_n is the stochastic process and f_n^A the observable. To see this, introduce for a given stochastic process $(f_n)_{n\in\mathbb{N}}$ the random variable

$$
\begin{aligned}
h : \Omega &\mapsto \mathbb{R}^{\mathbb{N}} \\
\omega &\mapsto h(\omega) = (f_0(\omega), f_1(\omega), \dots) \qquad (12.9)
\end{aligned}
$$

which has the distribution P_h. Next consider the shift map

$$
\begin{aligned}
T_+ : \mathbb{R}^{\mathbb{N}} &\mapsto \mathbb{R}^{\mathbb{N}} \\
(x_0, x_1, x_2, \dots) &\mapsto (x_1, x_2, x_3, \dots) \qquad (12.10)
\end{aligned}
$$

and its nth iterate T_+^n. Next take as the observable A the projection onto the first component

$$
\begin{aligned}
A : \mathbb{R}^{\mathbb{N}} &\mapsto \mathbb{R} \\
(x_0, x_1, x_2, \dots) &\mapsto x_0 \qquad (12.11)
\end{aligned}
$$

We then obtain

$$\begin{aligned} f_n^A(h(\omega))) &= A(S_n(f_0(\omega), f_1(\omega), \ldots)) \\ &= A(f_n(\omega), f_{n+1}(\omega), \ldots) \\ &= f_n(\omega) \end{aligned} \tag{12.12}$$

which shows that every stochastic process can be thought of as being generated by an observable of a dynamical system.

If $(f_n)_{n\in\mathbb{N}}$ is a stationary process, then P_h is an invariant measure of the shift map, $P_h(T_+^{-1}(B) = P_h(B)$ for all $B \in \mathcal{B}^{\mathbb{N}}$. It is sufficient to show this for cylinder sets. Since $T_+^{-1}(C(B_0, \ldots, B_N)) = \mathbb{R} \times B_) \times \cdots \times B_N \times \mathbb{R} \times \ldots$ is also a cylinder set, we have indeed

$$\begin{aligned} P_h(T_+^{-1}(C(B_0, B_1, \ldots, B_N)) &= P_h(\mathbb{R} \times B_0 \times B_1 \times \cdots \times B_N \times \mathbb{R} \times \ldots) \\ &= P_0(f_0^{-1}(\mathbb{R}) \cap f_1^{-1}(B_0) \cap \cdots \cap f_{N+1}^{-1}(B_N) \cap f_{N+2}^{-1}(\mathbb{R}) \cap \ldots) \\ &= P_0(\Omega \cap f_1^{-1}(B_0) \cap \cdots \cap f_{N+1}^{-1}(B_N) \cap \Omega \cap \ldots) \\ &= P_0(f_1^{-1}(B_0) \cap \ldots) \cap f_{N+1}^{-1}(B_N)) \\ &= P_0(f_0^{-1}(B_0) \cap \ldots f_N^{-1}(B_N) \ldots) \\ &= P_h(C(B_0, \ldots, B_N)) \end{aligned} \tag{12.13}$$

where the second to last equality follows from the stationarity of the stochastic process.

If the measure P_h is an ergodic measure under the shift map, then we call the stochastic process ergodic. In this case, any observable $A : \mathbb{R}^{\mathbb{N}} \mapsto \mathbb{R}$ satisfies

$$\lim_{N\to\infty} \frac{1}{N} \sum_{n=0}^{N-1} f_n^A = \langle A \rangle \tag{12.14}$$

where $f_n^A = A \circ T_+^n$. If one chooses for A the projection onto the first component, then f_n^A becomes the stochastic process f_n and one finds that the stochastic process satisfies

$$\lim_{N\to\infty} \frac{1}{N} \sum_{n=0}^{N-1} f_n = \langle f_0 \rangle \tag{12.15}$$

An important application of these findings is the insight that the probabilistic strong law of large numbers can be viewed as a special case of this ergodic law. The probabilistic strong law of large numbers requires as sufficient conditions that the sequence f_n of random variables represents an iid process with finite mean. For such an iid process, the probability distribution of h is given by the product probability distribution $P_h = P_f \cdot P_f \cdot \ldots$. This product measure is an invariant and ergodic measure of the shift map T_+ and hence (12.15) follows.

If one chooses as the observable

$$\begin{aligned} A : \mathbb{R}^{\mathbb{N}} &\mapsto \mathbb{R} \\ x_0, x_1, \ldots &\mapsto x_0 \ldots x_K \end{aligned} \tag{12.16}$$

then

$$\begin{aligned} f_n^A(h(\omega)) &= A \circ S_+^n(f_0(\omega), f_1(\omega), \ldots) \\ &= A \circ (f_n(\omega), f_{n+1}(\omega), \ldots \\ &= f_n(\omega) f_{n+1}(\omega) \ldots f_{n+K}(\omega) \end{aligned} \tag{12.17}$$

and the stochastic process f_n satisfies

$$\lim_{N\to\infty} \frac{1}{N} \sum_{n=0}^{N-1} f_n f_{n+1} \ldots f_{n+K} = \langle f_0 f_1 \ldots f_K \rangle \tag{12.18}$$

almost surely.

12.3 Coarse-Graining

Coarse-graining is based on a partition of the state space. The subsets of the partition are often called bins, boxes, groups, or cells. In general, many trajectories emanate from one given box. They might end up in different boxes. Only knowing that the initial condition is in a given box is then not sufficient to determine in which box the system will end up. If the initial conditions are randomly distributed, then at most what we can expect is a transitional probability of going from one box to another box. This is indeed the case.

The index function. Given a dynamical system $(S_t)_{t\in T}$, broadened initial conditions with probability measure P_0 and a partition $X = \cup_{k=1}^{\infty} B_k$, one introduces the index function

$$I(x) := \sum_{k=1}^{\infty} k\chi_{B_k}(x) \tag{12.19}$$

The index function assigns to each point x the index of the box that x occupies. The index function is an observable, $I : X \mapsto \mathbb{N}$. For any point x and any ordered sequence of times $t_1 \ldots < t_N$ with $N \in \mathbb{N}$, the associated observable $f_t^I(x) = I(S_t(x))$ thus generates a sequence

$$x \to k_1, k_2, \ldots, k_N \tag{12.20}$$

where $k_n = I(S_{t_n}(x))$ is the index of the box that $S_{t_n}(x)$ occupies. If we consider broadened initial conditions with measure P_0, then the joint probabilities of these sequences are given by

$$p_{t_N,\ldots,t_1}(k_N,\ldots,k_1) = P_0(S_{t_N}^{-1}(B_{k_N}) \cap \cdots \cap S_{t_1}^{-1}(B_{k_1})) \tag{12.21}$$

This set of joint probabilities satisfies the compatibility conditions such as

$$p_{t_{N-1},\ldots,t_1}(k_{N-1},\ldots,k_1) = \sum_{k_N=1}^{\infty} p_{t_N,\ldots,t_1}(k_N,\ldots,k_1) \tag{12.22}$$

The set of all joint probabilities thus defines a probability measure P_h on the sample space $\mathbb{N}^T$ with the associated σ-algebra being the power set $\mathcal{P}(\mathbb{N}^T)$. One thus has arrived at a stochastic process and can investigate its properties. This is the first important result. Coarse-graining generates a stochastic process. The stochastic process depends on the partition and on the initial probability measure P_0.

Example: Bernoulli chain. Consider the doubling map and the partition $B_0 = [0, 1/2)$ and $B_1 = [1/2, 1)$. Then $k_n = 0$ if $x_n \in B_0$ and $k_n = 1$ if $x_n \in B_1$. Each initial value thus generates a sequence $k_1, \ldots, k_N, \ldots$. If P_0 is the uniform measure, we obtain

$$p_{N,\ldots,1}(k_N,\ldots,k_1) = 2^{-N} \tag{12.23}$$

independent of $(k_N,\ldots,k_1)$ (see Fig. 12.2). We thus arrived at the Bernoulli chain with $p(0) = p(1) = 1/2$ which is an iid process. A realization of this process is obtained by choosing an initial value x_0 and constructing the associated sequence. Random number generator on a computer often uses such algorithms.

Example: Golden mean process. A special case of the map $F(x) = \alpha x \ mod\ 1$ defined on the interval $I = [0, 1[$ is the golden mean map where $\alpha = (1+\sqrt{5})/2$ is the golden mean, satisfying $\alpha - 1 = 1/\alpha$. Use the partition $B_0 = [0, 1/\alpha[$ and $B_1 = [1/\alpha, 1[$. One then finds that $F(B_0) = I$ and $F(B_1) = B_0$ since $\alpha - 1 = 1/\alpha$ (see Fig. 12.3). Hence, if one arrives in the interval B_1 the next iteration will give a point in B_0. The symbolic sequence does not have two consecutive 1s in it. Using this partition and the uniform initial probability distribution, one finds (see Fig. 12.2)

$$p_{N,\ldots,1}(k_N,\ldots,k_1) = \pi(k_N|k_{N-1})\pi(k_{N-1}|k_{N_2})\ldots\pi(k_2|k_1)p_*(k_1) \tag{12.24}$$

where

$$\pi(k|l) = \begin{pmatrix} \frac{1}{\alpha} & 1 \\ \frac{1}{\alpha^2} & 0 \end{pmatrix} \tag{12.25}$$

and

$$p_* = (\frac{1}{\alpha}, \frac{1}{\alpha^2})^T \tag{12.26}$$

We thus arrive at the golden mean process, discussed in the section on Markov chains.

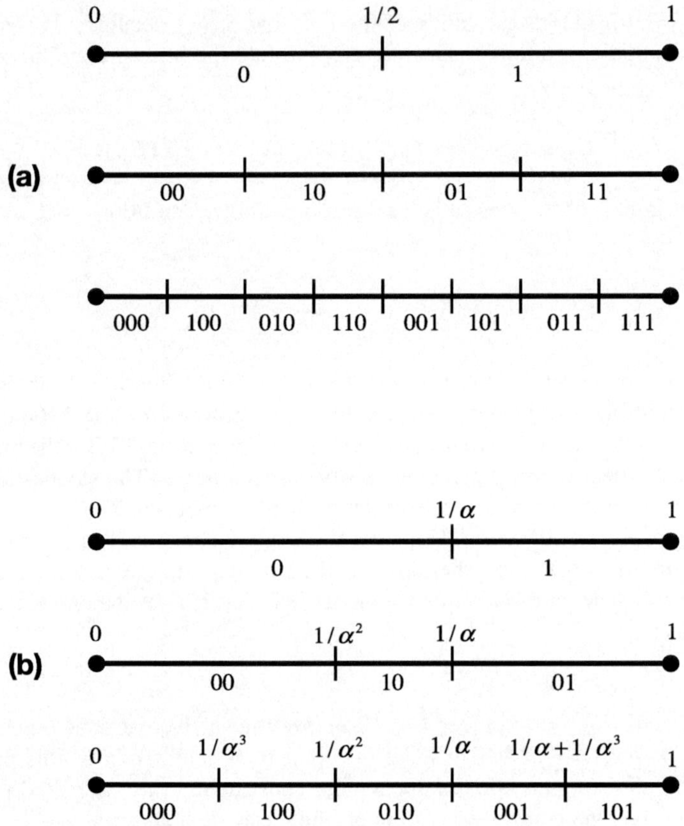

Fig. 12.2 The first three generations of initial conditions on the interval [0, 1[that lead to the indicated sequences for **a** the Bernoulli chain and **b** for the golden mean map

Behavior of entropy. Let us assume that the measure P_0 of the broadened initial conditions is given by a density ρ_0 so that we can apply the results of the chapter on broadened initial conditions. This initial density evolves as $\rho_t = F_t[\rho_0]$ where F_t is the Frobenius–Perron operator. Once we coarse-grain the state space, we can introduce the probability that the system is in box k at time t, given by

$$p_t(k) = P_0(S_t^{-1}(B_k)) = \int_{S_t^{-1}(B_k)} \mu(dx)\rho_0(x)$$
$$= \int_{B_k} \mu(dx)\rho_t(x) \quad k = 1, 2, \ldots \tag{12.27}$$

and the coarse-grained density

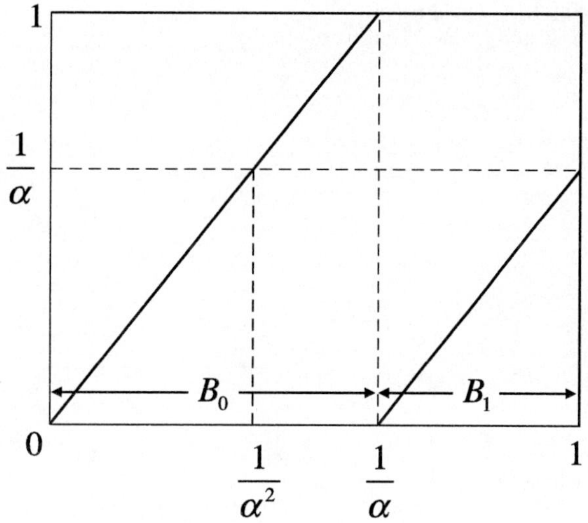

Fig. 12.3 The golden mean map

$$\rho_t^{cg}(x) := \sum_{k=1}^{\infty} \frac{p_t(k)}{\mu(B_k)} \chi_{B_k}(x) \tag{12.28}$$

Assume the dynamical system to be invertible and mixing. Mixing implies that there exists a unique stationary density ρ_* and the weak convergence

$$\lim_{t\to\infty} \int \mu(dx)\rho_t(x)g(x) = \int \mu(dx)\rho_*(x)g(x) \tag{12.29}$$

for all bounded measurable functions g. Choose $g(x) = \chi_{B_k}(x)$ and one finds

$$\lim_{t\to\infty} p_t(k) = \int_{B_k} \mu(dx)\rho_*(x) =: p_*(k) \tag{12.30}$$

$$\lim_{t\to\infty} \rho_t^{cg} = \rho_*^{cg} \tag{12.31}$$

Both probabilities converge point-wise to their respective stationary states. What does this result imply for the entropy?

The situation is most clear-cut for the Kullback–Leibler divergences which satisfy

$$D[\rho_1 \parallel \rho_2] \geq D[p_1 \parallel p_2] = D[\rho_1^{cg} \parallel \rho_2^{cg}] \tag{12.32}$$

and for whose evolution we find

$$\lim_{t\to\infty} D[p_t \parallel p_*] = \lim_{t\to\infty} D[\rho_t^{cg} \parallel \rho_*^{cg}] = 0 \tag{12.33}$$

Fig. 12.4 Evolution of Kullback–Leibler divergence

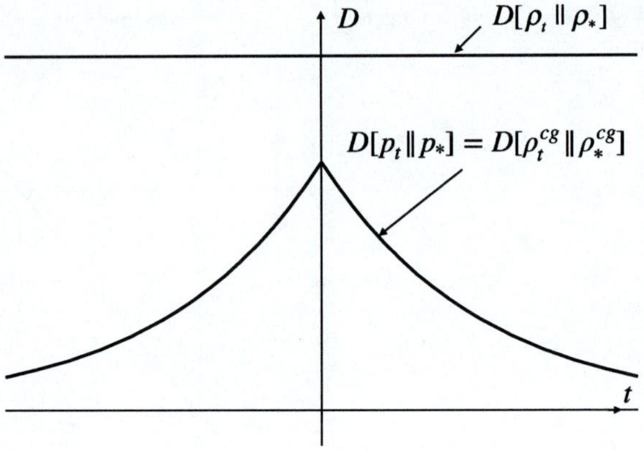

whereas

$$D[\rho_t \parallel \rho_*] = D[\rho_0 \parallel \rho_*] \tag{12.34}$$

from Sect. 11.7. Coarse-graining implies an approach toward a stationary state, even for invertible dynamical systems. However, since the system is invertible, all the above equations also hold in the limit $t \to -\infty$. The divergence approaches zero in both time directions. This situation is depicted in Fig. 12.4. Thus, coarse-graining does not help with explaining the second law of thermodynamics, as we will discuss further in Chap. 19. Rather the state at $t = 0$ has to be viewed as a fluctuation or an extreme.

The situation for the entropy is more muddled. We have the inequality

$$H[\rho^{cg}] \geq H[\rho] \tag{12.35}$$

but cannot relate these entropies to the entropy $H[p]$ since $H[p]$ is scale-invariant, whereas the entropy of continuous random variables is not. A second problem is that $H[\rho]$ is not necessarily constant if the flow is divergent. There are contributions to the entropy arising from volume changes which are pointed out in Sect. 11.7 for the over-damped oscillator. We do have the limits

$$\lim_{t\to\infty} H[p_t] = H[p_*] \tag{12.36}$$

$$\lim_{t\to\infty} H[\rho_t^{cg}] = H[\rho_*^{cg}] \tag{12.37}$$

but can, in general, not say how these entropies approach their final values. The situation becomes more structured when the stationary state is the density that maximizes the entropy, the uniform density for a state space with finite Lebesgue measure. This is the case for the baker transformation where the entropy increases toward its equilibrium value.

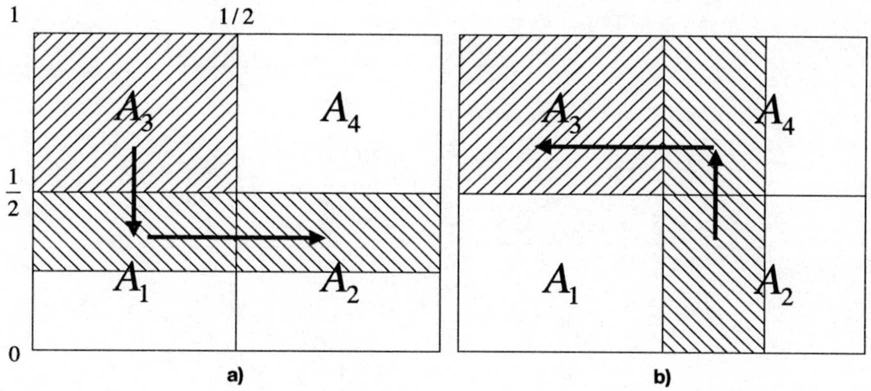

Fig. 12.5 Baker transformation

Example: Baker transformation. From the previous chapter, we have

$$H[\rho_*] = 0 \tag{12.38}$$

$$H[\rho_t] = H[\rho_0] \tag{12.39}$$

Consider the partition $A_1, \ldots, A_4$ of the unit square as depicted in Fig. 12.5.

Start out with

$$\rho_0(x, y) = \begin{cases} 4 \text{ if } (x, y) \in A_3 \\ 0 \text{ if } (x, y) \in A_1, A_2, A_4 \end{cases} \tag{12.40}$$

A forward iteration moves the right-hatched area A_3 into the left-hatched areas in (a) and a backward iteration moves the left-hatched areas into the right-hatched area A_3 in (b). We then have

$$p_0(k) = \begin{cases} 1 \text{ if } k = 3 \\ 0 \text{ if } k = 1, 2, 4 \end{cases} \tag{12.41}$$

and

$$p_1(k) = \begin{cases} 1/2 \text{ if } k = 1, 2 \\ 0 \quad\text{ if } k = 3, 4 \end{cases} \tag{12.42}$$

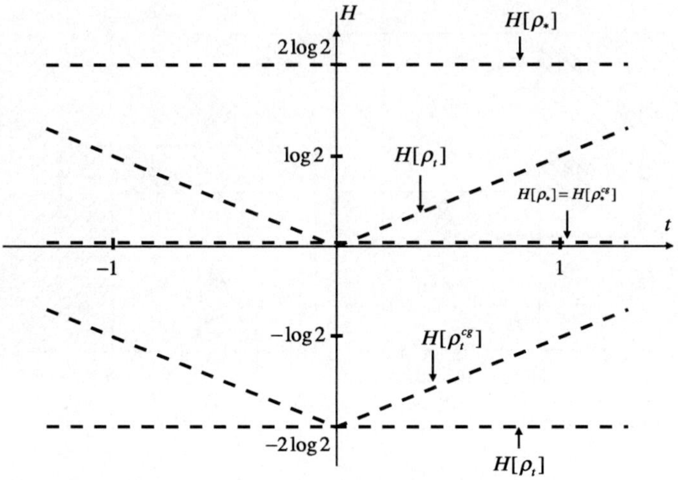

Fig. 12.6 Evolution of entropies $H[\rho_t]$, $H[p_t]$ and $H[\rho_t^{cg}]$ for the baker transformation

$$p_{-1}(k) = \begin{cases} 1/2 \text{ if } k = 2, 4 \\ 0 \quad \text{if } k = 1, 3 \end{cases} \tag{12.43}$$

and hence $H[p_0] = 0$ and $H[p_{-1}] = H[p_1] = \log 2$ and $H[p_*] = 2\log 2$. The entropy thus increases in both time directions toward the equilibrium value. Similarly, we find $H[\rho_0^{cg}] = -2\log 2$ and $H[\rho_1^{cg}] = H[\rho_{-1}^{cg}] = -\log 2$ and $H[\rho_*^{cg}] = 0$. Again, the entropy increases toward the equilibrium value in both time directions. These changes in entropy are depicted in Fig. 12.6.

There are many more examples that demonstrate the approach of the box probabilities and the coarse-grained densities toward their respective equilibrium densities and the accompanying increase of entropy, in both time directions in the case of invertible dynamical systems. This result also holds when the dynamics are not deterministic but stochastic as the following example of marbles in two bowls demonstrates.

Example: Marbles in two bowls. Consider K white marbles and K black marbles in two bowls such that K marbles are in each bowl. The dynamics is to take randomly one marble from each bowl and exchange them. In the language of statistical mechanics, the micro-state of the system is determined by which marble is in which bowl. There are $(2K)!/(K!K!)$ micro-states, each of them assumed to have the same probability. Each of these micro-states turns into K^2 different micro-states with equal probability. The one-step transition probability of going from one of these micro-states A to another micro-state B is thus

$$\pi^{micro}(A|B) = \begin{cases} \frac{1}{K^2} \text{ if B can be reached from A} \\ 0 \quad \text{otherwise} \end{cases} \tag{12.44}$$

The stochastic dynamics represents a stationary homogeneous Markov chain with the uniform distribution as the stationary one-point probability. If B can be reached from A, then A can be also reached from B. We thus have

$$\pi^{micro}(B|A) = \pi^{micro}(A|B) \tag{12.45}$$

The microscopic dynamics is in detailed balance.

The macro-state can be characterized by the number k of white marbles in bowl 1. Then there are $K-k$ black marbles in bowl 1 and $K-k$ white and k black marbles in bowl 2. To proceed, we need to find out in how many ways one can distribute the K different white marbles into two bowls such that k marbles are in the first bowl and $K-k$ marbles in the second bowl. This is a problem of combinatorics. Unordered selections of k objects from K objects are called k-combinations and their number is given by

$$\binom{K}{k} = \frac{K!}{k!(K-k)!} \tag{12.46}$$

Applying the same counting to the black marbles, the number of possible micro-states for the macro-state k is

$$C(k) = \left(\frac{K!}{k!(K-k)!}\right)^2 \tag{12.47}$$

Since there are $(2K)!/(K!K!)$ micro-states, the probability of the macro-state is given by

$$p_*^{macro}(k) = \left(\frac{K!}{k!(K-k)!}\right)^2 \frac{K!K!}{(2K)!} \tag{12.48}$$

This probability distribution has its largest value at $k_* = K/2$, the expected result. The value k_* may be interpreted as the equilibrium value, in the spirit of Boltzmann when applied to statistical mechanics.

Next consider the conditional probabilities to move from the macro-state k_n to the macro-state k_{n+1} which are given by

$$\pi^{macro}(k_{n+1}|k_n) = \begin{cases} \frac{(K-k_n)^2}{K^2} & \text{for } k_{n+1} = k_n + 1 \\ 2\frac{(K-k_n)k_n}{K^2} & \text{for } k_{n+1} = k_n \\ \frac{k_n^2}{K^2} & \text{for } k_{n+1} = k_n - 1 \end{cases}$$

and zero otherwise. These probabilities can be rationalized as follows: In order to go from k_n to $k_{n+1} = k_n + 1$, one has to take one black marble out of bowl 1. The probability for this is $(K-k_n)/K$ since there are $(K-k_n)$ black marbles in bowl 1. Simultaneously, one has to take one white marble out of bowl 2. The probability for this is $(K-k_n)/K$ since there are $(K-k_n)$ white marbles in bowl 2. The other probabilities follow by the same reasoning. The macro-process is a homogeneous Markov chain, also in detailed balance. Especially

$$\frac{\pi^{macro}(k_n + 1|k_n)}{\pi^{macro}(k_n - 1|k_n)} = \frac{(K - k_n)^2}{k_n^2} \tag{12.49}$$

If $k_n < K/2$, then this ratio is larger than one. It is thus more probable that the system will move toward the equilibrium value $k_* = K/2$ than away from it. This tendency is the more pronounced the larger N is and the further away from equilibrium you are. If $k_n > K/2$ then this ratio is smaller than one and again it is more probable that one moves toward the equilibrium value than away from it.

Consider where the system is most likely coming from. Using the generally valid formula

$$p_{n-1|n}(k_{n-1}|k_n) = \frac{p_{n|n-1}(k_n|k_{n-1})p_{n-1}(k_{n-1})}{p_n(k_n)} \tag{12.50}$$

we again find the ratio

$$\frac{p_{n-1|n}(k_n + 1|k_n)}{p_{n-1|n}(k_n - 1|k_n)} = \frac{(K - k_n)^2}{k_n^2} \tag{12.51}$$

upon substitution of $p_{n|n-1} = \pi^{macro}$ and $p_n = p_*^{macro}$. Again, if $n_k < N/2$, this ratio is larger than one. It is thus more probable that the system came from $k_{n-1} = k_n + 1$ than from $k_{n-1} = k_n - 1$. The state k_n is thus most likely an extreme. It came most likely from a state closer to equilibrium and it will move most likely move toward a state closer to equilibrium. We find that the system is fluctuating about its equilibrium state. This is an important point:

It does not matter whether or not the dynamics is deterministic or stochastic, coarse-graining does not break the time symmetry.

12.4 Elimination of Variables

Another way to generate stochastic processes from dynamical systems is the elimination of variables. Assume that the state space consists of two sets of variables (x, y) where y is the variable of interest and x is the ancillary variable to be eliminated. The function $(x, y) \mapsto y$ is an observable. Many trajectories y_t emanate from a given initial condition y_0 depending on the many possible initial conditions x_0. Assuming some distribution for the initial conditions of the ancillary variable, one obtains a stochastic process for the variable of interest by a construction similar to the one for coarse-graining. We demonstrate this construction by a few examples.

Example: Baker transformation. The x-component of the baker transformation is just the Bernoulli shift. Though elimination of a variable leads in this case to a dynamical system, note that the elimination changes the character of the dynamical system. The baker transformation is invertible. The Bernoulli shift is not. Consider next the y-component:

$$y_{n+1} = \begin{cases} \frac{1}{2} y_n & \text{for } x \in [0, \frac{1}{2}) \\ \frac{1}{2} y_n + \frac{1}{2} & \text{for } x \in [\frac{1}{2}, 1) \end{cases} \tag{12.52}$$

This can also be written as

$$y_{n+1} = \frac{1}{2} y_n + g_n \tag{12.53}$$

where

$$g_n = \frac{1}{2} \chi_{[1/2,1[}(x_n) \tag{12.54}$$

takes the value zero if $x_n \in [0, 1/2[$ and the value 1/2 if $x_n \in [1/2, 1[$. Now consider the Bernoulli shift with broadened initial conditions. Assume P_0 to be the uniform density. In the previous section on coarse-graining, we showed that g_n then becomes the Bernoulli chain $\mathcal{B}(0, 1/2; 1/2)$. We thus arrived at an autoregressive process AR(1) for y_n. Such processes are discussed in more detail in Sect. 13.2.

Example: Coupled harmonic oscillators. A continuous-time example is a set of two coupled harmonic oscillators. This example is most easily treated in complex notation (see Sect. A.4). Assume two linearly coupled oscillators a_0 and a_1 with Hamilton function

$$H = \omega_0 a_0 a_0^* + \omega_1 a_1 a_1^* - c_{01} a_0 a_1^* - c_{10} a_1 a_0^* \tag{12.55}$$

with frequencies ω_0 and ω_1 and coupling coefficient $c_{10} = c_{01}^*$. Hamilton's equations give

$$\dot{a}_1 = -i\omega_1 a_1 + i c_{01} a_0 \tag{12.56}$$

$$\dot{a}_0 = -i\omega_0 a_0 + i c_{10} a_1 \tag{12.57}$$

For given $a_0(t)$, Eq. (12.56) is solved by

$$a_1(t) = a_1(0) e^{-i\omega_1 t} + i \int_0^t dt' c_{01} a_0(t') e^{-i\omega_1 (t-t')} \tag{12.58}$$

Substituting this solution into Eq. (12.57) and introducing $\tilde{a}_0(t) := a_0(t) \exp\{i\omega_0 t\}$, one obtains the integro-differential equation

$$\dot{\tilde{a}}_0(t) = -\int_0^t dt' M(t-t') \tilde{a}_0(t') + F(t) \tag{12.59}$$

with memory function

$$M(t) = c_{10}c_{10}^* \exp\{-i(\omega_1 - \omega_0)t\} \tag{12.60}$$

and forcing function

$$F(t) = ic_{10}a_1(0)\exp\{-i(\omega_1 - \omega_0)t\} \tag{12.61}$$

If we assume that the initial amplitudes $a_1(0)$ are random variables, then we arrived at a stochastic integro-differential equation for a_0 with a stochastic forcing term $F(t)$. Since (12.61) is a linear relation between $a_1(0)$ and $F(t)$, the properties of this stochastic equation are most easily analyzed in terms of the moments of the distribution. Assume that the initial amplitudes have the first two moments

$$\begin{aligned} \langle a_1(0)\rangle &= 0 \\ \langle a_1(0)a_1^*(0)\rangle &= \sigma_1^2 \\ \langle a_1(0)a_1(0)\rangle = \langle a_1^*(0)a_1^*(0)\rangle &= 0 \end{aligned} \tag{12.62}$$

These initial moments are equivalent to assuming that the initial coordinate and momentum have zero mean, that the coordinate and momentum are uncorrelated, that the averaged initial potential and kinetic energies are equal, and that the averaged initial total energy has a constant value, proportional to σ_1^2. The first two moments of the forcing term are then

$$\begin{aligned} \langle F(t)\rangle &= 0 \\ \langle F(t)F^*(t')\rangle &= c_{10}c_{10}^*\sigma_1^2 \exp\{-i(\omega_1 - \omega_0)(t - t')\} \\ \langle F(t)F(t')\rangle &= \langle F^*(t)F^*(t')\rangle = 0 \end{aligned} \tag{12.63}$$

The covariance function $\langle F(t)F^*(t')\rangle$ is an oscillating function of $t - t'$.

The Langevin limit. The case of two coupled oscillators is easily generalized to the case that the oscillator a_0 is linearly coupled to a set of N oscillators a_i, $i = 1, \ldots, N$. These N oscillators constitute a heat bath for the oscillator a_0. There then appears a sum over all these heat bath oscillators in the expressions for the memory and forcing function. A Langevin equation can be obtained by taking a particular continuum limit.

Specifically, start with the memory function

$$M(t) = \sum_{i=1}^{N} c_{i0}c_{i0}^* \exp\{-i(\omega_i - \omega_0)t\} \tag{12.64}$$

Assume that the frequency spacing $\Delta\omega$ between the heat bath oscillators and the coupling coefficients c_{i0} approach zero such that $c_{i0}c_{i0}^*/\Delta\omega = |c(\omega)|^2$ remains finite. The sum over i can then be replaced by an integral and the memory function becomes

$$M(t) = \int_{-\infty}^{\infty} d\omega |c(\omega)|^2 \exp\{-i(\omega - \omega_0)t\} \tag{12.65}$$

Assuming that all the coupling coefficients are equal, $|c(\omega)|^2 = |c|^2$, we obtain

$$M(t) = |c|^2 \int_{-\infty}^{\infty} d\omega \exp\{-i(\omega - \omega_0)t\} = |c|^2 2\pi\delta(t) \tag{12.66}$$

Applying the same continuum limit to the forcing function results in

$$F(t) = \int_{-\infty}^{\infty} d\omega i c(\omega) a(\omega) \exp\{-i(\omega - \omega_0)t\} \tag{12.67}$$

Assume that each heat bath oscillator $a(\omega)$ satisfies the same statistical conditions (12.62) as the single amplitude a_1 in our two oscillator example and assume furthermore that the heat bath oscillators are not coupled to each other, $\langle a(\omega)a^*(\omega')\rangle = \sigma^2(\omega)\delta(\omega - \omega')$, then we again obtain $\langle F(t)\rangle = 0$ and the covariance function

$$\langle F(t)F^*(t')\rangle = \int_{-\infty}^{\infty} d\omega |c(\omega)|^2 \sigma^2(\omega) \exp\{-i(\omega - \omega_0)(t - t')\} \tag{12.68}$$

Assuming again constant coupling coefficients and constant variance results in

$$\begin{aligned}\langle F(t)F^*(t')\rangle &= |c|^2\sigma^2 \int_{-\infty}^{\infty} d\omega \exp\{-i(\omega - \omega_0)(t - t')\} \\ &= |c|^2\sigma^2 2\pi\delta(t - t')\end{aligned} \tag{12.69}$$

Under these approximations, the equation for the oscillator becomes the Langevin equation

$$\dot{\tilde{a}}_0(t) = -\gamma\tilde{a}_0(t) + F(t) \tag{12.70}$$

with a damping constant $\gamma = 2\pi|c(\omega_0)|^2/2$ and a stochastic forcing term that has zero mean and a covariance proportional to $\delta(t - t')$. The factor 1/2 in the damping constant was added to account for the convention that $\int_{-\infty}^{0} dx f(x)\delta(x) = f(0)/2$. The Langevin equation is more fully discussed in Sect. 13.4.

This derivation demonstrates what it takes to arrive at a Langevin equation or more generally at a white noise forcing term, namely an infinite number of uncorrelated forcing terms distributed equally over all frequencies, reflecting the fact that the variance of white noise is distributed equally over all frequencies.

12.5 Adiabatic Elimination of Fast Variables

The adiabatic elimination of fast variables is an important analysis tool to arrive at reduced dynamical systems. It assumes that the dynamical system consists of a slow-responding and a fast-responding subsystem. It is the difference in response times that allows the elimination of the fast variables. We first describe the technicalities of this elimination which requires some attention to notation.

Fast and slow variables. By suitable scaling, the evolution equations for a two-time-scale system can be cast into the generic form

$$\begin{aligned}\dot{x}(t) &= \frac{1}{\epsilon}u(x(t), y(t)) \\ \dot{y}(t) &= v(x(t), y(t))\end{aligned} \tag{12.71}$$

where x is the fast-responding subsystem and y is the slow-responding subsystem. The parameter $\epsilon \ll 1$ is the ratio of the response times. The factor $1/\epsilon$ in front of u ensures that x is the fast-responding system. Equation (12.71) constitute a coupled system of ODEs that has to be solved simultaneously for the functions $x(t)$ and $y(t)$. We envision a situation where we are not interested in the details of the evolution of the fast variable x but only in the evolution of the slow variable y. As an example, we may envision x to describe fast weather variables and y slow climate variables.

Heuristic approach. The heuristic approach assumes that the fast variable rapidly adjusts to a quasi-steady equilibrium value for which $\dot{x} = 0$. During the rapid adjustment, the slow variable y can be regarded as constant such that the final adjusted state can be calculated for each y from the condition

$$u(x, y) = 0 \tag{12.72}$$

Assume that this equation has a unique solution x for each y which we denote by $x_y(\infty)$. The somewhat convoluted notation will be rationalized later. When this solution is substituted into the evolution equation for the slow variable, one obtains

$$\dot{y}(t) = v(x_{y(t)}(\infty), y(t)) \tag{12.73}$$

which is a closed equation for the function $y(t)$. The fast variable $x(t)$ has been eliminated. It is given by $x_y(\infty)$. One says that the fast variable has been slaved by the slow variable. This slaving of the fast variable constitutes a reduction. Instead of the need to solve the original problem simultaneously for $x(t)$ and $y(t)$, one can approach the problem successively, first solving (12.72) for $x_y(\infty)$ and then solving (12.73) for $y(t)$.

Analogous numerical algorithm. The more general approach mimics the numerical solution of the system (12.71) where the slow variable is solved with a large time step ΔT and the fast variable with a small time step $\Delta t = \Delta T/M$ where $M = O(1/\epsilon)$. At the nth time step of the slow variable, the fast variable is calculated by

$$\hat{x}^n_{m+1} = \hat{x}^n_m + \Delta t u(\hat{x}^n_m, \hat{y}_n) \tag{12.74}$$

for $m = 0, \ldots, M$. The slow variable y_n is held constant during this time stepping of the fast variable. The time stepping of the fast variable is subjected to the initial condition $\hat{x}^0_0$ and the continuity condition $\hat{x}^{n+1}_0 = \hat{x}^n_M$.

The slow variable is time-stepped by averaging over the fast variable

$$\hat{y}_{n+1} = \hat{y}_n + \Delta T \frac{1}{M} \sum_{m=0}^{M-1} v(\hat{x}^n_m, \hat{y}_n) \tag{12.75}$$

with given initial condition $\hat{y}_0$.

The method of averaging. The continuous limit of this numerical scheme can then be obtained by taking the limits $\Delta t \to 0$, $\Delta T \to 0$, and $M \to \infty$ such that $M\Delta t = T \to \infty$. The result is the 'method of averaging' which consists of the following steps:

(1) Solve the fast problem

$$\frac{d}{d\tau}\hat{x}(\tau) = u(\hat{x}(\tau), \hat{y}) \tag{12.76}$$

for each fixed value of $\hat{y}$ and denote the result by $\hat{x}_{\hat{y}}(\tau)$ where τ is the fast time variable.

(2) Substitute this result into the function $v(x, y)$ which results in $v(\hat{x}_{\hat{y}}(\tau), \hat{y})$.

(3) Average the equation for the slow variable over the fast time scale τ which results in

$$\frac{d}{dt}\hat{y}(t) = \lim_{T\to\infty} \frac{1}{T} \int_0^T d\tau v(\hat{x}_{\hat{y}(t)}(\tau), \hat{y}(t)) \tag{12.77}$$

which is a closed equation for the function $\hat{y}(t)$.

We put a hat on the functions x and y to indicate that these are not solutions to the original problem but of a different problem that is supposed to approximate the original problem. The important point is that the slow variable $\hat{y}$ acts as a parameter in the fast problem. Thus, the solution for the fast problem has to be indexed by $\hat{y}$. Again, the coupled problem for $x(t)$ and $y(t)$ has been reduced to two successive problems. A proper derivation of the method of averaging using a two-time scale expansion is given in Appendix D.3.

When the method of averaging is applied to the case of our two linearly coupled oscillators, one finds

$$\hat{a}_1(\tau) = \hat{a}_1(0)e^{-i\omega_1\tau} + \frac{c_{01}\hat{a}_0}{\omega_1}[1 - e^{-i\omega_1\tau}] \tag{12.78}$$

Time averaging yields

$$\frac{d}{dt}\hat{a}_0(t) = -i\omega_0\hat{a}_0 + i\frac{c_{10}c_{01}}{\omega_1}\hat{a}_0 \tag{12.79}$$

The slow oscillator thus experiences a frequency shift. If one takes the limit, its exact value depends on how ω_1 and the coupling coefficients c_{10} and c_{01} approach infinity such that $c_{01}c_{10}/\omega_1$ remains finite. Equation (12.79) is also obtained by the heuristic approach although the fast variable does not approach rapidly a quasi-equilibrium state. On the contrary, the fast variable oscillates. For nonlinear problems, the method of averaging is, in general, *not* identical to the heuristic approach. The average in Eq. (12.77) will generally be very different from the heuristic quasi-equilibrium value. The method of averaging can thus deal with situations in which the fast component exhibits long-time oscillations and provides a formalism for studying nonlinear effects of these fast oscillations on the slow variables.

If we replace the harmonic oscillator with an over-damped oscillator $i\omega_1 \to \gamma_1$, the fast problem is indeed a rapid relaxation to an equilibrium value. Then we encounter the fact that the time average only exists in the forward time direction, $T \to +\infty$, but not in the backward time direction, $T \to -\infty$. The integral diverges. 'Rapid relaxation to a quasi-equilibrium state' is thus a one-way street and the method of averaging approximates the evolution of the slow variable in only one time direction. In the oscillator example, the time averages exist in both directions and are identical. It might also happen that the two time averages exist but are different.

Finally, note that the method of averaging relies on asymptotic scale separation and cannot address problems in which a continuous spectrum of scales plays a crucial role.

Ergodicity. The method of averaging requires that we solve for each state $\hat{y}$ the fast problem (12.76)–for each climate state we have to calculate the evolution of the weather. This calculation is a tall order. Some relief is obtained when we assume that the $\hat{x}_{\hat{y}}$ represent ergodic dynamical systems. Then there exist ergodic measures $\mu_{\hat{y}}$ for each $\hat{y}$ and the time integral in Eq. (12.77) can be replaced by an ensemble average

$$\lim_{|T|\to\infty} \frac{1}{T}\int_0^T d\tau v(\hat{x}_{\hat{y}}(\tau), \hat{y}) = \int \mu_{\hat{y}}(dx) v(x, \hat{y}) = \langle v(\hat{y})\rangle_{\hat{y}} \tag{12.80}$$

for almost all initial conditions. The function $\hat{x}_{\hat{y}}(\tau)$ is thus replaced by a random variable x with distribution $\mu_{\hat{y}}$. It is stressed that the distribution depends on $\hat{y}$. The statistics of weather depends on the climate state. This dependence is the reason for the subscript $\hat{y}$ on the cornered brackets. Equation (12.80) thus becomes

$$\frac{d}{dt}\hat{y}(t) = \langle v(\hat{y}(t))\rangle_{\hat{y}(t)} \tag{12.81}$$

which is a deterministic evolution equation. The slow component $\hat{y}$ evolves deterministically. One may identify $\hat{y}$ with the mean of y, $\hat{y} = \langle y\rangle_{\hat{y}}$. The mean climate state thus evolves deterministically. It is stressed that the ergodic measures $\mu_{\hat{y}}$ do not represent any subjective uncertainty imposed by the investigator. They are determined or 'chosen' by the dynamics of the system $\hat{x}_{\hat{y}}$.

The deviation $y' := y - \hat{y}$ from the mean state is governed by

$$\frac{d}{dt}\, y'(t) = v(x_{\hat{y}(t)+y'(t)}, \hat{y}(t) + y'(t)) - \langle v(\hat{y}(t))\rangle_{\hat{y}(t)} \tag{12.82}$$

Here, $\hat{y}(t)$ is a given function, obtained by solving the deterministic problem (12.81), and x_y a family of random variables with distributions μ_y. Equation (12.82) is thus a stochastic differential equation and its solution y' a stochastic process. The joint distribution of the family x_y of random variables is, however, not fully determined. The ergodicity of the dynamical system $x_y(\tau)$ only determines the marginal one-point distributions of x_y. The two-point joint distribution, such that the random variables x_y and $x_{y'}$ take certain values, and higher point joint distribution are not determined. Here, additional assumptions need to be incorporated. One such set of assumptions and simplifications leads to the stochastic climate model of Hasselmann (1976).

12.6 Summary

When random initial conditions are imposed on a dynamical system, then the evolution of observables represents a stochastic process. This fact is used to generate stochastic processes in practice. We considered in detail two observables, those that represent coarse-graining and the elimination of variables. These examples demonstrate that one can 'derive' many of the prototypical examples of stochastic processes, such as the Bernoulli chain, the autoregressive process, and the Langevin equation from such observables.

We also considered the effect of coarse-graining on the evolution of the Kullback–Leibler divergence. For invertible mixing systems, the non-grained density converges weakly toward a stationary density but the Kullback–Leibler divergence remains the same. The coarse-grained density, on the other hand, converges point-wise to its stationary density and the Kullback–Leibler divergence converges to zero. It converges, however, in both time directions to zero. Coarse-graining does not break the time symmetry.

The fast-responding variables can be eliminated from a two-time-scale system by the method of averaging, also called adiabatic elimination. If these fast-responding subsystems are ergodic, then one encounters the remarkable situation that the probability distribution of the initial conditions does not need to be prescribed by the investigator but is determined or 'chosen' by the dynamics.

Reference

Hasselmann, K.: Stochastic climate models. Part I: Theory. Tellus **28**, 473–485 (1976)

Chapter 13
Stochastic Dynamical Systems

13.1 Introduction

Randomness is introduced into dynamical systems for many reasons. A dynamical system might depend on some not well-determined parameters. Assuming this parameter to be a random variable results in a system where many trajectories emanate from a point in state space, one for each realization of the random variable, and one can analyze the spread of these trajectories to study how sensitive the dynamical system is to the exact value of the parameter.

Here, we introduce stochastic processes into a dynamical system to account for noise or random forcing, such as the 'random' forcing of suspended ink particles by the fluid molecules in Brownian motion. This introduction of a stochastic process converts the dynamical system to a stochastic dynamical system. Solutions are not trajectories but stochastic processes.

When adding noise to a dynamical system an important distinction needs to be made between additive and multiplicative noise. We demonstrate the difference for the simple case that the underlying dynamical system is linear. In the case of an iterative map, the result is an autoregressive process where a stochastic process is added to a (linear) regression formula or a multiplicative process where the coefficient of the (linear) regression formula is made a stochastic process.

When integrating differential equations with added noise, one encounters the problem that standard estimates of integrals fail and one needs to develop a new calculus which differs from ordinary calculus; the most satisfying of these new calculi is the Ito calculus. The continuous-time analogue of the autoregressive process is the Langevin equation which models Brownian motion. A continuous-time example of a multiplicative process is geometric Brownian motion.

P. Müller, *Handbook of Dynamics and Probability*,
https://doi.org/10.1007/978-3-030-88486-4_13

13.2 Autoregressive Processes

One of the simplest discrete-time systems is obtained from the linear regression formula $x_{n+1} = ax_n + b$. Assuming the forcing term b to be a stochastic process leads to the autoregressive process AR(1). Assuming a to be a stochastic process leads to a multiplicative process.

The AR(1) process. The AR(1) process is defined by the recursion formula

$$f_{n+1} = af_n + g_n \tag{13.1}$$

where $g_n(\omega)$ is a given stochastic process. This recursion formula has the solution

$$f_n = a^n f_0 + \sum_{m=1}^{n} a^{m-1} g_{n-m} = a^n f_0 + \sum_{k=0}^{n-1} a^{n-k-1} g_k \tag{13.2}$$

for $n \geq 0$ which defines a discrete-time stochastic process f_n if the g_n and f_0 are prescribed. The given process g_n is called noise, and it enters here as additive noise. We discuss the AR(1) in some detail because it is the prototype of a discrete-time stochastic process with additive noise g_n. The analysis becomes particularly simple when the noise process is an iid process. We also fold in random initial conditions and assume that they are statistically independent of the noise.

Probability distributions. Assuming the state space to be $X = \mathbb{R}$, the joint probability density of the AR(1) process is given by

$$\begin{aligned} \rho_{f_{n+1}, f_n, \ldots, f_0}(x_{n+1}, x_n, \ldots, x_0) &= \langle \delta(x_{n+1} - f_{n+1})\delta(x_n - f_n) \cdot \ldots \cdot \delta(x_0 - f_0) \rangle \\ &= \rho_g(x_{n+1} - ax_n)\rho_g(x_n - ax_{n-1}) \cdot \ldots \cdot \rho_{f_0}(x_0) \end{aligned} \tag{13.3}$$

where we used the recursion formula $f_{n+1} = af_n + g_n$, the assumption that the noise g_n is an iid process and independent of the initial conditions. One thus finds for the conditional density

$$\rho_{f_{n+1}|f_n, \ldots, f_0}(x_{n+1}|x_n, \ldots, x_0) = \rho_g(x_n - ax_{n-1} = \rho_{f_{n+1}|f_n}(x_{n+1}|x_n) \tag{13.4}$$

It only depends on the previous outcome. The AR(1) process is, not surprisingly, a Markov process with transition probability density

$$\rho_{f_{n+1}|f_n}(y|x) = \rho_g(y - ax) \tag{13.5}$$

independent of n. The AR(1) process is thus a homogeneous Markov process. The one-point densities evolve according to

$$\rho_{f_{n+1}}(y) = \int_{\mathbb{R}} dx \rho_g(y - ax)\rho_{f_n}(x) \tag{13.6}$$

This formula can generally not be inverted to give ρ_{f_n} in terms of $\rho_{f_{n+1}}$. The AR(1) process is in general a one-sided process. Where has this time asymmetry sneaked in? The basic dynamical system $x_{n+1} = ax_n + b$ is invertible for $a \neq 0$. The added noise is an iid stochastic process with no time direction singled out. The time asymmetry was introduced by the innocuous-looking assumption that the initial conditions are statistically independent of the noise in the future. This 'natural' assumption singles out the forward time direction. We will discover that such assumptions are the cause in most situations where the time symmetry is broken.

Moments of AR(1) process. Because of the linear nature of the AR(1) process, its moments are easily calculated. For the calculation of the first two moments, it suffices to assume that the noise has covariance function $\langle g_n g_m \rangle = \sigma_g^2 \delta_{nm}$, and is not correlated with the initial condition, $\langle f_0 g_n \rangle = 0$ for all $n \geq 0$. For notational simplicity, we assume that the noise process has zero mean, $\mu_g = 0$. Start the process with initial mean μ_{f_0} and initial variance $\sigma_{f_0}^2$. The mean and variance of the AR(1) process then evolve according to

$$\begin{aligned} \mu_{f_n} &= a\mu_{f_{n-1}} \\ &= a^n \mu_{f_0} \end{aligned} \tag{13.7}$$

$$\begin{aligned} \sigma_{f_n}^2 &= a^2\sigma_{f_{n-1}}^2 + \sigma_g^2 \\ &= a^{2n}\sigma_{f_0}^2 + \sigma_g^2 \frac{1-a^{2n}}{1-a^2} \end{aligned} \tag{13.8}$$

The cross-covariance function between the process and the noise is given by

$$C_{f_m, g_n} = \begin{cases} a^{m-n-1}\sigma_g^2 \text{ for } m > n \\ \\ 0 \text{ for } 0 \leq m \leq n \end{cases} \tag{13.9}$$

and the auto-covariance function by

$$C_{f_m, f_n} = a^{m+n}\sigma_{f_0}^2 + \sigma_g^2 \frac{a^{|m-n|} - a^{m+n}}{1-a^2} \tag{13.10}$$

Of course, $C_{f_n, f_n} = \sigma_{f_n}^2$. As the AR(1) process evolves, mean and variance change, the process develops correlation with the past noise, and the process becomes correlated with itself.

In the case $|a| > 1$, the mean and variance grow geometrically, the mean as a^n and the variance asymptotically as a^{2n}. The case $a = 1$,

$$f_n = f_{n-1} + g_{n-1} \tag{13.11}$$

recovers the random walk II on a continuum. The mean and variance evolve as

$$\langle f_n \rangle = \langle f_0 \rangle \tag{13.12}$$

$$\sigma_{f_n}^2 = \sigma_{f_0}^2 + n\sigma_g^2 \tag{13.13}$$

The mean remains constant and the variance increases linearly, by σ_g^2 with each time step. The cross- and auto-covariance functions become

$$C_{f_m, g_n} = \begin{cases} \sigma_g^2 \text{ for } m > n \\ 0 \text{ for } 0 \le m \le n \end{cases} \tag{13.14}$$

and

$$C_{f_m, f_n} = \sigma_{f_0}^2 + \begin{cases} n\sigma_g^2 \text{ for } m > n \\ m\sigma_g^2 \text{ for } 0 \le m \le n \end{cases} \tag{13.15}$$

If $a = 1$, then f_n is the sum of iid random variables. The central limit theorem can be applied giving the Gaussian as the limiting distribution.

The case $|a| < 1$ is more interesting. There exists a stationary process f_n^* with mean, variance, and auto-covariance

$$\begin{aligned} \mu_{f_n^*} &= 0 \\ \sigma_{f_n^*}^2 &= \frac{\sigma_g^2}{1 - a^2} \\ C_{f_{n+k}^*, f_n^*} &= a^{|k|} \frac{\sigma_g^2}{1 - a^2} \end{aligned} \tag{13.16}$$

independent of n. Furthermore, we find that any AR(1) process approaches these values in the limit $n \to \infty$. In the case $|a| < 1$, the mean of an AR(1) process thus decays to zero, its variance is arrested at the finite value $\sigma_g^2/(1 - a^2)$, and its auto-covariance function decays with lag k geometrically, as $a^{|k|}$. Any AR(1) process also converges to the stationary process f_* as n approaches infinity. If we had not assumed $\mu_g = 0$, then $\mu_{f_n^*} = \mu_g/(1 - a)$.

For any realization $\omega \in \Omega$ of the stationary AR(1) process, we find the time mean

$$\bar{f}_*(\omega) = \lim_{N\to\infty} \frac{1}{N} \sum_{n=1}^{N} f_n^*(\omega)$$
$$= a \lim_{N\to\infty} \frac{1}{N} \sum_{n=1}^{N} f_{n-1}^*(\omega) + \lim_{N\to\infty} \frac{1}{N} \sum_{n=1}^{N} g_{n-1}(\omega) \tag{13.17}$$

Since g_n is an iid process, the second term equals μ_g and we thus find the ergodic property

$$\bar{f}_*(\omega) = \frac{\mu_g}{1-a} = \langle f_* \rangle \tag{13.18}$$

The time mean equals the ensemble mean almost surely. If ρ_g is the Gaussian distribution $N(0, \sigma_g^2)$, then the stationary process has the Gaussian distribution $N(0, \sigma_g^2/(1-a^2))$.

We have formulated the AR(1) process on $X = \mathbb{R}$. We might have as well chosen the discrete lattice $X = \mathbb{Z}$. Assuming $a = 1$ and g_n to be the Bernoulli chain $B(-1, 1; 1/2)$ and the initial condition $f_0(\omega) = 0$ almost surely then recovers our random walk I on a discrete lattice. This is an important point:

The random walks I and II can alternatively be viewed as AR(1) processes with $a = 1$.

AR(K) processes. AR(1) processes are the simplest examples of discrete-time stochastic dynamical processes with additive noise. They are a subclass of a more general class of processes which are defined by the recursion formulas

$$f_{n+1} = \sum_{k=1}^{K} a_k f_{n-k+1} + g_n \tag{13.19}$$

where again g_n is a given iid process and K is called the order of the autoregressive process. An autoregressive process of order K is denoted by AR(K). For $K = 1$ and $a_1 = a$, we recover our AR(1) process. For $K = 2$, we find explicitly

$$f_{n+1} = a_1 f_n + a_2 f_{n-1} + g_n \tag{13.20}$$

Autoregressive processes can generate stationary processes if all (complex and real) roots y_k of the characteristic polynomial $P(y) = 1 - \sum_{k=1}^{K} a_k y^k$ satisfy $|y_k| > 1$. For an AR(1) process, we recover $|a_1| < 1$ and $a_2 \pm a_1 < 1$, $|a_2| < 1$ for an AR(2) process. If f_n is an AR(K) process, then $h_n = (f_n, f_{n-1}, \ldots, f_{n-K+1})$ is an AR(1) vector process.

13.3 Multiplicative Processes

While AR(1) processes and the associated random walks are well-studied and relatively well-understood, this is not the case for multiplicative processes because of the special role of extreme events.

Multiplicative processes. Multiplicative processes can be obtained from the simple linear iteration formula $x_{n+1} = ax_n$ by assuming that the coefficient a is a random process g_n. One then obtains the recursion formula

$$f_n = g_n f_{n-1} \tag{13.21}$$

or

$$f_n = (\Pi_{m=1}^n g_m) f_0 \tag{13.22}$$

where g_m is a prescribed sequence of random variables and f_0 the initial condition. The logarithm of f_n is given by

$$\log f_n = \log (\Pi_{m=1}^n g_m) f_0 = \sum_{m=1}^{n} \log g_m + \log f_0 \tag{13.23}$$

and represents an AR(1) process. Multiplicative processes are used in finance to model the growth of an investment portfolio f_n under the assumption that the growth rate g_m fluctuates in some manner. To stay within this interpretation, we assume that $f_0 > 0$ and $g_m > 0$ and hence $f_n > 0$. Without loss of generality, we also assume $f_0 = 1$. Otherwise, one has to replace f_n by f_n/f_0 in all the formulas below.

The theory becomes again particularly simple when the g_n are a sequence of iid random variables. In this case, the mean and variance of f_n are given by

$$\mu_{f_n} = \mu_{g_1}^n \tag{13.24}$$

$$\sigma_{f_n}^2 = (\sigma_{g_1}^2 + \mu_{g_1}^2)^n - \mu_{g_1}^{2n} \tag{13.25}$$

Asymptotically, if $\mu_{g_1} > 1$, the mean and variance approach infinity; if $\mu_{g_1} = 1$, the mean stays at 1 and the variance approaches infinity; if $\mu_{g_1} < 1$, the mean approaches zero and the variance approaches zero or infinity depending on whether $\sigma_{g_1}^2$ is smaller or larger than $1 - \mu_{g_1}^2$. Though these are straightforward results, they do not reveal the intricacies of multiplicative processes. For these, we turn to the binomial process.

Binomial multiplicative process. Assume g_n to be a Bernoulli chain $B(x_1, x_2; p)$ with $x_1, x_2 > 0$. Each of the mutually independent random variables g_n takes the value x_1 with probability p and the value x_2 with probability $(1 - p)$. The random variable f_n then takes the values $x_m = x_1^m x_2^{n-m} > 0$ for $m = 0, \ldots, n$. The probability distribution of f_n is the binomial distribution

$$p_{f_n}(x_m) = \binom{n}{m} p^m (1-p)^{n-m} \tag{13.26}$$

The expectation value of f_n is given by

$$\langle f_n \rangle = \sum_{m=0}^{n} \binom{n}{m} p^m (1-p)^{n-m} x_1^m x_2^{n-m} = [px_1 + (1-p)x_2]^n \tag{13.27}$$

where the second equality is the binomial formula. The most probable value x_n^{mp} of the distribution, on the other hand, is given by

$$x_n^{mp} = [x_1^p x_2^{(1-p)}]^n \tag{13.28}$$

since the binomial distribution takes its maximum at $m = np$. One finds that

$$\langle f_n \rangle \geq x_n^{mp} \tag{13.29}$$

The proof makes use of the fact that $\log f_n$ is an AR(1) process with a binomial distribution for which the mean and most probable value are the same, $\langle \log f_n \rangle = \log x_n^{mp}$. Thus

$$x_n^{mp} = e^{\langle \log f_n \rangle} \tag{13.30}$$

and, of course,

$$\langle f_n \rangle = \langle e^{\log f_n} \rangle \tag{13.31}$$

Jensen's inequality then implies the inequality $\langle f_n \rangle \geq x_n^{mp}$.

The mean can be much larger than the most probable value. Take the specific example of $x_1 = 1/2$ and $x_2 = 2$. At each time step, you lose half of your investment with probability p or double it with probability $(1-p)$. The values x_m range from 2^{-n} to 2^n and are geometrically spaced. The mean is then $\mu_{f_n} = (2 - 3p/2)^n$ and plotted in Fig. 13.1 for $n = 4$ as a function of p. For small values of p, the mean is close to the maximum value 2^n. The expectation value is dominated by values on the tail of the distribution. The most probable value $x_n^{max} = 2^{n(1-2p)}$ is also shown in Fig. 13.1. In the interval $1/2 < p < 2/3$, one arrives at the particular situation that $\mu_{f_n} \to \infty$ as $n \to \infty$ whereas $x_n^{mp} \to 0$ in the same limit.

Asymptotic behavior. The asymptotic behavior of f_n for large n can be obtained by noting that $\log f_n$ is the sum of iid random variables, $\log f_n = \sum_{m=1}^{n} \log g_m$. Applying the central limit theorem, one thus finds that $\log f_n$ has the Gaussian distribution

$$\rho_{\log f_n}(y) \approx \frac{1}{\sqrt{2\pi n \sigma^2_{\log g_1}}} \exp\left\{ -\frac{(y - n\mu_{\log g_1})^2}{2n\sigma^2_{\log g_1}} \right\} \tag{13.32}$$

as its limiting distribution and f_n hence the log-normal distribution

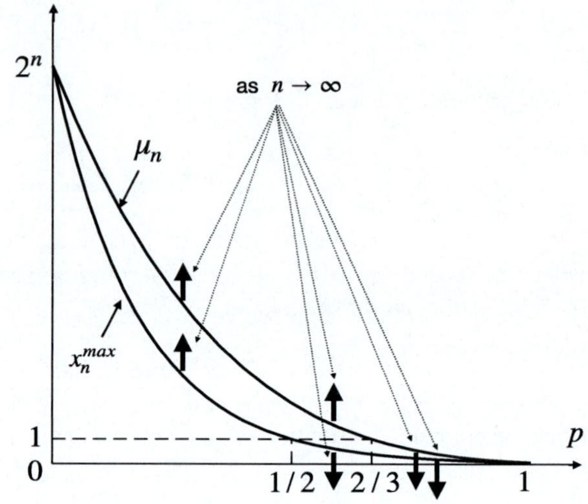

Fig. 13.1 Mean and most probable value as a function of p for $n = 4$

$$\rho_{f_n}(x) \approx \frac{1}{x\sqrt{2\pi n\sigma^2_{\log g_1}}} \exp\left\{-\frac{(\log x - n\mu_{\log g_1})^2}{2n\sigma^2_{\log g_1}}\right\} \tag{13.33}$$

with expectation

$$\langle f_n\rangle^{Gaussian} \approx \exp\left\{n(\mu_{\log g_1} + \frac{\sigma^2_{\log g_1}}{2})\right\} \tag{13.34}$$

This derivation of the asymptotic distribution is straightforward but care must be taken when calculating moments with this distribution. Consider the binomial multiplicative process for which $\mu_{\log g_1} = \log x_1^{mp}$ and $\sigma^2_{\log g_1} = p(1-p)(\log x_1/x_2)^2$. Then the expectation (13.34) differs from the expectation (13.27). The reason for this discrepancy can be seen by an alternative derivation of the log-normal distribution which starts from the fact that the random variable

$$h_n := \frac{1}{\log x_1/x_2}(\log f_n - n\log x_2) \tag{13.35}$$

has the support $m = 0, \ldots, n$ and the binomial distribution $p_b(m; m, p)$. Applying the Gaussian approximation to the binomial distribution, one arrives at

$$\rho^{Gaussian}_{\log f_n}(y) = \frac{1}{\sqrt{2\pi np(1-p)(\log x_1/x_2)^2}} \exp\left\{-\frac{(y - n\log x_1^{mp})^2}{2np(1-p)(\log x_1/x_2)^2}\right\} \tag{13.36}$$

and hence at the log-normal distribution (13.33) for f_n.

As pointed out by Redner (1990), the discrepancy is then due to the fact that the Gaussian approximation is a parabolic expansion about the maximum $m = np$ of the binomial distribution whereas for the calculation of the expectation, one should perform a parabolic expansion about the maximum of the binomial distribution times $x_1^m x_2^{n-m}$.

Growth rates. Another peculiarity of multiplicative processes is obtained when one considers growth rates. If we write the averaged process as $\langle f_n \rangle = e^{\tilde{\gamma}_n n}$ with growth rate $\tilde{\gamma}_n$, then we find in the limit $n \to \infty$ that the growth rate of the averaged process is given by

$$\begin{aligned}\tilde{\gamma} := \lim_{n\to\infty} \tilde{\gamma}_n &= \lim_{n\to\infty} \frac{1}{n} \log \langle f_n \rangle \\ &= \log \langle g_1 \rangle \end{aligned} \tag{13.37}$$

since $\langle f_n \rangle = \langle g_1 \rangle^n$ because g_n is an iid process. The time-averaged growth rate of a single realization, on the other hand, is given by

$$\begin{aligned}\bar{\gamma} :&= \lim_{n\to\infty} \frac{1}{n} \log f_n \\ &= \lim_{n\to\infty} \frac{1}{n} \sum_{m=1}^{n} \log g_m = \langle \log g_1 \rangle \end{aligned} \tag{13.38}$$

almost surely since $\log g_n$ is an iid process for which the strong law of large number holds. Jensen's inequality implies

$$\bar{\gamma} = \langle \log g_1 \rangle \leq \log \langle g_1 \rangle = \tilde{\gamma} \tag{13.39}$$

The inequality $\bar{\gamma} \leq \tilde{\gamma}$ has been the ruin of many men who based their gambling or investment strategies on expectation values rather than on time averages. Time averages were also Peter's (2011) means to resolve the St. Petersburg paradox, as is discussed in Sect. (16.9).

13.4 Langevin Equation

Two issues need to be addressed when one wants to extend the concept of discrete-time autoregressive and multiplicative processes to continuous-time stochastic processes:

- 1. We need to find a continuous-time noise process that acts as a driver for these processes.
- 2. We need to define a differential calculus for processes whose paths are not differentiable, as we demonstrated for the Wiener process.

This section addresses these two issues in a heuristic manner, the following chapter in a more rigorous manner.

Continuous limit of AR(1) process. Start from the AR(1) process $f_{n+1} = (1 - a) f_n + g_n$ where g_n is an iid time-discrete process with a $N(0, \sigma_1^2)$ distribution. Write the AR(1) process as

$$f_{(n+1)\Delta t} = (1 - a) f_{n\Delta t} + g_{n\Delta t} \tag{13.40}$$

and assume that the noise term scales as

$$g_{n\Delta t} = \sqrt{\Delta t} g_n \tag{13.41}$$

The rational for this choice is that it is the variance of the AR(1) process that grows linearly in t so the forcing must scale as the square root of t. In the limit $\Delta t \to 0$, $n \to \infty$ and $a \to \infty$ such that $t = n\Delta t$ and $\gamma = a\Delta t$ remain finite one arrives at

$$\frac{d}{dt} f_t = -\gamma f_t + g_t \tag{13.42}$$

where the process

$$g_t = \lim_{\Delta t \to 0, n \to \infty} \frac{g_{n\Delta t}}{\Delta t} = \lim_{\Delta t \to 0} \frac{g_n}{\sqrt{\Delta t}} \tag{13.43}$$

has mean and covariance function

$$\begin{aligned} \langle g_t \rangle &= 0 \\ \langle g_t g_{t'} \rangle &= \lim_{\Delta t \to 0} \frac{\langle g_n g_{n'} \rangle}{\Delta t} \\ &= \lim_{\Delta t \to \infty} \frac{\sigma_1^2 \delta_{nn'}}{\Delta t} \\ &= \sigma_1^2 \delta(t - t') \end{aligned} \tag{13.44}$$

and is hence the white noise process $\dot{w}_t$.

Langevin equation. The equation

$$\frac{df_t}{dt} = -\gamma f_t + \dot{w}_t \tag{13.45}$$

is called the Langevin equation. It has the explicit path-wise solution

$$f_t = f_0 e^{-\gamma t} + \int_0^t ds \dot{w}_s e^{-\gamma(t-s)} \tag{13.46}$$

which by linearity is a Gaussian process.

The Langevin equation was originally introduced to model Brownian motion. The stochastic process f_t is the velocity of the ink particle. The stochastic forcing $\dot{w}_t$ models the random impacts of fluid molecules on the ink particle, and the term $-\gamma f_t$ with $\gamma > 0$ models the viscous drag on the ink particle. The quantitative analysis assumes $\gamma = 6\pi\eta a$ where η is the kinematic viscosity and a the radius of a spherically symmetric ink particle and infers the variance of the forcing from the requirement that in thermal equilibrium $\langle mv^2/2\rangle = kT/2$.

A more recent application of the Langevin equation is the stochastic climate model of Hasselmann (1976). It asserts that changes in the climate state f_t are caused by random weather fluctuations $\dot{w}_t$. Contrary to previous climate models that ascribe climate changes to instabilities in the climate system, the stochastic climate model requires 'damping mechanisms' to avoid run-away climate changes.

13.5 Stochastic Differential Equations

The most rigorous way to deal with stochastic processes driven by white noise is to only consider the increment or integral equations since these might be well-defined even for non-differentiable processes. In the case of the Langevin equation, one thus considers the increment equation

$$df_t = -\gamma f_t dt + dw_t \tag{13.47}$$

or the integral equation

$$f_t - f_{t_0} = -\gamma \int_{t_0}^{t} dt' f_{t'} + \int_{t_0}^{t} dw_{t'} \tag{13.48}$$

The increment form exhibits most clearly the difference between deterministic and stochastic forcing. A deterministic forcing term would lead to an increment proportional to dt whereas the stochastic increment is proportional to dw_t. The problem is that the white noise process is so irregular that standard estimates of increments fail. This irregularity causes the variance of $w(t)$ to grow more slowly, proportional to t as shown in Sect. 9.6 on the Wiener process and not proportional to t^2 as it would for deterministic forcing.

Stochastic differential equations. A straightforward generalization of the Langevin equation is

$$df_t = A(f_t, t)dt + B(f_t, t)dw_t \tag{13.49}$$

or

$$f_t - f_{t_0} = \int_{t_0}^{t} dt' A(f_{t'}, t') + \int_{t_0}^{t} dw_{t'} B(f_{t'}, t') \tag{13.50}$$

where A and B are functions $\mathbb{R}^2 \mapsto \mathbb{R}$. If the function B depends on f_t, then the noise becomes multiplicative, as opposed to additive.

Equation (13.49) is the prototype of an ordinary stochastic differential equation. There are many subtleties when working with such stochastic differential equations, all due to the fact that the stochastic forcing term grows slower than dt. Thus, ordinary calculus rules do not apply. Without going into detail, we just present two points.

Ito integral. The first issue is the definition of integrals $\int_{t_0}^{t} dw_{t'} g(t')$. In the spirit of a Riemann–Stieltjes integral, one might divide the interval $[t_0, t]$ into N subintervals $t_0 < t_1 < \cdots < t_{N-1} < t_N$ and approximate the integral by the sum

$$S_N := \sum_{n=1}^{N} g(\tau_n)[w_{t_n} - w_{t_{n-1}}] \tag{13.51}$$

where $\tau_n \in]t_{n-1}, t_n[$. Unfortunately, this sum depends on the way one chooses τ_n. Take, for example, $g_t = w_t$ and τ_n such that $g(\tau_n) = \alpha g(t_n) + (1 - \alpha)g(t_{n-1})$. Then the expectation of the sum is given by

$$\begin{aligned}\langle S_N \rangle &= \sum_{n=1}^{N} \langle [\alpha w_{t_n} + (1 - \alpha) w_{t_{n-1}}][w_{t_n} - w_{t_{n-1}}] \rangle \\ &= 2D \sum_{n=1}^{N} [\alpha t_n - \alpha t_{n-1} + (1 - \alpha)t_{n-1} - (1 - \alpha)t_{n-1}] \\ &= 2D \sum_{n=1}^{N} \alpha(t_n - t_{n-1})\end{aligned} \tag{13.52}$$

and depends on the choice of α. The choice $\alpha = 0$ corresponds to the Ito integral and has desirable properties. First, from a mathematical point of view, one can prove that the Ito stochastic integral defined by

$$\int_{t_0}^{t} dw_{t'} g(t') := \lim_{N \to \infty} \sum_{n=1}^{N} g(t_{n-1})[w_{t_n} - w_{t_{n-1}}] \tag{13.53}$$

exists in the second mean or mean square limit as defined in Sect. 7.9, provided $g(t)$ is a 'non-anticipating' function. Non-anticipating functions g_t are stochastic processes such that $g(t)$ is at any time t statistically independent of the process $w_s - w_t$ for $s > t$. Non-anticipating functions at time t are independent of the behavior of the Wiener process at times larger than t, a physically very reasonable class.

The choice $\alpha = 0$ is also the reasonable choice in many practical situations. Think of w_{t_n} as the price of an asset and of $g(t_n)$ as the number of units you hold of this asset. Then the integrand $g(t_{n-1})[w_{t_n} - w_{t_{n-1}}]$ will be your gain or loss. For these and other reasons, the Ito integral is mathematically the most satisfying calculus, but physical or financial reasoning might lead to other choices. One such different choice is $\alpha = 1/2$ which leads to the Stratonovich stochastic integral, not covered here. In both cases, one has to develop a whole new calculus that determines the rules for manipulating stochastic differential equations.

Ito chain rule. One important rule in the Ito calculus is

$$\int_{t_0}^{t} dw_{t'}dw_{t'}g(t') = 2D\int_{t_0}^{t} dt'g(t') \tag{13.54}$$

This formula is usually used in the form $dw_t dw_t = 2Ddt$ with the implicit understanding that this substitution only makes sense when integrating. One application of this rule is the chain rule. Consider a function $g : \mathbb{R} \mapsto \mathbb{R}$ and its composition $h_t = g \circ f_t$ with f_t. Assume that f_t satisfy the stochastic differential Eq. (13.49). The Ito calculus then asserts that

$$\begin{aligned} dh_t &= g(f_t + df_t) - g(f_t) \\ &= g'(f_t)df_t + \frac{1}{2}g'' df_t df_t + \ldots \\ &= g'(f_t)[A(f_t, t)dt + B(f_t, t)dw_t] \\ &\quad + \frac{1}{2}g''(f_t)[A^2(f_t, t)dt^2 + 2A(f_t, t)B(f_t, t)dtdw_t + B^2(f_t, t)dw_t dw_t] + \ldots \\ &= [g'(f_t)A(f_t, t) + g''(f_t)DB^2(f_t, t)]dt + g'(f_t)B(f_t, t)dw_t \end{aligned} \tag{13.55}$$

where the primes denote derivatives with respect to x and $dw_t dw_t = 2Ddt$ has been substituted. The Ito expression differs from the usual chain rule by the second term in the first bracket.

Fokker–Planck equation. The stochastic process f_t defined by the stochastic differential Eq. (13.49) is a Markov process. Heuristically, one might argue: since the driving white noise is delta-correlated the conditional probability $\rho_{t_2|t_1}$ cannot depend on the values of f_t at previous times $t < t_1$. As discussed in Sect. 9.5, the probability density function $\rho(x, t)$ of the stochastic process f_t is therefore governed by the Kramers–Moyal expansion of the master equation

$$\partial_t \rho_{t|t'}(x|y) = \sum n = 1^{\infty} \frac{(-1)^n}{m!} \frac{\partial^m}{\partial x^m}[a_t^{(n)}(x)\rho_{t'|t}(x|y)] \tag{13.56}$$

where the moments are given by

$$a_t^{(n)}(x) = \lim_{\Delta t \to 0} \frac{1}{\Delta t} < (f_{t+\Delta t} - f_t)^n > |_{f_t = x} \tag{13.57}$$

The subscript is to indicate that the process has to start sharp at $f_t = x$. Expansion of the coefficients in the stochastic differential Eq. (13.49) results in

$$\begin{aligned} A(f_{t_1}, t_1) &= A(f_t, t_1) + A'(f_t, t_1)(f_{t_1} - f_t) + \ldots \\ B(f_{t_1}, t_1) &= B(f_t, t_1) + B'(f_t, t_1)(f_{t_1} - f_t) + \ldots \end{aligned} \tag{13.58}$$

where the prime on A and B denotes differentiation with respect to the first argument. With this expansion one obtains

$$f_{t+\Delta t} - f_t = \int_t^{t+\Delta t} dt_1 [A(f_t, t_1) + A'(f_t, t_1)(f_{t_1} - f_t) + \ldots] + \int_t^{t+\Delta t} dw_{t_1} [B(f_t, t_1) + B'(f_t, t_1)(f_{t_1} - f_t) + \ldots] \quad (13.59)$$

Iteration of this formula leads to

$$f_{t+\Delta t} - f_t = \int_t^{t+\Delta t} dt_1 [A(f_t, t_1) + A'(f_t, t_1)(\int_t^{t_1} dt_2 A(f_t, t_2) + \ldots) + \ldots] + \int_t^{t+\Delta t} dw_{t_1} [B(f_t, t_1) + B'(f_t, t_1)(\int_t^{t_1} dt_2 A(f_t, t_2) + \ldots) + \ldots] \quad (13.60)$$

Further iteration leads to an infinite sum of terms. All these terms only contain f_t which according to the definition (13.57) has the sharp value x for all realizations of the stochastic process. The only random terms thus arise from dw and are products of the form $< dw_{t_1} dw_{t_2} \ldots >$. Since the white noise is assumed to be Gaussian, the expectation of these products can be expressed in terms of the first moment $< dw_{t'} >= 0$ and the second moment $< dw_{t'} dw_{t''} >= 2D_0 \delta(t' - t'')$. Furthermore, only terms proportional to Δt contribute to $a_t^{(n)}$. Taking all this into account and sifting carefully through all those terms, one obtains for the first moment

$$a_t^{(1)}(x) = \lim_{\Delta t \to 0} \frac{1}{\Delta t} [\int_t^{t+\Delta t} dt_1 A(f_t, t_1) + 2D \int_t^{t+\Delta t} dt_1 B'(x, t_1) \int_t^{t_1} dt_2 B(x, t_2) \delta(t_2 - t_1)] \quad (13.61)$$

Since $\int_t^{t_1} dt_2 f(t_2) \delta(t_2 - t_1) = 0$ in the Ito calculus and equal to $\frac{1}{2} f(t_1)$ in the Stratonovich calculus, one obtains

$$a_t^{(1)}(x) = A(x, t) \quad (13.62)$$

in the Ito calculus and

$$a_t^{(1)}(x) = A(x, t) + D_0 B(x, t) b'(x, t) \quad (13.63)$$

in the Stratonovich case. The extra term in the Stratonovich calculus is called the noise-induced drift. Doing the same careful sifting, one obtains for the higher moments

$$a_t^{(2)}(x) = 2D_0 B^2(x, t) \quad (13.64)$$

independent of which calculus is used and

$$a_t^{(n)}(x) = 0 \quad n \geq 3 \quad (13.65)$$

The probability density function for the stochastic differential Eq. (13.49) is thus governed by the Fokker–Planck equation

$$\partial_t \rho(x,t) = -\partial_x (A(x,t)\rho(x,t)) + D_0 \partial_x^2 (B^2(x,t)\rho(x,t)) \tag{13.66}$$

in the Ito case. The Stratonovich case is obtained by adding the noise-induced drift term.

The Langevin equation is the special case $A(x,t) = -\gamma x$ and $B = 1$ and results in the Fokker–Planck equation

$$\partial_t \rho(x,t) = \partial_x (\gamma x \rho(x,t)) + D_0 \partial_x^2 \rho(x,t) \tag{13.67}$$

and the Langevin equation turns out to be the differential formulation of the Ornstein–Uhlenbeck process, discussed in Sect. 9.5. As shown there, the solution for initial condition $\rho(x,0) = \delta(x - x_0)$ is a Gaussian

$$\rho_{x,t} = \frac{1}{\sqrt{2\pi\sigma_t^2}} \exp\left[\frac{(x - x_0 - \mu_t)^2}{2\sigma_t^2}\right] \tag{13.68}$$

with mean

$$\mu_t = x_0 e^{-\gamma t} \tag{13.69}$$

and variance

$$\sigma_t^2 = \frac{D_0}{\gamma}[1 - e^{-2\gamma t}] \tag{13.70}$$

Geometric Brownian motion. Consider the stochastic differential equation

$$df_t = f_t[\gamma dt + dw_t] \tag{13.71}$$

where γ is a deterministic growth rate and dw_t the increment of the Wiener process. The noise is multiplicative. The process is called geometric Brownian motion. It is the continuous version of the multiplicative discrete process considered in Sect. 13.3. Using Ito's formula, the geometric Brownian motion can be written as

$$d \log f_t = [\gamma - D]dt + dw_t \tag{13.72}$$

which is easily integrated to yield the solution

$$f_t = f_0 \exp\left[(\gamma - D)t + w_t\right] \tag{13.73}$$

Geometric Brownian motion is a non-stationary and hence non-ergodic process. The growth rate for the ensemble-averaged process is

$$\tilde{\gamma} = \frac{d}{dt} \log\left\langle \frac{f_t}{f_0} \right\rangle \tag{13.74}$$

The expectation of f_t/f_0 is given by

$$\left\langle \frac{f_t}{f_0} \right\rangle = e^{(\gamma - D)t} \langle e^{w_t} \rangle = e^{\gamma t} \tag{13.75}$$

since $\langle e^{w_t} \rangle = e^{Dt}$. The growth rate for the ensemble-averaged process is thus γ. This differs from the time mean growth rate of a single realization which is given by

$$\bar{\gamma} = \lim_{T \to \infty} \frac{1}{T} \int_0^T dt \frac{d}{dt} \log \frac{f_t}{f_0} = \gamma - D + \lim_{T \to \infty} \frac{1}{T} w_T \tag{13.76}$$

Since w_T/T has a $N(0, 2D/T)$ distribution, the last term almost surely converges to zero and the time mean growth rate becomes $\bar{\gamma} = \gamma - D$, again smaller than the growth rate of the averaged process.

Differential equations can be driven by other than white noise processes, such as birth and death processes. Stochastic terms can also be added to partial differential equations. Solutions of stochastic differential equations show behavior that is often impossible (or at least difficult) to achieve with deterministic equations. Patchiness, i.e., the clustering into disconnected communities, can, for example, be obtained relatively easily by combining a random walk with a random birth and death process. Overall, stochastic differential equations and the resulting stochastic processes offer a tool to describe and understand complex environmental systems. They ascribe irregular behavior to stochastic elements in the system in contrast to dynamical systems where irregular behavior is caused by nonlinearities in the system.

13.6 Summary

In this chapter, we added a stochastic process to a dynamical system in order to account for noise or random forcing in the dynamical system. This addition turns the dynamical system into a stochastic dynamical system. The most prominent examples are autoregressive and multiplicative processes in the discrete-time case. They are driven by independent identically distributed stochastic processes. In the case of autoregressive processes, the stochastic process is added to a simple linear recursion formula. In the case of multiplicative processes, the stochastic process replaces the coefficient of the same simple linear recursion formula. The random walks I and II turn out to be special autoregressive processes. Multiplicative processes are non-ergodic processes and show very peculiar behavior. The growth rate of the average process differs from the time-averaged growth rate of a single realization. Continuous stochastic dynamical systems are derived from stochastic differential equations which are driven by white noise. The most famous example is the Langevin equation which describes the Ornstein–Uhlenbeck process. Analyzing stochastic differential equations beyond the Langevin equation requires the introduction of a new calculus that is different from ordinary calculus since the increments are so irregular that standard estimates fail. The most satisfying calculus is the Ito calculus which has rules that differ from ordinary calculus. The general result is that we could extend all the processes and concepts of discrete-time processes to continuous-time processes at the price of introducing this modified calculus.

Overall, stochastic dynamical systems are an efficient tool to actually construct stochastic processes. More on stochastic dynamical systems can be found in Arnold (1974), Karatzas and Shreve (1994), and Oksendal (1992).

References

Arnold, L.: Stochastic Differential Equations: Theory and Applications. Wiley (1974)
Hasselmann, K.: Stochastic climate models. Part I: Theory. Tellus **28**, 473–485 (1976)
Karatzas, I., Shreve, S.E.: Brownian Motion and Stochastic Calculus. Springer (1994)
Oksendal, B.: Stochastic Differential Equations. Springer (1992)
Peters, O.: The time resolution of the St. Petersburg paradox. Phil. Trans. Roy. Soc. A **369**, 4913–4921 (2011)
Redner, S.: Random multiplicative processes. An elementary tutorial. Am. J. Phys. **58**, 267 (1990)

Part IV
Probability in Scientific Reasoning

The concept of probability is introduced to deal with random or uncertain phenomena and with incomplete knowledge. This part addresses three interrelated questions:
1. Is there anything in the real world that corresponds to the mathematical concept of probability?
2. If so, how does one determine or measure such probabilities?
3. How should one use probabilities when making inferences under uncertainty?

The connection between probability theory and real-world events raises many unresolved foundational issues. Though probability theory was developed to quantify the outcomes of games of chance and other uncertain phenomena, the resulting axiomatic theory is a mathematical theory and therefore belongs to the Platonic world, when one adopts the philosophical distinction between the Platonic, physical, and mental worlds, the latter two worlds here referred to as the real world. Philosophers have been thinking about these worlds and their interrelations since the times of Plato. A major sub-discipline is the philosophy of science where philosophers have searched for entities in the real world that correspond to various mathematical concepts. In the case of probability, this search goes under the name of the interpretation of probability. The relationship between the mathematical concept of probability and phenomena in the real world is far from trivial. It requires that one identifies the basic concepts of probability theory such as probability spaces, conditional probabilities, statistical independence, and random variables with entities in the real world. What might outcomes, events and probabilities be in the real world? What might a probability measure correspond to in the real world? What do we mean by saying an event A has the probability *P(A)*. Answers to these questions are not straightforward. Various interpretations have been suggested, including the physicalist's, frequentist's, logicist's, and subjectivist's interpretation which we review but there is no consensus among philosophers (and lay people) about the proper interpretation of probability in the real world.

A more pragmatic approach is the utilitarian interpretation, adopted here. The utilitarian interpretation views probability theory as a model for certain phenomena. If the model works, fine. If it does not, one has to dismiss it. Technically, 'to work' means that the model has to pass the epistemological loop. The utilitarian interpretation does not assert that real-world phenomena have probabilities in any sense but only that it is a useful description of certain phenomena for certain purposes.

Probability theory does not assign probabilities to events. It only tells us how to manipulate probabilities. Of course, if one uses probabilities as a model then one needs to correctly follow its rules. The assignment of probabilities is the task of statistics. Statistics analyzes data, measurements, or observation, called a sample. The sample is assumed to be a realization of a random variable. This is the basic assumption of statistics. Statistics is to infer properties of the probability distribution of this random variable. A standard problem is to determine the mean and variance of the distribution. 'Traditional' statistics which subscribes to the frequentist's interpretation of probability provides estimators for the estimation of these free parameters and tests for the testing of hypotheses about these parameters. Probability theory objectively determines the properties of these estimators and tests, how they will perform when applied repeatedly but it does not tell us anything about the particular estimate or test result obtained from a sample. There is also no objective way to select a particular estimator or test in a given situation. Estimators and tests require the specification of parameters such as of acceptable error bars. The selection and specification have to be based on and rationalized by criteria extraneous to probability theory. Some of these extraneous concepts such as utility and risk may be quantified but only transfer the problem. Instead of specifying error bars one has to specify risks. Overall, 'tradotional' statistics requires subjective choices by the investigator, and different statisticians might differ in their choices.

Some of the shortcomings of the traditional statistics are overcome by Bayesian statistics or inference. It explicitly acknowledges the fact that one cannot make any inferences without assumptions. It forces the statistician to quantify his prior knowledge and updates prior probabilities to posterior probabilities when new evidence becomes available, such as by taking a sample. Bayesian inference can be applied to a much broader range of problems than the traditional approach. Bayesian inference works with conditional probabilities. At its heart is Bayes theorem. Bayesian inference also allows a logicist's or subjectivist's interpretation, not just the frequentist's interpretation.

We conclude with a general discussion of the epistemological framework and strategies that govern skillful inferences under uncertainty. Data assimilation will serve as an example to illuminate the framework and strategies when probabilistic inference is applied to a complex problem with many different uncertainties.

Chapter 14
Interpretations of Probability

14.1 Introduction

Probability is a mathematical construct that satisfies Kolmogorov's axioms. Probability theory tells us how to manipulate given probabilities according to the logical implications of these axioms. What things in the real world might correspond to this mathematical construct is a contentious issue. Applying probability theory to the real world requires first to identify 'events' to which one can assign probabilities. We consider four main classes of 'events' or four interpretations of probability (Guttman 1999):

(1) The physicalist's interpretation: Probability is an objective property of the (physical) system at hand, much like the mass or material of a coin. This interpretation has to answer the question of why the system at hand has the disposition (or propensity) to assume certain states with a certain probability. This is a view sometimes taken in quantum mechanics. The problem is how to measure this probability.

(2) The frequentist's interpretation: The frequentist's interpretation is a generalization of Laplace's view which is the oldest and most common interpretation. Laplace's view states that the probability of an event is the ratio of all positive outcomes to all possible outcomes. More generally, the probability is the relative frequency of a specific event in a sequence of events. Probability is, thus, a formal feature of sequences of events. Probability becomes a proportion.

(3) The logicist's interpretation: Here, probability theory is the extension of propositional logic to the case where there is insufficient knowledge. Probability is, thus, a logical relation between propositions. It is, thus, a logical and mathematical property. Probability becomes plausibility. Since these plausibilities depend heavily on the state of the logician's knowledge, the logicist's interpretation is formulated in terms of conditional probabilities and uses Bayesian algorithms.

P. Müller, *Handbook of Dynamics and Probability*,
https://doi.org/10.1007/978-3-030-88486-4_14

(4) The subjectivist's interpretation: Probabilities describe the degrees of belief of a person. This view has its roots in decision-making with insufficient knowledge and in betting strategies. The subjectivist's interpretation is also formulated in terms of conditional probabilities and uses Bayesian algorithms.

These interpretations progress from locating probability in the physical world to locating it in the measurement procedure, to locating it in sequences of numbers, and to locating it in the mental realm. We use somewhat uncommon names for these interpretations, first to have a convenient label and secondly to stress that it is a human being who makes these interpretations.

A few disclaimers are in order. First of all, there are other interpretations. Second, the various interpretations are not mutually exclusive. Thirdly, the names of these four interpretations are simply used to indicate different interpretations and might not correspond strictly to whatever philosophers associate with terms like physicalism or subjectivism. In practice, scientists use different interpretations depending on the problem at hand and the purpose of the investigation or they amalgamate different interpretations. The challenge for all interpretations is to demonstrate that they satisfy Kolmogorov's axioms.

We stay away from these foundational issues and use what might be called the utilitarian interpretation. Probabilities and the associated theory are used as a model for certain phenomena in the real world. Whether or not probability provides a good model depends on whether or not the model works or, more technically, survives the epistemological loop.

14.2 The Physicalist's Interpretation

The physicalist assumes that probability is an objective property of the physical system at hand. The physicalist's view is the one that one encounters most often in physics and other natural sciences. Specifically, the physicalist assumes that the state space X of the system is endowed with a probability measure P. If $f : X \mapsto \mathbb{R}$ is an observable, then the probability of an event $B \in \mathcal{B}$ is given by

$$P_f(B) := P(f^{-1}(B)) \tag{14.1}$$

where $f^{-1}(B)$ is the set of all states that imply the event B. The probability of obtaining head in a coin tossing experiment is, thus, assumed to be the fraction of all initial states that lead to the result of head. Kolmogorov's axioms are trivially satisfied by this definition since P is a probability. The physicalist's answer to the question of what is probability is thus: probability is a probability measure on the state space of a dynamical system. Ergodicity is often invoked to justify this view since it assures the existence of probability measures for certain systems. Such ergodic probability measures are indeed a property of a dynamical system, perhaps not quite on par with properties like mass or color but a property of the system nevertheless.

A similar interpretation is also encountered in quantum mechanics. There observables are represented by self-adjoint operators A on the Hilbert space $\mathcal{H}$. The possible values of an observable are the eigenvalues $a^{(i)}$, $i \in I$, of the operator A. If the system is in the normalized state $|x\rangle$, then the probability of obtaining the value $a^{(j)}$ of an observable A is given by

$$P(A = a^{(j)} || x\rangle) = \langle x|a^{(j)}\rangle\langle a^{(j)}|x\rangle \tag{14.2}$$

where $|a^{(j)}\rangle$ is the normalized eigenvector corresponding to the eigenvalue $a^{(j)}$ and where we have used Dirac's notation. This quantum-mechanical probability does not derive from measure theory but from a generalization of the Pythagorean theorem.

The main question is of course how does one determine, assign, or measure this probability measure of the system. The standard answer is that one measures the observable repeatedly, independently, from identical versions of the system, or in any other way that assures that one can regard these measurements as independent realizations and can apply the strong law of large numbers which states that the relative frequency of the occurrence of an event will converge to the probability of the event almost surely. This measurement technique for probabilities brings us to the second interpretation.

14.3 The Frequentist's Interpretation

The frequentist's interpretation takes observations or measurements as the only thing that is real. The frequentist's interpretation is the most common and oldest interpretation. It attempts to determine probabilities from observed outcomes with no reference to some underlying pre-existing probability distribution.

When one tosses a coin repeatedly, one obtains a binary sequence of numbers, or more generally, when one repeats an experiment, one obtains a sample s_n, $n = 1, \ldots, N$ where the index n denotes or counts the successive outcomes. Note that the sample is a sequence of numbers. At this point, it is unclear what random variable and what underlying probability space one should associate with this sequence. Most importantly one has to define what one means by a random sample, randomly repeating an experiment, randomly drawing from a population. One has first to construct these quantities. The frequentist does this construction in the following way.

Relative frequencies. The frequentist presumes a mechanism f that produces (real) numbers $s_1, s_2, \ldots$. The outcome is uncertain and differs from realization to realization. The mechanism f is to mimic a random variable but a random variable is not yet defined. It is assumed that all realizations are equivalent and independent, that the uncertainties are the same for each realization, and that the value s_n provides no knowledge about the value of the next realization s_{n+1} or any other realization. This verbiage is to indicate that the different realizations are statistically independent but, again, this concept is not yet defined. The output is, however, assumed not to be completely irregular. It is assumed that the relative frequency

$$RF_f([a, b]) := \lim_{N\to\infty} \frac{|\{s_{n+k} \in [a, b]; k \leq N\}|}{N} \tag{14.3}$$

exists for any interval $[a, b]$ and is independent of n. Here, vertical bars ($|\cdot|$) denote the number of elements of a set, i.e., in our case, the number of realizations with values in the interval $[a, b]$. The function RF_f is a real-valued function with values between zero and one that operates on intervals. It is further asserted that RF_f is σ-additive. It can, hence, be extended to operate on the Borel σ-algebra and is assumed to satisfy all the axioms of a probability measure. A mechanism satisfying these conditions is called a random variable, named f, with probability distribution RF_f. The probability distribution describes the 'uncertainty' of the mechanism. This is the frequentist's construction of probability.

The above, somewhat pedantic, definition with the requirement that RF_f be independent of n makes explicitly the fact that it is only the tail of the sequence that matters. A more common and less pedantic definition is

$$RF_f(B) := \lim_{N\to\infty} \frac{1}{N} \sum_{n=1}^{N} \chi_B(s_n) \tag{14.4}$$

where s_n is again the sample and $B \in \mathcal{B}$.

The frequentist's interpretation of the statement that the probability of obtaining head is $P(h)$ when tossing a coin is thus: if one throws the coin an infinite number of times, in the above-defined random way, then the relative frequency with which one obtains 'head' is $P(h)$. Probability is a relative frequency or proportion. The frequentist, thus, locates probability in the measurement procedure, not in the underlying system, as the physicalist does. There is an even stronger frequentist interpretation that locates probability only in sequences of numbers, not in the system and measurement procedure that produces these numbers. Whether subscribing to the measurement or sequence interpretation, the frequentist views probabilities as statements about possible, potential outcomes and not about a specific outcome, or as it is often phrased, the frequentist considers populations and not individuals.

Frequentism has been justified by the strong law of large numbers. If the sequence $s_1, s_2, \ldots$ is regarded as a realization of an infinite sequence of iid random variables for which the mean exists, then the strong law of large numbers indeed states that

$$RF(B) = \lim_{N\to\infty} \frac{1}{N} \sum_{n=1}^{N} \chi_B(s_n) = P(B) \tag{14.5}$$

almost surely. Nevertheless, this justification and the above construction made a few assumptions and are problematic for a number of reasons.

Practical problems. There are practical issues:

(1) Not all events to which one might apply probabilistic reasoning are repeatable. Events can be unrepeatable for a variety of reasons. How would one asses the

probability that the author will break his bones if he jumps out of his office window or that the earth is warming or that Mr. Smith committed a certain crime? The whole universe constitutes only a single case event. The frequentist's approach is limited to a situation where repetition is at least in principle possible.

(2) The definition is not always practical, even in a coin tossing experiment; though one might address the perhaps interesting question of what wears out first, the coin tosser or the coin, mind or matter.

(3) The limit $N \to \infty$ cannot be carried out in reality. The world does not provide infinite sequences. As we pointed out, it is the tail of the sequence that matters whereas in reality one only has its beginning. Frequentism, thus, considers a hypothetical problem. This criticism can of course always be raised when one takes limits. One usually argues that finite sequences approximate the limit, and one then has to assess the errors and the rate of convergence.

(4) Relative frequencies are dependent on context. The author is a member of many different groups. He is male, lives in a city, and walks to work. There are different relative frequencies with which males, city dwellers, and pedestrians die. Which of these relative frequencies applies to the author? One needs to consider the intersection. But the author is a member of many more groups. In the end, there is the individual.

Formal problems. The major problem with the frequentist's view is its internal inconsistency. It purports that it only applies to sequences (finite or infinite) but in its justification assumes that probabilities can be assigned to single events. The verbiage of independence and equivalence covers the fact that a single event is assumed to have a probability. All the standard inferences from an 'independent' sequence of coin tosses assume that each toss has a probability and that it is the same for each toss.

Additionally, the definition of the relative frequency requires taking a limit and is, therefore, prone to all the subtleties of this operation. First, the limit simply might not exist but furthermore:

(1) The definition of relative frequency is not invariant under reordering of the sequence. The relative frequency of even numbers in the sequence $(1, 2, 3 \ldots)$ is 1/2 whereas it is 2/3 in the reordered sequence $(1, 2, 4, 3, 6, 8, 5, .)$.

(2) Infinite subsets s_{n_m} of the sequence s_n do not necessarily converge to the same limit.

$$\lim_{N\to\infty} \frac{1}{N} \sum_{n=1}^{N} \chi_A(s_n) = \lim_{M\to\infty} \frac{1}{M} \sum_{m=1}^{M} \chi_A(s_{n_m}) \tag{14.6}$$

does not hold in general. The relative frequency of even numbers in the sequence of all natural numbers is 1/2 whereas in the subset of even numbers, it is one.

(3) Relative frequencies are not necessarily σ-additive as the following counter-example proves. Consider a partition of the positive real line $\mathbb{R}_+ = \cup_{m=1}^{\infty} A_m$ such that $A_m(n) = \delta_{mn}$ for all $n, m \in \mathbb{N}$ and again the sequence $(1, 2, 3, \ldots)$ consisting of all integers. Then

$$
\begin{aligned}
RF(\cup_{m=1}^{\infty} A_m) &= \lim_{N\to\infty} \frac{1}{N} \sum_{n=1}^{N} \lim_{M\to\infty} \sum_{m=1}^{M} \chi_{A_m}(s_n) \\
&= \lim_{N\to\infty} \frac{1}{N} \sum_{n=1}^{N} \lim_{M\to\infty} \sum_{m=1}^{M} \delta_{nm} \\
&= 1
\end{aligned} \tag{14.7}
$$

whereas

$$
\begin{aligned}
\sum_{m=1}^{\infty} RF(A_m) &= \lim_{M\to\infty} \sum_{m=1}^{M} \lim_{N\to\infty} \frac{1}{N} \sum_{n=1}^{N} \chi_{A_m}(s_n) \\
&= \lim_{M\to\infty} \sum_{m=1}^{M} \lim_{N\to\infty} \frac{1}{N} \sum_{n=1}^{N} \delta_{nm} \\
&= 0
\end{aligned} \tag{14.8}
$$

This counter-example reflects the fact that limits and summations cannot always be interchanged. The order of limits will also become an issue when we consider multiplicative processes in Section (13.3).

One way to deal with some of the formal issues is to simply require that the mechanism that produces the sample produces sequences such that the relative frequencies are invariant under reordering and sub-sampling. Instead of simply requiring these properties, one might infer them from symmetry arguments.

Relative frequencies

$$
FF(B) := \frac{1}{N} \sum_{n=1}^{N} \chi_A(s_n) \tag{14.9}
$$

obtained from finite sequences $s_1, \ldots, s_N$, of course, do not suffer from problems caused by taking limits. They are σ-additive. The above counter-example does not work since the last factor in (14.8) remains at the finite value of $1/N$ for finite N. Though the relative frequencies $FF(B)$ satisfy all the axioms of probability theory, they are nothing but proportions. One simply takes a tally. The relative frequencies $FF(B)$ are not suitable to define a probabilities.

The attempt of frequentists to *define* probabilities via relative frequencies must be distinguished from the approach taken in statistics. There one takes probabilities for granted. The sequence $s_1, \ldots, s_N$ is *assumed* to be a realization of a (finite) sequence of iid random variables. The relative frequency $FF(B)$ is viewed as an estimate of a presumed probability $P(B)$. The relative frequency becomes an estimate of this probability.

14.4 The Logicist's Interpretation

The logicist's interpretation starts from the recognition that binary logic, the basis of classical propositional logic, is limited in its applications. Take the logical implication $A \rightarrow B$ in binary logic. This implication being true implies: if A is true, then B must be true, and if B is false, then A must be false. It does not say anything about the other two cases, A being false or B being true. Let "if it rains the lawn is wet" be a true implication, then I know that the lawn is wet when it rained and when the lawn is dry it has not rained. When it has not rained, my lawn might be either dry or wet; the yard person might have watered the lawn. Similarly, when the lawn is wet, it might or might not have rained; again the yard person might have watered the lawn. The logicist addresses these latter two cases.

The logicist's interpretation goes back to Laplace's (1812) original conception of probability theory as a precise mathematical theory to be used for reasoning in situations with incomplete, limited, or uncertain information, such as in the above example. It generalizes his law of insufficient reason which states: If there are $n > 1$ mutually exclusive, collectively exhaustive, and indistinguishable possibilities and no other information, then each possibility should be assigned a probability equal to $1/n$. Laplace's emphasis was on reasoning. The axiomatization of probability theory by Kolmogorov, based on measure-theoretical concepts and in part inspired by Max Born's view of probability in quantum mechanics, emphasized measuring and counting and distracted from Laplace's original view. R. T. Cox (1961) resurrected Laplace's view by trying to put it on an axiomatic basis. The name most associated with the logicist's approach is Jaynes' (2006) book: "Probability Theory: The Logic of Science". Probability theory is viewed as the generalization of propositional logic. It gives a precise mathematical formulation for logical reasoning in circumstances where one has incomplete or insufficient knowledge. The technical approach is as follows:

Plausibilities. The logicist views an event A as a proposition. In classical propositional logic, a proposition is either true or false (see Appendix E.18). Since information is incomplete, the logicist only assesses the conditional plausibility $PL(A|B)$ of a proposition A conditioned on the available evidence B. According to Cox, these plausibilities should satisfy three 'desiderata':

1. The plausibility $PL(A|B)$ of a proposition A is a real number between 0 and 1 and depends on the information B that one has related to the proposition.
2. The plausibilities should vary sensibly with the amount of information B that one has about the proposition.
3. If plausibilities can be arrived at in different ways, the result ought to be the same. In particular, anyone with the same information at hand should assign the same number to the plausibilities.

Various sets of precise (and simple and intuitive) axioms for these plausibilities have been put forward. One such set is as follows:

(1) Plausibilities are transitive

$$PL(A|D) > PL(B|D), PL(B|D) > PL(C|D) \Rightarrow PL(A|D) > PL(C|D) \tag{14.10}$$

This transitivity implies an ordering of plausibilities and allows plausibilities to be mapped onto real numbers.

(2) The plausibility of the negation $\neg A$ should be a function of $PL(A|B)$. There should exist a function G such that

$$PL(\neg A|B) = G[PL(A|B)] \tag{14.11}$$

Because a double negation is an affirmation, this function G must satisfy the functional equation $G \circ G = I$.

(3) The plausibility of the conjunction $A \wedge B$ of two propositions A and B given some evidence C is a function F of the plausibility of B given the evidence C and of the plausibility of A given the evidence B and C,

$$PL(A \wedge B|C) = F[PL(B|C), PL(A|B \wedge C)] \tag{14.12}$$

The conjunction is assumed to be a commutative and associative operation in this axiom. One, thus, could as well use $PL(A|C)$ and $PL(B|A \wedge C)$ as arguments of the function F. Dependencies on $PL(A|B)$ and $PL(B|A)$ are disregarded since the plausibility of A or B in the absence of C should not be relevant in a situation where one knows that C is true. The plausibilities $PL(C|A)$ and $PL(C|B)$ are also disregarded since C is assumed to be given. The plausibility $PL(A|C)$ is also not needed since in the above formulation one assesses the plausibility of $A \wedge B$ given C by first assessing the plausibility of B given C and then the plausibility of A given B and C. One can also demonstrate that other combinations of arguments fail the desiderata. Thus,

$$PL(A \wedge B|C) = F[PL(B|C), PL(A|C)] \tag{14.13}$$

does not work since the plausibility of A and B given C does not necessarily tell us anything about the plausibility of $A \wedge B$ given C. It is plausible that the next car you see has two doors. It is plausible that the next car you see is a sports car and it is plausible that the next car you see is a sports car with two doors. On the other hand, it is again plausible that the next car you see has two doors. It is also plausible that the next car you see is a taxi. It is, however, fairly implausible that the next car you see is a taxi with two doors.

The above axioms are intuitively reasonable. Jaynes demonstrates that these three axioms imply the sum and product rule

$$PL(A|C) + PL(\neg A|C) = 1 \tag{14.14}$$

$$PL(A \wedge B|C) = PL(A|B \wedge C)PL(B|C) \tag{14.15}$$

and by convention, the rules $PL(A|C) = 0$ if A is known to be false and $PL(A|C) = 1$ if A is known to be true. The sum and product rules are sufficient for performing all finite compound logical operations since the rule for the logical disjunction $A \vee B$ can be obtained from the rule of negation and conjunction by $A \vee B = \neg(\neg A \wedge \neg B)$. These sum and product rules become the sum and product rule of probability theory when A, B, and C are sets and the conjunction $\wedge$ becomes the intersection $\cap$, the negation $\neg A$ the complement $X \setminus A$, and the disjunction $\vee$ the union $\cup$.

The proof that the functions G and F are $G(x) = 1 - x$ and $F(x, y) = xy$ is given in Jaynes (2006) and is somewhat involved. There are two points that make Jaynes' approach somewhat unsatisfactory: (i) Jaynes' proof requires continuity and differentiability. This requirement is very unsatisfying since the arguments of PL are propositions and propositions belong to the mathematics of discrete objects. (ii) One would expect that one can express one's plausibility on an arbitrary scale from a to b and that only ratios such as $PL(A|B)/PL(\neg A|B)$ are scale-invariant. Jaynes obtains the result that plausibilities are real numbers between zero and one from the fact that zero and one are the only idempotent real numbers under multiplication.

Nevertheless, Jaynes' assertion is that the sum and product rules of probability are the rules of logical reasoning in the presence of uncertainty. One caveat is that the proof only yields finite σ-additivity, so one has to avoid reasoning that requires an infinite number of operations.

There are approaches that start from different sets of (intuitively obvious) axioms which do not require continuity and allow arbitrary scales, most notably Dupreé and Tipler (2004). Nevertheless, the sum and product rules are the foundations for the logicist. It is still an open question whether or not there exist other consistent rules for reasoning in the face of uncertainty. One might, of course, just assert that reasoning under uncertainty should obey the sum and product rule, rather than deriving these rules from (more obvious) axioms. Obviousness is in the eye of the beholder.

This logicist's interpretation relieves the statistician from the task of finding an objective operative definition for the probability. It only requires that the investigator should not arbitrarily assign plausibilities but should do so in a logically consistent manner. The logicist's view also allows the statistician to assign plausibilities to quantities that are not elements of a σ-algebra of events. Thus, the investigator can talk about the plausibility that it will rain tomorrow and the plausibility that Mr. Smith murdered his wife (or Mrs. Smith her husband).

The logicists are Bayesians in that their plausibilities are conditioned on (prior) information. Their inferences use conditional probabilities. They use samples to change a prior plausibility into an a posteriori plausibility.

14.5 The Subjectivist's Interpretation

The subjectivist's interpretation has its roots in betting strategies and in the theory of decision-making. It can be applied to a gambling situation where the die has been thrown and has a definite face value but where one player does know the face value, whereas the other players might not know it. It, thus, acknowledges that different people might ascribe different probabilities to events, depending on different information that they have, and that probabilities might change in time depending on changing circumstances. Here, we discuss the justification of the subjectivist's approach using consistent betting strategies. A betting strategy is deemed consistent if it is one that does not make you a sure loser. Standard textbooks are Jeffrey's (2004) "Subjective Probability: The Real Thing" and DeFinetti's (1974, 1975) "Theory of Probability". A modern treatment is Kadane's (2020) "Principles of Uncertainty".

The Dutch book argument. The subjectivist's approach asks the question: For what price $PR(A)$ are you willing to buy or sell lottery tickets that pay \$1 if A turns out to be true. Again one would like to infer the sum and product rule. To infer these rules, one relies on the *Dutch book argument.* A 'book' is the set of all bets a bookmaker has accepted. A 'Dutch book' is a book where the bookmaker will lose whatever the outcome.

The sum rule. Following Jeffrey, imagine the situation that there are two sages among the candidates for an election. Consider the proposition H_1 that the first sage will be elected, the proposition H_2 that the second sage will be elected, and the proposition H that a sage will be elected. If you are willing to bet $PR(H_1)$ and $PR(H_2)$ on the first two propositions, why should you bet

$$PR(H) = PR(H_1) + PR(H_2) \tag{14.16}$$

on the third proposition? If you bet less on H, the bookmaker will sell you one ticket for the price $PR(H_1)$ and one ticket for the price $PR(H_2)$ and will buy from you one ticket for the price of $PR(H)$. If no sage gets elected, the bookmaker wins $PR(H_1) + PR(H_2) - PR(H) > 0$. If a sage gets elected, then again the bookmaker wins the same amount because the bookmaker has to pay you \$1 and you have to pay the bookmaker \$1. This is because if H occurs, either H_1 or H_2 must be true. If you bet more on H, the bookmaker will buy from you one ticket each for the price $PR(H_1)$ and $PR(H_2)$ and sell you one ticket for the price of $PR(H)$. Again the bookmaker comes out ahead independent of the outcome of the elections. Thus, in order not to become a sure loser, your bets should obey the above additivity rule. If you add the normalization that the price for a ticket of a sure event should be \$1 and for the impossible event \$ 0, then this becomes the sum rule.

The product rule. To arrive at the product rule, consider the ticket T_1 that pays \$1 if $A \wedge B$ occurs, $PR(A|B)$ if B does not occur, and \$0 if B occurs but not A (see Fig. 14.1). If you pay $PR(A|B)$ for this ticket, it pays \$1 if A and B become true and returns to you the price $PR(A|B)$ you paid for the ticket if B becomes false; the bet is called off. Why should this price obey the chain rule? For this consider the

Price	A true B true	A true B false	A false B true	A false B false
T_1 $PR(A\|B)$	\$1	$PR(A\|B)$	\$0	$PR(A\|B)$
T_2 $PR(A \cap B)$	\$1	\$0	\$0	\$0
T_3 $PR(A\|B)PR(\neg B)$	\$0	$PR(A\|B)$	\$0	$PR(A\|B)$

Fig. 14.1 Price and pay-out of the three tickets considered in the 'derivation' of the product rule

additional tickets T_2 and T_3 whose price and pay-out are also listed in Fig. 14.1. As can be seen, the two tickets T_2 and T_3 pay the same amount as ticket T_1 whatever the outcome. Therefore, the price of ticket T_1 should be the same as the sum of the prices of tickets T_2 and T_3 or

$$PR(A|B) = PR(A \wedge B) + PR(A|B)PR(\neg B) \tag{14.17}$$

which becomes the product rule.

Thus, in order to not be a sure loser, your betting rates should obey the sum and product rules, from which everything else follows.

As with the logicist's interpretation, the subjectivist's view does not require an objective operative definition for probability. It only requires that the investigator should not arbitrarily quantify his or her degree of belief but should do so in a logically consistent manner. The subjectivist can also formulate his or her belief about events that do not constitute a σ-algebra.

Subjectivists are also Bayesians in that they realize that degrees of belief depend on or are conditioned by prior information.

14.6 The Utilitarian Approach

Since there is no convincing interpretation of probability in the real world, we present a pragmatic view in this section. Probabilities and the associated theory are used as a model for certain phenomena in the real world. Whether or not probability provides a good model depends on whether or not the model works. This approach is the utilitarian approach. We discuss the concept of a model, its verification and validation, and its utility in general.

Models, theories, and laws. Models have been used in science since their beginnings. Today, a model denotes any device that serves to represent a system in terms of which a scientist tries to understand or anticipate that system. Models represent only limited aspects of the system which are obtained by processes such as simplification, abstraction, omission, reduction, and idealization. One distinguishes different kinds of models with qualifiers such as (in alphabetical order) analogical, cognitive, computational, conceptual, dynamical, empirical, kinematic, mathematical, probabilistic, scale, statistical, stochastic, and theoretical. Here, we are, of course, concerned with mathematical, especially with probabilistic models.

Models are something less than a theory. A theory is more abstract and more fundamental than a model. A theory of matter is more than a model of matter. A law is even more abstract and fundamental. Models are tools to be used rather than ontological statements to be assessed for truth or falsity. A probabilistic model might be useful, independent of whether or not one agrees with Einstein (Born 1971) that "He does not throw dice". Bohr's (1913) model of the atom with electrons circling around the nucleus is useful even though it is not a correct quantum-mechanical description. As with any tool, a model can be the source of valuable information and significant insight in the hands of the able scientist or the source of delusions and misconceptions in the hands of the dilettante.

One usually thinks of models as models of a system, but as important is the purpose of the model: whether it is aimed at understanding, predicting, or developing policies. Different maxims apply to different purposes, ranging from "as simple as possible" for the purpose of understanding to "as detailed as necessary" for the purpose of policy development. A particular model might be useful for one purpose but not for another one. The important point is:

Models are not statements about a system that are true or false but they are tools that are useful or not useful, adequate or not adequate for a specific purpose.

One usually constructs models of a system by observing the system. One then studies the logical and mathematical consequences of the model in order to apply it in new situations. This extrapolation to new situations has been rationalized in various ways. One rationalization involves the processes of denotation, demonstration, and interpretation (Hughes 1997). An element of the real world is denoted by an element in the model. The logical and mathematical consequences of the model allow us to demonstrate theoretical conclusions. These conclusions must be interpreted in the real world. Often, scientists forget the last step, the interpretation. Then they might have learned something about the model but not about the real world.

Another rationalization is Mary Hesse's (1970) use of analogs. The model shares some features with reality. These features are called positive analogs. Other model features are known to not occur in reality or reality has features that are not represented or misrepresented in the model. These are called negative analogs. And then there are features for which it is not known whether or not they correspond to reality. These are the neutral analogs. If one assumes that neutral analogs are positive analogs, then the model provides new hypotheses and possibly new information about reality.

All these processes, by which one asserts new knowledge, whether by extrapolating to new situations or by converting neutral into positive analogs, require a leap of faith. There is no way around it. Asserting the validity of model results in new regimes is an act that requires a leap of faith.

Verification and validation. The typical verification procedure for a model is the epistemological loop which consists of the three steps (see Fig. 14.2):

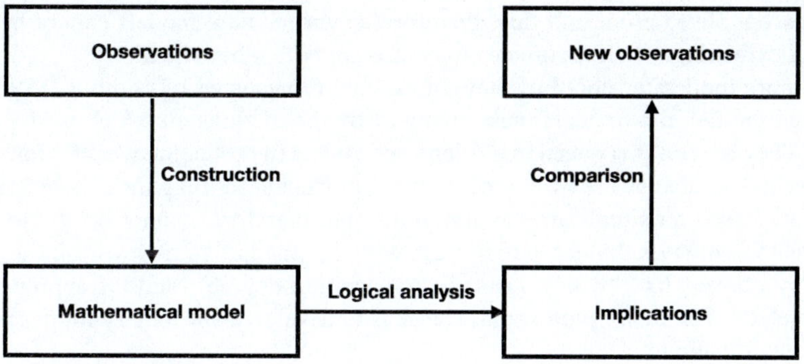

Fig. 14.2 The epistemological loop

1. Construction of the model from observations.
2. Logical and mathematical analysis of the model to arrive at new results.
3. Comparing these new results with new observations.

This verification procedure of models meets with various problems.

1. One problem arises from the rules of propositional logic: if proposition A implies proposition B, then non B implies non A. Let A correspond to the model and B to the model output. If B is true since the model output agrees with data, then one cannot say anything about the truth of the model. Another model might also produce output that agrees with the data. The situation is under-determined. If B is false since the model output disagrees with the data, then model A must be false as well. One cannot verify but only falsify propositions. One cannot establish the truth of propositions, be they hypotheses or models. This philosophical position has most prominently been advanced by the philosopher Karl Popper (1959). He also advocated that one must seek hypotheses that are falsifiable. Indeed, the more falsifiable a hypothesis is, the more valuable it is. If no negative instances are found, one gains trust in the truth of the hypothesis.

2. There are also practical problems with falsification. If the model output disagrees with the data, then there is something wrong but what? The data might be a fault; data might be suspect for many reasons. Or the hypothesis or model might be at fault and then it might be some principal or some auxiliary aspect of the model. And then there is the question: what exactly do we mean when we say the model output agrees with the data?

Since models cannot be verified, one has to settle for something less, which is called validation. A model is said to be valid if it is consistent with data and contains no obvious flaws or inconsistencies. Data are said to confirm or corroborate a model when they are consistent with it. In a much-referenced article Oreskes et al. (1994) take the extreme view that computer models, at least in environmental sciences, can neither be verified nor validated, for the above logical reasons and for the fact that environmental systems are never closed and can never be known completely due to their fractal structure. They conclude that computer models are just heuristic devices, not susceptible to proof and that, therefore, environmental models cannot be used for policymaking. Their arguments may also apply to other situations.

A more moderate school of thought as, e.g., represented by Morton (1993) has realized the fact that models make many of the predictions on which we base our lives. They are routinely used in airplane design and in producing weather forecasts. This school is also aware of the fact that the models describing these complex systems are based on simplifying assumptions and, therefore, cannot be strictly true. This school replaces the standard of truth with the standard of trustworthiness, qualified by phrases like “within limits”, “to a certain degree”, and “if appropriately interpreted”. The major philosophical task is to develop a vocabulary for describing these qualifications.

Why is the use of models controversial in climate sciences and not in weather forecast or aircraft design? Weather forecasts and airplanes work! The models have proven their skill many times over. This classical epistemological approach of hypothesis testing, formulating a hypothesis, comparing it to data, and then either accepting or rejecting it, works. The epistemological loop does not work for the climate system with its long time scales. Instead, as discussed in Müller (2010), new information, either in the form of new data, a new parameter estimate, or a new dynamical insight, is used not to test the model but to update the model, leading to an improved modified model in which we can have more trust. Updating a model is the Bayesian approach in statistics. One assigns a certain degree of belief to one’s model, the a priori probability. When a new piece of information arises, one updates the model and increases one’s degree of belief in it to a new a posteriori probability.

Even if a model has survived the epistemological loop repeatedly in the past, there is no guarantee that it will do so in the future. The well-known story is one of the turkeys whose model of the world is that humans are benevolent, providing them twice a day with delicious and nutritious feed; a model that has passed the epistemological loop many times, until a few days before Thanksgiving Day it turned out to be wrong.

Epistemological attitude. The utilitarian approach is the most versatile and malleable approach. It requires the following philosophical stance:

(1) One must distinguish between the Platonic world and the real world. Probability theory belongs to the Platonic world.
(2) When we apply probability theory to real-world problems, then we are using probability theory as a model. Probability and statistical theories are models of reality rather than statements about reality. Models are just tools to learn something about reality. They are not true or untrue; they might be useful or not useful for a specific purpose. Random or stochastic models share this property with any other model. Treating molecules in a gas as mass points and applying the deterministic laws of classical mechanics to them are useful for some purposes; treating their positions and momenta as realizations of a random variable is useful for other purposes. The ultimate justification of a probabilistic model, as with any model, is that it works and that it survives the epistemological loop. As with any tool, models can be used to great virtue or to great disaster.

(3) Treating a real-world process as a random process is a model assumption and does not exclude that the process is deterministic. In the real world, an event is a random event if it has no apparent cause. The adjective 'apparent' in this definition is to indicate that 'no cause' might be a subjective assessment, not an objective quality. The causal chain might be just too complicated to be treated in a deterministic manner and so the event is treated as random. The celebrated rolling of a die is treated as random, not because of an underlying indeterminacy but because we lack all the detailed information for its deterministic prediction (assuming that the laws of classical mechanics apply).

(4) One must carefully distinguish between the usage of concepts in probability theory and in the real world. Take the adjective 'independent'. In probability theory, statistical independence is a well-defined property of events or random variables. The major colloquial usage calls a drawing (or outcome) independent when it is obtained by a procedure or mechanism where all drawings are equivalent, do not affect each other, and where the result of one drawing does not provide any information about the next or any other drawing. Whether or not a given mechanism is well modeled by a set of statistically independent random variables is determined by how well this assumption survives the epistemological loop.

14.7 Summary

In this chapter, we considered four different interpretations of probability: the physicalist's, the frequentist's, the logicist's, and the subjectivist's interpretation. These different interpretations identify different entities to which probabilities can be assigned.

The physicalist assigns probabilities to subsets of the state space of a dynamical system. Probabilities are a property of the dynamical system. In quantum mechanics, the probability that an observable takes certain values depends on the state of the system.

The frequentist assigns probabilities to subsets of possible outcomes of an experiment. The probability is the relative frequency with which this outcome occurs in an infinite sequence of repeated experiments. Probability is, thus, a property of an infinite sequence of numbers.

The logicist assigns probabilities to logical propositions. The probability is the plausibility that a proposition A is true given that a proposition B is true. Probability theory, thus, provides the rules of propositional logic in the case of insufficient knowledge.

The subjectivist also assigns probabilities to the proposition. The probability is a person's degree of belief that proposition A is true given some knowledge B. Probability theory, thus, provides the rules to formulate a person's beliefs consistently, without becoming a sure loser.

According to these different interpretations, probability, thus, resides in the physical system, in infinite sequences, in logic, or in the mind.

These four interpretations attempt to 'derive' Kolmogorov's axioms from certain premises which are claimed to be self-evident. 'Deriving' countably infinite σ-additivity is a major problem and they usually settle for 'deriving' the sum and product rule. Logicists and subjectivists also assert that probabilities only make sense when conditioned on prior knowledge.

Because none of the interpretations is sufficiently general and satisfactory, we advocate the utilitarian approach. It leaves behind the question of whether or not there exist probabilities in the real world and asks instead whether or not probability might be an adequate model for events and phenomena in the real world, with adequate meaning surviving the epistemological loop.

References

Bohr, N.: On the constitution of atoms and molecules. Part I. Philos. Mag. **26**, 1–24 (1913)

Born, M.: The Born-Einstein Letters: Friendship, Politics and Physics in Uncertain Times. Translated by I Born, Macmillan, London, 235pp (1971)

Cox, R.T.: The Algebra of Probable Inference. The Johns Hopkins Press, Maryland (1961)

deFinetti, B.: Theory of Probability, Vol. I, II, (1974), (1975)

Dupreé, M.J., Tipler, F.J.: New axioms for rigorous Bayesian probability. Bayesian Anal. **1**(1), 1–8 (2004)

Guttman, Y.: The Concept of Probability in Statistical Physics. Cambridge University Press, Cambridge (1999)

Hesse, M.B.: Models and Analogies in Science. University of Notre Dame, Notre Dame (1970)

Jaynes, E.T.: Probability Theory. Cambridge University Press, The Logic of Science (2006)

Jeffrey, R.: Subjective Probability: The Real Thing. Cambridge University Press, Cambridge (2004)

Kadane, J.B.: Principles of Uncertainty. CRC Press (2020)

Morton, A.: Mathematical models: questions of trustworthiness. Brit. J. Phil. Sci. **44**, 659–674 (1993)

Müller, P.: Constructing knowledge with computer models. WIRE Clim. Chang. **1**, 565–580 (2010)

Oreskes, N., Shrader-Frechette, K., Belitz, K.: Verification, validation, and confirmation of numerical models in the Earth sciences. Science **263**, 641–646 (1994)

Popper, K.: The Logic of Scientific Discovery. Routledge Classics, London New York (1959)

Chapter 15
Parameter Estimation and Hypothesis Testing

15.1 Introduction

The task of determining probabilities is generally referred to as statistics. Statistics analyzes evidence, observations, data or samples. These samples are regarded as realizations of random variables. Regarding a sample as a realization of a random variable is the basic assumption of statistics. A major sub-discipline of statistics is statistical inference. It provides methods of how to infer properties of the assumed underlying probability distribution from the sample. Other sub-disciplines deal with problems such as what constitutes a good sampling strategy and an appropriate probability model. Given a probability model that is characterized by a set of parameters, statistical inference addresses two main problems: the estimation of parameters and the testing of hypotheses about these parameters.

In this chapter, we discuss *methods* for estimating parameters and testing hypotheses about these parameters. They are interpreted from a frequentist's point of view. Here we discuss only the properties of these methods. This part of parameter estimation and hypotheses testing is objective. We will defer questions concerning the selection of a method to the chapter on Bayesian inference since these are subjective decisions that need to be done by the statistician based on aspects that are extraneous to probability theory.

The major points will be:

(1) There are different methods to estimate parameters and test hypotheses about these parameters.
(2) One can objectively describe the properties of these different methods.
(3) Choosing a specific method will depend on extraneous arguments, such as what else one knows and what acceptable risks are.
(4) Statistics does not make statements about single realizations in the frequentist's interpretation.

P. Müller, *Handbook of Dynamics and Probability*,
https://doi.org/10.1007/978-3-030-88486-4_15

15.2 Sampling

The statistician analyses samples. What is a sample? Consider the drawing of balls from an urn that contains black and white balls. Suppose for definiteness an urn that contains white and black balls. Draw a ball from the urn, register its color, and put it back into the urn and then draw another ball. Do this N times. Denote the result "black ball" by 0, and the result "white ball" by 1. The resulting sequence $s = (s_1, \ldots, s_N)$ is thus a specific binary sequence of length N, consisting of the numbers 0 and 1. This is your sample. This sample is regarded as a specific realization of the random variable $f := (f_1, \ldots, f_N)$ where f_i is the random variable that describes the possible outcomes of the ith draw.

In the general case, we define as a sample s of size N

$$s = (s_1, \ldots, s_N) = (f_1(\omega), \ldots, f_N(\omega)) \tag{15.1}$$

a specific realization of a random variable $f := (f_1, \ldots, f_N)$. Here we introduced a new symbol s for the sample, although it is just a realization of a sequence of random variables which up until now we have denoted by $f(\omega)$. We introduce a new symbol because we want to stress that a sample is just a sequence of numbers. The ω on the right-hand side represents the specific sample being drawn. This ω in the argument of the random function is fixed and cannot be varied any more once the sample is drawn. Working with $f(\omega)$ instead of s might occlude this fact and one might be tempted to consider ω a variable argument in $f(\omega)$. The sample s is a sequence of numbers. It is a specific realization of the random variable f. The random variable f has a distribution. The sample does not have a distribution.

The sampling procedure or mechanism might also be of the kind that all drawings are equivalent and independent from each other such that the result of one drawing does not affect or is affected by any other drawing. In such circumstances, all components of the random variable $f = (f_1, \ldots, f_N)$ are *assumed* to be identically distributed and statistically independent. Its probability density function is thus given by

$$\rho_f(x_1, \ldots, x_N) = \prod_{i=1}^{i=N} \rho_{f_i}(x_i) \tag{15.2}$$

Statistical independence is the formal property that assures that successive drawings have all the properties that we require of 'random drawings'. In particular, the conditional density becomes $\rho_{f_i|f_j}(x_i|x_j) = \rho_{f_i}(x_i)$. The ith outcome does not depend on the jth outcome for $i \neq j$.

The assumptions of statistical independence and identical distribution cannot be proven. They must be assumed. They are adequate if the inferences survive the epistemological loop.

15.3 Estimator Versus Estimate

Let us now state the basic premise of estimation theory. Estimation theory starts from a sample $s = (s_1, \ldots, s_N)$. A specific sample is an element of the sample space S, containing all possible samples. The basic premise is that these observations are realizations of a sequence of random variables $f = (f_1, \ldots, f_N)$. The probability density function of f is assumed to belong to a family $\rho(x; \theta)$ characterized by a set of parameters $\theta = (\theta_1, \ldots, \theta_K)$. The parameters are elements of the parameter space Θ. Often one assumes that the random variables f_n are identical and independent. The important point is that the sample s is a realization of the random variable f. Examples of the parameter θ, considered below, are the mean, the variance, and the percentiles of the probability distribution.

Statistic. The problem of statistical estimation is to estimate the parameter θ from the sample s. To do this, one introduces a function E_θ from the sample to the parameter space

$$\begin{aligned} E_\theta : S &\mapsto \Theta \\ s &\mapsto E_\theta(s) \end{aligned} \tag{15.3}$$

This function is called a statistic.

Estimator. For understanding estimation theory, it is important to distinguish between numbers and random variables. We thus have to distinguish between two operations: when the specific sample s is chosen as an argument of E_θ, one obtains an estimate $E_\theta(s)$ of the parameter θ; when E_θ is composed with the random variable f, one obtains the random variable

$$E_\theta \circ f \tag{15.4}$$

called an estimator. An estimate is a number. The estimator is a random variable. The estimate is a realization of the estimator.

Example: Estimating the mean. Consider the basic task of estimating the mean of f_1 from a sample $s = (s_1, \ldots, s_N)$ assumed to be a realization of N independent identically distributed random variables. Typically, statistical inference uses the sample mean

$$\bar{s}_N := \frac{1}{N} \sum_{j=1}^{N} s_j \tag{15.5}$$

The function E_μ is then given by

$$E_\mu(s) = \bar{s}_N \tag{15.6}$$

When the method of calculating a mean from the sample is considered, then the estimator

$$E_\mu \circ f = \frac{1}{N}\sum_{j=1}^{N} f_j =: \bar{f}_N : \tag{15.7}$$

is introduced which is a random variable. The sample mean $\bar{s}_N$ has no uncertainty; it is just a number. The random variable $\bar{f}_N$, on the other hand, assigns a probability to all possible outcomes. It is a statement about how the method performs when applied repeatedly. The sample mean $\bar{s}_N$ is a realization of the random variable $\bar{f}_N$.

Since the estimator is a random variable, one can determine its expectation value which is

$$\langle \bar{f}_N \rangle = \langle f_1 \rangle = \mu_{f_1} \tag{15.8}$$

Thus, calculating the sample mean is a meaningful way of *estimating* the mean of f_1. The variance of $\bar{f}_N$ is given by

$$\langle (\bar{f}_N - \mu_{f_1})^2 \rangle = \frac{\sigma_{f_1}^2}{N} \tag{15.9}$$

and decreases with increasing N.

In the above discussion, we assumed that mean and variance exist. This does not need to be the case. Indeed, if we assume f_1 to be distributed according to the Cauchy distribution

$$\rho_C(x; x_0 = 0, a = 1) = \frac{1}{\pi}\frac{1}{1+x^2} \tag{15.10}$$

then we arrive at the peculiar situation that $\bar{f}_N$ is distributed according to the same distribution, $\rho_{\bar{f}_N} = \rho_{f_1}$. This is most easily proven by using the convolution theorem and noting that the characteristic function of the Cauchy distribution is $\hat{\rho}_C(k) = e^{-k}$. Taking additional measurements does not help. The central limit theorem does not apply.

The above line of argument, presented for the simplest of all cases, namely, the estimation of the mean, applies to *all* estimation problems. We present one more example, namely, confidence intervals.

Example: Estimating confidence intervals. When estimating a parameter θ one would like to have a confidence interval $[c_L, c_U]$ such that $P(c_L < \theta < c_U) = q$, where q is a certain given probability. As an explicit example, assume f_1 to be a normally distributed random variable with unknown mean μ_{f_1} and known variance $\sigma_{f_1}^2$. As we just discussed the estimator $E_\mu \circ f = \bar{f}_N$ of the mean has mean μ_{f_1} and variance $\sigma_{f_1}^2/N$. To arrive at the confidence interval for the unknown mean μ_{f_1}, we transform to the random variable

$$g := \sqrt{N}(\bar{f}_N - \mu_{f_1})/\sigma_{f_1} \tag{15.11}$$

which has a standard Gaussian distribution with zero mean and unit variance. Given a fraction q, we can thus determine numbers x_L and $x_U = -x_L$ such that $P_g(x \in$

$[x_L, x_U]) = q$. These numbers are tabulated for the standard Gaussian distribution for various values of q. Transforming back to $\bar{f}_N$ one finds

$$P_{\bar{f}_N}\left(\bar{s}_N - x_U \frac{\sigma_{f_1}}{\sqrt{N}} < \mu_{f_1} < \bar{s}_N + x_U \frac{\sigma_{f_1}}{\sqrt{N}}\right) = q \tag{15.12}$$

The functions

$$E_{c_L, c_U}(s) = \bar{s}_N \mp x_U \frac{\sigma_{f_1}}{\sqrt{N}} \tag{15.13}$$

thus provides estimates of the confidence interval $[c_L, c_U]$ of the mean μ_{f_1}.

To summarize, estimation theory introduces a function $E_\theta : S \mapsto \Theta$, called a statistic. This function becomes an *estimator* of the parameter when it is composed with the random variable $f = (f_1, \ldots, f_N)$; thus the estimator is the random variable $E_\theta \circ f$. The function becomes an estimate when the sample $s = (s_1, \ldots, s_N)$ is substituted, $E_\theta(s_1, ..., s_N)$. The estimator has a probability distribution, an expectation value $\langle E_\theta \circ f \rangle$ and a variance $\langle (E_\theta \circ f)^2 \rangle - \langle E_\theta \circ f \rangle^2$. Estimates and estimators are different entities. Estimates are realizations of a random variable.

15.4 Interpretation of Estimates

Estimation theory provides an estimate and the properties of the estimator. There are two problems: one is to interpret the estimate and the other to choose the estimator. Here we consider the first problem. The choice of the estimator is considered in the following sections.

The usual interpretation of an estimate is the frequentist's interpretation: When we repeatedly apply the estimation formula $\bar{s}_N = \frac{1}{N} \sum_{j=1}^{N} s_j$, then we draw samples from the random variable $\bar{f}_N$. The error of this operation $\bar{s}_N - \mu_{f_1}$ will be on average zero, and the standard deviation of this operation will be $\sigma_{f_1}/\sqrt{N}$. This is a correct assertion, but it does not imply anything about the error made, when *one* concrete sample is used to estimate the mean of the random variable f_1. Estimation theory makes statements about the accuracy of the *method* of estimating parameters, not about the error made when a single estimate is derived from a sample. When one estimate is calculated, then one realization of a random variable is drawn. This may be a number much larger or much smaller than the mean, and there is no way to find out how close it actually is to the mean.

The same interpretation applies to the confidence limits. When we repeat the sampling and calculate the confidence limits $(E_{c_L}(s), E_{c_U}(s))$ often enough then in $q \times 100$ percent of all cases the (changing) interval will contain the fixed parameter μ_{f_1}. Again, nothing is said about a single realization. When we estimate a realization $(E_{c_L}(s), E_{c_U}(s))$ of the random variable $(E_{c_L} \circ f, E_{c_U} \circ f)$, then we do not know whether it contains the parameter μ_{f_1} or not. The confidence interval does not provide us with confidence about the location of the true parameter but about the accuracy of the method. When we calculate a *confidence interval* for a parameter, we are not voicing our confidence about a number but about the process of calculating this number. In the chapter on Bayesian inference, we will see how one may actually gain confidence in the location of the parameter.

Often the law of large numbers is invoked to get around the distinction between estimates and estimators. The strong law of large numbers states that $\bar{f}_N$ converges almost surely against $\langle f_1 \rangle$ in the limit $N \to \infty$. The real world, however, only provides finite samples, $\bar{s}_N$ with $N < \infty$, and these have a finite standard deviation $\sigma_{f_1}/\sqrt{N} > 0$.

15.5 Maximum Likelihood Estimators

Here we consider a particular estimator, the maximum likelihood estimator, that illuminates some of the issues that go into the choice of an estimator.

Likelihood. Laypersons often use the words probability and likelihood interchangeably. In probability theory, likelihood has, however, a technical definition that differs from probability. The likelihood of an event B for a given event A is defined by $P(A|B)$, with the understanding that the argument before the bar is fixed, whereas the argument behind the bar varies. The likelihood is not a probability, e.g., $P(A|\Omega) = P(A) \neq 1$, hence the different name.

Maximum likelihood function. One systematic way to construct estimators is the maximum likelihood method. The parameters are determined such that the probability density function takes its maximum value for the sample. Let us assume that we have a sample $s = (s_1, \ldots, s_N)$ available to estimate a parameter θ. Let $\rho_n(x; \theta)$ be the known density function of f_n with the unknown parameter θ. If the f_n are assumed to be identically and independently distributed, then the joint probability density function of $f = (f_1, \ldots, f_n)$ is given by

$$\rho_f(x_1, \ldots, x_N; \theta) = \prod_{j=1}^{N} \rho_{f_1}(x_j; \theta) \tag{15.14}$$

Here $(x_1, \ldots, x_N)$ are the independent arguments and θ is a given parameter. If we insert the realization $(s_1, \ldots, s_N)$ into this formula, we obtain the likelihood function

$$L(\theta; s_1, \ldots, s_N) = \prod_{j=1}^{N} \rho_{f_1}(s_j; \theta) \tag{15.15}$$

where the sample is now given and θ is assumed to be the independent variable. It is called likelihood function, because it is the argument before the semicolon that is fixed and the argument behind the semicolon that varies and the argument behind the semicolon that is fixed. A maximum likelihood estimate of the parameter θ is the value θ_{max} that maximizes this likelihood function. This estimate depends on the realization s, $\theta_{max} = \theta_{max}(s_1, \ldots, s_N)$. By substituting the random variables for the realization, we obtain the maximum likelihood estimator

$$E_\theta^{ML} \circ f = \theta_{max}(f_1, \ldots, f_N) \tag{15.16}$$

Example: Gaussian random variables. Let all f_n be iid $N(\mu_{f_1}, \sigma_{f_1}^2)$ Gaussian random variables. The logarithm of the likelihood function then becomes

$$\begin{aligned} \log L(\mu_{f_1}, \sigma_{f_1}^2; s_1, \ldots, s_N) &= -N \log \sqrt{2\pi} - N \log \sigma_{f_1} - \sum_{n=1}^{N} \frac{(s_n - \mu_{f_1})^2}{2\sigma_{f_1}^2} \\ &= -N \log \sqrt{2\pi} - N \log \sigma_{f_1} - \frac{S_1^2}{2\sigma_{f_1}^2} - \frac{N(\bar{s}_N - \mu_{f_1})^2}{2\sigma_{f_1}^2} \end{aligned} \tag{15.17}$$

where

$$S_1^2 := \sum_{j=1}^{N} (s_j - \bar{s}_N)^2 \tag{15.18}$$

Only the combinations $\bar{s}_N$ and S_1^2 of the data enter the likelihood function. These two quantities thus constitute 'sufficient statistics'. Contours of this maximum likelihood function are shown in Fig. 15.1.

We may now distinguish between different epistemological situations and different purposes:

1. Both the mean and the variance are not known and one would like to estimate both. Then one needs to seek the maximum of the likelihood function in the two-dimensional $(\mu_{f_1}, \sigma_{f_1}^2)$-plane which is at

$$\mu_{f_1} = \bar{s}_N \tag{15.19}$$

$$\sigma_{f_1}^2 = \frac{S_1^2}{N} \tag{15.20}$$

2. The variance $\sigma_{f_1}^2$ is known and one would like to estimate the mean. One then needs to seek the maximum of the likelihood function along the line of the known value of the variance which is also at

$$\mu_{f_1} = \bar{s}_N \tag{15.21}$$

3. The mean μ_{f_1} is known and one would like to estimate the variance. One then needs to determine the the maximum of $\sigma_{f_1}^2$ along the line of the known value of the mean which is at

$$\sigma_{f_1}^2 = S_1^2/N + N(\bar{s}_N - \mu_{f_1})^2 \tag{15.22}$$

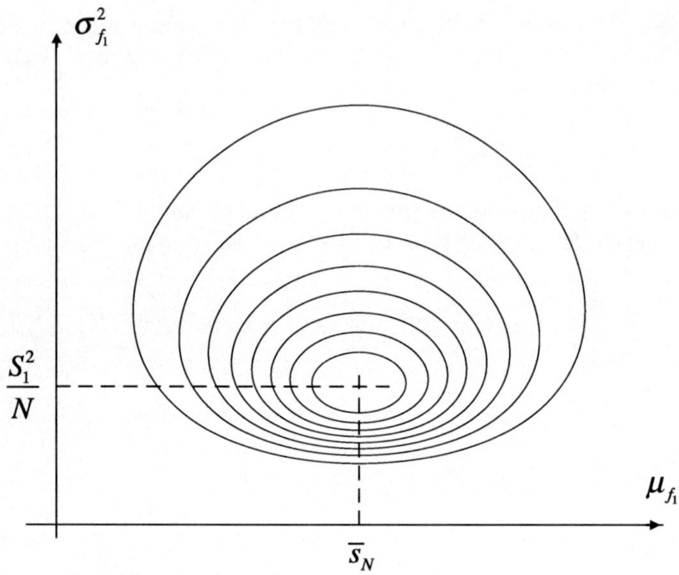

Fig. 15.1 Contours of the maximum likelihood function (15.17). The spacing between the contour lines is uniform in L, not in $\log L$

4. Yet another situation is that we do not know both the mean and the variance but only want to estimate the variance. One way to do this is to integrate the likelihood function over μ_{f_1}, or to marginalize μ_{f_1}, as the jargon is. The logarithm of the likelihood function then becomes

$$\log L(\sigma^2_{f_1}; s_1, \ldots, s_N) = -N \log \sqrt{2\pi} - N \log \sigma_{f_1} - \frac{S_1^2}{2\sigma^2_{f_1}} + \log \sqrt{\frac{2\pi\sigma^2_{f_1}}{N}} \tag{15.23}$$

The marginalization gives an extra term $\log \sigma_{f_1}$ so that the maximum likelihood estimate becomes

$$\sigma^2_{f_1} = \frac{S_1^2}{N-1} \tag{15.24}$$

This example illustrates an important point, which we will elaborate upon in the following sections, namely:

The choice of the estimator depends on what one knows or does not know.

15.6 Mean Square Error

There are no strict rules on how to design estimators. Different estimators have different properties. Whether or not these properties are desirable depends on the problem under consideration and on the purpose of the investigation. Here we address the question of how to compare different estimators.

Mean square error. One way to compare different estimators for the same parameter θ is the mean square error (MSE)

$$m_{E_\theta \circ f}(\theta_t) := \langle (E_\theta \circ f - \theta_t)^2 \rangle \tag{15.25}$$

where θ_t is the true value of the parameter. The mean squared error may be split into two contributions

$$m_{E_\theta \circ f} = b^2_{E_\theta \circ f} + \sigma^2_{E_\theta \circ f} \tag{15.26}$$

where

$$b_{E_\theta \circ f} := \langle E_\theta \circ f \rangle - \theta_t \tag{15.27}$$

is called the bias. It represents the expected error of the estimator or the systematic error. The variance represents the irregular fluctuations of the estimator. If the bias is zero the estimator is called *unbiased*. The mean square error of an estimator depends, however, on both, the bias *and* the variance. When estimating the mean μ_{f_1} by the sample mean $\bar{s}_N$, then the bias is zero and the variance $\sigma^2_{f_1}/N$.

If the MSE of one estimator is smaller than the MSE of another estimator for all values of θ_t, then the first estimator is called more efficient than the second estimator. An estimator is called consistent if the MSE converges to zero for $N \to \infty$. To further illustrate these concepts, we consider the estimation of the variance.

Example: Estimation of variance. Consider the two estimates of the variance

$$S^2_N := \frac{1}{N} \sum_{j=1}^{N} (s_j - \bar{s}_N)^2 \tag{15.28}$$

$$S^2_{N-1} := \frac{1}{N-1} \sum_{j=1}^{N} (s_j - \bar{s}_N)^2 \tag{15.29}$$

that we encountered in the section on maximum likelihood estimators. The biases of these estimators are

$$\begin{aligned} b_{S^2_N \circ f} &= -\frac{\sigma^2_{f_1}}{N} \\ b_{S^2_{N-1} \circ f} &= 0 \end{aligned} \tag{15.30}$$

If f_1 is a Gaussian random variable (such that fourth order moments factorize into product of second order moments), then the variances of the respective estimators are

$$\begin{aligned}\sigma^2_{S^2_N \circ f} &= \frac{2(N-1)}{N^2}\sigma^4_{f_1}\\ \sigma^2_{S^2_{N-1} \circ f} &= \frac{2}{N-1}\sigma^4_{f_1}\end{aligned} \tag{15.31}$$

and the mean square errors are

$$\begin{aligned}m_{S^2_N \circ f} &= \frac{2N-1}{N^2}\sigma^4_{f_1}\\ m_{S^2_{N-1} \circ f} &= \frac{2}{N-1}\sigma^4_{f_1}\end{aligned} \tag{15.32}$$

The estimator $S^2_{N-1} \circ f$ is unbiased; the estimator $S^2_N \circ f$ is biased but more efficient than the unbiased estimator since

$$\frac{m_{S^2_N \circ f}}{m_{S^2_{N-1} \circ f}} = \frac{(2N-1)(N-1)}{2N^2} < 1 \tag{15.33}$$

for all $N > 1$. Both estimators are consistent.

Note that the mean square error depends on the true (or to-be-estimated parameter) $\sigma^2_{f_1}$. We were able to compare the two mean square errors because they happen to be both proportional to $\sigma^4_{f_1}$, which is a consequence of our assumption of underlying Gaussian random variables. This dependence of the mean square error on the unknown true parameter generally complicates the comparison of estimators. Consider the two estimators resulting from $E^{(1)}_\theta = c_1$ and $E^{(2)}_\theta = c_2$. Their mean square errors are $(c_1 - \theta_t)^2$ and $(c_2 - \theta_t)^2$. None of the mean square errors is smaller than the other for all values of θ_t.

The dependence of the mean square error on the to-be-estimated parameter also complicates attempts to find estimators that minimize the mean square error. If one varies the mean square error functional with respect to the estimator, then one obtains

$$\begin{aligned}\delta\langle(E_\theta \circ f - \theta)^2\rangle &= \delta \int_S ds (E_\theta(s) - \theta)^2 \rho_f(s)\\ &= \int_S ds 2(E_\theta(s) - \theta)\rho_f(s)\delta E_\theta(s) = 0\end{aligned} \tag{15.34}$$

which results in the optimal estimator $E_\theta(s) = \theta_t$ which is, of course, optimal but useless since it requires knowledge of the true parameter.

One way out of this problem is to average over a reasonable range of values of the true parameter. We will do so in the next section where the concept of the mean square error is shown to be a special case of the more general concept of a loss function.

15.7 Minimizing Risk

The mean square error is not the only measure to compare different estimators. Decision theory introduces concepts like loss and risk to provide a more general framework.

Decision function. Decision theory is based on a decision function which is a function that maps the sample space S into an action space A

$$D : S \to A \tag{15.35}$$

The decision function must partition the sample space. It must assign an action $a \in A$ to all possible outcomes $s \in S$. The function E_θ is a special case of a decision function. Its action consists of ascribing the value $E_\theta(s)$ to the parameter θ.

Loss function. Decision theory attempts to find an optimal decision function. To find such an optimal decision function, the investigator must quantify the losses that occur when action $a \in A$ is taken if the true parameter value is θ_t. This is done by a loss function

$$L : A \mapsto \mathbb{R} \tag{15.36}$$

which maps the action space into the real numbers and describes the loss if action a is taken. Alternatively, one may specify a gain function that describes the gain of the action. Composing the decision and loss function one arrives at the risk function

$$R := L \circ D : S \mapsto \mathbb{R} \tag{15.37}$$

Risk. In evaluating the effectiveness of a specific decision function, one must again recognize that one cannot assess how well it performs for a specific sample but only how well it performs on average. One must again make the important distinction between numbers and random variables. Instead of the sample s, we should consider the random variables f of which the sample is a realization. The function $R \circ f$ is then also a random variable. The risk is then defined as the expectation value $\langle R \circ f \rangle$.

A possible loss function is the quadratic function

$$L(a) = (a - \theta_t)^2 \tag{15.38}$$

and the risk function becomes

$$R(s) = (E_\theta(s) - \theta_t)^2 \tag{15.39}$$

and the risk becomes the *mean square error*

$$m_{E_\theta}(\theta_t) = \langle (E_\theta \circ f - \theta_t)^2 \rangle \tag{15.40}$$

Minimizing the risk. The overall idea is to determine an optimal decision function by minimizing the risk (or maximizing gain or utility). As we already pointed out, when discussing the mean square error, the minimization of the risk is not a well-defined variational problem since the risk depends on the value of the unknown true parameter θ_t. Since this is the root of the problem, one way out is to take an average over a reasonable range of θ_t with an averaging function $g(\theta_t)$. One then arrives at

$$\overline{\langle R\rangle} = \int\int d\theta_t ds g(\theta_t)(E_\theta(s) - \theta_t)^2 \rho_f(s;\theta_t) \tag{15.41}$$

where the expectation value is written as an integral over s with probability density $\rho_f(s;\theta_t)$. If this functional is varied with respect to $E_\theta(s)$, we obtain

$$\delta\overline{\langle R\rangle} = \int\int d\theta_t ds g(\theta_t) 2(E_\theta(s) - \theta_t)\rho_f(s;\theta_t)\delta E_\theta(s) \tag{15.42}$$

This variation vanishes for

$$E_\theta(s) = \frac{\int d\theta_t g(\theta_t)\theta_t \rho_f(s;\theta_t)}{\int d\theta_t g(\theta_t)\rho_f(s;\theta_t)} \tag{15.43}$$

The variation thus leads to a well-defined solution. If we introduce $\rho(s|\theta_t) := \rho_f(s;\theta_t)$ and $\rho(\theta_t) := g(\theta_t)$, then $E_\theta(s)$ becomes the mean value of the distribution

$$\rho(\theta_t|s) = \frac{\rho(s|\theta_t)\rho(\theta_t)}{\int d\theta_t \rho(s|\theta_t)\rho(\theta_t)} \tag{15.44}$$

which we will encounter as the posterior distribution in Bayesian inference. Thus, minimizing the mean square error corresponds to taking the mean value of the posterior distribution.

Similarly, if we take as the loss function the absolute error $L(a) = |a - \theta_t|$, then the variation of the averaged risk leads to

$$\delta\overline{\langle R\rangle} = \int ds[\int_{-\infty}^{E_\theta(s)} d\theta_t g(\theta_t)\rho(s;\theta_t) - \int_{E_\theta(s)}^{\infty} d\theta_t g(\theta_t)\rho(s;\theta_t)]\delta E_\theta(s) \tag{15.45}$$

and vanishes if

$$\int_{-\infty}^{E_\theta(s)} d\theta_t g(\theta_t)\rho(s;\theta_t) = \int_{E_\theta(s)}^{\infty} d\theta_t g(\theta_t)\rho(s;\theta_t) \tag{15.46}$$

which implies that we should take for $E_\theta(s)$ the median of the posterior distribution for which

$$\int_{-\infty}^{E_\theta(s)} d\theta_t \rho(s|\theta_t)\rho(\theta_t) = \int_{E_\theta(s)}^{\infty} d\theta_t \rho(s|\theta_t)\rho(\theta_t) = \frac{1}{2} \tag{15.47}$$

The absolute error penalizes larger deviations more and smaller deviations less than the mean square error.

Finally, if we only care whether or not we hit the target and not by how much (a miss is a miss), then we might choose as the loss function

$$L(a) = \begin{cases} 0 \text{ if } |a - \theta_t| < \epsilon \\ \frac{1}{2\epsilon} \text{ otherwise} \end{cases} \tag{15.48}$$

The vanishing of the variation then requires

$$\int d\theta_t \frac{1}{2\epsilon}[\delta(\theta_t - (E_\theta(s) + \epsilon)) - \delta(\theta_t - (E_\theta(s) - \epsilon))]\rho(s|\theta_t)\rho(\theta_t) = 0 \tag{15.49}$$

In the limit $\epsilon \to 0$ the difference of the delta functions divided by 2ϵ becomes the derivative of the delta function. Integrating by parts then results in the function E_θ that assigns to every value s the value θ_t at which the posterior distribution $\rho(\theta_t|s)$ takes its maximum. If $\rho(\theta_t)$ does not change much around this maximum, then one can instead choose the function E_θ that assigns to every s the value θ_t where the likelihood function $\rho(x|\theta_t)$ takes its maximum and one arrives at the maximum likelihood estimator.

The concept of a loss function has proven to be helpful in rationalizing the choices of estimators. It also proves the important point:

The choice of the estimator depends on the preferences of the investigator, here formulated as the kind of losses that he or she can tolerate.

A major caveat in using the loss function is that it is based on expectation values, the average over a reasonable range of the unknown parameter. The discussion of the St. Petersburg paradox in Sect. 16.9 is an example that expectation values might be misleading.

There are other ways to compare decision functions. One is the minimax criterion. A decision function is called the minimax decision function D_0 if the maximum of $\langle L \circ D_0 \circ f \rangle$ over θ_t is smaller than the maximum of $\langle L \circ D \circ f \rangle$ for all competing decision functions D.

15.8 Hypothesis Testing

Another widely used inference approach is the *test of a hypothesis*. The basic idea is to introduce a probabilistic model and to examine whether the sample (the evidence) is consistent with this model. To do so, the sample is compared with the expected statistics of the model. If this comparison leads to the assessment that the evidence is unlikely to emerge from the model, the *null hypothesis* 'sample complies with the model' is *rejected*. If the assessment finds no contradiction, then the null hypothesis is

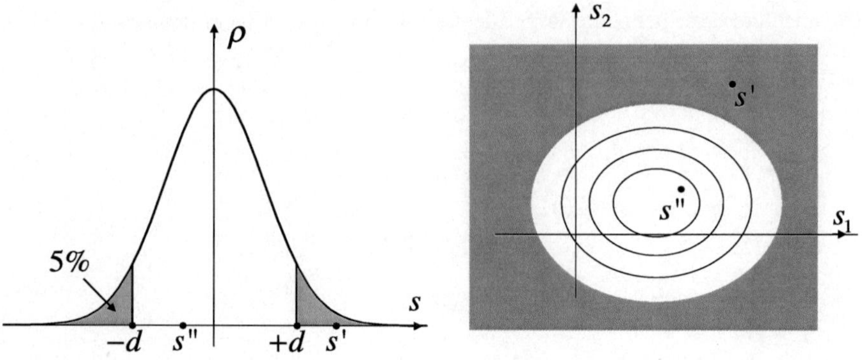

Fig. 15.2 Sketch of the probability density for a univariate and bivariate case. The rejection area is hatched. The null hypothesis is rejected for s', and it is not for s''

not rejected, but it is also not accepted. Instead, a weaker statement is made, namely, that the evidence is not inconsistent with the model. It may be that at a later time when more evidence has been gathered, the null hypothesis will be rejected.

Hypothesis testing. To introduce the concept more formally, we consider a simple prototypical case. Let f be a random variable with density function ρ_f. This is the statistical model upon which the test is based. We have one realization or sample s, but it is unknown whether s has been drawn from f or not. The null hypothesis, usually denoted by H_0, is that s is drawn from f. One then determines the smallest possible range A of outcomes of f so that $P(A) = \tilde{p}$ with some specified, normally large probability $\tilde{p}$. If the sample s lies in A, we consider s to be consistent with f even if we admit that it may be drawn from a distribution very similar to that of f. However, if s does not lie in A, then we consider this a sufficiently unlikely event under the null hypothesis—and we *decide* to reject the null hypothesis.

When f is a univariate random variable with a normalized Gaussian distribution ρ_G, then one chooses $A = [-d, d]$ where d is defined by

$$\int_{-d}^{d} dx\ \rho_G(x) = \tilde{p} \tag{15.50}$$

When f is a bivariate Gaussian distribution, the 'region of non-rejection' A is an ellipse (given by the covariance matrix) so that the integral of the probability density function over this area is $\tilde{p}$. The situation is sketched for both the uni- and bivariate cases in Fig. 15.2.

Significance level. The number $1 - \tilde{p}$ is often called the *significance level* and gives the acceptable probability to erroneously reject the null hypothesis. It must be selected subjectively. Its choice depends on the risk of a false rejection that one is willing to accept. Obviously, this choice depends on the context within which the decision is made. Instead of a risk of $1 - \tilde{p}$, also the term 'testing at a significance level of $1 - \tilde{p}$' is used. When the null hypothesis is rejected, the finding is declared 'significant at the $1 - \tilde{p}$ level'. Note that the use of the term 'significance' is different here from its colloquial use, where it represents 'the quality of being important'. The term 'significance' is also not uniformly used in the statistical community. Sometimes $\tilde{p} = 95\%$ is referred to as '95% significant' or 'the null hypothesis is rejected with a 95% confidence level'. If $\tilde{p} = 95\%$ we will say that the result is 'significant at the 5% level' or 'the null hypothesis is rejected with a 5% risk of error'.

The probability of correctly rejecting the null hypothesis is called the *power* and is not known. It is at least as large as the significance level and may be arbitrarily close to the significance level, since the correct model to describe the sample s may be very close to the tested model f. Generally, the concept of 'power' is theoretically useful, but the power can in most practical situation not be determined because it depends on the unknown true parameter. However, the power of all reasonable tests increases with increasing sample size, for a given deviation from the null hypothesis.

Formal definition. Formally a test is a decision function

$$D : S \mapsto \{0, 1\} \tag{15.51}$$

where the result '0' means acceptance of the hypothesis and '1' means the rejection of the hypothesis. A test thus partitions the sample space S into two domains

$$\begin{aligned} A_0 &:= D^{-1}(\{0\}), \\ A_1 &:= D^{-1}(\{1\}). \end{aligned} \tag{15.52}$$

If $s \in A_0$ one 'accepts' the hypothesis. If $s \in A_1$ one rejects the hypothesis. If two tests with different test statistics and decision functions lead to the same partition of the sample space, the tests are called equivalent.

Example: Test of the mean. A widely used test of a null hypothesis is the test whether an observed sample $s_1, \ldots, s_N$ derived from a sequence of iid random variables is consistent with the assumption that the random variables have a mean value of μ_0. Formally: $H_0 : \mu_{f_1} = \mu_0$ with some number μ_0. Assume that f_1 is a normally distributed random variable and that its variance $\sigma^2_{f_1}$ is known. If the null hypothesis is true, then the random variable

$$g := \frac{\sqrt{N}(\bar{f}_N - \mu_0)}{\sigma_{f_1}} \tag{15.53}$$

has the normalized Gaussian distribution $N(0, 1)$. One thus chooses as a test statistics

$$T(s) = \frac{\sqrt{N}(\bar{s}_N - \mu_0)}{\sigma_{f_1}} \tag{15.54}$$

and rejects the null hypothesis with a risk of $\tilde{p}$ if the test statistics $|T(s)| > d$, where d is given by (15.50). The decision function is thus

$$D(x) = \begin{cases} 0 \text{ for } x \in [-d, d] \\ \\ 1 \text{ for } x \notin [-d, d] \end{cases} \tag{15.55}$$

Our basic assertion that estimation theory does not provide estimates but information about estimation methods also holds for this second main branch of statistical inference, the testing of hypotheses. When we reject a null hypothesis at a 5% significance level, we are stating that our testing method will give wrong results in 5% of all cases when we repeatedly use this test.

15.9 Type I and II Errors

The line of argument in the previous section involves various choices. One is the choice of $\tilde{p}$. The second choice is the rejection region. We have assigned it to the fringes of the distribution. This choice is perhaps sensible. It does minimize the length of the acceptance region. The third and major choice is the test itself. Are some test statistics better than others. These choices can be analyzed quantitatively if one introduces alternative hypotheses.

Alternative hypothesis. As with estimation, the problem is how to choose the test statistics T or how to partition the sample space into an acceptance region A_0 and a rejection region A_1. One way to address this problem is to introduce alternative hypotheses. To introduce this concept more formally, we again assume that we have a sample s which is the realization of a random variable f with a probability density function $\rho_f(x; \theta)$ parameterized by θ. The goal is to test some hypothesis H_0 about θ versus some alternative hypothesis H_1. One distinguishes between simple (or point) and composite hypotheses. Simple hypotheses are of the form $H_0 : \theta = \theta_0$ versus $H_1 : \theta = \theta_1 (\neq \theta_0)$. Composite hypotheses are of the form $H_0 : \theta \in \Theta_0$ versus $H_1 : \theta \in \Theta_1$ where Θ_0 and Θ_1 are subsets of the parameter space Θ. Typical examples of composite hypotheses are $H_0 : \theta \leq \theta_0$ versus $H_1 : \theta > \theta_0$ or $H_0 : \theta = \theta_0$ versus $H_1 : \theta \neq \theta_0$.

Type I and II errors. A test consists of a partition of the sample space into a region A_0 where one accepts the null hypothesis H_0 and the complementary region A_1 where one rejects the null hypothesis. There are two types of errors that can occur

$$\alpha := P(x \in A_1 | H_0) \tag{15.56}$$

$$\beta := P(x \in A_0 | H_1) \tag{15.57}$$

The error α is also called the type I error and is the error that we reject the null hypothesis H_0 although it is true. The error β is called the type II error and is the error that we accept the null hypothesis H_0 although it is not true. A_1 is also called the critical region for testing H_0 and α is the called the size of the critical region. The type I error is our significance level $1 - \tilde{p}$ and $1 - \beta$ is the power.

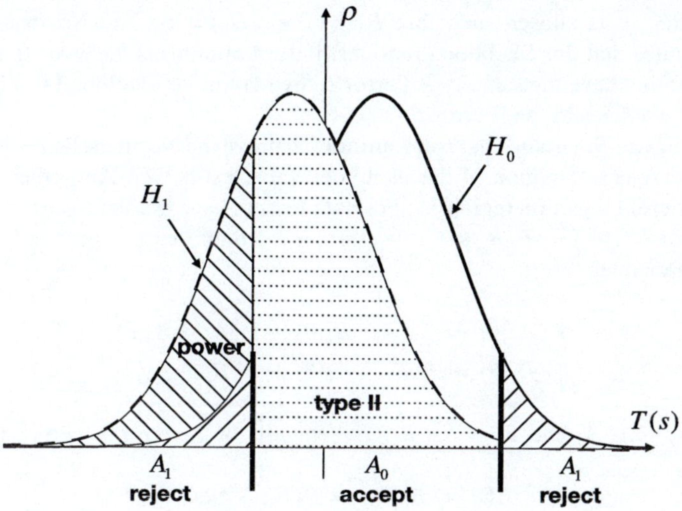

Fig. 15.3 Test of a simple hypothesis. The probability distributions of the test statistics $T(s)$ are sketched for the two hypotheses H_0 and H_1. The right-hatched area is the type I error or significance level α. The stippled area is type II error. The left-hatched area is the power

The situation is sketched for a simple hypothesis in Fig. 15.3.

As it is obvious from the definition of the errors, they cannot be minimized simultaneously. The error α can only be reduced by making A_1 smaller which will make its complement A_0 larger and will hence increase the error β. One, therefore, usually assumes a fixed value for α and searches for tests that minimize β or maximize the power $1-\beta$.

15.10 Likelihood Ratio Tests

A test that minimizes the error β for given α is the likelihood ratio test.

Likelihood ratio test. Consider the simple hypothesis $H_0 : \theta = \theta_0$ versus $H_1 : \theta = \theta_1$ first. The test is then based on the likelihood ratio

$$\lambda(s) := \frac{L(\theta_1; s)}{L(\theta_0; s)} = \frac{\rho_f(s; \theta_1)}{\rho_f(s; \theta_0)} \tag{15.58}$$

which depends on the sample. If the sample has a higher probability for θ_0 then for θ_1 then the likelihood ratio is smaller than 1 and vice versa. This result suggests the test

$$\begin{aligned} &if\ \lambda(s) \le c \text{ accept } H_0 \\ &if\ \lambda(s) > c \text{ reject } H_0 \end{aligned}$$

The constant c is chosen such that $P(\lambda(x) > c|H_0) = \alpha$. The Neyman–Pearson Lemma states that this likelihood ratio test indeed minimizes the error β, meaning: of all tests that have the same type I error α the maximum likelihood test is the one that has the smallest type II error β.

Proof of the Neyman–Pearson Lemma. To prove the Neyman–Pearson Lemma, denote the rejection region of the likelihood ratio test by A_1. Any other test must have a different rejection region B_1. For both tests to have the same error α, we have to require $P(A_1|\theta_0) = P(B_1|\theta_0)$. It is useful to decompose these probabilities in the following manner:

$$\begin{aligned} P(A_1|\theta) &= P(A_1 \cap B_1|\theta) + P(A_1 \cap B_0|\theta) \\ P(B_1|\theta) &= P(B_1 \cap A_1|\theta) + P(B_1 \cap A_0|\theta) \end{aligned} \tag{15.59}$$

where $A_0 = S \setminus A_1$ and $B_0 = S \setminus B_1$. The equality of the error α then implies

$$P(A_1 \cap B_0|\theta_0) = P(B_1 \cap A_0|\theta_0) \tag{15.60}$$

For the likelihood ratio test to be the most powerful test, we have to prove that the error β of the maximum likelihood test is smaller than or equal to the error β of any other test

$$P(A_0|\theta_1) \leq P(B_0|\theta_1) \tag{15.61}$$

or equivalently

$$P(A_0 \cap B_1|\theta_1) \leq P(B_0 \cap A_1|\theta_1) \tag{15.62}$$

according to (15.59). Taking into account that

$$P(C|\theta) = \int_C ds L(\theta; s) \tag{15.63}$$

is the probability that a sample s falls for given θ into a region C of the sample space, this assertion can then be proven in the following manner:

$$\begin{aligned} P(A_0 \cap B_1|\theta_1) &= \int_{A_0 \cap B_1} ds L(\theta_1; s) \\ &= \int_{A_0 \cap B_1} ds \lambda(s) L(\theta_0; s) \\ &\leq c \int_{A_0 \cap B_1} ds L(\theta_0; s) \\ &= cP(A_0 \cap B_1|\theta_0) \\ &= cP(A_1 \cap B_0|\theta_0) \\ &= c \int_{A_1 \cap B_0} ds L(\theta_0; s) \end{aligned}$$

$$= c \int_{A_1 \cap B_0} ds \frac{1}{\lambda(s)} L(\theta_1; s)$$
$$\leq \int_{A_1 \cap B_0} ds L(\theta_1; s) = P(A_1 \cap B_0 | \theta_1) \tag{15.64}$$

Example: Test of simple hypothesis. For a Gaussian random variable with known variance, the likelihood ratio test for $H_0 : \mu = \mu_0$ versus $H_1 : \mu = \mu_1 (> \mu_0)$ rejects the null hypothesis if

$$\lambda(s) = \exp\left\{-\frac{1}{2\sigma^2}\sum_{i=1}^{N}(s_i - \mu_1)^2 + \frac{1}{2\sigma^2}\sum_{i=1}^{N}(s_i - \mu_0)^2\right\}$$
$$= \exp\left\{\frac{(\mu_1 - \mu_0)\sum_{i=1}^{N} s_i}{\sigma^2} + \frac{N(\mu_0^2 - \mu_1^2)}{2\sigma^2}\right\} > c \tag{15.65}$$

Taking the logarithm, this inequality can be rewritten

$$\bar{s}_N > \frac{\sigma^2 \log c}{N(\mu_1 - \mu_0)} + \frac{1}{2}(\mu_1 + \mu_0) =: a \tag{15.66}$$

The value of a is determined by the condition $P(\bar{s}_N > a | H_0) = \alpha$. Equation (15.66) implies the test statistics. One needs only to determine $\bar{s}_N$ and find out whether or not it is smaller or larger than a. If $\bar{s}_N < a$, one accepts H_0. If $\bar{s}_N > a$, one rejects H_0. The error β is given by $\beta = P(\bar{s}N < a | H_1)$. If $\mu_1 < \mu_0$, the inequality has to be reversed in (15.66). The situation is sketched in Fig. 15.4. A numerical example is given in the next chapter.

Example: Test of composite hypothesis. Next consider the composite hypothesis $H_0 : \theta \in \Theta_0$ versus $H_1 : \theta \in \Theta_1$. The likelihood ratio is then given by

$$\lambda(s) = \frac{\sup_{\theta \in \Theta_1} L(\theta; s)}{\sup_{\theta \in \Theta_0} L(\theta; s)} \tag{15.67}$$

As an example, consider again our Gaussian case and test $H_0; \mu = \mu_0$ versus $H_1 : \mu \neq \mu_0$ with known variance σ^2. The likelihood ratio test then rejects the null hypothesis if

$$\lambda(s) = \frac{\sup_{\mu \neq \mu_0} \exp\{-\frac{1}{2\sigma^2}\sum_{i=1}^{N}(s_i - \mu_1)^2\}}{\exp\{\frac{1}{2\sigma^2}\sum_{i=1}^{N}(s_i - \mu_0)^2\}} > c \tag{15.68}$$

The supremum is taken for $\mu = \bar{s}_N$. The likelihood ratio test thus becomes

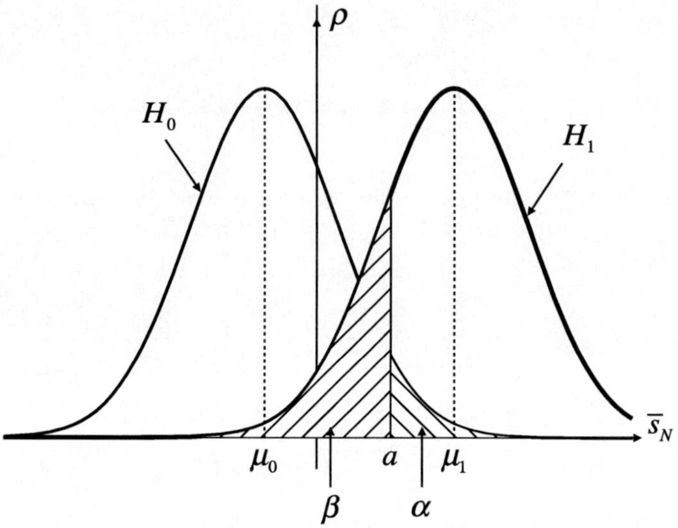

Fig. 15.4 Likelihood ratio test of the simple hypothesis $H_0 : \mu = \mu_0$ versus $H_1 : \mu = \mu_1 (> \mu_0)$ for a Gaussian variable with known variance. The left-hatched area is the type I error α; the right-hatched area is the type II error β

$$\lambda(s) = \exp\left\{-\frac{1}{2\sigma^2}\sum_{i=1}^{N}(s_i - \bar{s}_N)^2 + \frac{1}{2\sigma^2}\sum_{i=1}^{N}(s_i - \mu_0)^2\right\}$$
$$= \exp\{\frac{N(\bar{s}_N - \mu_0)^2}{2\sigma^2}\} > c \tag{15.69}$$

and upon taking the logarithm one obtains the likelihood ratio test

$$\frac{N(\bar{s}_N - \mu_0)^2}{\sigma^2} > 2\log c \tag{15.70}$$

or

$$|\frac{\sqrt{N}(\bar{s}_N - \mu_0)}{\sigma_{f_1}}| > d \tag{15.71}$$

which is indeed the test asserted in (15.54). Thus, assigning the rejection region symmetrically to the fringes of the $N(0, 1)$ distribution is indeed the optimal way, leading to the smallest error of the second kind.

Similarly, if σ^2 is not known, one has to consider the likelihood ratio

$$\lambda(s) = \frac{\sup_{\mu_1 \in \mathbf{R}, \sigma^2 \in \mathbf{R}_+} \sqrt{2\pi\sigma^2}^{-N/2} \exp\{-\frac{1}{2\sigma^2}\sum_{i=1}^{N}(s_i - \mu_1)^2\}}{\sup_{\sigma^2 \in \mathbf{R}_+} \sqrt{2\pi\sigma^2}^{-N/2} \exp\{\frac{1}{2\sigma^2}\sum_{i=1}^{N}(s_i - \mu_0)^2\}} \tag{15.72}$$

The suprema are taken for $\mu = \bar{s}_N$ and $\sigma^2 = S_N^2$. The likelihood ratio test then yields

$$\frac{N(\bar{s}_N - \mu_0)^2}{S_N^2} > 2 \log c \tag{15.73}$$

or

$$\left| \frac{\sqrt{N}(\bar{s}_N - \mu_0)}{\sqrt{\frac{1}{N-1} \sum_{n=1}^{N} (s_n - \bar{s})^2}} \right| > d \tag{15.74}$$

which is called the t-test since the test statistics is distributed according to the Student's t-distribution with $N - 1$ degrees of freedom. Its density function can be derived from the transformation rule for probability densities and its values are listed in text books on statistics.

15.11 Summary

This chapter considered parameter estimation and testing with a frequentist's interpretation. The starting point is a sample, which is a sequence of numbers. The sample is considered a realization of a random variable. An estimate is given by a statistic, a function from the sample space, the space of all possible samples, into the parameter space, the space of all possible parameters. Substituting into the statistic, the specific observed sample leads to an estimate of the parameter. Composing the statistic with the random variable results in an estimator which is a random variable. An estimate is a realization of the estimator. One can determine the properties of the estimator such as its mean, variance, mean square error, and bias. These properties tell you something about the estimator, not something about the observed sample. These properties are different for different estimators. Whether or not these properties are desirable is not addressed. Which of the estimators to choose is ultimately a decision of the statistician and might depend on the circumstances and purpose of the investigation or the preferences of the statistician.

Similarly, hypothesis tests are mappings from the sample space into the two-element space: hypothesis accepted or hypothesis rejected. Again one has to distinguish two situations. When a specific observed sample is substituted into this mapping a test result is obtained: either reject or accept. When the mapping is composed with the random variable of which the observed sample is a realization then the test becomes a random variable whose properties one can study.

The frequentist's interpretation of estimators and tests is that if one repeats the sampling over and over again, then a certain fraction of the resulting estimates has certain properties. The theory does not make statements about a single sample.

Chapter 16
Bayesian Inferences

16.1 Introduction

Bayesian inference also addresses the problem of parameter estimation and hypothesis testing. It differs from the frequentist view of the previous chapter in various respects:

(1) Bayesian inference uses conditional probabilities. This reflects the Bayesian premise that all inferences are conditional on evidence. These conditional probabilities are not necessarily assumed to represent causal relations or the arrow of time.
(2) Bayesian inference updates *prior* probabilities to *posterior* probabilities due to *new additional evidence*.
(3) Bayesian inference can deal with more general situations such as *backward probabilities*.
(4) Bayesian inference allows the logicist's or subjectivist's interpretation of probability.

We first introduce the Bayes formula and then the Bayesian paradigm, then consider the standard problems of parameter estimation and hypothesis testing, and finally compare the Bayesian approach to the frequentist's approach.

We conclude with a discussion of the general epistemological framework that defines a probabilistic inference and of the aspects that enter and are extraneous to probability theory.

For the standard formalism of Bayesian statistics; see *Cowan* (2007) for a concise introduction or *Jaynes* (2003) for more detail.

P. Müller, *Handbook of Dynamics and Probability*,
https://doi.org/10.1007/978-3-030-88486-4_16

16.2 Bayes Theorem

The theorem is named after the eighteenth century English minister Thomas Bayes, who is said to have used conditional probabilities for the first time in probabilistic reasoning. His works were edited and published posthumously by his friend Richard Price. For this reason, the theorem is also referred to as the Bayes–Price theorem.

Bayes theorem. The probability that both events A and B occur (i.e., $A \cap B$) can be written in two different ways as

$$P(A \cap B) = P(B|A)P(A) = P(A|B)P(B), \tag{16.1}$$

where $P(X|Y)$ is the conditional probability of X given Y and $P(X)$ is the probability of X. From (16.1), it follows that

$$P(A|B) = \frac{P(B|A)P(A)}{P(B)}, \tag{16.2}$$

Since $P(A|B) + P(\neg A|B) = 1$ we find

$$P(B) = P(B|A)P(A) + P(B|\neg A)P(\neg A) \tag{16.3}$$

and hence

$$P(A|B) = \frac{P(B|A)P(A)}{P(B|A)P(A) + P(B|\neg A)P(\neg A)} \tag{16.4}$$

which is Bayes' theorem. If the state space Ω is partitioned as $\Omega = \cup_{j=1}^{N} A_j$ with $A_j \cap A_k = \emptyset$ for $j \neq k$, then

$$P(B) = \sum_{j=1}^{N} P(B|A_j)P(A_j) \tag{16.5}$$

and Bayes' theorem generalizes to

$$P(A_i|B) = \frac{P(B|A_i)P(A_i)}{\sum_{j=1}^{N} P(B|A_j)P(A_j)} \tag{16.6}$$

Bayes' theorem is the basis for the calculation of conditional probabilities. It allows to calculate the probability of $P(A|B)$ from the probability of $P(B|A)$ given $P(A)$. The probability $P(A|B)$ when calculated from $P(B|A)$ is often called the backward probability. There are two important points:

In general $P(A|B) \neq P(B|A)$ which should be obvious.

The calculation of $P(A|B)$ from $P(B|A)$ requires the knowledge of $P(A)$.

The second point has important implications as demonstrated in the following example.

Example: Medical tests. Consider a test for a medical condition. Assume that the test gives a correct positive result for 99% and a false negative result for 1% of the people that have the condition, and gives a false positive result for 5% and a correct negative result for 95% of the people who do not have the condition. Let A denote the case that the person has the medical condition and B the case that the test result is positive. Then the situation is characterized by the conditional probabilities

$$\begin{aligned} P(B|A) &= 0.99 \\ P(\neg B|A) &= 0.01 \end{aligned} \tag{16.7}$$

$$\begin{aligned} P(B|\neg A) &= 0.05 \\ P(\neg B|\neg A) &= 0.95 \end{aligned} \tag{16.8}$$

If a person's test result is positive, what are the chances that he/she has the medical condition? We are interested in $P(A|B)$. According to Bayes theorem, the answer depends on the probability $P(A)$. Assume that the individual is part of a population where 1:1,000 people have the medical condition

$$P(\neg A) = 0.999 \tag{16.9}$$

$$P(A) = 0.001 \tag{16.10}$$

Bayes formula then gives as the conditional probability

$$P(A|B) = \frac{P(B|A)P(A)}{P(B|A)P(A) + P(B|\neg A)P(\neg A)} = \frac{0.99 \times 0.001}{0.99 \times 0.001 + 0.05 \times 0.999} \approx 0.02 \tag{16.11}$$

and

$$\frac{P(A|B)}{P(\neg A|B)} = \frac{P(B|A)P(A)}{P(B|\neg A)P(\neg A)} \approx \frac{0.99}{0.05} \approx 20 \tag{16.12}$$

A person who tested positive for the condition has only a 2% probability of actually having the condition. What has happened by the test is that the original probability $P(A)$ of having the condition has been updated to the new probability $P(A|B)$ of having the condition by using the result of the test. Although the test is highly accurate, the incidence of the condition is so low in the population that most of the persons who test positive do not have the condition. Nevertheless, by testing positive, the probability has increased from 1:1,000 to 1:50. This updating of the probability is the essence of the Bayesian approach.

The Bayesian theorem also allows to calculate backward probabilities and make predictions which are difficult or impossible to obtain within the frequentist's approach, as demonstrated in the following example.

Picking balls from urns. Consider $L+1$ urns $U_0, \ldots, U_L$. Each urn contains L balls. Urn U_l contains l black balls and $(L-l)$ white balls. Pick one urn at random and draw N times a ball from this urn, replacing the ball after each draw. The result may be represented as a binary sequence $s = (s_1, \ldots, s_N)$ where $s_n = 1$ corresponds to the drawing of a black ball and $s_n = 0$ to the drawing of a white ball. Regard this binary sequence as a realization of a Bernoulli chain. Assume that $n \leq N$ black balls were drawn. What is the probability that one picked urn U_l? To answer this question, we apply Bayes formula in the form

$$p(U_l|n) = \frac{p(n|U_l)p(U_l)}{\sum_{k=0}^{L} p(n|U_k)p(U_k)} \tag{16.13}$$

The conditional probability $p(n|U_l)$, i.e., the probability to draw n black balls from urn U_l is given by the binomial formula

$$p(n|U_l) = \binom{N}{n} p_l(1)^n(1-p_l(1))^{N-n} \tag{16.14}$$

with $p_l(1) = l/L$. As the probability of picking an urn assume the uniform distribution $p(U_l) = 1/(L+1)$. With this information, one can calculate the backward probability that one picked urn U_l. This probability cannot be obtained by repeating the experiment which would only yield a sequence of values $n_1, n_2, \ldots$.

Having calculated the backward probability, one can also address the question: what is the probability that the next ball that I draw is also black? The answer is

$$\begin{aligned} p(s_{N+1} = 1|n) &= \sum_{k=0}^{L} p(s_{N+1} = 1|n \textit{ and } U_k)p(U_k|n) \\ &= \sum_{k=0}^{L} \frac{k}{L} p(U_k|n) \end{aligned} \tag{16.15}$$

In these two examples Bayes formula updates probabilities from some prior value to some posterior value by incorporating the result of some data: in the medical example from $P(A)$ to $P(A|B)$ given the test result B, in the urn example from $P(U_l)$ to $P(U_l|n)$ given the observation that n black balls have been drawn in N drawings. This updating is the Bayesian approach and is formalized in the next section.

16.3 The Bayesian Paradigm

The Bayesian formulas are also applicable to situations where the outcomes or events represent logical statements, as used in the logicist's and subjectivist's interpretation of probability. The Bayesian theorem then becomes Bayesian inference and has developed its own technical terms, explicated next.

Bayesian inference. Bayesian inference uses the formula

$$P(H|E) = \frac{P(E|H)P(H)}{P(E)} \tag{16.16}$$

where

- H is the hypothesis. It may be an event as formalized in probability theory but it may also be a statement or proposition, which is either true or false but whose truth or falsity is not known and can only be assigned a probability in the sense of the logicist or subjectivist.
- E is the evidence. It is data or observations collected in order to assess the probability of the hypothesis.
- $P(E|H)$ is the conditional probability of the evidence given that the hypothesis is true. It is called the likelihood function if it is considered as a function of H for given fixed E.
- $P(H)$ is the prior probability of the hypothesis. It is the probability of H before the evidence is taken into account.
- P(E) is the probability of obtaining the evidence E no matter what hypothesis is true. It is usually written in the form

$$P(E) = \sum_{n=1}^{N} P(E|H_n)P(H_n) \tag{16.17}$$

  where the H_n are a complete set of mutually exclusive hypotheses. $P(E)$ may be viewed as a normalization constant.
- $P(H|E)$ is the posterior probability given the evidence. It reflects the fact that the probability ought to change when new evidence is collected.

It is important to realize that the above recipe requires an a priori probability $P(H)$ that the hypothesis is true. The Bayesian formula can be applied repeatedly as further evidence becomes available. The factor $P(E|H)/P(E)$ represents the impact of the evidence on our degree of belief. The impact is large if $P(E)$ is small or if $P(E|H)$ is large. The impact is small when $P(E)$ is large or $P(E|H)$ is small. Though this statement is correct, one must realize that $P(E)$ and $P(E|H)$ are not independent since $P(E) = P(E|H) + P(E|\neg H)$. A set of independent variables are $P(H)$, $P(\neg E|H)$, and $P(E|\neg H)$. In terms of these independent variables, the posterior probability is given by

$$P(H|E) = \frac{[1 - P(\neg E|H)]P(H)}{[1 - P(\neg E|H)]P(H) + P(E|\neg H)[1 - P(H)]} \tag{16.18}$$

The dependencies of $P(H|E)$ on these three variables are shown in Fig. 16.1.

Fig. 16.1 Dependence of the posterior probability $P(H|E)$ on the prior probability $P(H)$, on the probability $P(\neg E|H)$ that the hypothesis does not yield the evidence, and on the probability $P(E|\neg H)$ that alternative hypotheses yield the evidence

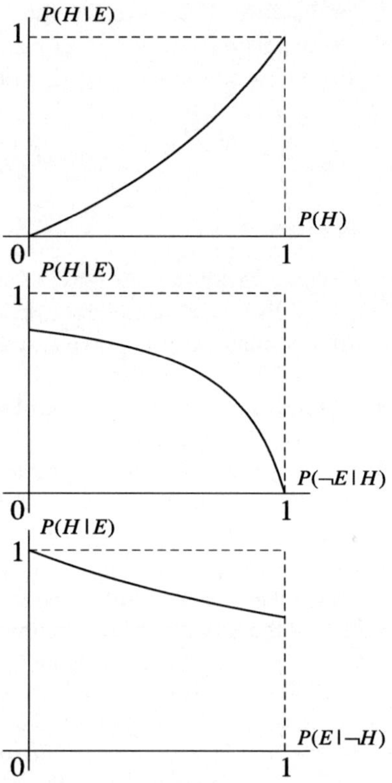

Since $P(E)$ may be viewed as a normalization constant, the Bayesian formula may also be written as

$$P(H|E) \propto P(E|H)P(H) \tag{16.19}$$

If there is no prior expectation, i.e., with a flat $P(H)$, then $P(H|E) \propto P(E|H)$ so that the data completely determine the expected probability of H. Conversely, with very limited new data, $P(E|H)$ might be a rather flat function of H and $P(H|E)$ would differ little from some prior $P(H)$. The more likely situation is that both the prior $P(H)$ and the data-determined $P(E|H)$ are localized in H, but peak at different values of H. As shown in Fig. 16.2, the subsequent $P(H|E)$ is shifted away from $P(H)$ and toward $P(E|H)$. The narrower that $P(E|H)$ is (as a consequence of increasing amounts of data), the more will the posterior distribution shift toward it, whereas a narrow prior $P(H)$ will allow only a small shift if $P(E|H)$ is broad.

All this is really just a mathematical way of stating that any reasonable person will modify his or her expectation based on evidence, with the change of position being large if the data are convincing, and small if the person has a narrow prior view! The subjective element remains, but Bayes' approach allows for the subjectivity to be quantified. If people with different views can be persuaded to state their prior

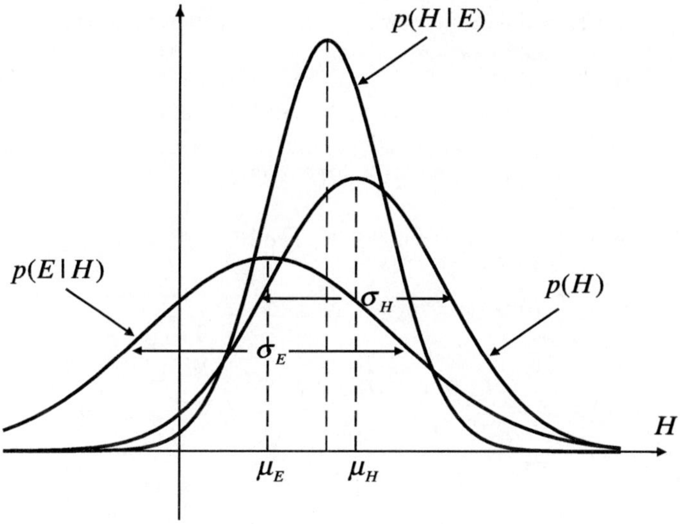

Fig. 16.2 The figure shows the case where the prior $P(H)$ has a $N(\mu_H, \sigma_H^2)$ and $P(E|H)$ a $N(\mu_E, \sigma_E^2)$ distribution, where $\sigma_E^2 = 2\sigma_H^2$. The posterior distribution is then a Gaussian distribution with mean $\mu = \frac{\mu_H\sigma_E^2 + \mu_E\sigma_H^2}{\sigma_E^2 + \sigma_H^2}$ and variance $\sigma^2 = \frac{\sigma_E^2\sigma_H^2}{\sigma_E^2 + \sigma_H^2}$

probability distribution functions, then they should accept the Bayesian shift required by new data.

Monty Hall game. Perhaps the most famous example of Bayesian inference is the Monty Hall game. It is named after the host Monty Hall of the TV show Let's make a deal. There are three doors on the stage. Behind one door is the prize, a brand new car. Behind the other two doors are goats. The rules of the game are:

(1) The player can choose one door.
(2) The host, who knows behind which door the prize is, then opens one of the other doors to show you that there is a goat behind that door.
(3) The player is then allowed to reconsider its initial choice and switch to the other remaining door.

The question is: Should the player switch?

The first answer is: you should at least reconsider. You have a new piece of information, namely, that the prize is not behind a certain door. You should also switch. The Bayesian reasoning is as follows:

Denote by H_i the hypothesis that the prize is behind door number i. Your prior probabilities should be $P(H_1) = P(H_2) = P(H_3) = 1/3$. Assume you chose door number 1. If the prize is behind door number 1, the host then opens door number 2 or 3. Presume the host has no preference then the likelihood is

$$P(D = 2|H_1) = P(D = 3|H_1) = 1/2 \tag{16.20}$$

If the prize is not behind door number 1, then the host has no choice which door to open. Therefore, the likelihoods are

$$P(D=2|H_2) = P(D=3|H_3) = 0 \tag{16.21}$$

$$P(D=2|H_3) = P(D=3|H_2) = 1 \tag{16.22}$$

Assume that the host opens door number 3. Then Bayes formula gives

$$P(H_1|D=3) = \frac{P(D=3|H_1)P(H_1)}{P(D=3)} \tag{16.23}$$

$$P(H_2|D=3) = \frac{P(D=3|H_2)P(H_2)}{P(D=3)} = 2P(H_1|D=3) \tag{16.24}$$

and of course

$$P(H_3|D=3) = P(D=3|H_3)P(H_3)/P(D=3) = 0 \tag{16.25}$$

The probability that the prize is behind door number 2 is twice the probability that the prize is behind door number 1. Therefore, you should switch.

If this result is counter your intuition, imagine a situation where there are a million doors. You have chosen door number 1 and the host opens all other doors to show you that there is no prize behind them, except one other. Where do you think the prize is?

The above result can of course also be obtained by calculating the probability of the various possible outcomes. Without loss of generality let the prize be behind the door number 1. The possible scenarios and their probabilities are given in Fig. 16.3.

If one does not switch, the scenarios (1, 2) and (1, 3) win, with a combined probability of 1/3. If one switches, then the scenarios (2, 3) and (3, 2) win with a combined probability 2/3. Thus, one should change the initial guess.

If one were to give all the possible scenarios equal probability, then one arrives at the result that it does not matter whether or not one switches. Such an assignment (given in Fig. 16.4) corresponds to the case that the host does not know where the prize is.

Since the prize is behind door number 1, the scenarios (2, 1) and (3, 1) reveal the prize. You have lost, independent of whether or not you switch. The two cases (1, 2) and (1, 3) where you win when you do not switch have combined probability $1/3$ and the two cases (2, 3) and (3, 2) where you win when you switch have also combined probability $1/3$. Switching does not improve your chance. The information that was used in the original problem is that the host knows where the prize is.

Fig. 16.3 Monty Hall game. The table lists the possible scenarios and their probabilities if the prize is behind door number 1. The first factor 1/3 arises from the player having three different options. The second factor 1/2 or 1 arises from the host having two or one option

player points to door \ host opens door	1	2	3
1		$\frac{1}{3}\cdot\frac{1}{2}=\frac{1}{6}$	$\frac{1}{3}\cdot\frac{1}{2}=\frac{1}{6}$
2			$\frac{1}{3}\cdot 1=\frac{1}{3}$
3		$\frac{1}{3}\cdot 1=\frac{1}{3}$	

Fig. 16.4 The table lists the possible scenarios and their probabilities if the prize is behind door number 1. The first factor 1/3 arises from the player having three different options. The second factor 1/2 arises from the host having two options

player points to door \ host opens door	1	2	3
1		$\frac{1}{3}\cdot\frac{1}{2}=\frac{1}{6}$	$\frac{1}{3}\cdot\frac{1}{2}=\frac{1}{6}$
2	$\frac{1}{3}\cdot\frac{1}{2}=\frac{1}{6}$		$\frac{1}{3}\cdot\frac{1}{2}=\frac{1}{6}$
3	$\frac{1}{3}\cdot\frac{1}{2}=\frac{1}{6}$	$\frac{1}{3}\cdot\frac{1}{2}=\frac{1}{6}$	

16.4 Bayesian Estimators

Though more general than the frequentist's approach, Bayesian inference can also be applied to the estimation and testing problem.

Bayesian estimates. Bayesian estimation methods assign a probability distribution to the parameter θ, not because it is a random variable but because its value is unknown. The known distribution $\rho_f(x;\theta)$ with a fixed constant θ is thus replaced by the known conditional distribution $\rho(x|\theta)$. Using Bayes formula, one then obtains

$$\rho(\theta|x) = c\rho(x|\theta)\rho(\theta) \tag{16.26}$$

where $\rho(\theta)$ is the prior distribution of the parameter θ and the factor c can be obtained from the normalization condition $\int d\theta\rho(\theta|x) = 1$. Substituting the sample s, one obtains

$$\rho(\theta|s) = c\rho(s|\theta)\rho(\theta) \tag{16.27}$$

which is the posterior distribution of the parameter θ after sampling in terms of the prior distribution of the parameter θ before sampling.

The Bayesian approach requires more input than, e.g., the maximum likelihood estimator. It needs the prior distribution of the parameter. It also delivers more: a distribution of the parameter after sampling, not just a value. If a value is needed, one may choose the expectation value of the posterior distribution, or the value at which the posterior distribution has its maximum, or some other well-defined parameter of the posterior distribution. These values depend on the sample s. By substituting the random variable for the realization, one obtains an estimator whose properties can be compared with the properties of other estimators.

Example: Coin tossing. Consider a binary sample $s = (s_1, \ldots, s_N)$ of length N regarded as a realization of a Bernoulli chain $B(0, 1; p)$ with unknown p. Assume that the number of zeros in this sample is n. Application of Bayes' formula gives

$$\rho(p|n; N) = \frac{p(n|p; N)\rho(p)}{p(n; N)} \tag{16.28}$$

The likelihood function is given by

$$p(n|p; N) = \binom{N}{n} p^n (1-p)^{N-n} \tag{16.29}$$

and the normalization constant by

$$\begin{aligned} p(n; N) &= \binom{N}{n} \int_0^1 dp p^n (1-p)^{N-n} \\ &= \binom{N}{n} \frac{\Gamma(n+1)\Gamma(N-n+1)}{\Gamma(N+2)} \\ &= \binom{N}{n} \frac{N!(N-n)!}{(N+1)!} = \frac{1}{N+1} \end{aligned} \tag{16.30}$$

upon using the formula

$$B(a, b) = \int_0^1 dx x^{a-1}(1-x)^{b-1} = \frac{\Gamma(a)\Gamma(b)}{\Gamma(a+b)} \tag{16.31}$$

for the beta integral. Assuming a uniform prior distribution, $\rho(p) = 1$, one obtains the posterior distribution

$$\rho(p|n;N) = (N+1)\binom{N}{n} p^n (1-p)^{N-n} \tag{16.32}$$

This distribution has its maximum at $p = n/N$.

What is the probability that the next draw will also give a zero? It is given by

$$\begin{aligned} p(s_{N+1} = 1|s) &= \int_0^1 dp p(s_{N+1} = 1|p)\rho(p|s) \\ &= \int_0^1 dp p \rho(p|s) \\ &= (N+1)\binom{N}{n} \int_0^1 dp p^{n+1}(1-p)^{N-n} \\ &= (N+1)\binom{N}{n} \frac{(n)!(N-n)!}{(N+2)!} = \frac{n+1}{N+2} \end{aligned} \tag{16.33}$$

which is a special case of the law of succession by Laplace. Laplace's law of succession considers K exhaustive mutually exclusive alternatives $A_1, \ldots, A_K$. If N random drawings are taken and the alternative A_i occurs n_i times such that $\sum_{i=1}^{K} n_i = N$ then the probability that A_i will occur in the next drawing is

$$P(A_i|n_i, N, K) = \frac{n_i + 1}{N + K} \tag{16.34}$$

Our result (16.33) is the special case K = 2.

This Bayesian prediction is the basis of most learning algorithms.

16.5 Bayesian Hypothesis Testing

The Bayesian approach can also be applied to hypothesis testing in a straightforward manner. A hypothesis is just another proposition that can be assigned a probability in the sense that we do not know whether it is true or false. Applying Bayes formula requires the likelihood function $P(E|H)$, the prior $P(H)$, and the probability $P(E)$ of the evidence. Again, one usually considers the ratio of two alternative hypotheses. As an example, consider again a binary sample s of length N resulting from a Bernoulli experiment. The number of zeros in this sample is n. Compare the probability of two hypotheses H_0 and H_1 which is given by

$$\frac{P(H_1|n;N)}{P(H_0|n;N)} = \frac{P(n|H_1;N)P(H_1)}{P(n|H_0;N)P(H_0)} \tag{16.35}$$

and only requires the prior probabilities $P(H_0)$ and $P(H_1)$. Take for example H_0 to be $p = p_0$ and H_1 to be the hypothesis $p = p_1$. Assuming equal prior probabilities of these two hypotheses, the ratio then becomes

$$\frac{P(H_1|n;N)}{P(H_0|n;N)} = \frac{p_1^n(1-p_1)^{N-n}}{p_0^n(1-p_0)^{N-n}} \tag{16.36}$$

If one takes as the alternative hypothesis H_1, the hypothesis that p can take any value then

$$P(n|H_1;N) = \binom{N}{n}\int_0^1 dp\,P^n(1-p)^{N-n} = \frac{1}{N+1} \tag{16.37}$$

and the ratio becomes

$$\frac{P(H_1|n;N)}{P(H_0|n;N)} = \frac{n!(N-n)!}{(N+1)!}\frac{1}{p_0^n(1-p_0)^{N-n}} \tag{16.38}$$

A numerical example is given in the next section.

16.6 Comparison of Frequentist and Bayesian Approach

The Bayesian approach is much more versatile than the frequentist's approach. It can assign probabilities to parameters and propositions that reflect the statistician's uncertainty about the true value of the parameter or the proposition. The Bayesian approach is, therefore, much broader in its applications. However, even when applied to the same situation, the two approaches lead to different inferences. We demonstrate this fact for the simple coin tossing experiment, the Bernoulli chain. Both approaches assume as the underlying probabilistic model that the outcomes are governed by the binomial distribution

$$p_B(n|p;N) = \binom{N}{n}p^n(1-p)^{N-n} \tag{16.39}$$

For numerical values, we consider a sample of this Bernoulli process that has $N = 10$ and $n = 6$. The various approaches give the following results:

1. Estimation. The Bayesian estimation with a uniform prior gives the posterior distribution

$$\rho(p|n=6;N=10) = 11\binom{10}{6}p^6(1-p)^4 \tag{16.40}$$

If the frequentist uses the maximum likelihood estimate, he/she would estimate the probability $p = \mu_{f_1}$ by the sample mean which results in

$$E_p = \frac{n}{N} = \frac{6}{10} \tag{16.41}$$

which corresponds to the maximum of the Bayesian posterior distribution. If the frequentist uses an estimate that minimizes the mean square error, the estimate would be $E_p = (n+1)/(N+2) = 7/12 < 6/10$, which corresponds to the expectation value of the posterior distribution. If the frequentist uses the estimator that minimizes the absolute error, then the estimate would be the median of the posterior distribution, given by solving

$$(N+1)\binom{N}{n}B(p, n+1, N-n+1) = \frac{1}{2} \tag{16.42}$$

for p. Here

$$B(x; a, b) = \int_0^x dx x^{a-1}(1-x)^{b-1} \tag{16.43}$$

is the incomplete beta function. The (complete) beta function is $B(a, b) = B(1; a, b)$.

2. Prediction. Whatever estimate for E_p is chosen, the frequentist would predict that the next toss of the coin would yield a zero with probability E_p. The Bayesian approach gives the probability that the next coin toss results in a zero as

$$p(s_{11} = 1|n = 6; N = 10) = 7/12 < 6/10 \tag{16.44}$$

Though the differences might be numerically small in this case, different predictions by the frequentist and Bayesian are contentious issues in debates about the foundation of statistics.

3. Simple hypothesis. To test the simple hypothesis $H_0 : p = 1/2$ versus the hypothesis $H_1 : p = 2/3$, the frequentist might perform the likelihood ratio test (15.59). The likelihood ratio test rejects the null hypothesis if

$$\lambda(n = 6) = \frac{p_B(n = 6; N = 10, p = 2/3)}{p_B n = 6; N = 10, p = 1/2)} > c \tag{16.45}$$

or $n > n_0$. Since

$$\sum_{m=8}^{10} p_B(m; N = 10, p = 1/2) = \left(\frac{1}{2}\right)^{10} (1 + 10 + 45)) \approx 0.056 \tag{16.46}$$

the frequentist would reject the hypothesis H_0 for all outcomes $m = 8, 9, 10$ with a type I error of $\alpha \approx 0.056$. The type II error β is then

$$\beta = \sum_{m=0}^{7} p_B(m; N = 10, p = 2/3) \approx 0.70 \tag{16.47}$$

To test the hypothesis $H_0 : p = 1/2$ versus the hypothesis $H_1 : p = 2/3$ the Bayesian would consider the ratio

$$\frac{P(H_1|n=6;N=10)P(H_1)}{P(H_0|n=6;N=10)P(H_0)} = \frac{p_B(n=6|p(1)=2/3;N=10)P(H_1)}{p_B(n=6|p(1)=1/2;N=10)P(H_0)}$$
$$= \frac{2^{16}}{3^{10}}\frac{P(H_1)}{P(H_0)} \approx 1.11\frac{P(H_1)}{P(H_0)} \tag{16.48}$$

If the Bayesian does not have any prior preference for one of the two hypotheses, $(P(H_1) = P(H_0))$, then the Bayesian would find the hypothesis $H_1 : p = 2/3$ more probable than the hypothesis $H_0 : p = 1/2$. However, if the Bayesian's prior belief is that the sample was three times as likely generated by a fair coin rather than a bent coin, $P(H_0)/P(H_1) = 3$, then the Bayesian would find the hypothesis H_0 more likely after the sampling.

4. Composite hypothesis. To test the hypothesis $H_0 : p = 1/2$ versus $H_1 : p \neq 1/2$ the frequentist may again perform the likelihood ratio test. Since

$$\begin{aligned} &p_B(n=0;N=10,p=1/2) + p_B(n=1;N=10,p=1/2) \\ &= p_B(n=10;N=10,p=1/2) + p_B(n=9;N=10,p=1/2) \\ &= 2^{-10} + 10 \times 2^{-10} \approx 0.011 \end{aligned} \tag{16.49}$$

the frequentist would reject the hypothesis $p = 1/2$ for all samples with $n = 0, 1, 9, 10$ with a type I error of 0.022. Especially, the frequentist would accept the null hypothesis for $n = 6$.

To test the hypothesis $H_0 : p = 1/2$ versus the hypothesis $H_1 : p \neq 1/2$, the Bayesian would consider the ratio

$$\frac{P(H_1|n=6;N=10)P(H_1)}{P(H_0|n=6;N=10)P(H_0)} = \frac{4!6!2^{10}}{11!}\frac{P(H_1)}{P(H_0)} \approx 0.443\frac{P(H_1)}{P(H_0)} \tag{16.50}$$

For equal prior probabilities of $P(H_0)$ and $P(H_1)$, the Bayesian would thus infer that the hypothesis H_0 is more probable.

Though there are numerical differences, the major difference between the frequentist and Bayesian approaches is the epistemological setting. The Bayesian has to assign a prior probability to parameters and hypotheses. The frequentist has to choose an estimator or a test. The frequentist's choice depends on what kind of error he/she would like to minimize and what amount of error he/she can tolerate. Prior knowledge also enters in that certain parameters of the assumed distributions are assumed to be known. The frequentist gets definite answers, an estimate of a parameter or either an acceptance or rejection of a hypothesis. The Bayesian gets probability distributions in both cases. The major task for both of them, the frequentist and the Bayesian, is how to interpret their respective results and what to do with them, which is the topic Sect. 16.8 on statistical inference.

16.7 Choice of Prior

Bayesian inference requires a prior that expresses the knowledge before the new data is taken into account. There is no easy rule on how to turn verbal knowledge into quantitative information. A general principle is Laplace's law of insufficient reason: If there is an exhaustive set of N mutually exclusive alternatives $H_1, \ldots, H_N$ then $P(H_i) = 1/N$ for $i = 1, \ldots, N$, or Bernoulli's principle of indifference: if A is true m times in the set of $H_1, \ldots, H_N$ and $N - m$ times false, then $P(A) = m/N$. We have made use of this principle on a number of occasions. The principles of indifference become the principle of maximum entropy which states: choose that probability distribution as a prior which maximizes the entropy under the constraints that prior knowledge provides. The principle of maximum entropy yields the distribution that is the least biased consistent with the prior knowledge. The principle of maximum entropy looks like a reasonable principle for choosing a prior, but it faces a fundamental problem. Suppose you are totally ignorant about the parameter θ that you want to estimate, no constraints. The principle of maximum entropy then tells you to choose the uniform distribution as the prior distribution for the parameter θ. But you are equally ignorant about the distribution of $1/\theta$ and the principle of maximum entropy tells you in this case to choose the uniform distribution for $1/\theta$, which differs from the uniform distribution for θ. Similarly suppose you do not know the position of a particle in a sphere. Do you assume a uniform distribution in Cartesian coordinates or in spherical coordinates. Again the two distributions differ. You do not really know how to formulate your ignorance about the position. We have previously stated as an important point that there is no inference without assumptions. Here we have to sharpen this point:

Ignorance is not a fallback option

The general hope is, however, that Bayesian inference is applied repeatedly. The posterior distribution becomes the prior once another set of data becomes available. It is assumed that in such repeated application of Bayesian inference, one ultimately approaches the true distribution, somewhat independent of where exactly one started. This is the premise of all learning. One improves from where ever one starts. An example of such continuous updating is discussed in Sect. 16.10 on data assimilation.

16.8 Statistical Inference

Statistical inference has five major ingredients:

(1) The epistemological framework.
(2) The evidence, often given in form of a sample.
(3) The selection of a probability model that allows the evidence to be regarded as a realization of a random variable.
(4) The selection of the inference method, such as an estimate or a test.

(5) The interpretation of the result.

We have actually not addressed the issues surrounding these ingredients in great detail. Instead, we have derived the probabilistic properties of various estimators and tests. This derivation was a straightforward and objective exercise, making use of probability theory, and is helpful to the statistician but does not relieve him or her of making important subjective choices and specifications. Here we discuss some of the broader issues that one faces when making statistical inferences.

The epistemological framework. When making statistical inferences, one needs to formulate exactly the question that one is asking or, more technically, one needs to assess the epistemological situation, including the epistemological starting point, what one does and does not know, and the epistemological end point, what one would like to know.

Take the coin tossing experiment. It resulted in a binary sequence of length N that contains $0 \leq n \leq N$ heads. One may ask the following questions:

(1) What is the number of different possible binary sequences of length N that contain n heads?
(2) What is the probability to observe a sequence that contains n heads?
(3) What is the probability that the result of a coin toss is head?
(4) Is the coin fair?
(5) What is the probability of obtaining head when you toss the coin again.

Combinatorics provides the answer to the first question. There are 2^N possible different sequences. Among these different sequences, $\binom{N}{n}$ sequences contain n heads. Such counting exercises are non-controversial. They are a result of probability theory. If one assumes an underlying Laplacian probability space, all possible sequences are equally probable, then the answer to the second question is $\binom{N}{n}/2^N$, an answer already conditioned on an assumption. The answers to the last three questions depend on whether one uses a Bayesian or frequentist approach, what kind of prior or estimator one applies, and what test and what alternatives one invokes.

There might be subtle differences. Consider the situation that the investigator knows the probability $p(H)$ of obtaining head when he or she tosses a coin. This is the probability of obtaining head before tossing the coin. Contrast this with the situation that the coin has been tossed but the investigator does not know the result. This situation is slightly different from the previous situation. Alea iacta est. Formally, if $f : \Omega \mapsto \{H, T\}$ is the random variable describing the coin tossing and $\omega \in \Omega$ the particular value that has been realized by the coin toss, then the investigator tries to assess the correctness of the statement $\omega \in f^{-1}(H)$. This statement is a Boolean proposition that is either true or false. The tossing of the coin has changed the epistemological situation. Once the coin has been tossed, one has left the realm of many different possible outcomes to which one might apply the frequentist's interpretation of probability and has arrived at a situation where there is a definite result but the investigator does not know it, a situation to which one might apply the logicist's or subjectivist's view. This change of the epistemological situation is similar to taking a measurement in quantum mechanics.

Evidence. The evidence might be just given to the statistician, say in form of a sample. Or the statistician has to take the sample. Then questions of sampling strategies arise, taking into account factors such as costs and representativeness. How many cars do you want to crash until you are satisfied that the side beam can sustain a 30 mph side impact? How many pollsters should you hire and where do you place them to poll the popularity of a certain party. Also, the evidence must include all the ancillary information that one has, in a precise and quantitative way. This is not always easy. Being faced with six heads in a row, it should make a difference whether they were thrown by a well-known crook in town or by your long-term trusted friend, but how to account for this difference quantitatively?

Probability model. The basic assumption of statistics is that your evidence is the realization of a random variable. How to choose the probability distribution of this random variable? It might even be questioned to use statistical inference at all. This is demonstrated by a problem posed to Fermat and Pascal, known as the problem of points.

Example: The problem of points. Assume that two players have agreed to play a game where

– each player contributes equally to the stake,
– one player bets on tail, the other on head,
– a fair coin will be tossed successively until one player has won N times,
– then this player gets the entire stake.

Assume that for some reason, the game is halted with player A having won $n_A < N$ times and player B having won $n_B < N$ times. How should one split the stake fairly or justly? This is a question about fairness. One option might be that each player gets back his or her share of the stake. The bet is called off. Another option is that the player who has won more often gets the whole stake. Fermat and Pascal suggested another solution by counting the possible outcomes if the players would have continued. After having played $n_A + n_B$ rounds, player A would need to win $M_A = N - n_A$ rounds and player B would need $M_B = N - n_B$ more rounds. At most, the players would have to play $M = M_A + M_B - 1$ more rounds to reach the end of the game. The game might end earlier but for counting purposes, it is easier to imagine that the players would play all the M additional rounds. Among the 2^M possible outcomes of these non-played rounds, the player A and B would win the relative fractions

$$p_A = \frac{\sum_{n=M_A}^{M} \binom{M}{n}}{2^M} \qquad p_B = \frac{\sum_{n=0}^{M_A-1} \binom{M}{n}}{2^M} \tag{16.51}$$

Fermat and Pascal suggested that the stake should be split according to these fractions. If $N = 10$, $n_A = 8$, and $n_B = 6$, then $M_A = 2$ and $M = 5$ and the fractions are $p_A = 26/32$ and $p_B = 6/32$. Note that the result will be the same for $N = 1000$, $n_A = 998$ and $n_B = 996$. Only the number of rounds that the players need to win among the non-played rounds matter. We will not discuss further subtleties of this problem and the prominent role that it played in the development of probability theory. Here we only want to point out that Fermat and Pascal provided an ingenious solution to the problem of points but it does not give an answer to what one might regard as fair in this situation. What is fair is outside of statistics.

More generally, there are the ever elusive questions: Are there random occurrences or is everything determined? Is there a free will? These are very lofty questions. Consider again an observed sequence of zeros and ones. These are our observations. We may interpret this sequence as resulting from a deterministic dynamical system, the deleted digits of the Bernoulli shift, or as being a realization of a stochastic process, the Bernoulli chain as realized by the tossing of a coin, or perhaps the sequence might be a coded message. When applying the utilitarian approach which relies on the epistemological loop one would have to explore the logical consequences of the assumed models and make some testable predictions and compare them to a new set of observations. Now it all depends on the experimental set-up. If one is able to repeat the experiment, then one would predict that one obtains nearly the same sequence in the case of the deterministic model, perhaps with some differences in the latter part of the sequence due to the inability to reproduce the initial state of the Bernoulli shift exactly. In the case of the stochastic model, one would predict a sequence different from the original one but with similar statistical properties. If the sequence is a coded message and one can ask the sender to send it again, one would expect exactly the same sequence. The situation is not clear-cut when one cannot repeat the experiment. If one just gets another sequence with no ancillary information about how it is generated, then it becomes hard even impossible to discriminate between these different models. This is the problem that researchers face who are looking for intelligent life out there. Do those signals result from deterministic processes, are they just random noise or is somebody trying to tell us something.

Inference method. The choice of the appropriate inference method is perhaps the most difficult choice a statistician has to make. There is no correct inference method. Different methods have different properties and these might make one method more appropriate than another. In choosing a method, the statistician is forced to specify acceptable error bars and risks and significance levels. A 5% risk of error seems to be a standard number but the probability of throwing two consecutive sixes is 1/36 = 0,0277..., well below the 5% mark and its occurrence would lead to a rejection of the hypothesis that the die is fair, at the 5% significance level.

Overall, the selection of a suitable inference method is like buying a car. There are many models out there. In choosing one model, one should consider one's needs and finances but in the end, it is not a totally rational decision.

Interpretation. When interpreting results of statistical inferences, first recall the fact that the frequentist's interpretation of statistics does not say anything about a single realization but only about ensembles. When one estimate of the mean is calculated, one gets a number which is the realization of a random variable. This number might be smaller or larger than the mean of the random variable and there is no way to find out from this one realization. However, some knowledge has been gained, as is most obvious from the Bayesian approach. But even there the result is a posterior distribution which only has an ensemble interpretation. The interpretation of statistical inferences is a huge problem in case of climate change. What does it mean that global warming is caused by man-made greenhouse gas emissions with 95% probability and what actions should follow from it?

Consistency In addition to all these considerations, one should, of course, obey the rules of probability theory if one chooses to apply it. Gilovich et al. (2002) give many examples that our intuitive mind often commits fallacies when confronted with random phenomena. There are some basic rules that need to be obeyed such as

(1) A less inclusive event is less probable than a more inclusive event, $P(A) \leq P(B)$ if $A \subseteq B$. A statistician is more likely a professor than a mathematics professor.
(2) Despite its higher 'regularity' the ordered sequence $HHHH$ has the same probability as the more 'irregular' sequence $HTHT$ when obtained from tossing a fair coin; also, the sequence $HHTHT$ is less probable than the sequence $HTHT$.
(3) There is a psychological law of small numbers which states that we often overestimate the inferences that can be drawn from small samples.
(4) An extreme event is more likely to be followed by a more moderate event. This is a consequence of $P(|x| \leq x_0) = 1 - P(|x| > x_0)$. If x_0 is large so that $P(|x| > x_0)$ is small than $P(|x| \leq x_0)$ is close to one. This fact is often referred to as the regression toward the mean. An explicit example of this regression were the black and white marbles in the bowls discussed in Sect. 12.3. If one starts out from an extreme state, then it is more likely to move toward the equilibrium value than away from it.

16.9 Decisions

In real life, one often has to go beyond inference. One has to make decisions. Imagine yourself to be a coin manufacturer. Your buyer has requested fair coins with a tolerance of $0.45 \leq p(H) \leq 0.55$. You design a test. The test results are in. Now, you have to decide whether or not to ship the coins to the buyer. For a decision, you have to weigh in the financial and legal risks that you are willing to take. An inference is just a gain of knowledge. Making an incorrect inference has no tangible consequences except that one errs. A decision or action usually has some tangible consequences. There is 50% probability of rain tomorrow. Some kind of knowledge has been gained by such a result but how does it help one to make a decision. Should I take or not take an umbrella. It is one thing to be wrong about a prediction and another thing to be rained on. In decision theory, the consequences of decisions or actions are quantified by loss and risk functions. In economics, one usually uses the utility function.

Utility function. Consider the simple situation that you are offered either to play in a lottery that offers \$10,000,000 and \$1,000,000 both with equal probability or being paid \$5,000,000 for sure. The epistemic situation is clear-cut. The lottery offers you \$5,500,000 in the mean. How do you decide? Do you take the \$500,000 or do you gamble? Most of us, being risk avert, would take the half million for sure. How could one rationalize such a decision. These and similar questions can be addressed by introducing a utility function. A possible utility function in this situation is based on the argument that an extra dollar means less to a rich person than to a poor person. To account for this fact, one introduces a utility function $U(w)$ which depends on the wealth w of the player. The function $U(w)$ should be a concave function, accounting for the 'diminishing marginal utility of money'. One such choice is $U(w) = \log w$. If you own one million dollars and get an additional million, your gain in utility is $U(2) - U(1) = \log 2$, which is more than the gain in utility if you acquire another million when you already own 9 million, which is $U(10) - U(9) = \log(10/9)$. The expected gain in utility for the lottery is $(\log 10 + \log 1)/2$ which is smaller than the utility $\log 5$ of being paid 5 million for sure. So you should take the 5 million for sure. Of course, other choices of the utility function might better describe your preferences. More such examples can be found in Gilovich et al. (2002).

As a second example, consider a lottery where one million people pay 1 dollar for a lottery ticket that pays out 1 million dollars to one winner. One winner takes it all. The probability of winning is, however, extremely small, 1 over 1 million. The expectation is 1. Should one or should one not play the lottery? The expectation of the gain of the logarithmic utility function is

$$\langle \Delta U \rangle = \frac{1}{10^6}[\log(w(0) - 1 + 10^6) - \log w(0)] + \frac{10^6 - 1}{10^6}[\log(w(0) - 1) - \log w(0)] \tag{16.52}$$

which is negative, increasing monotonically from $-\infty$ for $w(0) = 1$ to zero for $w(0) \to \infty$. So, you should not play this lottery, if you subscribe to the logarithmic utility function.

The introduction of a utility function is also the way by which Daniel Bernoulli resolved the famous St. Petersburg paradox.

St. Petersburg paradox. The St. Petersburg paradox arises in the following game. A casino offers a game where a fair coin is flipped repeatedly. When the first tail appears the game is halted. If this first tail appears at the nth throw the player wins an amount 2^n. How much should a player pay the casino to play this game? For an answer, one might look at the expected gain. The gain is $g(n) = 2^n$ and has probability $p(n) = (1/2)^n$. The expected gain is thus

$$\langle g \rangle = \sum_{n=1}^{\infty} \left(\frac{1}{2}\right)^n 2^n = \sum_{n=1}^{\infty} 1 = \infty \tag{16.53}$$

The expected gain is infinite. Thus, a rational player should be willing to pay any amount to play this game. However, no one in his right mind would do so. This constitutes the paradox. One way out of this paradox is to acknowledge that any casino has only finite resources R and can sustain the game only for $N = \log_2 R + 1$ throws. Then the expected gain is

$$
\begin{aligned}
\langle g \rangle &= \sum_{n=1}^{\infty} \left(\frac{1}{2}\right)^n \min\{2^n, R\} \\
&= \sum_{n=1}^{N} 1 + R \sum_{n=N+1}^{\infty} \left(\frac{1}{2}\right)^n \\
&= N + \frac{R}{2^{N-1}}
\end{aligned}
\tag{16.54}
$$

which is finite. If the resources of the casino are \$ 10^9 the expected gain is only about \$ 32 .

Bernoulli suggested another resolution of the paradox. He introduced the wealth of the player

$$w(n) := w(0) - c + 2^n \tag{16.55}$$

where $w(0)$ is the initial wealth and c is the cost to play the game. As above, he noted that an extra dollar is worth less to a rich person than to a poor person and introduced the logarithmic utility $U(w) = \log w/w(0)$. Thus, instead of calculating the expected gain, Bernoulli calculated the expected gain in utility

$$\langle U \rangle = \sum_{n=1}^{\infty} \left(\frac{1}{2}\right)^n [\log w(n) - \log w(0)] < \infty \tag{16.56}$$

which is finite. The expected utility can be positive or negative. The maximal price c can be determined from the condition that the expected utility should be larger than zero, and depends on the initial wealth of the player. A millionaire ($w(0) = 1,000,000$) should be willing to pay up to \$20.88 and a person of lesser means ($w(0) = 100,000$) should be willing to pay up to \$10.95.

The concept of the utility function highlights again the fact that statistics is a fairly subjective business. Different stakeholders will assess the utility differently. Consider the stakeholders in public health, the main ones are the patients, the physicians, the employers, the insurance companies, the pharmaceutical companies, and the government. They all have different fiduciary, legal, moral obligations and will choose their utility function accordingly. The important point is

The utility function has to be chosen by the investigator.

A more satisfying way out of the St. Petersburg paradox was suggested by Ole Peters (2011), who notes that the increase in wealth is a non-ergodic process and calculates the growth rate of the time-averaged process as follows: Denote by

$$r_k := \frac{w - c + m_k}{w} \tag{16.57}$$

the factor by which the wealth of the player increases in one round of the game. Here w is the wealth of the player before he or she plays the round, c again the cost to play the game and m_k the payout of the round. The exponential growth rate then is $\gamma_k = \log r_k$. If K such rounds are played the factor becomes

$$r_K = \Pi_{k=1}^{K} r_k \quad (16.58)$$

and

$$\bar{r}_k = (\Pi_{k=1}^{K} r_k)^{1/K} \quad (16.59)$$

the average factor per round. If $s(n)$ is the number of rounds that have length n then this averaged factor can be written

$$\bar{r}_K = (\Pi_{k=1}^{\infty} r_n^{s(n)})^{1/K} \quad (16.60)$$

Since $s(n)/K$ approaches $p(n)$ in the limit $K \to \infty$, we obtain

$$\bar{r} = \Pi_{n=1}^{\infty} r_n^{p(n)} \quad (16.61)$$

and for the growth rate

$$\bar{\gamma} := \log \bar{r} = \sum_{n=1}^{\infty} (\frac{1}{2})^n [\log w(n) - \log w(0)] \quad (16.62)$$

which is identical to Bernoulli's result using a logarithmic utility function. This equality is just a coincidence since both approaches use the logarithm, Bernoulli to define a utility function, Peters to define the exponential growth rate. Using any other utility function would lead to a discrepancy. The main result, however, is that, instead of calculating the expectation of a rather arbitrary utility function, Peters calculates the growth rate of the time-averaged wealth of the player, a result that is the more satisfying since it is this time-average that matters to a player, not the average of an ensemble of players playing the game in parallel universes which yields an infinite ensemble-averaged growth rate.

16.10 Data Assimilation

Most of our examples have been standard examples, coin tossing, throwing dice, picking balls out of urns. They were chosen because they are technically simple. Statistics, of course, also applies to more complex problems such as weather and climate forecasts. As an example of such a complex problem, we consider data assimilation which combines many of the aspects of statistics.

Data assimilation refers to techniques that combine dynamical models with observational data in order to obtain a better product. The better product might be a better estimate of the observed field, a better estimate of a model parameter, or a better prediction. Observationalists use dynamical models to extract reliable information out of observational data. Data assimilation then becomes 'dynamically consistent interpolation'. Modelers use observations to force their models to stay close to reality and control the loss of predictability. Data assimilation then becomes what is called data analysis in operational weather forecasting. These and other data assimilation techniques have a common conceptual and mathematical background which comes from estimation and control theory. Here we describe the basic concepts of data assimilation.

Philosophically, data assimilation is the farthest away from the classical paradigm that separates measurements and observations from physical laws and theory. It not only realizes that models require data to yield useful predictions but also that data require models to extract useful information. The utility of data becomes model dependent.

Data assimilation consists of three basic components:

- a dynamical model,
- a set of observations, and
- an assimilation scheme.

Dynamical model. To fix ideas, let us think about a computer model for weather forecasting. The dynamical model is then a time-discrete stochastic process of the general form

$$f_{n+1} = A_n[f_n, p, \epsilon_n] \tag{16.63}$$

where n is the discrete-time index of the model, A_n a generally nonlinear operator that comprises the dynamics of the system, p a set of parameters, and ϵ_n a noise process which takes into account effects like:

- the operator A_n is inexact because it does not include all dynamical processes,
- the boundary conditions and external forcing fields are not exactly known,
- the spatial discretization requires uncertain parameterization of subgrid scale processes, and
- the discretization in time introduces truncation errors.

Observations. The second component of data assimilation is a set of observations. If the observations are combined at each time step into an observation vector s_n, then DA assumes that there exists an observation equation

$$s_n = B_n[f_n, \delta_n] \tag{16.64}$$

which relates the observation vector s_n to the state vector f_n. Here B_n is again a generally nonlinear operator and δ_n the observational error, representing instrumental, environmental, and sampling errors and other uncertainties.

A first problem arises because the time index n must be identical in the dynamical model (16.63) and the observation model (16.64). In weather forecasting, the time interval into which observations are binned is usually much longer than the time step of the numerical algorithm used to solve the dynamical model. In this case, the time step of data assimilation is thus given by the sampling procedure of the data and not by the original time step of the dynamical model. The collection of data into time bins also contributes to the error δ_n.

Assume the dimension of the state vector to be d_m and the dimension of the observation vector to be d_o. In weather forecasting, one is in a situation where one has sparse data, $d_o \ll d_m$. The observations represent a weak constraint on the model.

Assimilation scheme. In the case $d_o < d_m$, the operator B_n cannot be inverted to yield the state vector f_n. The state vector is under-determined. Additional constraints are needed to arrive at a unique solution. In data assimilation, the additional constraint is consistency with a dynamical model. The third component of data assimilation is, therefore, a blending scheme, where one combines the model prediction f_n with the observation s_n to form an improved estimate $\hat{f}_n$ of the state vector. There are two main blending schemes. For the discussion, assume that data and model predictions are available for times $n = 1, \ldots, N$.

A filtering problem combines past and present data to form a 'nowcast'. Formally, the nowcast $\hat{f}_n$ at time n is derived from all f_m and s_m with $m \leq n$. Such a nowcast can be used to provide optimal initial conditions for a forecast. Later, when data and forecast have become available for the time $n + 1$, the procedure can be repeated to estimate $\hat{f}_{n+1}$, and so forth.

A smoothing problem combines all data, past, present and future, to form an optimal estimate for all times. Formally, $\hat{f}_n$ for all $n = 1, \ldots, N$ is derived from all f_m and s_m with $m = 0, \ldots, N$.

Although both problems are estimation problems, the filtering problem has its methodological roots in estimation theory and the smoothing problem has its roots in control theory.

Given the distribution of the model error ϵ_n and the observation error δ_n one can for any given blending scheme calculate the statistics of the error of the estimate $\hat{f}_n$. Optimal blending schemes can then be determined that minimize this error. As we discussed, this is not as straightforward as it sounds

- Assigning quantitative error statistics to models and data involves considerable subjective judgements. What error is introduced into a dynamical model because it only represents approximate dynamics? What errors are introduced into an observational model because observations are not taken at grid points or represent proxy data?
- If nonlinear blending schemes are allowed for, then it is not sufficient to characterize the model and observation errors by just their means and covariances. Higher-order moments or the full probability distributions are needed.
- Any minimization requires the specification of a distance or norm, often in the form of cost, penalty or risk functions. As these names indicate, there is no unique objective way to arrive at such a distance. What is costly for one purpose might not be costly for another purpose.
- The minimization problem might not have a (unique) solution or the solution cannot be determined.

- Even if an optimal blending scheme can be determined it might be impractical, computationally excessive or suffer some other drawback.

For the above reasons, a variety of ad-hoc and heuristic data assimilation schemes have been developed. When using any of these schemes, it is paramount to

- understand a priori statistical assumptions that go into the scheme, and
- determine the error of the blended estimate.

Methodologically, data assimilation lies between validation and quality control. In validation, the model prediction and the observed data are compared without any feedback. In data assimilation, model prediction and observational data are combined to form an improved estimate. In the filtering problem, data are fed into the model to improve the forecast. In the smoothing problem, the model is fed into the data to extract dynamically consistent information. In both problems, the data are given. Control theory also determines which data need to be taken for optimal control. In our context, this would be the determination of an optimal observational sampling strategy.

16.11 Summary

Bayesian inference is a method to update the probability of a hypothesis from some prior value to a new posterior value due to new evidence. The inference is based on Bayes' formula which requires the prior probability, a probabilistic model to calculate the likelihood function which is the probability of the evidence given that the hypothesis is true, and the probability of the evidence no matter whether or not the hypothesis is true. If one compares different hypotheses, the probability of the evidence drops out.

Bayesian inference easily allows for the logicist's and the subjectivist's interpretation of probability, in addition to the frequentist's interpretation. It can thus be applied to situations where repetition is not possible, even in principle.

Bayesian inference requires a prior probability of the hypothesis. Bayesians argue that this requirement reflects the fact that one cannot do inference without assumptions. All statistical inferences are conditional on evidence.

To use probability theory skillfully in any given practical situation requires that one carefully evaluates the epistemological situation that one starts from and the goal that one aims at. These evaluations require choices whose justifications lie outside the realm of probability theory. One has to choose estimators, tests, and priors. One may justify these choices by minimizing risks or trying to be unbiased but these are principles outside probability theory.

Statistics is a human endeavor. The human cannot be taken out of it. One has to clearly delineate between results of probability theory and statistics and what people make of these results. Statistical theory (combinatorics) unequivocally and unambiguously determines the possible outcomes of a play of a fair roulette. People, however, react differently to this result. Some gamble. Others buy casino stocks.

References

Gilovich, T., Griffin, D., Kahnemann, D. (eds.): Heuristics and Biases. Cambridge University Press, Cambridge (2002)

Peters, O.: The time resolution of the St. Petersburg paradox. Phil. Trans. Roy. Soc. A **369**, 4913–4921 (2011)

Part V
Probability in Physics

Dynamics and probability are combined in statistical mechanics and in quantum mechanics. Statistical mechanics aims at deriving the macroscopic laws that govern matter such as equilibrium thermodynamics and fluid dynamics from the microscopic laws that govern its constituent atoms and molecules. Stationary measures and the convergence of initial measures are the basis for statistical mechanics. The stationary measures describe the thermodynamic equilibrium state. Convergence toward the stationary measure becomes the irreversible approach toward equilibrium. Statistical mechanics is successful on a technical level. Macroscopic properties are routinely derived from microscopic Hamilton functions or Hamiltonians but there are major foundational issues and it is not fully understood why statistical mechanics works. Here, we demonstrate that statistical mechanics can be viewed as a basic exercise in Bayesian inference. It is simply the best that we can do given the limited information about the system. Adding probability to dynamics does not change the basic character of the dynamics such as any time-symmetry in the dynamical evolution.

Quantum mechanics is also a theory where dynamics and probability are intertwined, at a very basic level. Even pure states only have a probabilistic interpretation. The probability in quantum mechanics derives from a generalization of the Pythagorean theorem, not from measure theory, and implies that quantum mechanics is governed by quantum logic, and not by classical logic. This difference between classical and quantum probabilities is emphasized, bypassing most of the technical aspects of quantum mechanics. As with statistical mechanics quantum mechanics works extremely well but it also faces numerous unresolved foundational problems. A Bayesian approach softens some of these foundational issues.

Chapter 17
Equilibrium Statistical Mechanics

17.1 Introduction

Statistical mechanics considers macroscopic systems that are governed by phenomenological laws such as the laws of thermodynamics or continuum mechanics. It attempts to derive these laws from the underlying microscopic system. Our main example of such a microscopic system will be a classical N-particle system governed by a Hamilton function. Statistical mechanics is most advanced for systems that are in thermodynamic equilibrium, and we will consider this case in this chapter, and the non-equilibrium case in the next chapter.

In general, the macroscopic state does not uniquely specify the microscopic state. One does not have sufficient information to exactly determine the properties of the microscopic system. This is no reason to despair. One still can make useful inferences. Given an ideal gas in a container and only given the number N of molecules, the volume $V = L^3$ of the container, and the total energy of $E = Nmv^2/2$ of the gas, where m is the mass of the molecule and v its speed, one might estimate the pressure as follows: one third of the molecules will move in any of the three directions and need time $\delta t = 2L/v$ for one round trip between the walls. At each encounter with a wall a molecule exchanges the momentum $2mv$ with the wall. The pressure, the force per unit area, that the molecules exert on a wall is then given by

$$p = \frac{1}{3}N\frac{2mv}{\delta t}\frac{1}{L^2} = \frac{2}{3}\frac{E}{V} \tag{17.1}$$

Statistical mechanics mimics this kind of inference. It first realizes that a macroscopic specification provides insufficient information to determine the microscopic state. At any given time there are many micro-states that are compatible with a given macro-state. It therefore applies probability theory, the logic that governs reasoning with insufficient information, to the problem. In the language of probabilistic inference, the ensemble of the many possible micro-states constitutes the many mutually exclusive but exhaustive possible alternatives that must be assigned a probability distribution. It

P. Müller, *Handbook of Dynamics and Probability*,
https://doi.org/10.1007/978-3-030-88486-4_17

is thus the first basic postulate of equilibrium statistical mechanics that this ensemble of micro-states must be described by a probability distribution on the state space.

The most powerful principle that we have at our disposal to choose such a unique probability distribution is the principle of maximum entropy since the probability distribution that maximizes Shannon's entropy functional, perhaps subject to some constraint, is the unique distribution that presupposes the least information about the system. It is thus the second basic postulate of equilibrium statistical mechanics that the probability distribution of the microscopic ensemble should be inferred from the principle of maximum entropy.

We thus present statistical mechanics as a statistical inference problem, estimating the conditional probability $P(A|B)$ that some 'event' A in the micro-world will be realized given a set B of macroscopic variables. This approach is advocated in Baierlein (1971), from which we took the above calculation of the pressure. We thus present equilibrium statistical mechanics as a motivated recipe. The recipe works and is the best we can do but it is not yet fully understood why it works. Such understanding meets with major foundational problem. Why is thermodynamic equilibrium characterized by a stationary probability distribution or ensemble? Why can some macro-variables be interpreted as expectations of observables? Why does a finite number of macro-variables suffices and are related as required by the laws of thermostatics. These and other issues are discussed in the Chap. 19 "Foundational issues".

Though equilibrium statistical mechanics has been covered by numerous authors we repeat it here to stress fundamental points;

(1) Statistical mechanics is an inference under insufficient information.
(2) The principle of maximum entropy (PME) implies most of the formal structure of thermostatics, except the thermodynamic limit.
(3) In the thermodynamic limit fluctuations diminish and statements become nearly deterministic.
(4) Though describing macroscopic systems statistical mechanics needs to invoke quantum-mechanical corrections.
(5) There is no need to solve the microscopic equations of motion. Thermostatics and equilibrium statistical mechanics are timeless theories.

First we review the basic structure of thermostatics.

17.2 Thermostatics

Here we briefly recapitulate the basic structure of equilibrium thermodynamics or thermostatic. A more detailed exposition is given in Appendix B.

Thermostatics is a macroscopic theory of matter that ignores its discrete molecular structure and assumes a continuum. It is concerned with the properties of systems in thermodynamic equilibrium. Thermodynamic equilibrium is the state that an isolated system, i.e., a system that does not exchange mass, momentum and energy with its surroundings, reaches after a sufficiently long time. Thermostatic only studies the system in thermodynamic equilibrium and does not say anything about the approach to equilibrium.

Thermodynamic variables. A thermostatic system is characterized by a relatively small set of macroscopic variables, called thermodynamic variables. Thermostatics studies the relationship between these variables. The choice of the variables depends on the control that we have over the system. Depending on this control, the same physical system might be represented by different thermodynamic systems.

For simplicity, we consider the case where the thermodynamic variables are the mass M, the volume V, the internal energy U, the entropy S, the pressure p, the temperature T and the chemical potential μ. This case represents a gas in a container, at rest and not exposed to an external field. Mass, volume and pressure have purely mechanical explanations. The internal energy U is defined by the first law of thermodynamics which equates changes in the internal energy to the amount of heat and mechanical and chemical work added to the system

$$dU = \delta Q + \delta W_{mech} + \delta W_{chem} \tag{17.2}$$

Thermostatics assumes that there are prescriptions to measure heat and the mechanical and chemical work. The entropy is the thermostatic entropy first introduced by Clausius. Changes in this thermostatic entropy are defined by restricting the second law to reversible processes

$$dS = \delta Q/T \tag{17.3}$$

Temperature and entropy do not have a mechanical explanation. They depend on the molecular structure of matter.

The main findings. The main phenomenological findings of thermostatics are that:

(1) The entropy S is a function of (U, V, M) and this entropy function $S(U, V, M)$ fully specifies the system.
(2) The entropy function is super-additive. When two systems A and B are brought into contact with each other then we find for the total system

$$S(U_A + U_B, V_A + V_B, M_A + M_B) \geq S(U_A, V_A, M_A) + S(U_B, V_B, M_B) \tag{17.4}$$

This property is a consequence of the second law.
Super-additivity is also expressed by the phrase "The entropy is a maximum in thermal equilibrium". For a proper interpretation of this phrase consider a system which is externally constrained to have the internal energy U, volume V and mass M but has no internal constraints. Compare this system with a set of subsystems that has the same external constraints but where internal constraints keep the subsystems from interacting. Then the entropy of the system with no internal constraints is equal or larger than the entropy of any system with internal constraints. This extremal property of the entropy implies that, without internal

constraints, all subsystems must have the same temperature, pressure and chemical potential and that the temperature, specific heat at constant volume, and isothermal compressibility are non-negative.

(3) The entropy function $S(U, V, M)$ is a continuous, differentiable, monotonically increasing function of the internal energy U. Thus $S(U_2, V, M) \geq S(U_1, V, M)$ if $U_2 \geq U_1$. The entropy function can thus be inverted to give the internal energy function $U(S, V, M)$. The internal energy function is a thermodynamic potential. Its derivative are the pressure p, the temperature T and chemical potential μ

$$p = -\left(\frac{\partial U}{\partial V}\right)_{S,M} \quad T = \left(\frac{\partial U}{\partial S}\right)_{V,M} \quad \mu = \left(\frac{\partial U}{\partial M}\right)_{S,V} \tag{17.5}$$

This finding is often written in the form

$$dU = -pdV + TdS + \mu dM \tag{17.6}$$

Instead of S, V and M one can change to other sets of independent variables by Legendre transformations and obtains equivalent thermodynamic potentials. We will later consider the free energy (that part of the internal energy that can be converted into mechanical work)

$$F(V, T, M) = U(V, T, M) - TS(V, T, M) \tag{17.7}$$

with the thermodynamic relations

$$p = -\left(\frac{\partial F}{\partial V}\right)_{T,M} \quad S = -\left(\frac{\partial F}{\partial T}\right)_{V,M} \quad \mu = \left(\frac{\partial F}{\partial M}\right)_{V,T} \tag{17.8}$$

(4) For sufficiently large systems the entropy, internal energy, volume and mass become extensive quantities that scale with the size of the system. Thus

$$U(\lambda S, \lambda V, \lambda M) = \lambda U(S, V, M) \tag{17.9}$$

The energy function $U(S, V, M)$ thus becomes a homogeneous function of degree one. Taking $\lambda = 1/M$ one can introduce the specific quantities $u := U/M$, $s := S/M$ and $v := V/M$ and obtains the specific internal energy function $u(s, v)$ with relations

$$p = -\left(\frac{\partial u}{\partial v}\right)_{s} \quad T = \left(\frac{\partial u}{\partial s}\right)_{v} \tag{17.10}$$

The chemical potential is then given by

$$\mu = u + pv - Ts \tag{17.11}$$

which follows from Euler's theorem for homogeneous functions.

Instead of dividing by the mass M, one can also divide by the volume V to get densities (quantities per unit volume).

The basic goal of equilibrium statistical mechanics is to derive the macroscopic energy function $U(S, V, M)$ from the properties of the underlying microscopic system. The basic assumption is: The macroscopic state is given by an invariant probability distribution ρ_* or ensemble on the state space of the microscopic system. This probability distribution can be derived from the principle of maximum entropy.

17.3 Implications of PME

Here we recapitulate those relations that are a consequence of the principle of maximum entropy (PME).

Principle of maximum entropy. If K observables g_k have certain given expectation values $\langle g_k \rangle$, $(k = 1, \ldots, K)$ then the maximum entropy principle states that the least biased density consistent with these constraints is

$$\rho_K^{max}(x; \lambda_1, \ldots, \lambda_K) = \frac{1}{Z_K(\lambda_1, \ldots, \lambda_K)} \exp\left\{-\sum_{k=1}^{K} \lambda_k g_k(x)\right\} \tag{17.12}$$

with partition function

$$Z_K(\lambda_1, \ldots, \lambda_K) := \int_X dx \exp\left\{-\sum_{k=1}^{K} \lambda_k g_k(x)\right\} \tag{17.13}$$

and Lagrange multipliers $(\lambda_1, \ldots, \lambda_K)$. One finds

$$\frac{\partial \log Z_K(\lambda_1, \ldots, \lambda_K)}{\partial \lambda_k} = -\langle g_k \rangle \quad k = 1, \ldots, K \tag{17.14}$$

and

$$\frac{\partial \log Z_K}{\partial \langle g_k \rangle} = 0 \quad k = 1, \ldots, K \tag{17.15}$$

The first set of equations constitutes a set of differential equations from which the Lagrange multipliers can be determined. The Lagrange multipliers are thus chosen such that the PME density reproduces the observed constraints or data.

If there are no constraints then the PME density is the uniform density

$$\rho_0^{max}(x) = \frac{\chi_X(x)}{Z_0} \tag{17.16}$$

where χ is the indicator function and the partition function is the volume of the state space, $Z_0 = V(X)$.

Entropy function. The entropy function is defined by

$$S_K := H[\rho_K^{max}] = \log Z_K(\lambda_1, \ldots, \lambda_K) + \sum_{i=1}^{K} \lambda_i \langle g_i \rangle \tag{17.17}$$

for which one finds

$$\frac{\partial S_K}{\partial \langle g_k \rangle} = \lambda_k \tag{17.18}$$

$$\frac{\partial S_K}{\partial \lambda_k} = 0 \tag{17.19}$$

Thus $S_K = S_K(\langle g_1 \rangle, \ldots, \langle g_K \rangle)$ and

$$dS_K = \sum_{k=1}^{K} \lambda_k d\langle g_k \rangle \tag{17.20}$$

The above formulas imply that $S_K = S_K(\langle g_1 \rangle, \ldots, \langle g_K \rangle)$ is the Legendre transform of $-\log Z_K(\lambda_1, \ldots, \lambda_K)$. One can thus use either the expectations $(\langle g_1 \rangle, \ldots, \langle g_K \rangle)$ as independent variables and work with the entropy function or the Lagrange multiplier $(\lambda_1, \ldots, \lambda_K)$ and work with the partition function or with any other combination by introducing the partial Legendre transform

$$\begin{aligned} F(\langle g_1 \rangle, \ldots, \langle g_M \rangle, \lambda_{M+1}, \ldots, \lambda_K) &:= \log Z_K(\lambda_1, \ldots, \lambda_K) + \sum_{i=1}^{M} \lambda_i \langle g_i \rangle \\ &= S_K(\langle g_1 \rangle, \ldots, \langle g_K \rangle) - \sum_{j=M+1}^{K} \lambda_j \langle g_j \rangle \end{aligned} \tag{17.21}$$

Parameter dependence. When the constraints $g = (g_1, \ldots, g_K)$ depend on parameters $r = (r_1, \ldots, r_M)$ such that $g = g(x; r)$ then the entropy function becomes dependent on r with the partial derivatives at constant $\langle g_k \rangle$ given by

$$\begin{aligned} \frac{\partial S_K}{\partial r_m} &= \frac{\partial \log Z_K}{\partial r_m} \\ &= -\frac{1}{Z_K} \int_X dx \exp\left\{ -\sum_{k=1}^{K} \lambda_k g_k(x) \right\} \sum_{l=1}^{K} \lambda_l \frac{\partial g_l}{\partial r_m} \\ &= -\sum_{k=1}^{K} \lambda_k \langle \frac{\partial g_k}{\partial r_m} \rangle \end{aligned} \tag{17.22}$$

and formula (17.20) generalizes to

$$dS_K = \sum_{k=1}^{K} \lambda_k \delta Q_k \tag{17.23}$$

where

$$\delta Q_k := d\langle g_k\rangle - \sum_{m=1}^{M} \langle \frac{\partial g_k}{\partial r_m}\rangle dr_m \tag{17.24}$$

In statistical mechanics this expression is taken as the definition of heat and its generalizations. In phenomenological thermostatics one assumes that one can measure these heats and defines the entropy via this equation.

PME thus distinguishes between observables g_k, their expectations $\langle g_k\rangle$, Lagrange multipliers λ_k, and parameters r_m.

Super-additivity. Consider two non-interacting systems A and B, each in their maximum entropy state $\rho_A^{max}(x_A)$ and $\rho_B^{max}(x_B)$. The maximum entropy state of the combined but still not interacting system is given by

$$\rho_{A,B}^{max}(x_A, x_B) = \rho_A^{max}(x_A)\rho_B^{max}(x_B) \tag{17.25}$$

and hence $S_{A,B} = S_A + S_B$. When these two systems are allowed to interact and to form a system D such that the constraints are additive, $\langle g_D\rangle = \langle g_A\rangle + \langle g_B\rangle$ then ρ_D^{max} leads to an entropy function

$$S_D \geq S_A + S_B \tag{17.26}$$

since $\rho_{A,B}^{max}$ is a possible probability distribution for the system D but not necessarily the one that maximizes the entropy. The equality sign in (17.26) holds if $\lambda_D = \lambda_A = \lambda_B$.

The PME entropy function S_K thus has the same formal properties as the thermostatic entropy S. It remains of course to be shown that the expectations $\langle g_k\rangle$ and Lagrange multiplier λ_k correspond to thermodynamic variables.

There are two things that PME does not provide. First, it needs to be shown that the PME probability distribution is a dynamically invariant distribution. Second, the thermodynamic limit has to be added, invoking additional properties of the system.

Different ensembles are obtained depending on which macroscopic variables are specified or controlled. We start out with the micro-canonical of a N-particle Hamiltonian system.

17.4 The Micro-canonical Ensemble

Let us consider a N-particle system governed by an autonomous Hamilton function $H(q, p; V, N)$ where $q = (q^{(1)}1, \dots, q^{(N)})$ and $p = (p^{(1)}, \dots, p^{(N)})$ are the $3N$-dimensional position and momentum vectors, V the volume, and N the number of particles. A typical example is

$$H(q, p; V, N) = \sum_{n=1}^{N} \frac{1}{2m} p^{(n)T} p^{(n)} + \frac{1}{2} \sum_{\substack{n,m=1\\ n\neq m}}^{N} m\Phi_{int}(q^{(nm)}) + \sum_{n=1}^{N} m\Phi_{ext}(q^{(n)}, t) \tag{17.27}$$

which describes N interacting particles of mass M exposed to an external field. The first term is the kinetic energy. The second term descibes the interaction of the particles governed by a pair interaction potential $\Phi_{int}(r)$. The pair interaction potential is assumed to be invariant under translation and rotation so that it only depends on the magnitude of the distance $q^{(nm)} := |q^{(n)} - q^{(m)}|$ between particles. This form of the pair potential is not essential for what follows. One has, however, to require stability and temperateness. Stability avoids configurations where particles congregate in a subregion of the configuration space. Temperateness assures that the interaction potential is sufficiently short ranged so that boundary effects can be neglected. Proper definitions can be found in Ruelle (1969). The typical pair potential energy shown in Fig. 17.1 satisfies all the properties that are implicitly assumed in the following. External potential energies $\Phi_{ext}(q^{(n)}, t)$ are also included and might represent external gravitational, electric or magnetic fields but also potentials that must be introduced to confine the system to a finite volume in physical space. Explicitly, we will only consider the ideal gas, the standard example in statistical mechanics, whose Hamilton function is given by first term only. Given such a Hamilton function, the microscopic state of the system is described by a point in the $6N$-dimensional phase space Γ. Observables g are real-valued functions on the phase space, $f : \Gamma \mapsto \mathbb{R}$. The Hamilton function is such an observable; it maps each point in phase space to its energy.

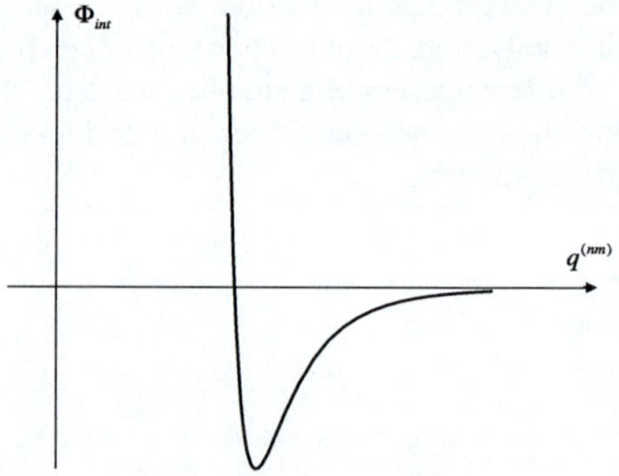

Fig. 17.1 Typical pair potential

Statistical mechanics must now relate the four macroscopic quantities U, S, V and M to the microscopic quantities that appear in the Hamilton function $H(p, q; V, N)$. This is no problem for the volume V which appears explicitly in the Hamilton function. It is also no problem for the mass which is given by $M = Nm$. It also does not require extraordinary ingenuity to find out how to connect the internal energy to the energy of the microscopic state. The major challenge is to give a microscopic interpretation of the thermostatic entropy S. This requires the introduction of a probability density function. It is the major accomplishment of Boltzmann and Gibbs to provide such an interpretation.

The micro-canonical ensemble. Consider a dynamical system in thermodynamic equilibrium. It is isolated, has the volume V, mass $M = Nm$ and Hamilton function $H(q, p; V, N)$. We assume that $\Gamma \subset R^{6N}$ and that R^{6N} is endowed with the usual Lebesgue measure $dqdp$. Since the Hamilton function does not explicitly depend on time, the energy E of the system is conserved. The micro-states are thus constrained to the $(6N-1)$-dimensional energy surface Γ_E defined by

$$\Gamma_E := \{(q, p)|(q, p) \in \Gamma \; H(p, q) = E\} \tag{17.28}$$

The micro-canonical ensemble is then the uniform distribution on the energy surface

$$\rho_0^{max}(q, p) = \frac{\chi_{\Gamma_E}(q, p)}{Z_0(E)} \tag{17.29}$$

where χ is the indicator function and $Z_0(E)$ the area of the energy hyper-surface Γ_E, called the partition function. The uniform density on the energy surface Γ_E is somewhat awkward to work with. One way out is to first calculate the volume

$$V(E) := \int_0^{H(p,q)=E} dqdp \tag{17.30}$$

Then the area of the energy surface is given by

$$Z_0(E) = \frac{dV(E)}{dE} \tag{17.31}$$

Another way out is to use the broadened ensemble

$$\tilde{\rho}_0^{max}(q, p) = \frac{\chi_{\Delta V}(q, p)}{Z_0(E)\Delta E} \tag{17.32}$$

where $Z_0(E)\Delta E$ is the phase space volume between the energy surfaces $H(p, q) = E$ and $H(p, q) = E + \Delta E$. This modification may also be rationalized by the fact that we cannot have a truly isolated system with constant energy but the energy will always be subjected to fluctuations. In any case one needs to demonstrate that results are independent of ΔE or that one can take the limit $\Delta E \to 0$. One may also define the micro-canonical ensemble by the density

$$\rho_0^{max}(p,q) := Z_0^{-1}(E)\delta(H(p,q)-E) \tag{17.33}$$

on the state space Γ with

$$Z_0(E) = \int_\Gamma dpdq\delta(H(p,q)-E) \tag{17.34}$$

The micro-canonical ensemble has the two (desirable) properties: (1) its density is the density that maximizes Shannon's entropy functional and (2) its measure is invariant under the flow S_t in phase space generated by Hamilton's equations. The proof of (2), which is given in Sect. 5.4, makes use of the facts that Hamiltonian flows are conservative (the velocity in phase space has zero divergence) and that the velocity normal to the energy surface is zero.

The micro-canonical ensemble is the only invariant density if and only if the system is ergodic, which is generally assumed. For a Hamiltonian system ergodicity implies that there are no constants of motion other than functions of the energy. If there were such other constants of motion, a properly normalized function of these constants would also be an invariant density that differs from the micro-canonical density. Momentum and angular momentum are not conserved because of the interaction with the 'rigid container'.

Note that the micro-canonical ensemble requires knowledge of just the Hamilton function but not of the solutions of the associated equations of motion. The ensemble is timeless.

So far we have only considered the PME in momentum space where it leads to the uniform distribution on the energy surface. In configuration space PME also leads to a uniform distribution. We thus have to multiply the partition function by V^N. If we resurrect the dependence of the Hamilton function on the volume V and mass $M = Nm$ the micro-canonical ensemble is the PME ensemble consistent with prescribed values of E,V, and M. The entropy function of this micro-canonical ensemble is given by

$$S_0(E,V,M) = H[\rho_0^{max}(E,V,M)] = \log Z_0(E,V,M) \tag{17.35}$$

For computational reasons one also works with

$$\tilde{S}_0(E,V,M) := \log\{Z_0(E,V,M)\Delta E\} \tag{17.36}$$

where ΔE is some small energy increment. This modification also assures that one can actually take the logarithm, since $Z_0(E)\Delta E$ is now a dimensionless quantity.

Thermostatics. Thermostatics now follows from the following two postulates:

Postulate 1: The internal energy U is given by

$$U = E \tag{17.37}$$

Postulate 2: The thermostatic entropy S is given by

$$S(U, V, M) = k_B S_0(E, V, M) \tag{17.38}$$

where we have introduced the Boltzmann constant k_B to conform with the notation in statistical mechanics. One has, of course, to prove that S has all the right properties listed in the section on thermostatics. If so, $U(S, V, M)$ becomes a thermodynamic potential, from which the rest of thermostatics follows. From an experimental point of view the above identifications 'work', for most systems.

Given the micro-canonical density one can calculate the probability densities of observables by the transformation rules of densities. Just to check the internal consistency of the micro-canonical ensemble we calculate the density of the energy

$$\begin{aligned} \rho_0(E') :&= \langle \delta(E' - H) \rangle \\ &= \frac{1}{Z_0(E)} \frac{1}{N!} \int_\Gamma \frac{dpdq}{(2\pi\hbar)^{3N}} \delta(E - H(p, q)) \delta(E' - H(p, q)) \\ &= \delta(E' - E) \end{aligned} \tag{17.39}$$

which is, as it is supposed to be, concentrated on the value E.

17.5 Quantum-Mechanical Corrections

Though the micro-canonical ensemble was formulated for a classical system one needs to add quantum-mechanical corrections. This is best demonstrated for a discrete system. Thus let W be the number of micro-states that are consistent with a specific macro-state. If there are no other constraints then PME asserts the uniform distribution

$$p_0(i) = 1/W \tag{17.40}$$

for $i = 1, \ldots, W$, which is the micro-canonical ensemble. All possible micro-states are equally probable. Then $S_0 = H[p_0]$ and the thermostatic entropy is given by the famous Boltzmann expression

$$S = k_B \log W \tag{17.41}$$

Let us extend this simple situation to N particles in K boxes or, more properly, to N particles that can occupy K different energy levels.

N particles in K boxes. Assume first that the particle are distinguishable. Then each particle can be in any of the K boxes and there are K^N possible micro-states. Therefore $W = K^N$ and

$$S = k_B N \log K \tag{17.42}$$

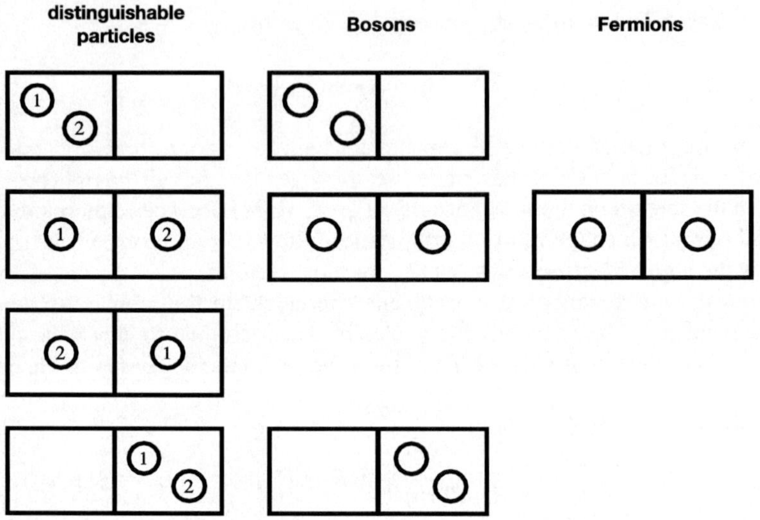

Fig. 17.2 Two particles in two boxes. There are 4 micro-states when the particles are distinguishable (left), only three when the particles are indistinguishable and can occupy the same state (Bosons) (middle) and only one when the particles are indistinguishable and cannot occupy the same state (Fermions) (right)

The entropy S is an extensive quantity if the number K of boxes remains constant, and it is super-additive

$$S_{A+B} - S_A - S_B = k_B N_A \log K_A + K_B/K_A + k_B N_B \log K_A + K_B/K_B \geq 0 \tag{17.43}$$

Statistical mechanics relies on the atomistic structure of matter and must conform with quantum mechanics. In quantum mechanics identical particles are indistinguishable. Furthermore one has to distinguish between Bosons that can occupy the same state and Fermions that cannot. In the case of Bosons one finds the Bose-Einstein statistics

$$W_{BE} = \binom{N+K-1}{N} \tag{17.44}$$

and for Fermions the Fermi-Dirac statistics

$$W_{FD} = \binom{K}{N} \tag{17.45}$$

where in the latter case K has to be larger than or equal to N. Consider as an example the case of two particles and two boxes depicted in Fig. 17.2.

In the classical limit $K \gg N \gg 1$ both these statistics approach

$$W_{cl} = \frac{K^N}{N!} \tag{17.46}$$

which can be proven by using Stirling's formula. To account for the quantum mechanical fact that particles are indistinguishable one has to divide the number of possible states by $N!$. This is often taken as obvious since $N!$ is the number of possible permutations of the particle labels and permutations of the particle label do not give a separate state in case of indistinguishable particles. It is stressed, however, that this is only true in the classical limit. If one has one box one has only one state. It is also noteworthy that making the particles indistinguishable reduces the entropy by an amount $\Delta S = -\log N!$. Intuitively, one feels that some sort of information is lost when particles become indistinguishable, but what S_0 measures is the number of distinguishable states and these are reduced, as can be seen by the example shown in Fig. 17.2.

Entropy in the classical limit. Since the classical limit requires $N \gg 1$ the entropy in the classical limit becomes

$$S(K, N) = k_B N \left(1 + \log \frac{K}{N}\right) \tag{17.47}$$

upon use of Stirling's formula. It is now an extensive function if the 'density' N/K remains constant. It remains super-additive since

$$\begin{aligned}\Delta S :&= S(K = K_A + K_B, N = N_A + N_B) - S(K_A, N_A) - S_B(K_B, N_B)\\ &= N(-\frac{N_A}{N}\log\frac{K_A}{K} - \frac{N_B}{N}\log\frac{K_B}{K} + \frac{N_A}{N}\log\frac{N_A}{N} + \frac{N_B}{N}\log\frac{N_A}{N}) \geq 0 \end{aligned} \tag{17.48}$$

because of Gibbs' inequality. The entropy difference is non-negative and zero only if the densities are equal, $N_A/K_A = N_B/K_B = N/K$. Super-additivity means that the number of distinguishable states in the combined system is larger than the product of the distinguishable states within each subsystem. The quantum mechanical correction does not destroy super-additivity.

Expansion and mixing. Consider the following processes (see Fig. 17.3).

(1) Expansion: An expansion process can be mimicked by changing the number of boxes from K to say $2K$. The change of entropy is then given by

$$\Delta S = N\left(1 + \log \frac{2K}{N}\right) - N\left(1 + \log \frac{K}{N}\right) = N \log 2 \geq 0 \tag{17.49}$$

Expansion increases the entropy since the number of available states increases.

(2) Mixing of two different kinds of particles A and B: In this case

$$\begin{aligned}\Delta S &= N_A(1 + \log\frac{2K}{N_A}) + N_B(1 + \log\frac{2K}{N_B}) - N_A(1 + \log\frac{K}{N_A}) - N_B(1 + \log\frac{K}{N})\\ &= (N_A + N_B)\log 2 > 0\end{aligned} \tag{17.50}$$

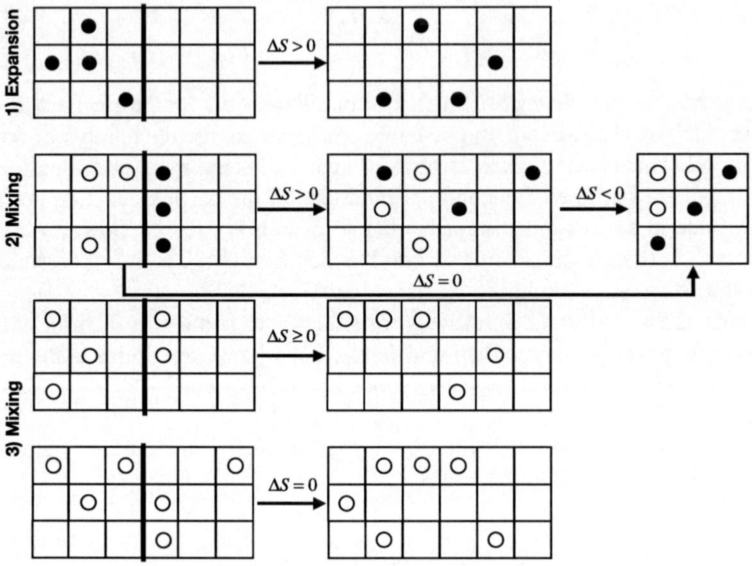

Fig. 17.3 Changes in entropy for expansion and mixing processes with indistinguishable particles and two different kinds of particles

Removing an internal constraint, the partition in this case, increases the entropy. This increase in entropy is called the entropy of mixing although it is due to the expansion. Indeed, if we compress the mixed system back to a volume K then the entropy decreases by the same amount and the total change is zero. Thus mixing two different kinds of particles at constant volume does not change the entropy.

(3) Mixing of indistinguishable particles: In this case

$$\Delta S = (N_A + N_B)(1 + \log \frac{2K}{(N_A + N_B)}) - N_A(1 + \log \frac{K}{N_A}) - N_B(1 + \log \frac{K}{N}) \tag{17.51}$$

which is the special case $K_A = K_B = K$ of (17.48). The entropy change is larger than zero when $N_A \neq N_B$ and equals zero when $N_A = N_B$. Removing the partition increases the entropy when the densities in the left and right part are different. The entropy remains the same when the densities are the same on both sides. Thus, neither removing a partition from a container with identical particles in equilibrium nor inserting a partition changes the entropy.
Note, to obtain this result we must use formula (17.47); we must both introduce the factor $1/N!$ and take the limit $N \to \infty$. If we were to use $S = \log K^N$ or $S_B = \log K^N/N!$ we would obtain the same result in the cases (1) and (2). However, using $S_B = \log K^N$ for the mixing of indistinguishable particles would result in

$$\Delta S = \log (2K)^N - \log K^{N_A} - \log K^{N_B} = N \log 2 > 0 \quad (17.52)$$

Removing the partition would increase the entropy and inserting it would decrease it, the famous Gibbs paradox. Including the factor $1/N!$ and using $S = \log K^N/N!$ decreases the entropy change to

$$\Delta S = N \log 2 - \log \frac{(N_A + N_B)!}{N_A! N_B!} > 0 \quad (17.53)$$

which is still larger than zero. The correction term is but one term in the binomial expansion of $(1+1)^N$. The entropy change approaches zero for $N_A = N_B$ only in the limit $N \to \infty$.

These results show the somewhat unexpected implications of the fact that particles are indistinguishable. In our minds we tend to label particles. If one empties a bucket of water in the ocean one believes that something is lost and that one can never retrieve the original state, though nothing is simpler than that. Just take a bucket of water from the ocean and you have obtained the original state. The argument that the molecules in the bucket are not the same molecules as in the original bucket is a fallacy which relies on our labeling of the molecules. This bucket example is just the same as removing and inserting a partition.

These examples also demonstrate that it is difficult to associate entropy with order. First there is no well-defined concept of order. Secondly, what one might perceive as an increase in disorder is not necessarily associated with an increase in entropy. In the expansion example one might argue that the state with all particles in the left box is more ordered than the state with the particles occupying the two boxes; and indeed the entropy is less for the state with all particles in the left box. The second example is already more problematic. The author, at least, finds the state with all the particles A in one box and all the particles B in another box more ordered than the state where both kinds of particles are in the same box. Yet the entropy of the two states does not differ. The important point is:

Counting must follow quantum mechanical rules where particles are indistinguishable. This leads to non-intuitive results.

Quantum mechanics requires that one adds a factor $1/N!$ to the partition function, reducing the phase space by a factor $1/N!$. A second quantum-mechanical factor $(2\pi\hbar)^{3N}$ needs to be added since the uncertainty relation requires the classical (q, p) phase space to be partitioned into cells of this size. The quantum mechanical phase space is of course a Hilbert space, not partitioned into cells. There the uncertainty relation states that the non-commutable observables q and p can only be measured simultaneously to this joint accuracy.

With these two quantum-mechanical corrections the micro-canonical partition function becomes

$$Z_0(E) := \frac{1}{N!} \int_\Gamma \frac{dpdq}{(2\pi\hbar)^{3N}} \delta(H(p, q) - E) \Delta E \quad (17.54)$$

where we also have added a factor ΔE so that the partition function becomes the phase space volume between the energy surface $H(q, p) = E$ and $H(q, p) = E + \Delta E$. This modification makes the partition function dimensionless, so that one can actually take the logarithm, and may be justified by the fact that there are no truly isolated systems with constant energy but rather there will always be energy fluctuations. In applications this ad hoc modification will not cause problems,

17.6 The Canonical Ensemble

We now assume that the system is in contact with a heat bath such that the expectation $\langle H \rangle$ is a given constant. In this case PME results in the density

$$\rho_1^{max}(q, p) = Z_1^{-1} \frac{1}{N!} \frac{1}{(2\pi\hbar)^{3N}} \exp(-\beta H(q, p)) \tag{17.55}$$

with partition function

$$Z_1(\beta) = \frac{1}{N!} \int_\Gamma \frac{dpdq}{(2\pi\hbar)^{3N}} \exp(-\beta H(q, p)) \tag{17.56}$$

where we have again included the quantum mechanical corrections. The entropy function becomes

$$S_1 = \log Z_1 + \beta \langle H \rangle \tag{17.57}$$

The Lagrange multiplier, denoted by β here to conform with statistical mechanics conventions, has to be determined from

$$\begin{aligned} \langle H \rangle &= Z_1^{-1}(\beta) \frac{1}{N!} \int_\Gamma \frac{dpdq}{(2\pi\hbar)^{3N}} H(q, p) \exp(-\beta H(q, p)) \\ &= -\frac{\partial \log Z_1(\beta)}{\partial \beta} \end{aligned} \tag{17.58}$$

This density ρ_1^{max} is called the canonical ensemble and was introduced by Gibbs. It can be viewed as the escort distribution of $e^{-H(q,p)}$. The partition function Z_1 is the Laplace transform of the partition function Z_0

$$Z_1(\beta) = \int_0^\infty dE e^{-\beta E} Z_0(E) \tag{17.59}$$

Since the energy is a constant of motion and Hamiltonian flows are non-divergent one finds that ρ_1^{max} is an invariant measure.

Thermostatics. The connection to equilibrium thermodynamics is obtained by the postulates:

Postulate 1: The temperature is given by

$$T = (k_B \beta)^{-1} \tag{17.60}$$

Postulate 2: The free energy is given by

$$F(T, V, M) = -k_B T \log Z_1(T, V, M) \tag{17.61}$$

Postulate 3: The thermostatic entropy is given by

$$S = k_B S_1 = k_B \log Z_1 + k_B \beta \langle H \rangle \tag{17.62}$$

It then follows that

$$\frac{\partial F}{\partial T} = -S \tag{17.63}$$

If we add

Postulate 4: The internal energy is given by

$$U = \langle H \rangle \tag{17.64}$$

one finds

$$U = F + TS = \langle H \rangle \tag{17.65}$$

Introduce infinitesimal changes in the mechanical and chemical work by

$$\delta W_{mech} := \langle \frac{\partial H}{\partial V} \rangle dV \tag{17.66}$$

$$\delta W_{chem} := \langle \frac{\partial H}{\partial M} \rangle dM \tag{17.67}$$

The PME relation (17.24) then becomes

$$\delta Q = dU - \delta W_{mech} - \delta W_{chem} \tag{17.68}$$

which is the first law and the PME relation (17.23) becomes

$$dS = \frac{\delta Q}{T} \tag{17.69}$$

which is the second law for reversible processes. Thus much of thermostatics follows from the canonical ensemble.

Additionally we find $S = k_B \log W$ with

$$W = \exp{(U - F)/k_B T} \tag{17.70}$$

Thus the number W of 'accessible states' is related to the amount of energy $U - F$ that cannot be converted into mechanical and chemical work.

The energy distribution is given by

$$\begin{aligned}\rho_1(E) :&= \langle\delta(E - H\rangle \\ &= \int_\Gamma dpdq \frac{1}{Z_1(T)} \exp\{-H(p,q)/k_B T\}\delta(E - H(p,q)) \\ &= \frac{1}{Z_1(T} \exp\{-E/k_B T\} \int_\Gamma dpdq\delta(E - H(p,q)) \\ &= \frac{Z_0(E)}{Z_1(T)} \exp\{-E/k_B T\} \end{aligned} \tag{17.71}$$

It has its maximum value at E_* where $\partial \log Z_0(E)/\partial E = \beta$ that is at that energy value where the micro-canonical ensemble has the same temperature as the canonical ensemble.

The micro-canonical ensemble represents an isolated system. The canonical ensemble represents a closed system in contact with a heat bath. Similarly one can introduce ensembles for systems that are in contact with a particle reservoir (the grand canonical ensemble) and ensembles for systems that are exposed to a constant pressure (the pressure ensemble). The thermodynamic potentials of these ensembles are related by Legendre transforms and the partition functions by Laplace transforms.

17.7 Macro-Variables

The micro-canonical ensemble is the PME ensemble for prescribed values of E, V, and M. The canonical ensemble is the PME ensemble for prescribed values of $\langle H\rangle$, V and M. What about other observables, once an ensemble has been specified? There are various approaches, one in the spirit of Boltzmann and one in the spirit of Gibbs.

Boltzmann's approach. Observables $g : \Gamma \to Y$ are generally many-to-one functions. A single value of y might thus be compatible with many micro-states. Assume the phase space Γ to be endowed with an ensemble P_*. The 'number' or 'volume' of micro-states that are compatible with each value y of the observable or macro-state is

$$W(y) := P_*(g^{-1}(y)) \tag{17.72}$$

Boltzmann regards as the equilibrium value y_* of the observable the value for which this 'number' or 'volume' is a maximum. The equilibrium value is the macro-state that can be realized by the largest 'number' or 'volume' of compatible micro-states. The equilibrium value y_* is the most probable value.

Apply this approach to our example of N particles distributed over K boxes. The micro-state is determined by specifying which particle occupies which box. There are K^N micro states. Each particle can be in any of the boxes. Assume a uniform distribution over all these micro-states, the micro-canonical ensemble. Take as the observable A the function that assigns to each micro-state the number of particles n_k that are in box k, $k = 1, \ldots, K$. Its probability distribution is the multinomial distribution

$$p(n_1, \ldots, n_K; N) = P(n_1, \ldots, n_K : N)\frac{1}{K^N} \tag{17.73}$$

where

$$P(n_1, \ldots, n_K : N) = \frac{N!}{n_1!\ldots n_K!} \tag{17.74}$$

is the number of ways one can distribute different N particles in K different boxes. The multinomial distribution has its maximum for $n_1 = \ldots = n_k = N/K$. The equilibrium state is thus the macro-state where N/K particles are in each box since this is the macro-state that can be realized in the largest number of ways.

Gibbs' approach. Gibbs identifies the observed value of an observable g with the expectation $\langle g \rangle$. The difference between the Boltzmann and Gibbs approach thus is: The observed value of an observable is the most probable value in the Boltzmann and the mean value in the Gibbs approach. In the above example these two approaches coincide. The multinomial distribution has the expectation $\langle n_k \rangle = N/K$. There are,of course, distributions for which mean and most probable value do not coincide. The difference becomes, however somewhat mute once the thermodynamic limit is applied.

17.8 The Thermodynamic Limit

It is amazing how much of thermostatics follows from PME, with appropriate postulates, and we have not even yet made use of the fact that statistical mechanics only 'works' for large systems. We have not yet made use of the fact that N is large, $O(10^{23})$, other than that the classical limit of quantum mechanics requires N to be large. In phenomenological thermostatics the thermodynamic limit states that the internal energy becomes an extensive variable that scales with the size of the system

$$U(\lambda S, \lambda V, \lambda N) = \lambda U(S, V, N) \tag{17.75}$$

Taking $\lambda = 1/M, 1/V, 1/N$ one gets specific quantities (quantities per unit mass), densities (quantities per unit volume) or quantities per molecule. The thermodynamic limit is a phenomenological result. It does not follow from PME. One has to explicitly demonstrate that the ensembles imply the thermodynamic limit for large N. This is the case for our example of N particles in K boxes where the entropy 17.42 turns out to be proportional to N in the classical limit. It can also be demonstrated for the ideal gas, but again requires the quantum-mechanical corrections. If the thermodynamic limit is asserted it has two important consequences: Fluctuations of observables diminish and the micro-canonical and the canonical ensemble become equivalent.

The diminishing of fluctuations. Consider first our discrete example of N particles in K boxes. The distribution of $(n_1, \ldots, n_K)$ is governed by the multinomial distribution which has its maximum at $n_1 = \cdots = n_K = N/K$ which is therefore sometimes called the 'high probability manifold'. Using Stirling's formula we find that the value at the maximum approaches 1 for large N. In this limit nearly all micro-states fall into the high probability manifold. Deviations from this high probability manifold become negligible. They become so dramatically. The probability that only half of the boxes are (uniformly) occupied is 2^{-N}, a very small number indeed for $N = 10^{23}$.

For a Hamiltonian N-particle system we have

$$\langle H \rangle = -\frac{\partial \log Z_1(\beta)}{\partial \beta} \tag{17.76}$$

and

$$\langle H^2 \rangle = \frac{1}{Z_1(\beta)} \frac{\partial^2 Z_1(\beta)}{\partial \beta^2} \tag{17.77}$$

The expectation of the fluctuations $\Delta H := H - \langle H \rangle$ thus becomes

$$\langle (\Delta H)^2 \rangle = -\frac{\partial \langle H \rangle}{\partial \beta} \tag{17.78}$$

which scales as $\langle H \rangle$. Thus if $\langle H \rangle$ scales with N then $\langle (\Delta H)^2 \rangle$ scales with N and we find

$$\frac{\langle (\Delta H)^2}{< H >^2} \sim \frac{1}{N} \tag{17.79}$$

The fluctuations vanish when N approaches infinity. One has of course to check whether or not the particular Hamilton function at hand scales with N. For the N particles in K boxes we could prove this.

Thus: If the thermodynamic limit holds then most of the micro-states lead to the same macro-state and fluctuations vanish. The most probable state becomes the mean state. Statements become deterministic, similar to the case when the strong law of large numbers holds.

Equivalence of micro-canonical and canonical distribution. So far there is no obvious connection between the micro-canonical and the canonical ensemble. Here we show that they become equivalent when the thermodynamic limit holds.

Consider a macroscopic system A that can exchange energy with a larger system B, called an energy reservoir, heat bath or thermostat. The combined system A+B has energy E and is described by the micro-canonical ensemble

$$\rho_{A+B}(q_A, p_A, q_B, p_B; E) = \frac{1}{Z_0^{A+B}} \delta(E - H(q_A, p_A, q_B, p_B) \tag{17.80}$$

where $Z_0^{A+B}(E)$ is the micro-canonical partition function of the combined system. The quantum mechanical corrections are neglected since they do not matter. The probability density function of system A is then given by

$$\rho_A(q_A, p_A) = \frac{1}{Z_0^{A+B}(E)} \int \mathrm{d}q_B \, \mathrm{d}p_B \, \delta(E - H(q_A, p_A, q_B, p_B)) \tag{17.81}$$

We now assume that the systems are large and the interactions short-ranged. Then the interactions at the interface between the two systems can be neglected and one has $H(q_A, p_A, q_B, p_B) = H_A(q_A, p_A) + H_B(q_B, p_B)$. Therefore

$$\begin{aligned}
\rho_A(q_A, p_A) &= \frac{1}{Z_0^{A+B}(E)} \int dq_B \, dp_B \, \delta(E - H_A(p_A, q_A) - H_B(p_B, q_B)) \\
&= \frac{Z_0^B(E - H_A(q_A, p_A))}{Z_0^{A+B}(E)} \\
&= \exp\left\{\frac{1}{k_B}[S_B(E - H^A(p_A, q_A)) - S_{A+B}(E)]\right\}
\end{aligned} \tag{17.82}$$

If the system B is assumed to be much larger than system A we have $E = (N_A + N_B)e \approx N_B e$ and hence in the thermodynamic limit

$$S^B(E - H^A) - S^{A+B}(E) = N_B s(e - H^A/N_B) - N_B s(e) \tag{17.83}$$

Taylor expansion of the first term results in

$$S_B(E - H_A) - S_{A+B}(E) = N_B s(e) - H_A s'(e) + \ldots - N_B s(e) = -\frac{1}{T} H_A \tag{17.84}$$

Therefore we have, except for the normalization constant

$$\rho_A(q_A, p_A) \sim e^{-\beta(e) H_A(p_A, q_A)} \tag{17.85}$$

with

$$\beta(e) := \frac{1}{k_B T} \tag{17.86}$$

The system A is characterized by the canonical ensemble.

17.9 The Ideal Gas of Particles

The textbook example for all these theoretical developments is the ideal gas. The ideal gas consists of N point particles whose interaction potential can be neglected when calculating the energy, but must of course be present for the system to approach thermodynamic equilibrium. It is a good approximation for a dilute mono-atomic gas with no internal degrees of freedom at high temperatures. Its Hamilton function is given by

$$H(q, p) = \frac{1}{2m} p^T p \tag{17.87}$$

where q and p are $3N$-dimensional vectors. The micro-canonical partition function is thus given by

$$Z_0(E, V, N) = \frac{1}{N!} \frac{1}{(2\pi\hbar)^{3N}} V^N \int_{\mathbb{R}^{3N}} dp \delta\left(E - \frac{1}{2m} p^T p\right) \Delta E \tag{17.88}$$

where we also included a uniform distribution in configuration space of volume V^N. We may evaluate this partition function as follows

$$Z_0(E, V, N) = \frac{1}{N!} \frac{1}{(2\pi\hbar)^{3N}} V^N \frac{(2mE)^{3N/2}}{2E} A(3N, R = 1) \Delta E \tag{17.89}$$

where $A(3N, R = 1)$ is the surface area of the unit sphere in $3N$ dimensions. This area can be calculated by first calculating the volume of a sphere with radius R in $3N$ dimensions which is given by

$$V(3N, R) = \int_{\mathbb{R}^{3N}} d^{3N}x \Theta\left(R^2 - \sum_{n=1}^{3N} x_n^2\right) = R^{3N} \frac{\pi^{3N/2}}{\Gamma\left(\frac{3N}{2} + 1\right)} \tag{17.90}$$

The surface area of this sphere is then given by the principle of Cavalieri

$$\begin{aligned} A(3N, R) &= \frac{dV(3N, R)}{dR} = 2R \int_{\mathbb{R}^{3N}} d^{3N}x \ \delta\left(R^2 - \sum_{n=1}^{3N} x_n^2\right) \\ &= 3N \frac{R^{3N-1} \pi^{3N/2}}{\Gamma\left(\frac{3N}{2} + 1\right)} \end{aligned} \tag{17.91}$$

Note that the ratio $A(3N, R)\Delta R / V(3N, R) = 3N\Delta R/R$ is order one for $\Delta R/R = 1/(3N)$. For high dimensional spheres most of the volume is in a thin shell at the surface. Do not peel an apple of high dimensions!

Entropy function. The entropy function $S_0(E, V, N) = k_B \log Z_0(E, V, N)$ then consists of four contributions:

(1) a volume contribution

$$k_B^{-1} S_V(V, N) := \log V^N = N \log v + N \log N \tag{17.92}$$

where we introduced $v := V/N$.

(2) an energy contribution

$$\begin{aligned} k_B^{-1} S_E(E, N) : &= N \log (2mE)^{3/2} - \log (2E) + \log F(3N, 1) + \log \Delta E \\ &\approx N \left[\log \left(\frac{4me\pi}{3} \right)^{3/2} + \frac{3}{2} \right] - \log \frac{2E}{3N\Delta E} \end{aligned} \tag{17.93}$$

where we introduced $e := E/N$, assumed $N \gg 1$, used the formula $\Gamma(N) = (N-1)!$, and Stirling's approximation $\log N! \approx N \log N - N$ of the factorials for large N.

(3) a first quantum-mechanical correction

$$k_B^{-1} S_{corr1}(N) := -\log N! \approx -N \log N + N \tag{17.94}$$

where we again used Stirling's formula, and

(4) a second quantum mechanical correction

$$k_B^{-1} S_{corr2}(N) := -\log (2\pi\hbar)^{3N} = -N \log (4\pi^2\hbar^2)^{3/2} \tag{17.95}$$

In the thermodynamic limit $N \to \infty$ the entropy should become an extensive quantity and hence a homogeneous function of first order of the extensive variables N, E, and V. The energy contribution satisfies this requirement if we neglect the term $-\log \frac{2E}{3N\Delta E}$. This may be safely done since it remains constant in the thermodynamic limit whereas the other terms grow with N. The volume contribution does not satisfy the thermodynamic limit, a fact referred to as Gibbs paradox. To resolve this paradox we must add the first quantum-mechanical correction which accounts for the fact that particles are indistinguishable. The sum of all four contributions gives

$$S(E, V, N) = k_B N \left[\log v + \log \left(\frac{me}{3\pi\hbar^2} \right)^{3/2} + \frac{5}{2} \right] \tag{17.96}$$

which satisfies the thermodynamic limit. Note that it is the first quantum-mechanical correction that cancels the 'bad' term in the volume contribution. It is the fact that particle are indistinguishable that resolves Gibbs paradox, as we demonstrated in the discrete case. The second quantum-mechanical correction needs to be applied because we assumed (q, p) as the microscopic state variables.

The above result is fairly robust. One can also integrate over all states with energies $E' \leq E$. One then obtains the partition function

$$\hat{Z}_0(E, V, N) = \frac{1}{N!} \frac{1}{(2\pi\hbar)^{3N}} V^N \int_{\mathbb{R}^{3N}} dp \Theta \left(E - \frac{1}{2m} p^T p \right)$$
$$= \frac{V^N}{N!(2\pi\hbar)^{3N}} (2mE)^{3N/2} V(3N, 1) \tag{17.97}$$

where $V(3N, 1)$ is the volume of the unit sphere in $3N$ dimensions. It leads to the same expression for the entropy, even without the need to neglect a non-growing term.

The above entropy expression implies the internal energy

$$U = \frac{3}{2} N k_B T \tag{17.98}$$

and the free energy

$$F = N k_b T [\log \frac{n}{n_\rho} - 1] \tag{17.99}$$

where $n = N/V$ is the number density and $n_\rho = (2\pi mkT/h^2)^{3/2}$ is the quantum concentration. This free energy expression is also obtained from the canonical partition function

$$Z_1(T, V, N) = \frac{1}{N!} \int_\Gamma \frac{dpdq}{(2\pi\hbar)^{3N}} \exp(-\frac{p^T p}{2mk_B T})$$
$$= -V^N \left(\frac{2\pi mkT}{h^2} \right)^{3N/2} \frac{1}{N!} \tag{17.100}$$

showing explicitly the equivalence of the micro-canonical and canonical ensemble in the thermodynamic limit.

It is also reassuring that after all this heavy machinery with quantum mechanical corrections we recover our result from the introduction, namely that the pressure of an ideal gas is given by

$$p = - \left(\frac{\partial U}{\partial V} \right)_{S,N} = \frac{2}{3} \frac{U}{V} \tag{17.101}$$

A very comforting result!

17.10 Summary

There are two descriptions of matter, the microscopic and macroscopic one. Here we considered the case that the microscopic description is given by the positions and momenta of N particles whose dynamics is governed by a Hamilton function and the macroscopic description by the values of a few observables, such as energy, volume and mass. A macro-state is generally compatible with many micro-states. Statistical mechanics bridges the gap between the microscopic and macroscopic description. It makes the following assertions:

- The macro-state can be described by an ensemble, i.e., by a probability distribution over the micro-states.
- The observed values of observables are the expectation values of the observables with respect to this ensemble.

The ensemble is chosen such that it satisfies the conditions:

- The density of the ensemble must be an invariant density of the dynamics.
- The density is the one that maximizes the information entropy subject to the controls that are exerted over the system.

The thermodynamic limit is required for statistical mechanics to work. The thermodynamic limit imposes certain restriction on the Hamilton function. In the thermodynamic limit most micro-states lead to the same value of observables. This is the reason for statistical mechanics to work. Most of the structure of thermostatic such that different thermodynamic potential are related by the Legendre transformations and that different partition functions are related by Laplace transformation follows from principle of the maximum entropy.

References

Baierlein, R.: Atoms and Information Theory. W.H. Freeman and Company, San Francisco (1971)
Ruelle, D.: Statistical Mechanics (Rigorous results). W.A. Benjamin Inc, Amsterdam (1969)

Chapter 18
Non-equilibrium Statistical Mechanics

18.1 Introduction

In this chapter we consider macroscopic systems that are not in equilibrium. Examples are fluids and gases as described by fluid dynamics, deformable bodies as described by the theory of elasticity, and electromagnetic fields and charges and currents as described by the macroscopic Maxwell equations.

As for equilibrium statistical mechanics the task is again to derive the macroscopic equations of motion from the microscopic equations, such as the equations of a N-particle system or the microscopic Maxwell equation. Our prototype problem will be the derivation of the fluid dynamical equations from a N-particle Hamiltonian.

When one tries to mimic the approach of equilibrium statistical mechanics one encounters the problem that fluid dynamics makes the continuum hypothesis. It replaces a finite discrete set of particles that have mass, momentum and energy by fields: the mass, momentum and energy densities as a function of position and time. The transition from discrete particle indexes to continuous fields might be viewed as just a technical issue for the spatial dependence, though it has its intricacies. The time dependence, however, constitutes a major challenge. It requires the explicit microscopic dynamics. In equilibrium statistical mechanics we never needed the solution of Hamilton's equations. Specification of the Hamilton function was sufficient since the energy is a constant of motion. The term "thermostatics" reflects this timeless non-dynamical characteristic of equilibrium systems.

As in equilibrium thermodynamics one must distinguish between the macroscopic level which deals with macroscopic quantities such as the mass, momentum and energy density and the microscopic level which deals with particles and their mass, momentum, and energy. These two levels are characterized by different space and time scales. There is a third level with scales in between, the mesoscopic level. Its most prominent representative is the Boltzmann equation which describes the time evolution of the probability distribution of one representative particle in momentum space. Most of the explicit calculations in non-equilibrium statistical mechanics are done on this mesoscopic level but are not considered here because they rely heavily

P. Müller, *Handbook of Dynamics and Probability*,
https://doi.org/10.1007/978-3-030-88486-4_18

on the specific physics of the problem. Rather we try to mimic the approach of equilibrium statistical mechanics and make the basic premises that the macroscopic state can be described by a probability distribution on the microscopic state space, that the macro-variables are observables, and that their observed values are the expectations with respect to this probability distribution. The question then is what are the appropriate probability distributions and how do they evolve in time. What replaces the equilibrium ensembles, the micro-canonical and the canonical ensemble in equilibrium statistical mechanics? Especially, we sketch how far one gets by applying the principle of maximum entropy.

Non-equilibrium statistical mechanics faces considerable technical difficulties. There is no simple textbook example in non-equilibrium statistical mechanics as the ideal gas in equilibrium statistical mechanic. Non-equilibrium statistical mechanics also faces the same foundational problems as equilibrium statistical mechanics such as problems with the probabilistic interpretation. It also faces additional issues such as:

- Why does a set of macro-variables exist, such as the mass, momentum and energy density in fluid dynamics, that obeys a closed set of evolution equations?
- How can irreversible macroscopic dynamics be obtained from reversible microscopic dynamics by simple ensemble averaging or any other operation? Specifically, the reversibility paradox or "Umkehreinwand" of Loschmidt (1876) and the recurrence paradox or "Wiederkehreinwand" of Zermelo (1896) have to be addressed.

These technical and foundational problems are so severe that the general consensus is that we do not yet have a generally accepted theory to predict the behavior of macroscopic non-equilibrium systems from the underlying microscopic dynamics. Where we have working theories we often do not fully understand why they work.

It is very important to distinguish between the continuum limit of a discrete system and an ensemble average of a discrete system. It is easy to derive the continuum limit of discrete systems, replacing discrete particle labels by a continuous label. This continuum limit does not introduce ireversibility into the system. It turns the N-particle system into the perfect fluid equations. There might appear new conservation laws in taking the continuum limit, such as Ertel's potential vorticity conservation, but it does not introduce ireversibility. Introducing an ensemble and averaging observables over this ensemble does allow for irreversibility, initial densities and observables converge toward their stationary states in mixing systems, but these operations do not fully break the microscopic time-symmetry since the convergence is in both time directions. When introducing observables and their average one also faces the closure problem. When attempting to close the problem one generally breaks the microscopic time-symmetry by asserting evolution in the forward time direction. The most prominent example is the closure of the Boltzmann equation by assuming molecular chaos in the past but not in the future. The same remarks apply when the microscopic system is not averaged but smoothed (Salmon 1998). Since the continuum limit does not introduce irreversibility the discrete atomistic structure of matter might be required for irreversibility.

18.2 Fluid Dynamics

Fluid dynamics is a phenomenological theory of macroscopic liquids and gases. The three key ingredients of irreversible fluid dynamics are: the balance equations for mass, momentum and energy, local thermostatic relations, and phenomenological flux laws.

The balance equations. The balance equations for mass, momentum and energy are

$$\partial_t \rho + \nabla^T(\rho u) = 0 \tag{18.1}$$

$$\partial_t(\rho u^T) + \nabla^T(\rho u u^T - \sigma) = -\rho \nabla^T \Phi_{ext} \tag{18.2}$$

$$\partial_t(\rho e) + \nabla^T(\rho e u + q) = tr(\sigma^T(\nabla u^T)) \tag{18.3}$$

where ρ is the density, u the fluid velocity, e the specific internal energy, σ the (symmetric) stress tensor, q the heat flux vector and Φ_{ext} an external potential. A phenomenological derivation of these balance equations is given in textbooks of fluid dynamics, e.g., Müller (2006).

The macroscopic balance equations are formulated in their flux form, the dependent variables are densities, the mass density ρ, the momentum density ρu and the internal energy density ρe. The balance equations state that the time rate of change of the density plus the divergence of a flux term equals a source term. The mass balance does not have a source term. The source term in the momentum equation describes the effect of external forces. The internal forces are represented by the stress tensor. It is a major tenet of fluid dynamics that internal forces can indeed be represented by the divergence of a stress tensor. The source term in the internal energy equation describes the conversion of kinetic to internal energy. The flux form of the macroscopic balance equations is especially convenient when considering integrated quantities such as the total mass, total momentum or total internal energy. The flux term then describes the flux across the boundary.

Local thermostatic relations. The balance equations have to be augmented by a thermostatic specification of the fluid. One assumes local thermodynamic equilibrium such that there exists a specific entropy function (entropy per unit mass)

$$\eta = \eta(e, v) \tag{18.4}$$

as a function of the specific internal energy e and specific volume v and that the specific entropy function determines the temperature T and pressure p via the thermodynamic relations

$$\frac{1}{T} = \left(\frac{\partial \eta}{\partial e}\right)_v \quad \frac{p}{T} = \left(\frac{\partial \eta}{\partial v}\right)_e \tag{18.5}$$

Note that the thermodynamic specification is not affected by a (uniform) motion of the fluid parcel and by its exposure to an external potential.

Phenomenological flux laws. The balance equations also contain the unknown stress tensor σ and heat flux q. They consist of reversible and irreversible (or molecular) contributions, $\sigma = \sigma^{rev} + \sigma^{mol}$ and $q = q^{rev} + q^{mol}$. There is no reversible heat flux, $q^{rev} = 0$. A basic linear theory assumes these irreversible fluxes to be proportional to the gradients of the velocity and temperature field. Assuming an isotropic fluid one arrives at the specific flux laws or constitutive relations for the stress tensor and heat flux

$$\begin{aligned} \sigma^{rev} &= -pI \\ \sigma^{mol} &= 2\rho\nu S + 3\rho\nu' N \\ q^{mol} &= -\rho c_p \lambda \nabla T \end{aligned} \tag{18.6}$$

where p is the thermostatic pressure, ν the kinematic shear viscosity, ν' the kinematic expansion viscosity, c_p the specific heat at constant pressure, λ the thermal conduction coefficient and where

$$\begin{aligned} S &:= D - \frac{1}{3} tr(D) I \\ N &:= \frac{1}{3} tr(D) I \end{aligned} \tag{18.7}$$

are the shear and normal components of the (symmetric) rate of deformation tensor

$$D := \frac{1}{2}[\nabla u^T + (\nabla u^T)^T] \tag{18.8}$$

The antisymmetric part, the vorticity tensor, does not induce stresses since it represents a uniform rotation of an infinitesimal fluid element.

The flux laws (18.6) can be viewed as the first term in an expansion about the state of thermal equilibrium. The phenomenological coefficients ν, ν' and λ are the new coefficients that need to be determined. They must also be assumed to depend on η and v, though in practice one writes them as functions of the pressure p and the temperature T.

The balance equations together with a given specific entropy function $\eta(e, v)$ and given phenomenological coefficients ν, ν' and λ completely describe the evolution of the fluid. The thermostatic pressure p, the temperature T and the specific heat c_p can be derived from the specific entropy function $\eta(e, v)$. The fluid dynamical equations represent a set of partial differential equations that we view here as a dynamical system, setting aside the formidable task of identifying the proper function spaces and related issues.

Irreversibility. The fluid dynamical equations are irreversible. The molecular fluxes are responsible for this irreversibility. Indeed, differentiating the relation $\eta(e, v)$ one finds

$$\partial_t(\rho\eta) + \nabla^T(\rho\eta u + \nabla^T(\frac{q^{mol}}{T})) = Q_\eta^{mol} \tag{18.9}$$

where

$$\begin{aligned} Q_\eta^{mol} &= q_{mol}^T \nabla(\frac{1}{T}) + \frac{1}{T} tr(D^T \sigma^{mol}) \\ &= \frac{1}{T}(2\rho\nu tr(S^T S) + 3\rho\nu' tr(N^T N)) + \frac{\rho c_p \lambda}{T^2}(\nabla T)^T(\nabla T) \end{aligned} \tag{18.10}$$

is the entropy production rate. The second law requires $Q_\eta^{mol} \geq 0$. Hence the phenomenological coefficients must satisfy $\nu, \nu', \lambda \geq 0$.

A state of thermodynamic equilibrium is achieved when $T=$ const., $D = 0$ and $\nabla_x p = -\rho\nabla_x\Phi_{ext}$ (hydrostatic equilibrium) and the entropy production is halted. More on fluid dynamics can be found in appendix C.

The key result is:

The fluid dynamical equations describe an irreversible evolution.

When attempting to derive the macroscopic fluid dynamical equations from the underlying microscopic physics the balance equations constitute the least problem. They just represent macroscopic formulations of the conservation of mass, momentum and energy which also hold on the microscopic level. There are some challenges: why is it that the internal forces can be represented by the gradient of a stress tensor. Local thermal equilibrium and the molecular or irreversible fluxes are the major problems. The basic problem is how to square the macroscopic irreversible behavior with the microscopic time-reversal invariance.

We first demonstrate that the continuum limit of a discrete N-particle system, replacing the discrete particle label by a continuous one, does not lead to irreversible behavior. It leads to the reversible perfect fluid equations.

18.3 The Continuum Limit of a N-particle System

The continuum limit is actually fairly trivial. It consists of replacing the discrete particle label by a continuous label. To prove that this replacement in the Hamilton function of a N-particle system leads to the perfect fluid equations we first demonstrate that the perfect fluid equations can be obtained from a Lagrangian density via Hamilton's principle and then that this Lagrangian density is the continuum limit of the N-particle Lagrangian.

The perfect fluid equations. The perfect fluid equations are obtained by ignoring the irreversible processes of friction and heat conduction. Setting the phenomenological coefficients ν, ν' and λ to zero the macroscopic balance equation become

$$\frac{D}{Dt}v = v\nabla_x^T u \tag{18.11}$$

$$\frac{D}{Dt}u = -v\nabla_x p - \nabla_x \Phi_{ext} \tag{18.12}$$

$$\frac{D}{Dt}e = -vp\nabla^T u \tag{18.13}$$

where $D/Dt = \partial_t + u^T\nabla_x$. They are written here in their advective form. These equations describe reversible fluid processes. There is no production of entropy. For this reason one usually replaces the specific internal energy e by the specific entropy η as a dependent variable and replaces Eq. 18.13 by

$$\frac{D}{Dt}\eta = 0 \tag{18.14}$$

These dynamical equations have then to be augmented by an equation of state $p = p(v, \eta)$ which characterizes the fluid.

We next convert to a Lagrangian description of the fluid, as opposed to the above Eulerian description. In the Lagrangian description, the fluid flow is described by the position $x(s, t)$ of a fluid particle as a function of its label s and time t. Usually the label s is taken to be the initial position. The Lagrangian description follows the paths of (marked) fluid particles. It is a field description but a field description in label space. The independent variables are s and t. In the Lagrangian description the Eqs. 18.11, 18.12, and 18.14 become

$$v(s,t) = v_0(s)J(s,t) \tag{18.15}$$

$$\ddot{x} = -\frac{1}{\rho}\nabla_x p - \nabla_x \Phi_{ext} \tag{18.16}$$

and

$$\eta(s,t) = \eta_0(s) \tag{18.17}$$

where $J(s,t) = \partial(x(s,t))/\partial(s)$ is the Jacobian. These equations have again to be augmented by an equation of state $p = p(v, \eta)$.

Hamilton's principle for the Lagrangian description. The Lagrangian equation of motion (18.16) can be obtained from Hamilton's principle applied to continuous systems. One has to specify the Lagrangian density $\mathcal{L}$ which is the kinetic minus potential energy density expressed in terms of the position x, the velocity $\dot{x}$, the displacement gradient tensor $F(s,t) = \nabla_s x^T(s,t)$ and the independent variables s and t. Hamilton's principle then states that

$$\delta \int dt \int\int\int \mathcal{L}(x, \dot{x}, F; s, t)d^3s = 0 \tag{18.18}$$

for variations $\delta x(s,t)$ that vanish on the boundary. The Euler-Lagrange equations are

$$\frac{\partial}{\partial t}\frac{\partial \mathcal{L}}{\partial \dot{x}_i} + \frac{\partial}{\partial s_j}\frac{\partial \mathcal{L}}{\partial F_{ji}} - \frac{\partial \mathcal{L}}{\partial x_i} = 0 \tag{18.19}$$

and are derived in the same way as the ones for a discrete particle system.

For a perfect fluid the appropriate Lagrangian density is

$$\mathcal{L} = \rho_0(s)[\frac{1}{2}\dot{x}^T\dot{x} - e(v,\eta) - \Phi_{ext}(x,t)] \tag{18.20}$$

where $e(v,\eta) = e(v_0(s)J(s,t), \eta_0(s))$ is the specific internal energy. The acceleration term and the external force term in (18.16) are straightforward to derive. To derive the pressure term one needs to notice that the only dependence on the displacement gradient tensor F is through $v(s,t) = v_0(s)J(s,t)$. Since $p = -\partial e(v,\eta)/\partial v$ one finds

$$\begin{aligned}\frac{\partial}{\partial s_j}\frac{\partial \mathcal{L}}{\partial F_{ji}} &= \frac{\partial}{\partial s_j}[\rho_0(s)pv_0(s)\frac{\partial J}{\partial F_{ji}}] \\ &= J(F^{-1})_{ij}\frac{\partial p}{\partial s_j} = J\frac{\partial p}{\partial x_i}\end{aligned} \tag{18.21}$$

where we have used (see appendix E.20)

$$\partial J/\partial F_{ji} = J(F^{-1})_{ij} \tag{18.22}$$

$$\partial^2 J/\partial F_{ji}\partial s_j = 0 \tag{18.23}$$

and where sums over repeated indexes are implied.

This continuous form of Hamilton's principle can easily be rationalized as the continuum limit of a discrete particle system. To obtain the Lagrangian density 18.20 one needs to replace in the Lagrangian of the N-particle system $\sum_n m_n$ by $\int\int\int \rho_0(s)d^3s$, $q^{(n)}$ by x, $p^{(n)}/m_n$ by $\dot{x}$ and associate the pair interaction potential energy with the internal energy. Further details on Hamilton's principle for fluids can be found in appendix C and in Müller (1995).

If the above reasoning via Hamilton's principle strikes the reader as too implicit consider the FPU system of Sect. 3.7 for a more explicit calculation.

The FPU system. The FPU system is a one-dimensional chain of $N+1$ particles that are coupled to each other by linear and quadratic next-neighbor restoring forces. For definiteness, assume that the two boundary particles $n=0$ and $n=N$ are fixed so that only $N-1$ particles can move. The Newtonian equations of motion are

$$m_n\ddot{p}_n = D[(q_{n+1}-q_n) - (q_n - q_{n-1}] + C[(q_{n+1}-q_n)^2 - (q_n - q_{n-1})^2] \tag{18.24}$$

where q_n is the displacement of the nth particle from its equilibrium position and m_n its mass and D is the linear and C the quadratic coupling coefficient. The notation is a little different because now we are also interested in knowing what happens to the coupling coefficients.

To perform the continuum limit assume that the particles are at positions $s_n^* = n\Delta s$ when the system is in the unperturbed equilibrium state where Δs is the equilibrium spacing between particles. Using this equilibrium spacing one may rewrite 18.24 as

$$\frac{m_n\ddot{q}_n}{\Delta s} = D\Delta s[\frac{(q_{n+1} - q_n)}{(\Delta s)^2} - \frac{(q_n - q_{n-1}}{(\Delta s)^2}] + C(\Delta s)^2[\frac{(q_{n+1} - q_n)^2}{(\Delta s)^3} - \frac{(q_n - q_{n-1})^2}{(\Delta s)^3}] \tag{18.25}$$

A continuous system can then be obtained in the limit $\Delta s \to 0$, $N \to \infty$, $n \to \infty$, $m_n \to 0$, $D \to \infty$ and $C \to \infty$ such that $n\Delta s \to s$, $N\Delta s \to L$, $m_n/\Delta s \to \rho(s)$, $D\Delta s \to \tilde{D}$, $C(\Delta)^2 \to \tilde{C}$ and $q_n(t) \to q(s,t)$, and results in

$$\rho(s)\partial_{tt}q(s,t) - \partial_s[\tilde{D}\partial_s q(s,t) + \tilde{C}(\partial_s q(s,t))^2] = 0 \tag{18.26}$$

where the term in square brackets is the stress which is a nonlinear function of the displacement gradient $\partial_s q(s,t)$. We included a quadratic restoring force to demonstrate that nonlinearities do not constitute a problem for this continuum limit.

The limit ($\Delta s \to 0$, $N \to \infty$) is the continuum limit. It is the discrete particle index n that becomes a continuous index s. The result of this continuum limit is a fairly well-defined PDE for a time-dependent displacement field $q(s,t)$ which still represents a time-reversal invariant system. By considering a continuum of particles we have not introduced any irreversible behavior. This is the important point:

The continuum limit does not break the time-reversal symmetry.

18.4 Average of a N-particle System

Averaging (or smoothing) the microscopic equations is a more promising procedure. It allows for irreversibility but faces two problems:

- What is the appropriate probability measure that one should use for the averaging.
- When considering observables one faces a closure problem. What is the appropriate closure hypothesis?

These problems are very challenging. Next we sketch how they have been approached for a N-particle Hamiltonian system. The state space Γ of such a system is a $6N$-dimensional space with elements (q,p) where $q = (q^{(1)}, \ldots, q^{)N)})^T$ and $p = (p^{(1)}, \ldots, p^{(N)})^T$ are $3N$-dimensional vectors and $q^{(n)}$ and $p^{(n)}$ are elements of the 3-dimensional position space and the 3-dimensional momentum space of the nth particle.

Microscopic densities. First we introduce microscopic mass, momentum and energy densities in physical space by

$$m_{y,t}(q,p) = \sum_{n=1}^{N} \delta(y - q_t^{(n)}(q,p))m \tag{18.27}$$

$$p_{y,t}(q,p) = \sum_{n=1}^{N} \delta(y - q_t^{(n)}(q,p))p_t^{(n)}(q,p) \tag{18.28}$$

$$h_{y,t}(q,p) = \sum_{n=1}^{N} \delta(y - q_t^{(n)}(q,p))h_t^{(n)}(q,p) \tag{18.29}$$

where $p_t^{(n)}$ is the momentum and $h_t^{(n)}$ the energy of the nth particle and $y \in V$ where V is the three-dimensional volume to which the fluid is confined. These densities are, of course, very spiky. They are all of the form

$$f_{y,t}(q,p) := \sum_{n=1}^{N} \delta(y - q_t^{(n)}(q,p))f_t^{(n)}(q,p) \tag{18.30}$$

and represent a family of observables indexed by the position $y \in V$ from a dynamical systems point of view. The values of the densities depend on the trajectories $q_t^{(n)}(q,p)$, $p_t^{(n)}(q,p)$ which in turn depend on the initial conditions (q,p), as indicated.

Microscopic balance equations. Taking the time derivative of the microscopic densities gives

$$\partial_t m_{y,t}(q,p) = -\frac{\partial}{\partial y}[\sum_{n=1}^{N} \delta(y - q_t^{(n)}(q,p))m\dot{q}_t^{(n)}(q,p) \tag{18.31}$$

$$\begin{aligned}\partial_t p_{y,t}(q,p) = &-\frac{\partial}{\partial y}[\sum_{n=1}^{N} \delta(y - q_t^{(n)}(q,p))p_t^{(n)}(q,p)\dot{q}_t^{(n)}(q,p)] \\ &+\sum_{n=1}^{N} \delta(y - q_t^{(n)}(q,p))\dot{p}_t^{(n)}(q,p)\end{aligned} \tag{18.32}$$

$$\begin{aligned}\partial_t h_{y,t}(q,p) = &-\frac{\partial}{\partial y}[\sum_{n=1}^{N} \delta(y - q_t^{(n)}(q,p))h_t^{(n)}(q,p)\dot{q}_t^{(n)}(q,p)] \\ &+\sum_{n=1}^{N} \delta(y - q_t^{(n)}(q,p))\dot{h}_t^{(n)}(q,p)\end{aligned} \tag{18.33}$$

where one now has to substitute $\dot{q}_t^{(n)}(q,p)$, $\dot{p}_t^{(n)}(q,p)$ and $\dot{h}_t^{(n)}(q,p)$ from Hamiltonian equations. This starts to get tedious for the momentum density and even more so for the energy density. If one grinds through though, as is done in appendix D.4, one finds that the microscopic balance equations all take the form

$$\partial_t f(y,t) + \nabla_y^T j(y,t) = Q(y,t) \tag{18.34}$$

where the flux vector $j(y,t)$ is due to the internal interactions and the source term $Q(y,t)$ due to the external interactions. In writing the microscopic balance equations in this form we suppressed the dependence on (q,p), but it is there. From now on we will just work with this generic form. The remarkable result in deriving these microscopic balance equations is that the effect of the internal interactions can be represented by the divergence of a flux vector. This result imposes some restrictions on the Hamilton function. It is for example sufficient that the interaction potential is isotropic, as is assumed in the derivation of appendix D.4.

Macroscopic balance equations. To proceed we average the microscopic balance equation over the initial condition. The ensemble average simply leads to

$$\partial_t \langle f(y,t)\rangle + \nabla_y^T \langle j(y,t)\rangle = \langle Q(y,t)\rangle \tag{18.35}$$

since time and space differentiation commute with the ensemble average. One presumes that the averaged functions are smooth. The averaged equations have the same form as the balance equations of fluid dynamics and one can easily identify the mass and energy densities, the fluid velocity, the stress tensor and the heat flux in the fluid dynamical equations with corresponding terms in the averaged balance equation. Details are found in Appendix D.4. It is thus fairly straightforward to 'derive' the macroscopic balance equations from the microscopic dynamics. This should not be too surprising since the macroscopic balance equations are just the macroscopic manifestation of mass, momentum and energy conservation which holds on the microscopic level. Though not surprising it is an important point:

The balance equations hold for both the microscopic and macroscopic level.

Alternatively, one might smooth the microscopic balance in physical space with a smoothing function (Salmon 1998). Typically, an isotropic, infinitely differentiable smoothing function such as a Gaussian is chosen. Smoothing also commutes with time and space derivatives and hence the smoothed equations have the same structure as the averaged equations, with the cornered brackets replaced by an over-bar, indicating the smoothing operation. Smoothing replaces the delta functions by the smoothing function. This has some technical advantages. Having a smoothing function with a certain width one can invoke the condition that the range of the intermolecular forces is much smaller than the smoothing scale. This leads to some technical simplifications.

The tasks that remain are: First, one has to specify the ensemble over which to average. Second, one has to close the equations. Third, one has to derive the local thermodynamic equilibrium relations, specifically one has to define a thermodynamic entropy. And fourth, one has to derive the phenomenological flux laws. Next we demonstrate to what extent the principle of maximum entropy (PME) can help to handle these tasks.

18.5 Applying PME

What follows is just a general sketch, following partly Guttman (1999). To keep things simple we consider the case where we just have one scalar balance equation, say for the energy. In this case $f(y, t)$ would be the microscopic energy density and $j(y, t)$ the microscopic heat flux and we attempt to derive the equation for heat conduction.

The PME density. Recall that $f(y, t)$ is just an abbreviation for $f_{y,t}(q, p)$ and depends on the initial conditions $(q, p) \in \Gamma$, and so does $j(y, t)$. To derive a probability distribution for these initial conditions via PME we require that the ensemble is consistent with the macroscopic evolution of the system in the past up to the considered time t. More specifically, we assume that this macroscopic evolution represents deviations $\langle \dot{f}(y, t')\rangle$ from thermal equilibrium for $t_0 < t' < t$ and $y \in V$. The initial conditions then refer to some time t_0 way back in time, at $t_0 = -\infty$ if you so wish. The dot denotes the partial time derivative. The reason for representing the deviations by the time derivative will become clear below. Taking these deviations as constraints PME gives the initial density

$$\rho_{t_0}^{max}[q, p; \beta_0, \delta\beta, t] = \frac{1}{Z[\beta_0, \delta\beta, t]} \exp\{-\beta_0 H(q, p) - \int_V d^3y' \int_{t_0}^{t} dt' \delta\beta_{y',t'} \dot{f}_{y',t'}(q, p)\} \tag{18.36}$$

with partition function

$$Z[\beta_0, \delta\beta, t] = \int_\Gamma dqdp \exp\{-\beta_0 H(q, p) - \int_V d^3y' \int_{t_0}^{t} dt' \delta\beta_{y',t'} \dot{f}_{y',t'}(q, p)\} \tag{18.37}$$

The Lagrange multiplier β_0 denotes the equilibrium inverse temperature and $\delta\beta_{y,t}$ the perturbation due to the constraints. By choosing the time derivative $\dot{f}(y, t)$ the Lagrange multipliers $\delta\beta_{y,t}$ have the same dimension as β_0. These multipliers have to be determined from

$$\langle H\rangle = -\frac{\partial \log Z[\beta_0, \delta\beta, t]}{\partial\beta_0} \tag{18.38}$$

$$\langle \dot{f}_{y,t'}\rangle = -\frac{\delta \log Z[\beta_0, \delta\beta, t]}{\delta\beta_{y,t'}} \tag{18.39}$$

where the δ denotes the functional derivative. Since we have a continuous set of constraints PME gives a space-time integral of the product of Lagrange multiplier and observable in the exponent. The above formulas are a straightforward extension from a discrete set of constraints to a continuous set. Sums become integrals. Partial derivatives become functional derivatives.

Since the flow in Γ-space is non-divergent and invertible for a Hamiltonian system the PME density does not change along trajectories (see Sect. 11.3) and hence

$$\rho_{t'}^{max}[q, p; \beta_0, \delta\beta, t] = \rho_{t_0}^{max}(q_{-(t'-t_0)}(q, p), p_{-(t'-t_0)}(q, p); \beta_0, \delta\beta, t) \quad (18.40)$$

Note that this density depends on two times, the time t up to which the constraints are taken into account and the time t' at which the density is evaluated.

Information entropy. The information entropy of the PME density is given by

$$S_{t'}(\beta_0, \delta\beta, t) = -\int_\Gamma dqdp\rho_{t'}^{max}[q, p, \beta_0, \delta\beta, t]\log\rho_{t'}^{max}[q, p; \beta_0, \delta\beta, t] \quad (18.41)$$

This information entropy depends on the time t up to which the constraints are taken into account. In Sect. 10.8 it was proven that increasing the number of constraints decreases the entropy since more constraints have to be satisfied. We find the same result in the continuous case, namely

$$\frac{d}{dt}S_{t'}[\beta_0, \delta\beta, t] \leq 0 \quad (18.42)$$

which can again be proven by Gibbs inequality.

In Sect. 11.7 it was proven that the information entropy does not change for an non-divergent invertible flow and hence

$$\frac{d}{dt'}S_{t'}[\beta_0, \delta\beta, t] = 0 \quad (18.43)$$

Introducing random initial condition does not induce any irreversibility.

Observables. Given the PME density the expectation value of an observable $g_{y,t'}(q, p)$ becomes

$$\langle g_{y,t'}\rangle = \int_\Gamma dqdp\rho_{t_0}^{max}[q, p; \beta_0, \delta\beta, t]g_{y,t'}(q, p) \quad (18.44)$$

Note that there is no restriction on the time t' here. It can be a future time $t' > t$ or be a past time $t' < t$ though we only consider the case $t = t'$ where the constraints are satisfied up to the time that we consider. In particular we find for $g = f$

$$\langle f_{y,t}\rangle = \int_\Gamma dqdp\rho_{t_0}^{max}[q, p; \beta_0, \delta\beta, t]f_{y,t}(q, p) \quad (18.45)$$

and for $g = j$

$$\langle j_{y,t}\rangle = \int_\Gamma dqdp\rho_{t_0}^{max}[q, p; \beta_0, \delta\beta, t]j_{y,t}(q, p) \quad (18.46)$$

which should be the basis for deriving the phenomenological flux laws. Though formally straightforward the above formulas are very hard to evaluate since, among other things, they still contain the trajectories $q_t(q, p)$, $p_t(q, p)$ of the N-particle system. Also, when taking the time derivative of $\langle f_{y,t}\rangle$ there are two contributions, one from the updating, through $\rho_{t_0}^{max}$, and one from the dynamical evolution, through $f_{y,t}$.

Linearization. To make progress one introduces the thermodynamic equilibrium density and partition function

$$\rho^{equ}(q, p, \beta_0) = \frac{1}{Z^{equ}(\beta_0)} e^{-\beta_0 H(q,p)} \tag{18.47}$$

$$Z^{equ}(\beta_0) = \int_\Gamma dqdpe^{-\beta_0 H(q,p)} \tag{18.48}$$

and assumes that the perturbation, the second term in the exponential of the PME density 18.36, is small compared to the first term. To first order in this perturbation the PME density then becomes

$$\rho_{t_0}^{max}[q, p; \beta_0, \delta\beta, t] = \frac{\rho^{equ}(q, p, \beta_0)}{Z^{equ}[\beta_0]}[1 - \int_V d^3y' \int_{t_0}^t dt'\delta\beta_{y',t'}(\dot{f}_{y',t'}(q, p) - \langle \dot{f}_{y',t'}\rangle_{equ})] \tag{18.49}$$

where the subscript "equ" on the cornered bracket indicates that the ensemble average is to be taken with respect to the equilibrium density.

Using this linearized density the deviation of the expectation of any observable from its equilibrium value becomes

$$\langle g_{y,t}\rangle - \langle g_{y,t}\rangle_{equ} = -\int_V d^3y' \int_{t_0}^t dt'\delta\beta_{y',t'}C_{\dot{f},g}^{equ}(y', y, t', t) \tag{18.50}$$

where

$$C_{\dot{f},g}^{equ}(y', y, t', t) = \langle \dot{f}_{y',t'} g_{y,t}\rangle_{equ} - \langle \dot{f}_{y',t'}\rangle_{equ}\langle g_{y,t}\rangle_{equ} \tag{18.51}$$

is the equilibrium covariance function. The important gain by this approximation is that one has to evaluate the covariance function at thermal equilibrium. Still, the calculation of the covariance function requires the microscopic trajectories.

Taking again $g = f$ one finds

$$\langle f(y, t)\rangle - \langle f(y)\rangle_{equ} = -\int_V d^3y' \int_{t_0}^t dt'\delta\beta(y', t')C_{\dot{f},f}^{equ}(y', y, t', t) \tag{18.52}$$

The averaged perturbation energy density is proportional to the perturbation inverse temperature over the whole space-time domain. The factor of proportionality is the equilibrium covariance function between $\dot{f}$ and f.

And taking $g = j$ results in

$$
\begin{aligned}
\langle j_f(y,t)\rangle &= -\int_V d^3y' \int_{t_0}^{t} dt'\delta\beta(y',t')C^{equ}_{\dot{f},j}(y',y,t',t) \\
&= -\int_V d^3y' \int_{t_0}^{t} dt'\delta\beta(y',t')\langle j(y,t)\partial_{t'} f(y',t')\rangle_{equ} \\
&= \int_V d^3y' \int_{t_0}^{t} dt'\delta\beta(y',t')\langle j(y,t)\nabla^T_{y'} j(y',t')\rangle_{equ} \\
&= -\int_V d^3y' \int_{t_0}^{t} dt'\langle j(y,t) j^T(y',t')\rangle_{equ}\nabla_{y'}\delta\beta(y',t') \qquad (18.53)
\end{aligned}
$$

where we used the microscopic balance equation, integrated by parts and disregarded the boundary terms arising from the integration by parts and assumed the equilibrium heat flux to be zero, $\langle j_f(y,t)\rangle_{equ} = 0$. The heat flux is proportional to the gradient of the inverse temperature over the whole space-time domain. The factor of proportionality is the equilibrium auto-covariance function of the heat flux.

Substituting the expressions for the averaged energy density and the averaged heat flux into the averaged balance equation (18.35) results in a complicated integro-differential equation for the inverse temperature field, containing equilibrium covariance functions. Though complicated it is a closed equation for the temperature field, given the equilibrium covariance functions.

Fourier's law. As a particular simple application we 'derive' Fourier's law of heat conduction. For this we assume $f(y,t)$ to be the energy density $h(y,t)$. Fourier's law may then be obtained by making the drastic assumption that the energy density and flux are uncorrelated in the equilibrium state

$$C^{equ}_{\dot{f},f}(y',y,t',t) = S_{\dot{f}f}\delta(y-y')\delta(t-t') \qquad (18.54)$$

$$C^{equ}_{j,j}(y',y,t',t) = S_{jj}\delta(y-y')\delta(t-t') \qquad (18.55)$$

where $S_{\dot{f}f}$ and S_{jj} are the (cross)-spectral levels. The complicated integral equation then reduces to

$$\partial_t\beta(y,t) = \frac{S_{jj}}{S_{\dot{f}f}}\nabla^T_y\nabla_y\beta(y,t) \qquad (18.56)$$

from which Fourier's temperature equation $\partial_t T = D\nabla^T_y\nabla_y T$ can be obtained. Of course, one needs to demonstrate that some short memory and short distance approximations to the dynamics results in covariance functions well represented by such delta functions and that the resulting diffusion coefficient is positive. Nevertheless, the remarkable thing is that one obtains a closed irreversible equation for the temperature field.

Where does the closure and irreversibility come from in this case? First, averaging the microscopic flux law over an initial ensemble neither closes the equation nor does it break the time-symmetry. The closure is accomplished by the linearization. The breaking of the time-symmetry is due to the set up of the problem. The evolution of f

at the present time t is constrained by the values $\langle f(y, t')\rangle$ in the past, $t' < t$, not by its future values. This is consistent with the more standard derivations of irreversible equations in statistical mechanics where one usually finds that time-asymmetry is not derived but assumed. One breaks the time symmetry by assuming that certain approximations are valid in the forward time direction but not in the backward time direction. The most famous example of assuming such a time-asymmetry is the assumption of molecular chaos when solving the Boltzmann equation which assumes that molecules are uncorrelated before the collision but not afterwards. The important point is:

Time-asymmetry is introduced by setting up the problem in a forward time direction.

Local thermostatic relations. To complete the derivation of the fluid dynamical equations one would need to derive the local thermostatic relations. One would first need to identify the thermodynamic variables. To equate the inverse temperature $1/T$ with the Lagrange multiplier β and the internal energy density $e(y, t)$ with $\langle h(y, t)\rangle$ is perhaps straightforward. Even the free energy density might be introduced as the logarithm of the partition function. The real challenge is to demonstrate that these quantities satisfy the local thermostatic relations. This presumably requires local thermodynamic equilibrium which in turn requires: (1) The fluid can be divided into parcels that are sufficiently small from a macroscopic point of view to be labeled as points but are still sufficiently large from a microscopic point of view to contain a large enough number of molecules for equilibrium thermodynamics to apply. (2) A fluid parcel, when disturbed, reaches its thermodynamic equilibrium so rapidly that the adjustment appears to be instantaneous from a macroscopic point of view. One would like to see under which conditions the PME ensemble exhibits such properties but it seems to be unclear whether or not a general result such as 18.53 for the heat flux can be obtained.

18.6 Summary

The task of non-equilibrium statistical mechanics is to derive the equations governing macroscopic systems from the equations governing the microscopic dynamics. The prototype problem considered here is the derivation of the macroscopic fluid dynamical equations from an underlying microscopic N-particle Hamiltonian system. The continuum limit, replacing discrete particle labels by continuous ones, is fairly straightforward but results in the perfect fluid equations and hence does not introduce irreversibility. An averaging or smoothing procedure is more promising. It is tedious but somewhat straightforward to derive the macroscopic balance equations as ensemble averages of the microscopic balance equations. Such ensemble averaging does not break any time-symmetry of the underlying microscopic dynamics. It does, however, introduce the closure problem and closure might introduce time-asymmetry. We considered the ensemble that results from the principle of maximum entropy when one takes the past evolution as a constraint. A closure can be achieved by linearization about the local thermodynamic equilibrium. This closure results in a generalized Fourier law where the heat flux depends on the gradients of the temperature field over the whole past time-space domain. When one closes the averaged system in this way the time symmetry is broken since the problem is set up that way. The present evolution is constrained by the past not by the future. There are numerous technical and foundational problems that remain. Most importantly there is not yet a general derivation of the local thermostatic relations. Non-equilibrium statistical mechanics remains a challenging task.

References

Loschmidt, J.: Wien. Ber. **73**, 128 (1876)
Zermelo, E.: Ann. Phys. (Leipzig) **57**, 485 (1896)
Salmon, R.: Lectures on Geophysical Fluid Dynamics, vol. 8. Oxford University Press (1998)
Müller, P.: Ertel's potential vorticity in physical oceanography. Rev. Geophys. **33**(1), 67–97 (1995)
Müller, P.: The Equations of Oceanic Motions. Cambridge University Press (2006)

Chapter 19
Foundational Issues of Statistical Mechanics

19.1 Introduction

Statistical mechanics relates macroscopic properties to the underlying microscopic dynamics. Though statistical mechanics is highly successful on a technical level, properties of macroscopic systems are routinely determined from Hamilton functions or Hamiltonians, it is still an open question why statistical mechanics works? The basic problem is:

Macroscopic systems are dynamical systems that exhibit irreversible behavior. Left to themselves they approach irreversibly a stationary or equilibrium state. Viewed as a dynamical system this equilibrium state is an attractor with the whole state space as the basin of attraction. The problem arises because macroscopic matter actually consists of many microscopic constituents such as atoms or molecules. "Many" typically means something like 10^{23}. These microscopic constituents follow their own microscopic dynamics which exhibits quite different behavior. It is invertible and time-reversal-invariant and has no attractors. The basic question that arises from this macroscopic-microscopic distinction is: How can macroscopic behavior be squared with microscopic dynamics? There are three major questions:

1. It is a phenomenological and remarkable fact that macroscopic systems represent dynamical systems. There exist sets of macroscopic variables which evolve according to a dynamical map. Fluid mechanics is our prime example. In statistical mechanics these macroscopic variables are interpreted as observables or expectations of observables. These observables are many-to-one functions. The evolution of observables generally does not constitute a dynamical system. What is required about the microscopic dynamics or the observables such that these observables evolve as a dynamical system?
2. All macroscopic processes are irreversible. Reversibility is just an idealization. The second law implies that the thermodynamic (Clausius) entropy never decreases in an isolated system? Isolated systems return to the state of thermal

P. Müller, *Handbook of Dynamics and Probability*,
https://doi.org/10.1007/978-3-030-88486-4_19

equilibrium when perturbed from it. Macroscopic systems exhibit the thermodynamic arrow of time. Microscopic systems, on the other hand, are invertible systems, cannot have attractors because of their Hamiltonian nature, are time-reversal invariant (or more generally CPT-invariant), and are subjected to Poincaré's recurrence theorem. In microscopic dynamics there is no obvious arrow of time. How does one extract macroscopic irreversibility from such microscopic laws?
3. Statistical mechanics makes generous use of ensembles. A given macroscopic system is, presumably, in a well-defined microscopic state. Instead of considering this one microscopic state statistical mechanics considers the ensemble of all microscopic states that are compatible with the given macroscopic state. Macroscopic properties are then interpreted as ensemble averages. Why are such ensemble averages relevant for the specific micro-state that the system occupies? More succinctly, ensemble averages can be interpreted as averages over independent parallel universes. Why is an average over parallel universes relevant? Ensembles are or can be characterized by some probability measure. The question then is: what do these probability measures represent?

These are still open questions. Here we just recapitulate some definitions and results from dynamical systems and probability theory, give the standard interpretations of thermodynamic equilibrium states and the standard rationalizations for the irreversible approach to these equilibrium states. These interpretations and rationalizations are not fully satisfying and still leave open important questions.

19.2 The Arrow of Time Versus Time-Reversible Systems

Here we just recall a few basic definitions about the time behavior of dynamical systems and their relation to the arrow of time.

The arrow of time. The nature of time is one of the fundamental questions in science and philosophy. Time is generally represented by the real line. Time is an ordered phenomenon. One distinguishes between the past, the present and the future. Time flows from the past to the future. This is the arrow of time. We humans experience this flow of time. We are born, live and die. For us, the past is factual. We have documents and memory of the past. The future is open. It has not yet occurred. It is potential. This experience of the flow of time is often referred to as the anthropogenic or psychological arrow of time. This directionality of time may be taken as a basic experiential or phenomenological fact, not reducible to anything else, or may be related to some other arrow of time such as:

- the thermodynamic arrow of time, which equates the direction of time with the increase of entropy,
- the cosmological arrow of time, which equates the direction of time with the expansion of the universe, and
- the causal arrow of time, which asserts that causes precede effects.

To what extent these arrows are indeed different, related or synonymous is an open question but does not put into question the basic phenomenological fact:

Time always flows in one direction, from the past to the future.

The thermodynamic arrow of time manifests itself in irreversible processes and the second thermodynamic law.

Irreversible processes. The distinction between irreversible and reversible thermodynamic processes is a phenomenological statement about macroscopic processes. One can change the state of a system by external manipulations. A process is called irreversible if one is unable to return to the original state by reversing these external manipulations. If one is able to do so then the process is called reversible. Reversibility is an idealization. All macroscopic processes are irreversible. Their time-reversed counterpart is not observed. There are numerous examples from everyday life: we live from birth to death and not vice-versa, a rock rolls down a hill crashing a few bushes on its way down and not broken bushes pushing a rock uphill, milk and coffee mix but do not unmix. The word "unmix" is not even in the dictionary since there is no such process. Making a movie of irreversible processes and viewing the movie backwards is a sure way to entertain children, and adults.

Reversible processes are not only an idealization. They also throw out time. They are viewed as a succession of thermodynamic equilibrium states following each other in a 'quasi-static' way. Equilibrium thermodynamics is a timeless theory. When one plots the evolution of a reversible process one actually only plots different equilibrium states.

The second law. The distinction between irreversible and reversible processes is made quantitative by the second law of thermodynamics. The weak version states that the entropy of an isolated system never decreases

$$d_i S \geq 0 \tag{19.1}$$

where the subscript 'i' refers to internal processes. Irreversible processes are then defined by $d_i S > 0$ and reversible processes by $d_i S = 0$. There are no processes for which $d_i S < 0$.

There is also a strong version of the second law which adds the fact that the entropy is a maximum when the system is in thermodynamic equilibrium. The strong version then states that an isolated system is either in thermodynamic equilibrium or it approaches equilibrium and its entropy increases to the maximum value possible.

The dynamical equations that describe macroscopic phenomena incorporate irreversible behavior through processes such as molecular viscosity, diffusion and conduction, as in the fluid dynamical balance equations. These macroscopic dynamical equations represent dynamical systems that are non-invertible, have no periodic orbits and have one attractive fixed point, the state of thermal equilibrium, all in case of an isolated system.

Despite this directionality of time, the micro-systems considered in this book exhibit various kinds of symmetry with respect to time.

Invertible systems. A dynamical $(S_t)_{t\in T}$ is invertible if the index set T (= time) is $\mathbb{R}$ or $\mathbb{Z}$. Irreversible processes represent non-invertible systems.

Time-reversal-invariant systems. The general definition of time-reversal invariance for a physical system asserts the existence of a transformation $f : X \mapsto X$ such that

$$S_t \circ f \circ S_t = f \tag{19.2}$$

or

$$f \circ S_{-t} = S_t \circ f \tag{19.3}$$

for all $t \in \mathbb{R}$ and $x \in X$. More explicitly: if the state x evolves according to $S_t(x)$ then the state $f(x)$ evolves according to $S_t(f(x)) = f(S_{-t}(x))$, both forward in time. It is again stressed that time-reversal invariance is a bad name since it does not reverse time. Systems always evolve forward in time. It only relates different solution. Hamiltonian systems that do not explicitly depend on time and whose Hamilton functions satisfy $H(q, p) = H(q, -p)$ are time-reversal invariant with the transformation being $f : (q, p) \mapsto (q, -p)$. If the Hamiltonian depends on a rotation vector (as is the case when the equations of motion are formulated in a rotating frame of reference) then the rotation vector must also be reversed. Time-reversal invariance becomes CPT (charge, parity, time) invariance if one leaves the realm of classical mechanics.

As a simple example consider a free particle

$$\begin{aligned} \dot{q} &= p \\ \dot{p} &= 0 \end{aligned} \tag{19.4}$$

Let the system evolve forward in time $(q_0, p_0) \mapsto (q_t(q_0, p_0), p_t(q_0, p_0))$, then reverse the momentum and let the system again evolve forward in time. We then get back to the original position with the momentum reversed

$$\tilde{q}_t(q_t(q_0, p_0), -p_t(q_0, p_0)) = q_0 \quad \tilde{p}_t(q_t(q_0, p_0), -p_t(q_0, p_0)) = -p_0 \tag{19.5}$$

There is another solution $(\tilde{q}_t, \tilde{p}_t)$ in state space for which the trace in physical space will be traversed in opposite direction (see Fig. 19.1). The important point is that this time-reversal invariance does not imply that time is reversed but only that the system exhibits a symmetry that relates two separate solutions. Not all systems are time-reversal invariant. The over-damped oscillator is not.

Time-reversed dynamical systems. Given a dynamical system $(S_t)_{t \in T}$ one defines the time-reversed system by

$$\tilde{S}_t := S_{-t} \tag{19.6}$$

The time-reversed system is the time mirror of the original system. The time reversed system $\tilde{S}_t$ is a different dynamical system that evolves forward in time as the original system would evolve backward in time. It is the movie played backwards. The time-reversed system is a different dynamical system.

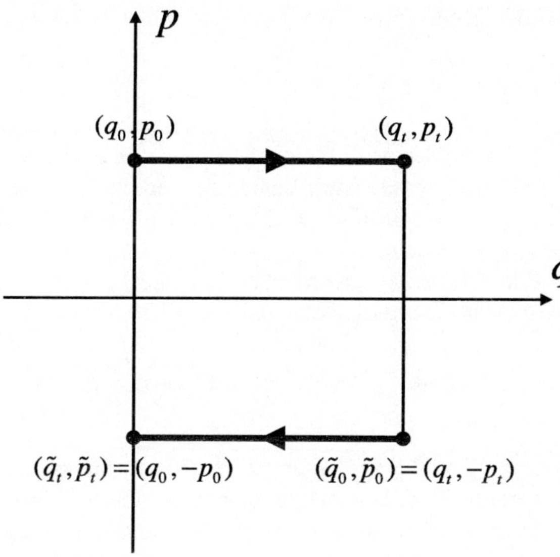

Fig. 19.1 Free particle. The time reversal invariance implies that to each solution (q_t, p_t) there exists a solution $(\tilde{q}_t, \tilde{p}_t)$ that traverses physical space in the opposite direction

Time-reversible stochastic processes. Given a stochastic process f_t the time-reversed process $\tilde{f}_{t,\tau}$ is defined by requiring that the joint probability densities satisfy

$$\tilde{\rho}_{t_N,\ldots,t_0} = \rho_{\tau-t_N,\ldots,\tau-t_0} \tag{19.7}$$

for all N, all $(t_N, \ldots, t_0)$ and all $\tau \in T$. As for time-reversed dynamical systems, the time-reversed process is a different stochastic process that is a time mirror of the original process with the time τ determining the time origin of the mirrored process. If the original process is stationary then the time-reversed process is stationary.

A reversible stochastic process is one for which all time reversed processes are equal to the original process, which implies that

$$\rho_{t_N,\ldots,t_0} = \rho_{\tau-t_N,\ldots,\tau-t_0} \tag{19.8}$$

holds for all N, all $(t_N, \ldots, t_0)$ and all $\tau \in T$. A reversible process is stationary since

$$\rho_{t_N+\tau,\ldots,t_0+\tau} = \rho_{-t_N,\ldots,-t_0} = \rho_{t_N,\ldots,t_0} \tag{19.9}$$

where the latter equality results from the special choice $\tau = 0$. A reversible process is thus a process which looks the same in the forward and backward direction. In case of dynamical systems, the requirement that the time-reversed system equals the original system can only be satisfied for the trivial system $S_t = I$ for all t.

The major conclusion to take from this discussion is:

The adjective 'reversible' is a statement about a dynamical system or stochastic process and not a statement about time.

19.3 Entropy

Entropy is the key quantity that connects the microscopic and macroscopic world. Some of the mystique and confusion related to entropy arises from the fact that there are different kinds of entropy, all well-defined in their respective contexts. These entropies are equivalent under certain circumstances and are not in others. We distinguished four different kinds of entropies, here characterized by the scientists who introduced them.

The Clausius entropy. The Clausius or thermodynamic entropy is defined for systems in thermodynamic equilibrium by

$$S_C^B - S_C^A := \int_A^B \frac{\delta Q}{T} \tag{19.10}$$

where δQ is the heat added reversibly to the system and T the temperature. The integral is along a reversible path in the thermostatic state space. Only entropy differences are defined in this way. An absolute value can be obtained by imposing the third law of thermodynamics which states that the entropy is zero at zero absolute temperature. The Clausius entropy is a thermodynamic variable. It depends only on the thermodynamic state of the system.

It is important to note that the Clausius entropy is only defined for systems in thermodynamic equilibrium or for aggregates of such systems, such as fluids in local thermodynamic equilibrium.

The Shannon entropy. The Shannon or information entropy is defined for probability distributions, in the discrete case by

$$H[p] := -k_B \sum_{i \in I} p(i) \log p(i) \tag{19.11}$$

and in the continuous case by

$$H[\rho] := -k_B \int_X dx \rho(x) \log \rho(x) \tag{19.12}$$

The Shannon entropy is a functional of the probability distribution. It measures the randomness of the distribution or the amount of missing information associated with the probability distribution p or ρ. It does not describe disorder, a not well-defined and subjective concept. The Shannon entropy has no dimension. We give it some dimension through a factor k_B in front of the definition to conform to some specific units. As discussed in Sect. 10.6 there are some problems with the continuous case which can be alleviated by considering the relative entropy or Kullback-Leibler divergence

$$D[\rho_1 \parallel \rho_2] := \int_X dx \rho_1(x) \log \frac{\rho_1(x)}{\rho_2(x)} \tag{19.13}$$

The information entropy is a concave functional and satisfies the properties discussed in Sect. 10.6. Two important properties are that coarse graining and smoothing never decrease the information entropy.

The Boltzmann entropy. The Boltzmann entropy connects microscopic and macroscopic dynamics. It presumes a microscopic state space X with an invariant measure μ_* and a set of macroscopic variables which are observables $f : X \mapsto Y$. For given values y of the observables the Boltzmann entropy is then defined by

$$S_B(y) := k_B \log [\mu_*(f^{-1}(y))] \tag{19.14}$$

where $f^{-1}(y)$ is the set of micro-states compatible with the macro-state y. This set has a certain measure. The Boltzmann entropy is the logarithm of this measure. If $\mu_*(X)$ is finite then $P(y) = \mu_*(f^{-1}(y))/\mu_*(X)$ defines a probability measure on Y.

The Boltzmann entropy requires specification of a measure μ_*. If the state space is discrete and one chooses the counting measure then $W(y)$ becomes the number of possible micro-states that is consistent with the macro state y and one obtains Boltzmann's famous formula

$$S_B = k_B \log W \tag{19.15}$$

The entropy is the logarithm of the number of compatible micro-states.

If one chooses the Lebesgue measure or volume measure μ_L for Hamiltonian flows then the entropy is the logarithm of the volume of compatible states. A technical problem often encountered is that the pre-image of $f^{-1}(y)$ might be a lower-dimensional set in X with zero volume. One then takes the pre-images of the balls $B(y; \epsilon) := \{y' | y' \in Y, d(y, y') < \epsilon\}$ with small but arbitrary ϵ and needs to demonstrate that results are independent of ϵ.

As demonstrated in Sect. 17.4 the Boltzmann entropy is obtained from the principle of maximum entropy when the value of y is used as a constraint.

There is also another entropy that was introduced by Boltzmann and is defined by

$$\hat{H}_B[\rho_1] := \int dp^{(1)} \rho_1(p^{(1)}) \log \rho_1(p^{(1)}) \tag{19.16}$$

where $\rho_1(p^{(1)})$ is the 1-particle distribution in momentum space. Note there is no minus sign and irreversibility becomes $dH_B/dt \leq 0$, the famous H-theorem. These different sign conventions have caused some confusion. We do not consider $\hat{H}_B$.

The Gibbs entropy. Gibbs entropy also connects microscopic and macroscopic dynamics. It also presumes a microscopic state space X and a set of macroscopic observables $f : X \mapsto Y$ but this time with given expectation values $\langle f \rangle$. One then applies the principle of maximum entropy (PME) and asks which probability distribution $\rho_{max}(x; \lambda)$ maximizes the Shannon entropy functional, with the expectation values $\langle f \rangle$ as constraints. This probability distribution depends on the Lagrange multipliers λ corresponding to the constraints. Substituting this distribution into the information entropy then gives Gibbs entropy

$$S_G(\lambda) := H[\rho_{max}(\lambda)] \tag{19.17}$$

which depends on the Lagrange multipliers λ.

Thus the Shannon or information entropy is a functional of a probability distribution and the Boltzmann and Gibbs entropies are the entropy functions that are obtained when the PME distributions are substituted into the Shannon functional for the case that the constraints are values or expectations of observables. The Shannon entropy depends on the probability distribution, the Boltzmann entropy on the value of the observable and the Gibbs entropy on the Lagrange multiplier corresponding to the expectation of the observable.

All the entropies are extensive or additive. The Clausius entropy is an extensive thermodynamic variable, in the thermodynamic limit. The Shannon, Boltzmann and Gibbs entropies are additive for independent subsystems.

These formal considerations are all fine. One would, of course, like to interpret entropy. What does entropy mean? We do not easily sense entropy as we, for example, sense temperature which is also not a trivial quantity. In all definitions entropy measures the amount of randomness, as perhaps most succinctly expressed by Boltzmann's definition: entropy is the logarithm of the number of micro-states compatible with the macro-state. In thermodynamics the connection between randomness and entropy manifests itself in that the free energy $F = U - TS$ measures the amount of energy that can be converted into work, into macroscopic (organized) motion. The higher the Clausius entropy the less microscopic motion can be converted into macroscopic motion, with the temperature being the controlling factor.

Though the Shannon entropy can be written as an expectation $H[\rho] = -\langle \log \rho \rangle$ one must realize that the probability density is not an observable. Densities and observables evolve differently. Thus one should not view entropy as some 'stuff' that can be distributed. One should also not view entropy as a measure of order. Order is not defined for a probability distribution. What we subjectively perceive as order has to do with the arrangements of things in space. If one were to define such order quantitatively it will turn out to be something different from entropy. Entropy can be compared to other characteristics of the probability distribution such as the variance. One finds for example that entropy and variance scale differently for discrete distributions, $H[p_{af}] = H[p_f]$ versus $\sigma^2_{af} = a^2\sigma^2_f$.

19.4 Dynamical Constraints

Consider invertible non-divergent flows, such as generated by Hamiltonian dynamics, with random initial conditions. The Lebesgue measure is then an invariant measure. When random initial conditions are realized by densities then they evolve according to (see Sect. 11.3)

$$\rho_t(x) = \rho_0(S_{-t}(x)) \tag{19.18}$$

since the flow is non-divergent. Consequently the information entropy does not change

$$H[\rho_t] = H[\rho_0] \tag{19.19}$$

If ρ_* is a stationary density then

$$P_*(A) = \int_A \mu_L(dx)\rho_*(x) \tag{19.20}$$

is an invariant probability measure. If the state space has a finite Lebesgue measure then the uniform density

$$\rho_*(x) = \frac{\chi_X(x)}{\mu_L(X)} \tag{19.21}$$

is a stationary density. If P_* is the only invariant measure then the system is ergodic. In this case arbitrary initial densities converge towards the stationary density in the Cesaro sense

$$\lim_{T\to\pm\infty} \frac{1}{T}\int_0^T dt\langle\rho_t, f\rangle = \langle\rho_*, f\rangle \tag{19.22}$$

where f is any bounded observable and the scalar product given by $\langle\rho, f\rangle = \int_X \mu_L(dx) f\rho(x)f(x)$. The convergence is in both time directions since the system is invertible.

If the system is mixing then one has the weak convergence,

$$\lim_{t\to\pm\infty}\int \mu_L(dx)\rho_t(x)f(x) = \int \mu_L(dx)\rho_*(x)(x) \tag{19.23}$$

for all bounded observables f. This convergence may be rewritten in terms of expectations as

$$\lim_{t\to\pm\infty}\langle f_t\rangle = \langle f_*\rangle \tag{19.24}$$

Expectations of observables converge towards their stationary value, again in both time directions. This is an important result, utilized in many 'derivations' of irreversible behavior. It specifically manifests itself in coarse-graining.

Consider a partition $X = \cup_{k=1}^{\infty} B_k$ and introduce the probability of each box

$$p_t(k) = \int_{B_k} \mu_L(dx)\rho_t(x) \quad k = 1, 2, \ldots \tag{19.25}$$

Assume a mixing system and choose $g(x) = \chi_{B_k}(x)$ in (19.23). One then finds convergence for the probabilities

$$\lim_{t\to\pm\infty} p_t(k) = \int_{B_k} \mu(dx)\rho_*(x) =: p_*(k) \tag{19.26}$$

for the entropy

$$\lim_{t\to\pm\infty} H[p_t] = H[p_*] \tag{19.27}$$

and for the Kullback-Leibler divergence

$$\lim_{t\to\pm\infty} D[p_t \parallel p_*] = 0 \tag{19.28}$$

Similarly the coarse-grained density

$$\rho_t^{cg}(x) := \sum_{k=1}^{\infty} \frac{p_t(k)}{\mu(B_k)} \chi_{B_k}(x) \tag{19.29}$$

also converges to the coarse-grained invariant density

$$\lim_{t\to\pm\infty} \rho_t^{cg} = \rho_*^{cg} \tag{19.30}$$

and we again find

$$\lim_{t\to\pm\infty} H[\rho_t^{cg}] = H[\rho_*^{cg}] \quad , \quad \lim_{t\to\pm\infty} D[\rho_t^{cg} \parallel \rho_*^{cg}] = 0 \tag{19.31}$$

In general, we cannot say how the entropy approaches its final value. Only if the stationary density is the density that maximizes the entropy, the uniform density for a state space with finite Lebesgue measure, can we say that the entropy increases towards its equilibrium value. This is, however, a somewhat minor point The major point is: when the system is invertible all the above equations also hold in the limit $t \to -\infty$. The time symmetry is thus not broken by considering averages of observables. It is also not broken by the continuum limit as discussed in Sect. 18.3 and by reversible stochastic dynamics as discussed in 12.3.

Ensemble-averaged observables exhibit time-directed behavior in invertible mixing system. This behavior should not yet be called irreversible behavior. Though the system will move toward its equilibrium in the future, it moved away from its equilibrium in the past. Irreversibility requires that one dismisses the past evolution for some reason. Irreversible behavior requires that one breaks the time-symmetry.

19.5 Irreversibility and Microscopic Dynamics

Whether macroscopic variables are interpreted as values of observables or as expectations of observables the macroscopic laws represent processes where initial states approach irreversibly the thermal equilibrium state in isolated systems. The microscopic dynamics exhibits usually quite different dynamics. In the case of Hamiltonian systems the dynamics is invertible and time-reversal-invariant and has no attractors.

It is indeed hard to find irreversible behavior in microscopic dynamics. One failed attempt were the calculations of Fermi, Pasta and Ulam. Starting out with a supposedly non-equilibrium state with most of the energy in just one mode the energy spread out into all the other modes and seem to reach a state of thermal equilibrium, only to return after further integration back to its initial state. The basic question that arises from this macroscopic-microscopic distinction is: How can irreversible macroscopic behavior be obtained from microscopic dynamics? Why do macroscopic systems show irreversible behavior?

The first basic question is: Why is it that certain sets of macroscopic variables evolve in time according to a closed set of equations? What kind of closure hypothesis needs to be applied?

And then there are the two famous objections to squaring macroscopic irreversibility with microscopic dynamics:

1. The microscopic dynamics supposedly satisfies all the requirements of Poincaré's recurrence theorem which states that every set of positive measure in a space of finite measure is revisited by almost all of its points infinitely often. How does this square with macroscopic irreversibility? This question is the "Wiederkehreinwand" (recurrence objection) of Zermelo (1896)
2. The microscopic laws are time-reversal-invariant (or CPT-invariant). How does one square microscopic time-reversal invariance with macroscopic irreversibility? This is the "Umkehreinwand" (time-reversal objection) of Loschmidt (1876).

The answer given to the recurrence objection is generally that the recurrence times are larger than the times required to approach equilibrium. Recurrence times are hard to estimate but are assumed to grow exponentially with the number of particles, reaching times much larger than the age of the universe for just a few hundreds of particles. The basic response to the recurrence objection thus is: true but physically irrelevant.

The answer given to the time-reversal objection is more complicated. First, one must realize that one can move along a trajectory in the state space of a dynamical system in only one direction, in the forward time direction, not in the reverse direction. Second, time-reversal invariance only states that each forward trajectory has a complementary forward trajectory that is related to the time-reversed system. Which trajectory a system occupies depends on the initial condition. The system might just be on a microscopic trajectory that exhibits irreversible macroscopic behavior, or it might not be. As we will demonstrate the response to the time-reversal objection is: you have to introduce observables and random initial conditions to arrive at a closed set of macroscopic equations that exhibit irreversible behavior. Introducing just observables is not enough. Consider the baker transformation which is a time-reversal invariant system. Its first component taken as an observable is a Bernoulli shift which is, admittedly, a non-invertible exact system. Its second component, however, does not constitute a closed dynamical system. One also needs random initial conditions.

Closure. To understand what is required to close the ensemble-averaged evolution equations for observables consider the irreversible expansion of an ideal gas from the

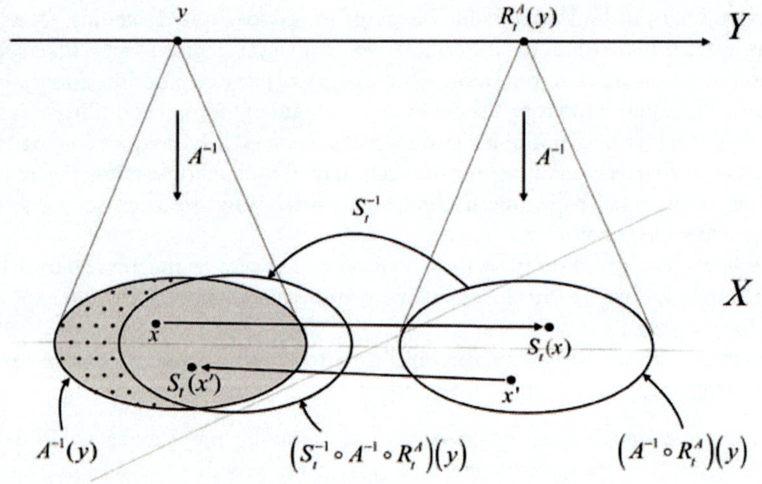

Fig. 19.2 Sketch of the pre-images $A^{-1}(y)$, $(A^{-1} \circ R_t^A)(y))$, and $(S_t^{-1} \circ A^{-1} \circ R_t^A)(y))$. The plain shaded area represents the 'good' initial conditions B_{good}, the stippled shaded area the 'bad' initial conditions B_{bad}. Time reversibility implies that for every point $x \in B_{good}$ that ends up at time t in $(A^{-1} \circ R_t^A)(y)$ there exists a point $x' \in (A^{-1} \circ R_t^A)(y)$ that ends up in B_{good} at time $2t$

left side of a container into a vacuous right hand side. There one argues that the expansion requires that most of the micro-states compatible with a volume $V/2$ evolve towards micro-states compatible with a volume V. This argument can be generalized to a time-reversible microscopic dynamical system $S_t : X \mapsto X$ that is ergodic with an invariant probability measure P_* as follows. One assumes that the macroscopic variables are represented by a set of observables $A : X \mapsto Y = \mathbb{R}^N$. As discussed in Sects. 2.6 and 12.2 these observables evolve according to $R_t^A := A \circ S_t \circ A^{-1}$ which is in general a multi-valued function that describes a probabilistic evolution. Under what circumstances will this function describe a deterministic evolution? For an answer assume that R_t^A is such a deterministic (single-valued) function and consider the following pre-images (see Fig. 19.2):

$$A^{-1}(y) = \{x | x \in X, A(x) = y\} \tag{19.32}$$

$$(A^{-1} \circ R_t^A)(y) = \{x | x \in X, A(x) = R_t(y)\} \tag{19.33}$$

$$(S_t^{-1} \circ A^{-1} \circ R_t^A)(y)) = \{x | x \in X, S_t(x) \in A^{-1}(R_t(y))\} \tag{19.34}$$

When we initially constrain the system to be in a macroscopic state y we only know that the microscopic system is in a microscopic state $x \in A^{-1}(y)$ but we do not know in which specific one. Each of the possible microscopic initial condition x evolves according to $S_t(x)$. The 'good' microscopic initial conditions $x \in A^{-1}(y)$ are those that lead to a point in $(A^{-1} \circ R_t^A)(y))$ and are thus given by

$$B_{good}:=A^{-1}(y)\cap(S_t^{-1}\circ A^{-1}\circ R_t^A)(y))) = \{x|x\in X, x\in A^{-1}(y), S_t(x)\in(A^{-1}\circ R_t^A)(y)\} \tag{19.35}$$

The 'bad' microscopic initial conditions are the complement

$$B_{bad}:=A^{-1}(y)\setminus(S_t\circ A^{-1}\circ R_t^A)(y)) = \{x|x\in X, x\in A^{-1}(y), S_t(x)\notin(A^{-1}\circ R_t^A)(y)\} \tag{19.36}$$

The number of bad initial condition should be much smaller than the number of good initial conditions. This requirement assures that the macroscopic system evolves from y to $R_t^A(y)$ in most cases. Nearly all microscopic initial conditions compatible with a value y of the observables then evolve to microscopic states at time t that are compatible with a value $R_t^A(y)$ of the observables. In terms of our measure, this requires that

$$P_*(B_{bad}) \ll P_*(B_{good}) \tag{19.37}$$

Under this condition the function $A\circ S_t\circ A^{-1}$ evolves nearly deterministically.

Irreversibility. Consider again the expansion of the gas. If after expansion, one were able to invert all the momenta of the gas molecules then the gas would return to its original state with all the molecules being in the left hand side of the container again. For this reversal to be highly unlikely or for the expansion to be irreversible one needs to require that the number of micro-states originating from the left hand side of the container is a small fraction of all the micro-states compatible with V. This is indeed the case since the ratio of the respective states is proportional to $(V/2)^N/V^N = 2^{-N}$ for an ideal gas. This argument can also be generalized. For the sake of argument, assume that the observables are not affected by the coordinate transformation f that renders the microscopic dynamics time-reversal invariant, i.e., $A(f(x)) = A(x)$. For every point $x\in B_{good}$ that ends up as $S_t(x)\in(A^{-1}\circ R_t^A)(y)$ we therefore have a point $x' = f(S_t(x))\in(A^{-1}\circ R_t^A)(y)$ that ends up back in B_{good} at time $2t$. In order for the expansion to be irreversible one needs to require that the number of micro-states in $(A^{-1}\circ R_t^A)(y)$ that return to B_{good} is small. Since for every point in B_{good} there is a point in $(A^{-1}\circ R_t^A)(y)$ that returns the number of these states is the number of states in B_{good}. In order for the macroscopic evolution to be irreversible we must thus require

$$P_*(B_{good}) \ll P_*((A^{-1}\circ R_t^A)(y))) \tag{19.38}$$

Iterating this argument one finds: in order for the macroscopic observables to evolve irreversibly the number of micro-states compatible with the values of the observables must increase. More and more micro-states must become compatible with the macro-state. The system moves to more and more probable states until it reaches thermal equilibrium, the most probable state. Conversely, as we move further and further into the future the volume of compatible micro-states at time $t = 0$ becomes an ever smaller fraction of the future compatible micro-states.

Of course, we know from dynamical systems theory that in mixing systems expectations converge weakly to the equilibrium state

$$\lim_{t\to\infty} \langle \rho_t, f \rangle = \langle \rho_*, f \rangle \tag{19.39}$$

for all bounded observables f. The above arguments demonstrate what is required for this approach to be nearly deterministic and irreversible. The arguments especially demonstrate that an irreversible approach towards equilibrium is possible, not that it happens. It only asserts that most initial conditions lead to irreversible behavior, not all of them. Thus

The approach towards equilibrium is a probabilistic phenomenon. The approach towards equilibrium is more probable than to move away from it. Deviations from irreversibility or the second law are not impossible but only highly improbable.

Entropies. We must again distinguish between different entropies. The density ρ_0 that describes the random initial conditions evolves according to $\rho_t(x) = \rho_0(S_{-t}(x))$ with respect to the microscopic dynamics. The information entropy hence does not change

$$H[\rho_t] = H[\rho_0] \tag{19.40}$$

or more exactly the Kullback-Leibler divergence does not change under the dynamical evolution of an invertible dynamical system

$$\frac{d}{dt} D[\rho_t \parallel \rho_*] = 0 \tag{19.41}$$

When at time t the macroscopic evolution of the system is in the state $R_t^A(y)$ then the number of compatible micro-states is $P_*((A^{-1} \circ R_t^A)(y))$. The Boltzmann entropy is therefore

$$S_B(R_t^A(y)) := k_B \log P_*((A^{-1} \circ R_t^A)(y)) \tag{19.42}$$

Since the irreveribility argument requires that this number has to increase in time this Boltzmann entropy will increase in time, until it reaches its equilibrium value.

Symmetry-breaking. There is a caveat in the above arguments. Assume you have set up a container with the (ideal) gas such that a bell rings when the gas only occupies the left hand side of the container. You would actually have to wait a long time for this bell to ring. However, when it happens what would you predict? You should predict (correctly) that the gas will most likely evolve to a state where it again occupies the whole container, according to Boltzmann's rationalization. You may also ask ask where this peculiar state came from. Since the system is time-reversal-invariant you should also apply the above rationalization backwards in time. This peculiar state most likely came from a state where the gas occupied the whole container since there are many more micro-states in that state than in any other state that will result in the gas being only in the left hand side of the container. The gas being in the left part of the container is just a statistical fluke. It most likely came from the equilibrium state and will most likely evolve towards the equilibrium state.

Formally: if you observe the system to be in a macro-state y at time $t = 0$ then most of the compatible micro-states $x \in A^{-1}(y)$will end up in the 'equilibrium manifold'

$$\lim_{t\to\infty} S_t(x) \in (A^{-1} \circ R^A_\infty)(y) \tag{19.43}$$

For time-reversal-invariant systems there exists a transformation f such that $S_{-t} \circ f = f \circ S_t$. Since we also required $f(x) \in A^{-1}(y')$ if $x \in A^{-1}(y')$ to hold for arbitrary $y' \in Y$ one finds

$$\lim_{t\to\infty} S_{-t}(x) \in (A^{-1} \circ R^A_\infty)(y) \tag{19.44}$$

as well. Most compatible micro-states come from the equilibrium manifold. This same behavior is found in Sect. 12.3 for a reversible stochastic dynamics, namely the exchange of marbles in two bowls. Thus

The time symmetry is neither broken by considering macro-states nor by considering ensembles. One cannot derive irreversible macroscopic laws from reversible microscopic dynamics, without invoking a symmetry breaking assumption.

So what is a possible symmetry breaking assumption? When considering systems that start out in a non-equilibrium state they will (when left to themselves) most probably approach equilibrium forward in time. We can now break the time symmetry by not applying this reasoning backward in time because we **assume** that the system is on its forward approach to equilibrium coming from an even less probable state in the past. In principle there are documents from this past. Only the future is open. The past is factual, and in principle known. When applying this symmetry breaking assumption to the universe as a whole one would have to infer that the universe started out at a very improbable state at the big bang and since then moves through ever more probable states towards the state of heat death. Application of the locally established second law to the whole universe needs of course justification.

In conclusion: From the microscopic point of view irreversibility is a probabilistic phenomenon. Macroscopic irreversible behavior results from highly improbable microscopic initial conditions. Deviations from irreversibility or the second law are still possible but highly improbable. This probabilistic interpretation does not resolve the conflict between microscopic time-reversal symmetry and macroscopic irreversibility. What resolves the conflict is that the problem is treated as an initial value problem solved forward in time. Thus

Irreversibility is introduced by setting up the problem in a forward time direction. Irreversibility resides in the initial conditions.

19.6 Ensemble

A specific state of a macroscopic system is generally compatible with many micro-states. Since one can only control the macro-variables one does not know which micro-state the system occupies. One is thus forced to consider all compatible micro-states. In statistical mechanics this is done by introducing an ensemble. It is a set of copies of the same macroscopic system that differ only in their microscopic state but not in their macroscopic state. It consists of all micro-states compatible with the macroscopic state. It is usually characterized by a probability distribution. The ensemble members evolve independently of each other. The average over the ensemble or the ensemble mean is thus an average over independent copies of the same macroscopic system or as one often formulates it, an average over many parallel universes. Why is such an ensemble average relevant. The system is only in one micro-state, though we do not know in which one.

The frequentist's argument. The frequentist's interpretation of probability is that the probability of a specific outcome of an experiment is the relative frequency with which this outcome occurs when one repeats this experiment (infinitely) often. The frequentist thus admits that probability does not say anything about a single experiment but only about many repetitions of the same experiment. This is a comfortable fall-back option.

The ergodic argument. A more ambitious justification for the use of ensemble means comes from the ergodic theorem which states that the time mean equals the ensemble mean for ergodic systems,

$$\bar{f} := \lim_{T\to\infty} \frac{1}{T} \int_0^T dt f(S_t(x)) = \langle f \rangle \tag{19.45}$$

One then argues that any measurement of a macroscopic variable f is actually a time average over an interval Δt

$$\bar{f}_{\Delta t} := \frac{1}{\Delta t} \int_0^{\Delta t} dt f(S_t(x)) \tag{19.46}$$

that is short from a macroscopic point of view but long from a microscopic point of view such that the measurement result is given by the ensemble mean. The error is of course hard to estimate for other than the simplest systems. One such simple system is a set of N uncoupled oscillators whose trajectories are given by

$$\theta_n(t) = \theta(0) + \omega_n t \quad n = 1, \ldots, N \tag{19.47}$$

This is the system that yielded the extraordinary long recurrence times in Sect. 5.8. Take as the observable the mean

$$f(\theta_1, \ldots, \theta_N) = \frac{1}{N} \sum_{n=1}^{N} \theta_n \tag{19.48}$$

If the frequencies are incommensurable then the uniform density

$$\rho_*(\theta_1, \ldots, \theta_N) = (2\pi)^{-N} \tag{19.49}$$

is the only invariant measure and the system hence ergodic. The ensemble mean is thus

$$\langle f \rangle = \pi \tag{19.50}$$

Elementary integration yields

$$\bar{f}_{\Delta t} = \pi(1 + \epsilon) \tag{19.51}$$

where the error term ϵ is smaller than $2\pi/\omega_{min}\Delta t$. For the error term to be small the measurement interval Δt must be much larger than the largest period of the oscillator, a perhaps reasonable assumption in certain contexts. The ergodic argument thus requires that the measurement interval is large from a microscopic point of view, and that the system is ergodic. Proving ergodicity is an unresolved problem for most systems. Sinai showed that hard spheres in a box constitute an ergodic system. The KAM theorem states that a slightly distorted integrable Hamiltonian system on the other hand is not ergodic. The tori that are only slightly distorted evolve independently and preserve their measure. The uniform Lebesgue measure of the complete state space is not the only invariant measure. Thus, ergodicity is an assertion for most more complicated systems where it is unknown what is required of the microscopic dynamics and of the observables for the ergodic argument to work.

The large N argument. A different argument invokes the law of large numbers in its various forms. Probabilistic statements then become nearly deterministic. The N-particle Hamiltonian system is the standard example. As discussed in Sect. 17.8 the fluctuations about the mean may become negligible in the thermodynamic limit $N \to \infty$ and statements become nearly deterministic. Again it is hard to prove if and how fluctuations decrease as the system becomes larger.

19.7 Summary

Here we addressed the basic question of how to square microscopic behavior with macroscopic behavior. Microscopic dynamics is invertible and time-reversal invariant and has no fixed points. Macroscopic is irreversible and approaches an equilibrium state. This discrepancy is generally resolved in several steps steps:

1. One introduces random initial conditions or an ensemble of microscopic systems. This step makes all further steps inherently probabilistic. In most cases the system will exhibit a certain behavior.
2. One identifies macro-variables with observables. The observed value of a macro-variable is either the value of the observable (the Boltzmann view) or the expectation of the observable (the Gibbs view).
3. The combination of random initial conditions and observables introduces some time-directed behavior. One can show for invertible mixing systems that ensemble-averaged observables converge toward their equilibrium values, with the caveat

that they do so in both time directions, into the future and into the past. This is not yet irreversible behavior.

4. Observables are generally many-to-one functions and their evolution does not constitute a dynamical system. One faces a closure problem.
5. When closing the evolution equation for the ensemble-averaged observables one usually invokes many diverse assumption. One key assumption, often not explicitly stated, is a breaking of the time-symmetry. One formulates the closure so that it is only applicable in the forward time direction, into the future, and not in the reverse time direction, into the past. It is this breaking of the time-symmetry that introduces real irreversible behavior.

The rationalization of irreversible behavior that emerges from these steps is the following: every value of the observable has a certain number of compatible micro-states. This number is the larger the closer the value is to its equilibrium value. When the system is far away from equilibrium there is only a small number of compatible micro-states. As these micro-states evolve in time most of them will, in a mixing system, hit the larger sets compatible with observables closer to equilibrium. Though originating from a low probability region, that correspond to a non-equilibrium value of the observable, they will eventually find themselves in a high probability region, corresponding to the equilibrium value of the observable.

This standard rationalization of irreversible behavior does not exclude rare deviations and presumes an arrow of time.

References

Zermelo, E.: Ann. Phys. (Leipzig) **57**, 485 (1896)
Loschmidt, J.: Wien. Ber. **73**, 128 (1876)

Chapter 20
Quantum Mechanics

20.1 Introduction

Any discussion of dynamics and probability must include a discussion of quantum mechanics where dynamics and probability are intertwined in a peculiar way. This chapter cannot, of course, cover quantum mechanics in any substantial way. Instead, we just present with a broad brush the pragmatic rules and some foundational issues of quantum mechanics. Practitioners of quantum mechanics generally agree on the pragmatic rules but disagree on many of the foundational issues. Probability is intrinsic to quantum mechanics. Pure states only have a probabilistic interpretation which derives from a generalization of the Pythagorean theorem rather than from a probability measure. The broadened states of classical mechanics correspond to mixed states in quantum mechanics. There is an additional type of state in quantum mechanics, an entangled state, which does not have any equivalent in classical mechanics. Entangled states derive from the fact that the state space of a composite system is given by the tensor product of the subsystems rather than by the Cartesian product as in classical physics. In quantum mechanics, identical particles are also indistinguishable which led to the non-intuitive counting rules in the chapter on equilibrium statistical mechanics. The differences between quantum and classical mechanics become most succinct when considering the logic applied to propositions in both fields.

20.2 The Pragmatic Rules

Quantum mechanics can be characterized by five pragmatic rules, here formulated for a finite-dimensional system:

(1) The state of a quantum system is given by a vector in a Hilbert space which is a complete unitary vector space (see Appendix E.4 for details). The Hilbert space is generally denoted by $\mathcal{H}$. Following Paul Dirac, the elements $x \in \mathcal{H}$ are

P. Müller, *Handbook of Dynamics and Probability*,
https://doi.org/10.1007/978-3-030-88486-4_20

denoted by ket-vectors $|x\rangle := x$ and the adjoint vectors by bra-vectors $\langle x| := x^\dagger$ (see Appendix E.10 for the rationale and rules of this notation). States are often called wave functions.

(2) Observables are represented by self-adjoint operators A on the Hilbert space $\mathcal{H}$. There are observables such as momentum and energy which have counterparts in classical mechanics, but there are also observables such as spin which do not have any counterparts in classical mechanics. For finite-dimensional systems, such operators can be represented by self-adjoint matrices.

(3) The possible values of an observable are the eigenvalues $\alpha^{(n)}$, $n = 1, \ldots, N$ of the operator A.

(4) If the system is in the normalized state $|x\rangle$, then the probability of obtaining the value $\alpha^{(n)}$ of an observable A is given by

$$Prob(A = \alpha^{(n)} || x\rangle) = \langle x|a^{(n)}\rangle\langle a^{(n)}|x\rangle \tag{20.1}$$

where $|a^{(n)}\rangle$ is the normalized eigenvector corresponding to the eigenvalue $\alpha^{(n)}$. This probabilistic rule is referred to as the intrinsic probability in quantum mechanics. The probability is here denoted by the symbol $Prob$, firstly because the symbol P is reserved for projection operators in quantum physics, and secondly because this intrinsic quantum-mechanical rule does not satisfy Kolmogorov's axioms though it is interpreted as the relative frequency with which the result $\alpha^{(n)}$ occurs in a sequence of repeated measurements.

(5) The time evolution of the quantum-mechanical state is given by the Schrödinger equation

$$i\hbar\frac{d}{dt}|x_t\rangle = H|x_t\rangle \tag{20.2}$$

where H is a self-adjoint operator on $\mathcal{H}$ called the Hamiltonian and $\hbar$ is Planck's constant. The Schrödinger equation induces the linear dynamical system

$$S_t(|x\rangle) = U_t|x\rangle \tag{20.3}$$

where

$$U_t := e^{-iHt/\hbar} \tag{20.4}$$

is a unitary operator, an operator that leaves the length of a vector unchanged. Thus, the dynamical evolution corresponds to a uniform rotation in the Hilbert space. The technical aspects of constructing explicit solutions may be demanding but from a dynamical systems point of view, a quantum-mechanical system is just a straightforward linear system, just a rotation.

(6) When a measurement is performed and a specific outcome is observed, the state changes, the wave function collapses to a new state in which the outcome has probability 1. This collapse is not described by the Schrödinger equation.

The pair $\{\mathcal{H}, H\}$ is called a quantum-mechanical system. In the following sections, we will elaborate on the probabilistic aspects of quantum physics. To avoid mostly technical problems, we will restrict ourselves in the following to Hilbert spaces of finite dimension N. The formalism is then one of matrix algebra. The extension to countably infinite dimensions is a straightforward extension of matrix algebra. The extension to non-countably infinite dimensions encounters sometimes technical problems.

20.3 Pure States

If one has maximal and complete information about the state of a quantum-mechanical system, then the system is said to be in a pure state, given by a normalized vector $|x\rangle \in \mathcal{H}$.

Since both the state and the eigenvector are normalized in the probability law (20.1), the probability of measuring the eigenvalue $\alpha^{(n)}$ is given by the cosine squared of the angle between the vector $|x\rangle$ and the vector $|a^{(n)}\rangle$. Furthermore, since the eigenvectors of a self-adjoint operator are orthonormal, $\langle a^{(m)}|a^{(n)}\rangle = \delta_{nm}$, and form a complete basis, $\sum_{n=1}^{N} |a^{(n)}\rangle\langle a^{(n)}| = I$, the probabilities sum to one

$$\sum_{n=1}^{N} Prob(A = \alpha^{(n)}||x\rangle) = \langle x|I|x\rangle = 1 \tag{20.5}$$

which is the generalized Pythagorean theorem. The quantum-mechanical probabilities thus have their roots in the generalization of this theorem, not in measure theory. As a special case, one finds

$$Prob(A = \alpha^{(n)}|B = \beta^{(m)}) = \langle b^{(m)}|a^{(n)}\rangle\langle a^{(n)}|b^{(m)}\rangle \tag{20.6}$$

The probability that the observable A takes the value $\alpha^{(n)}$ when the system is in a state $|b^{(m)}\rangle$ where an observable B has the value $\beta^{(m)}$ is given by the cosine squared of the angle between the two eigenvectors.

The expectation value of an observable A is given by

$$\begin{aligned}\langle A\rangle :&= \sum_{n=1}^{N} \alpha^{(n)} Prob(A = \alpha^{(n)}||x\rangle)\\ &= \sum_{n=1}^{N} \alpha^{(n)} \langle x|a^{(n)}\rangle\langle a^{(n)}|x\rangle\\ &= \langle x|A|x\rangle\end{aligned} \tag{20.7}$$

The Schrödinger equation implies that the state vector evolves in time as $|x_t\rangle = U_t|x_0\rangle$ with the unitary operator $U_t = \exp\{-itH/\hbar\}$. Observables evolve as

$$A_t = U_t^* A_0 U_t \tag{20.8}$$

The expectation value $\langle A_t\rangle$ of an observable A at time t can thus be written in two forms:

$$\begin{aligned}\langle A_t\rangle &= \langle x_t|A_0|x_t\rangle \\ &= \langle x_0|A_t|x_0\rangle\end{aligned} \tag{20.9}$$

The first line is the Schrödinger representation—the state evolves in time; the second line is the Heisenberg representation—the observable evolves in time.

The above formulas are often written in a form that introduces the trace of an operator which is defined by

$$tr\{A\} := \sum_{n=1}^{N} \langle e^{(n)}|A|e^{(n)}\rangle \tag{20.10}$$

for any orthonormal basis $e^{(1)}, \ldots, e^{(N)}$ and the self-adjoint operators

$$P_n := |e^{(n)}\rangle\langle e^{(n)}| \quad P_x := |x\rangle\langle x| \tag{20.11}$$

which are the projection operator onto the subspaces spanned by the eigenvector $|e^{(n)}\rangle$ and by the vector $|x\rangle$. The probabilities and expectation values then take the form

$$Prob(A = \alpha^{(n)}||x\rangle) = tr\{P_x P_n\} \tag{20.12}$$

and

$$\langle A\rangle = tr\{P_x A\} \tag{20.13}$$

In this form, the formulas can easily be generalized to mixed states considered in the next section.

Example: Binary quantum system. The simplest quantum system is a binary system where measurements can only have two possible outcomes. The spin s or angular momentum of an electron is such a binary system. In any direction, it can only take the values $s = \pm 1/2\hbar$, corresponding to a clockwise or anticlockwise rotation in that direction. Another example is the polarization of a photon which can also have only two values. In quantum information theory, such a system is called a qubit, with the possible outcomes 0 and 1. The Hilbert space of such a binary system is the two-dimensional space $\mathcal{H} = \mathbb{C}^2$. An orthonormal basis in this space are the vectors

$$|0\rangle = \begin{pmatrix} 1 \\ 0 \end{pmatrix} \tag{20.14}$$

$$|1\rangle = \begin{pmatrix} 0 \\ 1 \end{pmatrix} \tag{20.15}$$

The space $\mathcal{L}(\mathbb{C}^2)$ of linear operators on this Hilbert space can be represented by complex 2×2 matrices. The observable "spin in direction" $n = (n_x, n_y, n_z)$ with $\| n \| = 1$ is given by the matrix

$$S_n = n_x S_x + n_y S_y + n_z S_z \tag{20.16}$$

where

$$S_x = \begin{pmatrix} 0 & 1 \\ 1 & 0 \end{pmatrix} \; S_y = \begin{pmatrix} 0 & -i \\ i & 0 \end{pmatrix} \; S_z = \begin{pmatrix} 1 & 0 \\ 0 & -1 \end{pmatrix} \tag{20.17}$$

are the Pauli matrices. The spin observable has the eigenvalues ± 1 where the eigenvalue $+1$ corresponds to the spin in positive n-direction and the eigenvalue -1 to the spin in negative n-direction.

The two-dimensional Hilbert space $\mathbb{C}^2$ of such binary quantum systems can be represented in our three-dimensional physical space $\mathbb{R}^3$ as follows: choose as a basis the eigenvectors $|0\rangle = (1, 0)^T$ and $|1\rangle = (0, 1)^T$ of the spin matrix S_z. Any normalized vector $|x\rangle$ can then be written in the form

$$|x\rangle = \cos\theta/2|0\rangle + \sin\theta/2(\cos\varphi + i \sin\varphi)|1\rangle \tag{20.18}$$

where $\theta \in [0, \pi]$ and $\varphi \in [0, 2\pi]$. Interpret θ as the co-latitude and φ as the longitude of points on the unit sphere, with the z-axis pointing upwards. Then every pure quantum state can be represented by a point on this unit sphere, called the Bloch sphere in this context.

When the binary system is in the state $|x\rangle$, then the probability of observing a spin in the upward or downward direction is given by $\cos^2\theta/2$ or $\sin^2\theta/2$.

20.4 Mixed States

If one does not know the exact state of the quantum system but only that it is in states $|x^{(1)}\rangle, \ldots, |x^{(M)}\rangle$ with probabilities $p_1, \ldots, p_M$, then one obtains a mixed state. Here, p_m is the epistemic probability that the system is in state $|x^{(m)}\rangle$. This epistemic probability p must be distinguished from the intrinsic probability considered in the previous section. The mixed state corresponds to the broadened states in classical physics. This mixed state is characterized by the density matrix

$$\rho := \sum_{m=1}^{M} p_m P_m \tag{20.19}$$

where $P_m = |x^{(m)}\rangle\langle x^{(m)}|$ is the projection operator onto the subspace spanned by $|x^{(m)}\rangle$. The density matrix has the following properties: $\langle x|\rho|x\rangle \geq 0$ which reflects the fact that the epistemic probabilities p_m are non-negative, $tr\rho = 1$ which reflects the fact that the epistemic probabilities sum to one, and $\rho^* = \rho$ which reflects the fact that projection operators are self-adjoint. Pure states are those for which all p_m are zero except one which implies $\rho^2 = \rho$. Classical mechanics has only epistemic probabilities.

The probability that an observable A takes the value $\alpha^{(n)}$ when the system is in the mixed state ρ can then be written as

$$Prob(A = \alpha^{(n)}|\rho) = \sum_{m=1}^{M} p_m Prob(A = \alpha^{(n)}||x^{(m)}\rangle) = tr\{\rho P_n\} \tag{20.20}$$

and the expectation value of the observable becomes

$$\langle A\rangle = \sum_{n=1}^{N} a^{(n)} Prob(A = \alpha^{(n)}|\rho) = tr\{\rho A\} \tag{20.21}$$

Density matrices evolve in time according to

$$\rho_t = U_t \rho_0 U_t^* \tag{20.22}$$

which differs from the evolution of observables (20.8). Expectation values can again be written in the two equivalent forms:

$$\begin{aligned}\langle A_t\rangle &= tr\{\rho_0 A_t\} \\ &= tr\{\rho_t A_0\}\end{aligned} \tag{20.23}$$

Example: Binary quantum system. The space $\mathcal{L}(\mathbb{C}^2)$ of linear operators on $\mathcal{H} = \mathbb{C}^2$ is the set of all complex 2×2 matrices. The Pauli matrices together with the unit matrix I constitute a basis for such complex matrices. In this basis, the density matrix ρ of a binary system is given by

$$\rho = \frac{1}{2}(I + a_x S_x + a_y S_y + a_z S_z) \tag{20.24}$$

where $a = (a_x, a_y, a_z)$ must satisfy $\parallel a \parallel \leq 1$ and is called the Bloch vector. One finds indeed $tr\{\rho\} = 1, \rho = \rho^*$, and $\langle x|\rho|x\rangle \leq 1$ if the point (x, y, z) is a point of the Bloch sphere. The eigenvalues of the density matrix are $1/2(1\pm \parallel a \parallel)$. The state is pure, $\rho = \rho^2$, if the point x lies on the surface of the Bloch sphere. All the points in the interior of the Bloch sphere can be written as a convex combination of pure states.

20.5 Entangled States

Another difference between classical and quantum mechanics arises when composite systems are considered. In classical mechanics, the states of a composite system lie in the Cartesian product (direct sum) of the subsystems whereas in quantum mechanics they lie in the tensor product of the subsystems.

In classical mechanics, we thus have the following situation: If X_A and X_B are the state vector spaces of the two subsystems of dimensions N_A and N_B, then the state space of the composite system is given by the direct sum $X = X_A \oplus X_B$ with dimension $N_A + N_B$. If the states of the single particles are given by $(q^{(1)}, p^{(1)}) \in \mathbb{R}^2$ and $(q^{(2)}, p^{(2)}) \in \mathbb{R}^2$, then the state of the two-particle system is given by $(q^{(1)}, p^{(1)}, q^{(2)}.p^{(2)}) \in \mathbb{R}^4$. The direct sum is also a vector space with the addition and scalar multiplication satisfying the rules given in Appendix E.4.

In quantum mechanics, on the other hand, the composite states lie in the tensor product $\mathcal{H} = \mathcal{H}_A \otimes \mathcal{H}_B$ of dimension $N_A N_B$. The tensor product is also a vector space with the addition and scalar multiplication satisfying the rules given in Appendix E.4. If $|e_A^{(n)}\rangle$, $n = 1, \ldots, N_A$ is a basis in $\mathcal{H}_A$ and $|e_B^{(m)}\rangle$, $m = 1, \ldots, N_B$ is a basis in $\mathcal{H}_B$, then the tensor products $|e_A^{(n)}\rangle \otimes |e_B^{(m)}\rangle := |e_A^{(n)}\rangle\langle e_B^{(m)}|$ form a basis of $\mathcal{H}$. (In matrix notation, $|e_A^{(n)}\rangle \otimes |e_B^{(m)}\rangle = e_A^{(n)} e_B^{(m)\dagger}$ where the dagger denotes the adjoint. The space $\mathcal{H}$ may therefore be considered as the space of all $N_A \times N_B$ matrices. Thus, every vector $|x\rangle \in \mathcal{H}$ can be written as

$$|x\rangle = \sum_{n=1}^{N_A} \sum_{m=1}^{N_B} x_{nm} |e_A^{(n)}\rangle\langle e_B^{(m)}| \tag{20.25}$$

with $N_A N_B$ components x_{nm}.

If $|x_A\rangle = \sum_{n=1}^{N_A} x_n |e_A^{(n)}\rangle$ and $|x_B\rangle = \sum_{m=1}^{N_B} x_m |e_B^{(m)}\rangle$, then the tensor product of these vectors takes the form

$$|x_A\rangle\langle x_B| = \sum_{n=1}^{N_A} \sum_{m=1}^{N_B} x_n x_m^* |e_A^{(n)}\rangle\langle e_B^{(m)}| \tag{20.26}$$

We now apply these formulas to quantum systems and assume that the bases are orthonormal and the states are normalized. If a quantum state $|x\rangle$ can be written in the form (20.26), it is called separable. However, there exist states $|x\rangle$ which cannot be written in this form since there exist component matrices x_{nm} which cannot be written as the product of two vectors. Such states are called entangled states. The state of one subsystem cannot be described independently of the other subsystem.

Using density matrices, these entangled states can be characterized as follows: The density matrix of a pure state $|x\rangle$ in $\mathcal{H}$ is given by

$$\rho = |x\rangle\langle x| \tag{20.27}$$

The density matrix ρ_A of the subsystem A is obtained by 'tracing out' the subsystem B

$$\begin{aligned}\rho_A &= tr_B\{\rho\} \\ &= \sum_{n,m=1}^{N_A,N_B} x_{nm}|e_A^{(n)}\rangle\langle e_B^{(m)}| \sum_{k=1}^{N_B} |e_B^{(k)}\rangle\langle e_B^{(k)}| \sum_{n',m'=1}^{N_A,N_B} x_{n'm'}^{*}|e_B^{(m')}\rangle\langle e_A^{(n')}| \\ &= \sum_{k=1}^{N_B} p_k|x_A^{(k)}\rangle\langle x_A^{(k)}|\end{aligned} \tag{20.28}$$

where

$$|x_A^{(k)}\rangle = \sum_{n=1}^{N_A} \frac{x_{nk}}{\sqrt{p_k}}|e_A^{(n)}\rangle \tag{20.29}$$

and

$$p_k = \sum_{n=1}^{N_A} x_{nk}x_{nk}^{*} \tag{20.30}$$

The factor p_k assures that the vectors $|x_A^{(k)}\rangle$ are normalized to one. The normalization of $|x\rangle$ assures that $\sum_{k=1}^{N} p_k = 1$. The subsystem A is in a mixed state. If the system is in the separable state $|x\rangle = |x_A\rangle\langle x_B|$ and hence $x_{nm} = x_n x_m$, then the subsystems are in a pure state

$$\rho_A = |x_A\rangle\langle x_A| \tag{20.31}$$

Thus, entangled states are those for which the composite state is pure whereas the subsystems are in a mixed state.

Example: The binary quantum system. Consider the case that the two subsystems are binary quantum systems such as a pair of electrons. The state space is then the Hilbert space of all complex 2×2 matrices. A possible basis is formed by the matrices

$$|0\rangle\langle 0| = \begin{pmatrix} 1 & 0 \\ 0 & 0 \end{pmatrix}$$
$$|0\rangle\langle 1| = \begin{pmatrix} 0 & 1 \\ 0 & 0 \end{pmatrix}$$
$$|1\rangle\langle 0| = \begin{pmatrix} 0 & 0 \\ 1 & 0 \end{pmatrix}$$
$$|1\rangle\langle 1| = \begin{pmatrix} 0 & 0 \\ 0 & 1 \end{pmatrix} \tag{20.32}$$

It is, however, computationally more convenient to work in a four-dimensional vector space with basis vectors

$$|0\rangle\langle 0| = (1, 0, 0, 0)^T$$
$$|0\rangle\langle 1| = (0, 1, 0, 0)^T$$
$$|1\rangle\langle 0| = (0, 0, 1, 0)^T$$
$$|1\rangle\langle 1| = (0, 0, 0, 1)^T \tag{20.33}$$

Linear operators are then 4×4 complex matrices. The matrix S of the vertical spin is given by

$$S = S_A \otimes I_B + I_A \otimes S_B \tag{20.34}$$

where S_A, S_B are the Pauli matrices S_z and I_A, I_B the identity matrices of the first and second systems. In the basis (20.33), S takes the form

$$S = \begin{pmatrix} 2 & 0 & 0 & 0 \\ 0 & 0 & 0 & 0 \\ 0 & 0 & 0 & 0 \\ 0 & 0 & 0 & 2 \end{pmatrix} \tag{20.35}$$

This matrix has three eigenvalues:$s = +2$, $s = 0$, and $s = -2$. The eigenvalue $s = +2$ has the eigenvector $|0 >< 0|$ which represents the state where both spins point upwards. The eigenvalue $s = 0$ is degenerate and has two independent eigenvectors $|0 >< 1|$ and $|1 >< 0!$ which represent the two states where one spin points upwards and the other downwards. The eigenvalue $s = -2$ has the eigenvector $|1 >< 1|$ which represents the state where both spins point downwards.

Assume one has prepared a composite system to be in the state

$$|x\rangle = \frac{1}{\sqrt{2}}(|0\rangle_A\langle 1|_B - |1\rangle_A\langle 0|_B) \tag{20.36}$$

This is a pure but entangled state. Its total spin is 0. If one measures the spin of one electron and obtains the value $+1/2$, then the "wave function collapses" to the state $|0\rangle_A\langle 1|_B$ which is a separable state where the measured electron has spin $+1/2$ and the other electron has spin $-1/2$. A measurement of the spin of the other electron will thus yield the value $-1/2$, even if the electrons are far separated in space. This situation gives rise to the Einstein–Podolsky–Rosen paradox.

20.6 Quantum Probability

An alternative formulation of quantum physics goes under the name of quantum probability. It starts from the fact that observables form an algebra $\mathcal{A}$ over a unitary vector space. Such algebras satisfy for $A, B, C \in \mathcal{A}$ and $a, b, c \in \mathbb{C}$

$$
\begin{aligned}
(AB)C &= A(BC) && (20.37)\\
A(bB + cC) &= bAB + cAC && (20.38)\\
(aA + bB)C &= aAC + bBC && (20.39)\\
IA &= AI && (20.40)
\end{aligned}
$$

where I is the identity (matrix). If $AB = BA$ then the algebra is commutative and is called Abelian. Furthermore, the algebra contains an adjoint operation $\mathcal{A} \mapsto \mathcal{A}$, $A \mapsto A^*$ that satisfies

$$
\begin{aligned}
(A^*)^* &= A && (20.41)\\
(aA + bB)^* &= a^*A^* + b^*B^* && (20.42)\\
(AB)^* &= B^*A^* && (20.43)
\end{aligned}
$$

and

$$\| A^*A \| = \| A \|^2 \tag{20.44}$$

which is equivalent to $A^*A = 0 \Rightarrow A = 0$ for matrices. In general, an algebra with these properties is called a C^*-algebra and is the basis for a general approach to quantum probability. Here, we conform with the quantum-mechanical convention and denote the adjoint operation by a star rather than a dagger.

Within this approach, the state of the system is a linear mapping

$$
\begin{aligned}
\tilde{\rho} : \mathcal{A} &\mapsto \mathbb{C}\\
A &\mapsto \tilde{\rho}(A) && (20.45)
\end{aligned}
$$

with the following properties:

$$\tilde{\rho}(aA + bB) = a\tilde{\rho}(A) + b\tilde{\rho}(B) \tag{20.46}$$
$$\tilde{\rho}(A^*) = \tilde{\rho}(A)^* \tag{20.47}$$
$$\tilde{\rho}(AA^*) \geq 0 \tag{20.48}$$
$$\tilde{\rho}(I) = 1 \tag{20.49}$$

The pair $\{\mathcal{A}, \tilde{\rho}\}$ is called the quantum probability space. The states formed are convex sets. Extreme points are pure states. A projection operator P denotes a possible observation and $\tilde{\rho}(P)$ the probability to obtain this observation. Here, we denoted the state in quantum probability by $\tilde{\rho}$, which is a linear mapping from $\mathcal{A}$ into $\mathbb{C}$, to distinguish it from the state or density matrix ρ in the Hilbert space formulation, which is an element of $\mathcal{L}(\mathcal{H})$. However, when ρ is a state in the Hilbert space formulation, the mapping

$$\begin{aligned} \tilde{\rho} : \mathcal{A} &\mapsto \mathbb{C} \\ A &\mapsto tr\{\rho A\} \end{aligned} \tag{20.50}$$

is a state in the quantum probability formulation. The Hilbert space formulation is thus subsumed in the theory of quantum probability.

The C^*-algebra formalism can also be applied to classical systems. Consider the space $\mathcal{L}^1(\Omega)$ of all integrable functions from a probability space $\{\Omega, \mathcal{A}, P\}$ into the complex numbers. For $f \in \mathcal{L}^1(\Omega)$, define the adjoint operation point-wise by $f^*(\omega) = f(\omega)^*$. Then $\mathcal{L}^1(\Omega)$ is an Abelian C^*-algebra and the mapping

$$\begin{aligned} \tilde{\rho} : \mathcal{A} &\mapsto \mathbb{C} \\ A &\mapsto \int_\Omega A dP \end{aligned} \tag{20.51}$$

a state. Thus, Kolmogorov's probability theory can also be expressed by a C^*-algebra. The main difference is that the linear operators on a Hilbert space form in a non-commutative C^*-algebra whereas the integrable functions on a probability space form a commutative (or Abelian) C^*-algebra. This difference is also the key difference between classical and quantum propositional logic, discussed next.

20.7 Classical Versus Quantum Logic

The differences between classical and quantum physics come into sharp focus when the propositional logic of both theories is considered. Consider first classical mechanics.

Classical propositional logic. Starting point is an observable $f : X \mapsto \mathbb{R}$. The space of all observables $C(X, \mathbb{R})$ has an algebraic structure with

$$(\nu_1 f_1 + \nu_2 f_2) = \nu_1 f_1(x) + \nu_2 f_2(x) \tag{20.52}$$

$$(f_1 f_2)(x) = f_1(x) f_2(x) \tag{20.53}$$

An observable f has a value in $I \subset \mathbb{R}$ if

$$x \in f^{-1}(I) := \{y \in X | f(y) \in I\} \tag{20.54}$$

The proposition that the observable f has a value in I is thus associated with a subset $f^{-1}(I) \in X$ such that the proposition is true if $x \in f^{-1}(I)$ and false if $x \in X \setminus f^{-1}(I))$. One thus associates any logical proposition B with a subset $X_B \subset X$ such that B is true if $x \in X_B$ and B is false if $x \in X \setminus X_B$. A proposition B that is always true thus corresponds to $X_B = X$. A proposition that is always false corresponds to $X_B = \emptyset$. The logical operations of negation, conjunction, disjunction, and implication are then given by

$$X_{\neg B} = X \setminus X_B \tag{20.55}$$

$$X_{B \wedge C} = X_B \cap X_C \tag{20.56}$$

$$X_{B \vee C} = X_B \cup X_C \tag{20.57}$$

$$B \to C \equiv X_B \subset X_C \tag{20.58}$$

Instead of the subsets, X_B one can also introduce the corresponding indicator function χ_B. With both identifications, propositions satisfy the distributive rules

$$B \wedge (C \vee D) = (B \wedge C) \vee (B \wedge D) \tag{20.59}$$

$$B \vee (C \wedge D) = (B \vee C) \wedge (B \vee D) \tag{20.60}$$

which make the set of propositions a distributive Boolean algebra.

Quantum-mechanical propositional logic. If a quantum-mechanical system is in a pure state $|x\rangle$, then a measurement of an observable A will result in any of the eigenvalues $a^{(1)}, a^{(2)}, \ldots \in \mathbb{R}$. If $a^{(n_1)}, a^{(n_2)}, \ldots$ are the eigenvalues that fall into a subset $I \in \mathbb{R}$, then all states in the linear subspace $\mathcal{H}_I$ spanned by the corresponding eigenvectors $|a^{(n_1)}\rangle, |a^{(n_2)}\rangle, \ldots$ will result in an eigenvalue in I with probability one. Thus

$$P(A \in I || x\rangle) = 1 \tag{20.61}$$

if $|x\rangle \in \mathcal{H}_I$. If $|x\rangle \notin \mathcal{H}_I$ then $P(A \in I || x\rangle) < 1$. One thus identifies a proposition B with a linear subspace $\mathcal{H}_B$ such that the proposition B is true for all states $|x\rangle$ that lie in $\mathcal{H}_B$ and for which $P(B || x\rangle) = 1$ and false for all states that do not lie in $\mathcal{H}_B$ and for which $P(B || x\rangle) < 1$. One might have expected that a proposition B is false if $P(B || x\rangle) = 0$, but then there are states for which the proposition is neither true nor false. The negation of B is true for all states that lie in the orthogonal complement $(\mathcal{H}_B)^\perp$ and for which $P(B || x\rangle) = 0$. This definition has the following peculiar properties. The proposition B being false does not necessarily imply that

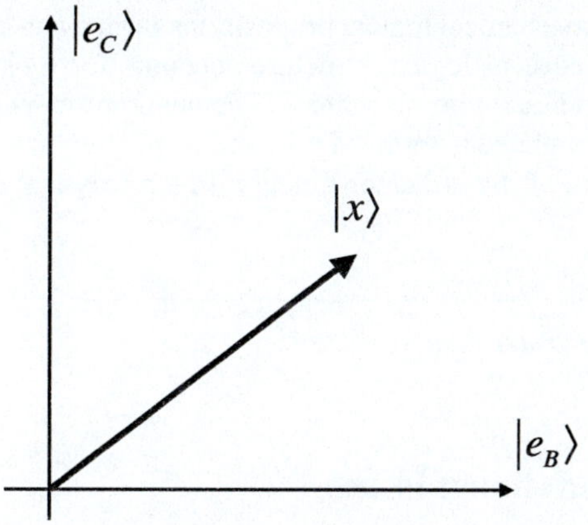

Fig. 20.1 A two-dimensional Hilbert space spanned by the vectors $|e_B\rangle$ and $|e_C\rangle$. The proposition B is true for all vectors $|x\rangle = c|e_B\rangle$ with $c \in \mathbb{C}$ and false for all other vectors. The negation $\neg B$ is true for all vectors $|x\rangle = c|e_C\rangle$ with $c \in \mathbb{C}$ and false for all other vectors. If $|x\rangle = c|e_B\rangle + c'|e_C\rangle$ with $c, c' \neq 0$ is a non-trivial linear combination, then both B and $\neg B$ are false. This is Schrödinger's cat. The proposition "the cat is alive" is false and the proposition "the cat is dead" is false

the proposition $\neg B$ is true. However, B being true implies $\neg B$ being false. Also, if $|x\rangle$ is a non-trivial linear combination of a vector from $\mathcal{H}_B$ and a vector from $(\mathcal{H}_B)^\perp$, then neither B nor $\neg B$ is true. A proposition B that is always true is given by the full space $\mathcal{H}$. A proposition B that is always false is given by $\mathcal{H}_B = \{|0\rangle\}$. These general properties are exemplified by the two-dimensional Hilbert space depicted in Fig. 20.1.

The logical connectives are given by

$$\mathcal{H}_{\neg B} = (\mathcal{H}_B)^\perp \tag{20.62}$$

$$\mathcal{H}_{B\wedge C} = \mathcal{H}_B \cap \mathcal{H}_C \tag{20.63}$$

$$\mathcal{H}_{B\vee C} = \mathcal{H}_B \oplus \mathcal{H}_C \tag{20.64}$$

$$B \to C \equiv \mathcal{H}_B \subset \mathcal{H}_C \tag{20.65}$$

Note that the disjunction is not $\mathcal{H}_B \cup \mathcal{H}_C$ because the union is not a linear subspace. Instead, the disjunction is the smallest linear subspace that contains both $\mathcal{H}_B$ and $\mathcal{H}_C$.

These logical connectives do not satisfy the distributive rules. A simple counterexample is the two-dimensional Hilbert space of Fig. 20.1. If a proposition D corresponds to a vector $|x\rangle$ that is a non-trivial linear combination of $|e_B\rangle$ and $|e_C\rangle$, then

$$\mathcal{H}_D \cap (\mathcal{H}_B \oplus \mathcal{H}_C) = \mathcal{H}_D \neq (\mathcal{H}_D \cap \mathcal{H}_B) \oplus (\mathcal{H}_D \cap \mathcal{H}_C) = |0\rangle \tag{20.66}$$

Thus, the quantum-mechanical logical propositions form a non-distributive Boolean algebra. This difference in logical structure accounts for most of the differences between classical and quantum mechanics. The quantum-mechanical propositional logic can also be formulated in terms of projection operators which are the quantum-mechanical analogue of the indicator function in the classical case. The important point is

Quantum logic derives from the fact that probability derives from the Pythagorean theorem, not from measure theory.

20.8 The Foundational Issues

Quantum mechanics is a theory with contentious foundations. While physicists generally agree on the pragmatic rules, there is wide disagreement on many conceptual issues. There are two main areas of disagreement, one is the interpretation of the intrinsic probability, the other concerns the state reduction in measurements.

Interpretation of the wave function. The discussion of the interpretation of the wave function mirrors to a large extent our discussion of the various interpretations of probability in Chap. 14, although it has its peculiar facets and jargon. There are two extreme positions which are called the realist and anti-realist views.

The realist claims that the individual system actually possesses values which are only revealed by measurement, as is assumed in classical physics. It is because of a lack of knowledge that we cannot predict the values. We have incomplete knowledge. There are hidden variables whose values we do not know or do not have access to. The intrinsic probability of quantum mechanics is thus an epistemic problem for the realist.

The extreme anti-realists claim that a single system does not possess any values. They claim that the basic entity of the theory is an ensemble of systems. Quantum theory predicts the probabilistic spread of measurements performed on this ensemble.

Between these two extremes, there are many other interpretations which differ on

- the existence of properties,
- the status of individual systems,
- our knowledge of properties,
- the role of individual and repeated measurements, and
- the meaning of probability.

With regard to the existence of properties, the positions range from the belief

- that an individual system has definite properties, some of them just hidden from us,
- that their values are intrinsically not sharp,
- that the system has propensities; the values are latent and come into being by the measurement, and
- that properties do not exist at all.

Measurements and state reduction. In classical physics, the measurement process is not regarded as a problem. Though one realizes that every measurement of the system distorts the system, it is assumed that in appropriate limits this distortion can be assumed to be negligible. The situation is different in quantum mechanics. There a measurement assumes the following steps:

(1) A system is assumed to be in the state $|x\rangle$, in a pure state for the sake of argument.
(2) When a measurement of an observable A is taken, then the wave function collapses into the state

$$\rho = \sum_{n=1}^{N} \langle x|a^{(n)}\rangle \langle a^{(n)}|a^{(n)}\rangle |a^{(n)}\rangle \langle a^{(n)}| \tag{20.67}$$

(3) When the result of the measurement is known, then the wave function further collapses into

$$\rho' = |a^{(n)}\rangle \langle a^{(n)}| \tag{20.68}$$

where the first reduction reflects the fact that one knows the probabilities that a measurement will result in any of the possible values, and the second reduction the fact that one knows which value was measured. A measurement irretrievably modifies the system. There is also the opposite act of state preparation. State preparation forces the system into a specific state after the preparation. Both these operations, measurement and state preparation, are not governed by the Schrödinger equation. They also involve interaction between the (microscopic) quantum-mechanical system and a macroscopic measuring/preparation device.

QBism. Quantum mechanics may be approached with a Bayesian mindset, which is the approach taken in QBism, where the 'Q' stands for quantum and 'B' for Bayesian. QBism makes the following assertions:

(1) The wave function $|x\rangle$ does not characterize the system; it summarizes our knowledge of the system. It does not state what the system is but what we know about the system. It is the map, not the territory.
(2) In Bayesian inference, the wave function acts as the prior. When new information becomes available, as is the case when one learns that a measurement has been taken, the prior needs to be updated, as in the first reduction above.
(3) When we additionally learn the value of the measurement, another updating is required, as in the second reduction above.

This Bayesian approach softens many of the quantum paradoxes. Take Schrödinger's cat. It is in a box and will die the moment a radioactive nucleus decays. As long as the box is closed, the cat seems to be in limbo, being neither dead nor alive. QBism does not ask what state the cat is in but only what we know about its state, which is that it is more likely to be dead the longer we wait, a statement which can be quantified by a trajectory in the Bloch sphere governed by the Schrödinger equation. Once the box is opened, we will know whether the cat is dead or alive. In QBism, the knowledge is also personal; different investigators may have different knowledge and hence ascribe different wave functions to the system.

20.9 Summary

Quantum mechanics distinguishes itself in many respects. The two aspects relevant here are

- When considered as a dynamical system, quantum-mechanical systems are linear dynamical systems. Their time evolution is governed by a unitary operator on the state space, which is a Hilbert space.
- Quantum mechanics has an intrinsic probability structure, absent in classical mechanics. The theory only predicts the probability of a measurement result. This intrinsic probability has its roots in the linear structure of quantum mechanics and takes the form of a generalized Pythagorean theorem rather than a normalized measure.

Quantum mechanics also allows for epistemic uncertainty, which is represented by a density matrix and is handled in the same way as in classical systems. Furthermore, quantum mechanics also allows for entangled states that are absent from classical mechanics.

In terms of propositional logic, quantum logic has the peculiar property that both a proposition and its negation can be false. This peculiar logical structure accounts for most of the counter-intuitive results of quantum mechanics, since our intuition is honed by classical propositional logic.

More on the foundations of quantum mechanics can be found in Isham (1995) and the dynamical systems view in Alicki and Fannes (2001).

References

Alicki, R., Fannes, M.: Quantum dynamical systems. Oxford University Press (2001)
Isham, C.J.: Lectures on Quantum Theory. Mathematical and Structural Foundations. Imperial College Press, London (1995)

Appendix A
Classical Mechanics

Classical mechanics considers the motion of point particles and rigid bodies in physical space under the action of forces. The concepts of point particles and rigid bodies are idealizations. Classical mechanics distinguishes itself from relativistic dynamics in that it models physical space as a three-dimensional Euclidean point space and time as a one-dimensional Euclidean point space. Here, we give a brief overview to provide the context for the examples used in the main text.

There are two general frameworks to formulate the dynamical laws: Newtonian and Hamiltonian dynamics.

A.1 Newtonian dynamics

The Newtonian formulation of the dynamics of a point particle is given by Newton's second law

$$m\ddot{x} = F(x, \dot{x}, t) \tag{A.1}$$

where $x = (x_1, x_2, x_3)^T$ is the particle position vector in physical space, $\dot{x} = (\dot{x}_1, \dot{x}_2, \dot{x}_3)^T$ the particle velocity vector, $\ddot{x} = (\ddot{x}_1, \ddot{x}_2, \ddot{x}_3)^T$ the acceleration vector, m the mass of the point particle, and F the force acting on the particle. The force may depend on the position, velocity, and explicitly on time. Newton's law states that the mass times the acceleration is given by the force exerted on the particle. Newton's law is a set of three second-order (in time) ODEs. This set of three second-order ODEs can be cast into a set of six first-order ODEs by the substitution $v = \dot{x}$ which yields

$$\dot{x} = v \tag{A.2}$$

$$\dot{v} = F(x, v, t) \tag{A.3}$$

P. Müller, *Handbook of Dynamics and Probability*,
https://doi.org/10.1007/978-3-030-88486-4

Given the initial position and initial velocity of the particle, Newton's law then determines the position and velocity of the particle at any other time. Hence, position and velocity comprise the state of the system and Newton's law determines the evolution of the state in time.

Newton's approach is used for macroscopic phenomena and can incorporate frictional effects. Generalization to many particles is straightforward. The vectors x, $\dot{x}$, $\ddot{x}$, and F then become 3N-dimensional vectors.

Instead of Cartesian coordinates, one may also introduce generalized coordinates, generally denoted by q, and Newton's law becomes $\ddot{q} = F(q, \dot{q}, t)$. Such generalized coordinates are useful if the motion is constrained. One example of such constrained motion is the pendulum.

Example: Pendulum. The equations of motion of a one-dimensional pendulum are given by

$$m\ddot{\theta} = -\frac{mg_0}{r_0}\sin\theta - \nu_0\dot{\theta} \tag{A.4}$$

where the generalized coordinate θ $(= q)$ is the angle that the pendulum makes with the vertical, g_0 the gravitational acceleration, r_0 the length of pendulum, and ν_0 the friction coefficient. The first term on the right-hand side describes the gravitational force and the second term a linear frictional force. If the gravitational restoring force is linearized ($\sin\theta \to \theta$) and friction is neglected, one obtains a harmonic oscillator equation

$$\ddot{\theta} = -\frac{g_0}{r_0}\theta \tag{A.5}$$

with frequency $\omega = \sqrt{g_0/r_0}$. If the gravitational force is neglected, one obtains the rotator

$$m\ddot{\theta} = -\nu_0\dot{\theta} \tag{A.6}$$

The limit that the (linear) gravitational restoring force is balanced by the frictional force is called over-damped oscillator and given by

$$\dot{\theta}(t) = -\gamma\theta(t) \tag{A.7}$$

with $\gamma := mg_0/r_0\nu_0$.

A.2 Hamilton's principle

Consider a point particle of mass m, position $q(t)$, and velocity $\dot{q}(t)$. If the forces acting on the particle are given by the gradients of its potential energy $U(q(t), t)$, then the Newtonian equations of motion take the form

$$m\ddot{q} = -\nabla_q U \tag{A.8}$$

Hamilton's principle asserts that these equations of motion can be derived from a variational principle. This assertion is true for many physical systems. Hamilton's principle is also called the principle of least action or the principle of stationary action. To formulate Hamilton's principle, one introduces the Lagrangian

$$L(q, \dot{q}, t) := \frac{1}{2} m \dot{q}^2 - U(q, t) \tag{A.9}$$

which is the difference between the kinetic and potential energies of the particle. Newton's equations (A.8) are then given by

$$\frac{d}{dt}\left(\frac{\partial L}{\partial \dot{q}}\right) - \frac{\partial L}{\partial q} = 0 \tag{A.10}$$

This equation results from the variational problem that states that the action integral

$$S := \int_{t_1}^{t_2} dt L(q(t), \dot{q}(t), t) \tag{A.11}$$

is an extreme for the actual motion between $q(t_1)$ and $q(t_2)$. Thus, $\delta S = 0$ for variations $\delta q(t)$ that vanish at the end points, $\delta q(t_1) = \delta q(t_2) = 0$. The proof is a standard exercise in classical mechanics and goes as follows:

$$\begin{aligned} \delta S &= \int_{t_1}^{t_2} dt \left(\frac{\partial L}{\partial q}\delta q + \frac{\partial L}{\partial \dot{q}}\delta \dot{q}\right) \\ &= \int_{t_1}^{t_2} dt \left(\frac{\partial L}{\partial q}\delta q + \frac{\partial L}{\partial \dot{q}}\frac{d}{dt}\delta q\right) \\ &= \int_{t_1}^{t_2} dt \left[\frac{\partial L}{\partial q} - \frac{d}{dt}\left(\frac{\partial L}{\partial \dot{q}}\right)\right]\delta q + \int_{t_1}^{t_2} dt \frac{d}{dt}\left(\frac{\partial L}{\partial \dot{q}}\delta q\right) \end{aligned} \tag{A.12}$$

upon integration by parts. The last term vanishes because there is no variation at the end points. Hence, the extreme is obtained when $q(t)$ satisfies (A.10) which are called the Lagrange–Euler equations of the variational principle. The variational principle can also be stated as $\delta S/\delta q(t) = 0$ if one introduces the functional derivative (see Appendix E.16).

One must distinguish between the Lagrangian L as a function of the variables $(q, \dot{q}, t)$ which is denoted by $L(q, \dot{q}, t)$) and the Lagrangian composed with the actual motion which is $L(q(t), \dot{q}(t), t)$). The derivatives with respect to q and $\dot{q}$ refer to the former, the derivative d/dt to the latter.

A.3 Hamiltonian Dynamics

The variational principle carries easily over to a formulation of the equations of motion where one introduces the momentum

$$p := \partial L/\partial \dot{q} \tag{A.13}$$

as a dependent variable. Using q and p as the dependent variables rather than q and $\dot{q}$ is the Hamiltonian formulation of the dynamics. One introduces the Hamilton function

$$H(q, p; t) := p\dot{q} - L(q, \dot{q}; t) \tag{A.14}$$

where $\dot{q}$ on the right-hand side needs to be expressed in terms of q and p. The Hamilton function is the energy of the system expressed in terms of the variables (q, p, t). The action then becomes

$$S = \int dt L = \int dt (p\dot{q} - H(p, q, t)) \tag{A.15}$$

Independent variation with respect to p and q then leads to the Lagrange–Euler equation

$$\frac{d}{dt}\left(\frac{\partial L}{\partial \dot{p}}\right) - \frac{\partial L}{\partial p} = 0 \tag{A.16}$$

$$\frac{d}{dt}\left(\frac{\partial L}{\partial \dot{q}}\right) - \frac{\partial L}{\partial q} = 0 \tag{A.17}$$

which imply Hamilton's equations

$$\dot{q} = \frac{\partial H}{\partial p} \tag{A.18}$$

$$\dot{p} = -\frac{\partial H}{\partial q} \tag{A.19}$$

which are also called the canonical equations of motion. These results also hold for $q = (q_1, \ldots, q_N)$ and $p = (p_1, \ldots, p_N)$.

The above form of Hamilton's equations is standard, but it must be realized that one has first to differentiate the Hamilton function to obtain the generalized velocity field

$$u_q(q, p, t) = \frac{\partial H(q, p, t)}{\partial p}$$
$$u_p(q, p, t) = -\frac{\partial H(q, p, t)}{\partial q} \tag{A.20}$$

into which one then has to substitute the functions $q(t)$ and $p(t)$ which results in the equations of motion

$$\begin{aligned}\dot{q}(t) &= u_q(q(t), p(t), t)\\ \dot{p}(t) &= u_p(q(t), p(t), t)\end{aligned} \tag{A.21}$$

An important property of Hamiltonian systems is that their flow in phase space is non-divergent

$$\nabla_q^T u_q + \nabla_p^T u_p = 0 \tag{A.22}$$

Hamilton's formulation is a considerable simplification of the dynamics since all relevant dynamical information is contained in one single scalar function, the Hamilton function $H(q, p, t)$. The peculiar structure of Hamilton's equations is exploited by introducing Poisson brackets and associated symplectic geometry.

Often, the Hamilton function has the form

$$H(q, p) = \frac{1}{2m}p^2 + U(q) \tag{A.23}$$

where the first term is the kinetic energy and U the potential energy that depends only on the generalized coordinates q and not on the momenta p. All our examples, the harmonic oscillator, the FPU system, and the N-particle system, are of this form.

A.4 Harmonic Oscillator

A particular simple example is the Hamilton function

$$H(q, p) = \frac{1}{2m}p^2 + \frac{1}{2}D^2q^2 \tag{A.24}$$

which represents a harmonic oscillator with mass m and spring constant D. Hamilton's equations are given by

$$\dot{q} = \frac{p}{m} \quad , \quad \dot{p} = -Dq \tag{A.25}$$

The frequency of the harmonic oscillator is $\omega = \sqrt{D/m}$. In this representation, the harmonic oscillator is characterized by two parameters, the mass m and the spring constant D. The harmonic oscillator is also obtained as the small excursion limit of the frictionless pendulum. Then the frequency is given by $\omega = \sqrt{g_0/r_0}$ where g_0 is the gravitational acceleration and r_0 the length of the pendulum. These parameters are very important for physicists. For example, physicists stress the fact that the frequency of the pendulum is independent of the mass.

In a dynamical system context, one often introduces other representations of the harmonic oscillator. Among them are as follows:

(1) In terms of the scaled variables

$$\tilde{q} := q\sqrt{m} \;\; \tilde{p} := \frac{p}{\sqrt{m}} \tag{A.26}$$

the Hamilton function takes the form

$$H(\tilde{q}, \tilde{p}) = \frac{1}{2}\tilde{p}^2 + \frac{1}{2}\omega^2\tilde{q}^2 \tag{A.27}$$

leaving the frequency as the only parameter characterizing the harmonic oscillator. Mass is eliminated though it determines the relation between the momentum p and the velocity $\dot{q}$. If only one harmonic oscillator is considered, then one might additionally scale time such that $\omega = 1$.

(2) A further scaling $\hat{q} := \sqrt{\omega}\tilde{q}$ and $\hat{p} := \tilde{p}/\sqrt{\omega}$ results in the Hamilton function

$$H(\hat{q}, \hat{p}) = \frac{1}{2}\omega\hat{p}^2 + \frac{1}{2}\omega\hat{q}^2 \tag{A.28}$$

(3) Polar coordinates are mimicked by introducing action $J \geq 0$ and angle θ variables

$$\tilde{q}q = \sqrt{\frac{2J}{\omega}}\sin\theta \tag{A.29}$$

$$\tilde{p} = \sqrt{2\omega J}\cos\theta \tag{A.30}$$

with inverse transformation

$$J = \frac{1}{\omega}[\frac{1}{2}\tilde{p}^2 + \frac{1}{2}\omega^2\tilde{q}^2] \tag{A.31}$$

$$\theta = \arctan\frac{\omega\tilde{q}}{p} \tag{A.32}$$

leading to the Hamilton function

$$H(\theta, J) = \omega J \tag{A.33}$$

(4) The two variables q and p are combined in the single complex amplitude

$$a := \frac{1}{\sqrt{2}}(\sqrt{m\omega}q + i\frac{p}{\sqrt{m\omega}}) \tag{A.34}$$

leading to the Hamilton function

$$H(a, a^*) = \omega a a^* \tag{A.35}$$

and Hamilton's equations

$$\begin{aligned} \dot{a} &= -i\frac{\partial H(a, a^*)}{\partial a^*} = -i\omega a \\ \dot{a}^* &= i\frac{\partial H(a, a^*)}{\partial a} = i\omega a^* \end{aligned} \tag{A.36}$$

In this representation, the solution is just given by $a(t, a_0) = a_0 e^{-i\omega t}$.

These scalings and transformations are introduced when the emphasis changes. Physics is more interested in how the properties of a physical system depend on the parameters of the system. Dynamical systems theory is more interested in the underlying mathematical structure of the governing equation independent of the concrete context.

Appendix B
Thermostatics

Thermodynamics is a phenomenological theory of matter in its macroscopic state. It ignores the discrete molecular structure of matter and regards matter as a continuum. Equilibrium thermodynamics studies systems in thermodynamic equilibrium. It is also called thermostatics. It does not say anything about the approach to equilibrium. Here, we briefly review some of the basic elements of equilibrium thermodynamics.

B.1 Thermodynamic Variables and Processes

An isolated system is one that cannot exchange mass, momentum, or energy with its surroundings. Such isolated systems have the tendency to move toward a state of thermal equilibrium. In the state of thermodynamic equilibrium, the system is characterized by a relatively small set of thermodynamic variables, such as its mass, its volume, its internal energy, its entropy, its chemical potential, its pressure, and its temperature. The entropy in thermostatics is a purely phenomenological quantity. It is neither time-dependent, nor microscopic, nor probabilistic. Specific values of the thermodynamic variables define the state of the system. The state is a point in the state space spanned by the thermodynamic variables. Mass, volume, and pressure have a mechanical interpretation whereas internal energy, entropy, chemical potential, and temperature are variables germane to thermodynamics. It is emphasized that some of the thermodynamic variables are only defined in thermodynamic equilibrium. A system not in thermodynamic equilibrium may not have a temperature.

There are also important quantities in thermodynamics which are not thermodynamic variables, notably heat and work. They play an important role in the first law of thermodynamics.

If the system under consideration is not isolated but interacts with its surroundings, then changes in the surrounding systems lead to changes in the system under consideration. Such changes are called *thermodynamic processes*. If these changes are

P. Müller, *Handbook of Dynamics and Probability*,
https://doi.org/10.1007/978-3-030-88486-4

so slow that the system under consideration always stays in thermodynamic equilibrium, then these changes are called *equilibrium processes* or *quasi-static processes*. They can be represented by a line in the state space spanned by the thermodynamic variables. Processes are called *reversible* if the system returns to its original state when the surrounding systems reverse to their original state. All other processes are called *irreversible*. The slow isothermal expansion of a gas in a cylinder by a slowly moving piston is reversible, whereas the free expansion into a vacuum by removing a partition is irreversible. All reversible processes are quasi-static processes, but there are quasi-static processes which are not reversible. The expansion of a gas into a vacuum by infinitesimal amounts is a quasi-static but irreversible process.

Thermodynamic processes are called adiabatic when the system is thermally isolated (i.e., it does not exchange heat with its surroundings), isothermal when the system is immersed in a 'heat bath' such that the temperature remains constant, and isentropic when the entropy remains constant.

Thermodynamic processes are subject to the first and second laws of thermodynamics.

B.2 First Law of Thermodynamics

Let δW_{mech} be the infinitesimal amount of *mechanical work* done on the system, δW_{chem} the infinitesimal amount of chemical work, and δQ the infinitesimal amount of *heat* received by the system, then the *first law of thermodynamics* states that the internal energy U changes by an infinitesimal amount

$$dU = \delta Q + \delta W_{mech} + \delta W_{chem} \tag{B.1}$$

The first law states that the internal energy is an energy reservoir that can be filled or depleted by exchanging energy with surrounding systems in two forms, in the form of heat which is energy in an 'unorganized' form, and mechanical and 'chemical' work which is energy in an 'organized' form. The name chemical work originates from systems with chemical reactions among different components. Here, it just means the work required to add mass to the system. If Q, W_{mech} and W_{chem} are defined and one knows how to measure them, then the first law implies the existence of a thermodynamic variable, namely the internal energy U, and a way to measure it, up to a constant. Note that Q, W_{mech}, and W_{chem} are not thermodynamic variables. Their values depend on the way a certain thermodynamic state is reached. Especially, $\oint \delta Q$ and $\oint \delta W$ do not vanish for *cyclic processes*, i.e., for processes that return to their starting state. To account for this fact, the infinitesimal changes of these quantities are denoted by the symbol δ. The symbol 'd' on the other hand is reserved for thermodynamic variables, i.e., for variables whose value depends only on the thermodynamic state, and not on the way the system reaches this state. The internal energy is such a thermodynamic variable and $\oint dU = 0$.

One also distinguishes between changes that are caused by interaction with surrounding external systems and changes caused by internal interactions. The externally caused changes are denoted by a subscript 'e', for external, and the internally caused changes by a subscript 'i', for internal. Writing

$$dU = d_e U + d_i U \tag{B.2}$$

the first law of thermodynamics then states $d_e U = \delta Q + \delta W_{mech} + \delta W_{chem}$ and

$$d_i U = 0 \tag{B.3}$$

There are no internal processes that can change the internal energy. This form results in the popular interpretation of the first law as stating that the energy of a system is conserved in the sense that it can only be changed by interaction with other systems.

Though $\oint \delta Q$ does not necessarily vanish for a cyclic process, it is found that $\oint \delta Q/T = 0$ for all cyclic reversible processes where T is the temperature. This finding implies the existence of a thermodynamic variable S, called the thermostatic (or Clausius) entropy, which is obtained by integrating

$$d_e S = \frac{\delta Q}{T} \tag{B.4}$$

along reversible paths. The entropy is also only defined up to an additive constant.

B.3 Second Law of Thermodynamics

Not all processes that conserve energy occur in nature. The second law of thermodynamics states that only processes occur for which

$$\oint \frac{\delta Q}{T} \leq 0 \tag{B.5}$$

The equality sign holds if and only if the process is reversible. Equivalent formulations are

$$dS \geq \frac{\delta Q}{T} \tag{B.6}$$

or

$$d_i S \geq 0 \tag{B.7}$$

if one writes $dS = d_e S + d_i S$ with $d_e S = \delta Q/T$. The entropy changes through heat exchange with its surroundings, the term $d_e S$, and through internal processes, the term $d_i S$. The internal processes never decrease the entropy. For reversible processes $d_i S = 0$, and for irreversible processes $d_i S > 0$. Hence, reversible adiabatic processes are isentropic and the entropy of thermally isolated systems never decreases. To determine whether or not a quasi-static process is reversible, one has to compare dS and $\delta Q/T$. If these two increments are equal, the process is reversible; otherwise, it is irreversible and dS will be larger than $\delta Q/T$.

Stronger forms of the second law are formulated for isolated systems. The weakest form is the law stated above, $S(t) \geq S(0)$ for all $t > 0$. The stronger forms are

$$\lim_{t\to\infty} S(t) = S^* \tag{B.8}$$

where S^* is a stationary state, possibly metastable and not necessarily unique. The strongest version of the second law is

$$\lim_{t\to\infty} S(t) = S^{max} \tag{B.9}$$

where S^{max} is the unique stationary state of maximum entropy. These stronger versions imply the approach to equilibrium.

B.4 Thermodynamic Potentials

Systems in thermodynamic equilibrium obey Gibbs' phase rule which states: a system consisting of N components and M phases is completely described by the specification of $N - M + 3$ thermodynamic variables. There exists a function of these $N - M + 3$ variables, called a thermodynamic potential, which determines all thermodynamic variables of interest. Gibbs' rule implies for a one-component, one-phase system that the state of the system is determined by the specification of three variables. The entropy S, volume V, and mass M may be chosen as these three independent variables. The thermodynamic potential can then be constructed in the following way: For reversible processes, the mechanical and chemical work and the heat are given by

$$\delta W_{mech} = -p\,dV \tag{B.10}$$

$$\delta W_{chem} = \mu\,dM \tag{B.11}$$

$$\delta Q = T dS \tag{B.12}$$

where p is the pressure and μ the chemical potential. The first law thus takes the form of a complete differential

$$dU = TdS - pdV + \mu dM \tag{B.13}$$

assuring the existence of an internal energy function $U(S, V, M)$ with

$$T = \left(\frac{\partial U}{\partial S}\right)_V \tag{B.14}$$

$$p = -\left(\frac{\partial U}{\partial V}\right)_S \tag{B.15}$$

$$\mu = \left(\frac{\partial U}{\partial M}\right)_{S,V} \tag{B.16}$$

provided all the points in the thermodynamic state space can be reached by a reversible process. The internal energy as a function of entropy, volume and mass constitutes a thermodynamic potential.

Instead of the independent variables (S, V, M), one can choose other sets of independent variables. This leads to alternative *thermodynamic representations*. The most common alternative representations are obtained by a Legendre transformation which is a mapping $f(x) \mapsto g(y)$ such that $y = \partial f/\partial x$ and $x = \partial g/\partial y$. Hence $g(y) = yx(y) - f(x(y))$. Thus, if one would like to introduce the pressure $p = -\partial U/\partial V$ instead of V as the independent variable, one finds as the thermodynamic potential the enthalpy

$$H(S, p, M) = U + pV \tag{B.17}$$

with $V = \partial H/\partial p$, $T = \partial H/\partial S$, and $\mu = \partial H/\partial M$. Similarly, one finds the free energy

$$F(T, V, M) := U - TS \tag{B.18}$$

with $S = -\partial F/\partial T$, $p = -\partial F/\partial V$, and $\mu = \partial H/\partial M$ and the free enthalpy

$$G(p, T, M) = U + pV - TS \tag{B.19}$$

with $V = \partial G/\partial p$, $S = \partial G/\partial T$, and $\mu = \partial H/\partial M$. This Legendre formalism easily extends to multi-component and multi-phase systems.

The energy function $U(S, V, M)$ is a continuous, differentiable monotonically increasing function of the entropy S. Thus, $U(S_2, V, M) \geq U(S_1, V, M)$ if $S_2 \geq S_1$. The energy function can thus be inverted to give the entropy function $S(U, V, M)$. Its derivatives are $\partial S/\partial U = 1/T$, $\partial S/\partial V = p/T$, and $\partial S/\partial M = -\mu/T$. The entropy function has the following important properties: When two systems '1' and '2' are brought into contact with each other, then we find for the total system

$$S(U_1 + U_2, V_1 + V_2, M_1 + M_2) \geq S(U_1, V_1, M_1) + S(U_2, V_2, M_2) \tag{B.20}$$

This property called super-additivity is a consequence of the second law.

B.5 Thermodynamic Limit

The thermodynamic variables U, M, V, and S are supposed to be *extensive* quantities. They are supposed to scale with the size of the system. They are supposed to be additive in thermodynamic equilibrium. If the system is divided into two subsystems having the internal energies U_1 and U_2, then the total system has the internal energy $U = U_1 + U_2$. This notion of extensiveness requires the neglect of boundary effects. The thermodynamic variables p, T, and μ, on the other hand, are *intensive* quantities. In thermodynamic equilibrium, their values are the same in any subsystem. This distinction is not always as clear as it may sound as the notion of partial pressures indicates.

To define the concept of extensiveness more formally, one introduces the thermodynamic limit. One assumes that the specific quantities $u := U/M$, $\eta := S/M$ and $v := V/M$ exist in the limit $M, U, S, V \mapsto \infty$. In this thermodynamic limit, the internal energy function must be a homogeneous function of the first order

$$U(\lambda S, \lambda V, \lambda M) = \lambda U(S, V, M) \tag{B.21}$$

Generally, a function $f(x, y)$ that satisfies the identity $f(tx, ty) = t^n f(x, y)$ for a fixed n is called a homogeneous function of degree n. Euler's theorem states that homogeneous functions of order n satisfy $x\partial f/\partial x + y\partial f/\partial y = nf(x, y)$. One thus finds

$$U = S\frac{\partial U}{\partial S} + V\frac{\partial U}{\partial V} + M\frac{\partial U}{\partial M} \tag{B.22}$$

or

$$U = TS - pV + \mu M \tag{B.23}$$

which is an identity between the thermodynamic variables, called *Euler's identity*. Note that the products in this identity are the product of an intensive and an extensive quantity.

Similarly, the entropy is an extensive quantity

$$S(\lambda E, \lambda V, \lambda M) = \lambda S(E, V, M) \tag{B.24}$$

in the thermodynamic limit.

For extensive quantities, one introduces *specific* variables which represent the amount per unit mass and *densities* which represent the amount per unit volume. The specific quantities and densities are intensive quantities.

Densities. To convert to a description in terms of densities, one introduces the mass density

$$\rho := \lim_{M,V\to\infty} \frac{M}{V} \tag{B.25}$$

and the entropy density

$$\tilde{\eta} := \lim_{S,V\to\infty} \frac{S}{V} \tag{B.26}$$

The internal energy density can then be expressed as

$$\tilde{e} := \lim_{E,V\to\infty} \frac{E}{V} = \tilde{e}(\tilde{\eta}, \rho) \tag{B.27}$$

with

$$T = \frac{\partial \tilde{e}}{\partial \tilde{\eta}} \tag{B.28}$$

$$\mu = \frac{\partial \tilde{e}}{\partial \rho} \tag{B.29}$$

The pressure p follows from Euler's identity and is given by

$$p = T\tilde{\eta} - \tilde{e} + \mu\rho \tag{B.30}$$

The intensive state of a system consisting of N components and M phases is thus determined by $N - M + 2$ variables. This is Gibbs' phase rule for intensive states.

Specific quantities. To convert to a description in terms of specific quantities, one introduces the specific volume

$$v := \lim_{V,M\to\infty} \frac{V}{M} \tag{B.31}$$

and the specific entropy

$$\eta := \lim_{S,M\to\infty} \frac{S}{M} \tag{B.32}$$

The specific internal energy function is then

$$e := \lim_{E,M\to\infty} \frac{E}{M} = e(v, \eta) \tag{B.33}$$

with

$$p = -\frac{\partial e}{\partial v} \tag{B.34}$$

$$T = \frac{\partial e}{\partial \eta} \tag{B.35}$$

The chemical potential μ follows from Euler's identity as

$$\mu = e - T\eta + pv \tag{B.36}$$

Appendix C
Fluid Dynamics

Fluid dynamics is a phenomenological theory of the macroscopic behavior of liquids and gases. Here, we

(1) list the key ingredients of irreversible fluid dynamics,
(2) discuss the Navier–Stokes equations which are often taken to be synonymous with fluid dynamics but describe a fairly restricted class of fluid motions,
(3) elaborate on the distinction between the Eulerian and Lagrangian descriptions in fluid dynamics, and
(4) point out that the equations of perfect fluids, those fluids that do not exhibit irreversible behavior, can straightforwardly be derived by applying the continuum limit to Hamilton's principle of the microscopic dynamics.

C.1 Irreversible Fluid Dynamics

Irreversible fluid dynamics has three key ingredients: local thermodynamic equilibrium, balance equations for mass, momentum, and energy, and phenomenological flux laws.

Local thermodynamic equilibrium. Fluid dynamical systems are not in thermodynamic equilibrium but are assumed to be in *local* thermodynamic equilibrium. Local thermodynamic equilibrium assumes that the fluid can be divided into parcels that are sufficiently small from a macroscopic point of view to be labeled as points but are still sufficiently large from a microscopic point of view to contain a large enough number of molecules for thermodynamics to apply. If, in addition, one assumes that a parcel, when disturbed, reaches its thermodynamic equilibrium so rapidly that the adjustment appears to be instantaneous from a macroscopic point of view, then the parcels can be regarded as being in thermodynamic equilibrium at all points x and all times t.

P. Müller, *Handbook of Dynamics and Probability*,
https://doi.org/10.1007/978-3-030-88486-4

The properties of a system in thermodynamic equilibrium are completely encoded in a thermodynamic potential. For a one-component, one-phase system, the internal energy function $e = e(\eta, v)$ which expresses the specific internal energy e in terms of the specific entropy η and the specific volume v is such a thermodynamic potential. In fluid dynamics, it is more convenient to use the specific entropy function

$$\eta = \eta(e, v) \tag{C.1}$$

with the thermodynamic relations

$$\begin{aligned} \frac{1}{T} &= \frac{\partial \eta}{\partial e} \\ \frac{p}{T} &= \frac{\partial \eta}{\partial v} \end{aligned} \tag{C.2}$$

for the temperature T and the pressure p.

Since only local thermodynamic equilibrium is assumed, the specific energy $e(x, t)$ and the specific volume $v(x, t)$ are (continuous) functions of position and time, and so will be the specific entropy and the temperature and the pressure. In addition, one allows the fluid parcels to move with a macroscopic velocity $u(x, t)$ and to be exposed to a prescribed external potential $\Phi_{ext}(x, t)$. These variables must be included in the description of the system. The important point is, however, that the thermodynamic state is not affected by a (uniform) motion of the fluid parcel and by its exposure to an external potential. The specific entropy function does not depend on these external state variables. The above thermodynamic specification remains valid at each point in space and time. The thermodynamic state of the system is still described by the specific entropy function (C.1).

What remains to be determined are the specific volume field $v(x, t)$, the velocity field $u(x, t)$, and the specific internal energy field $e(x, t)$.

Balance equations. The unknown fields $v(x, t)$, $u(x, t)$, and $e(x, t)$ are determined by the balance equations for mass, momentum, and energy

$$\partial_t \rho + \nabla^T(\rho u) = 0 \tag{C.3}$$

$$\partial_t(\rho u^T) + \nabla^T(\rho u u^T - \sigma) = -\rho \nabla^T \Phi_{ext} \tag{C.4}$$

$$\partial_t(\rho e) + \nabla^T(\rho e u + q) = tr(\sigma^T(\nabla u^T)) \tag{C.5}$$

where $\rho = v^{-1}$ is the mass density, σ the (symmetric) stress tensor, and q the heat flux vector. A phenomenological derivation of these balance equations can be found in textbooks of fluid dynamics, e.g., Müller (2006).

These balance equations are the balance equations in their flux form. They are formulated in terms of densities which are quantities per unit volume: the mass density ρ, the momentum density ρu, and the internal energy density ρe. The balance equations in their flux form state that the time rate of change of the density plus the divergence of a flux term equals a source term. The mass balance does not have such a source term. The source term in the momentum equation describes the effect of external forces. The internal forces are represented by the stress tensor. It is a major tenet of fluid dynamics that internal forces can indeed be represented by the divergence of a stress tensor. The source term in the internal energy equation describes the conversion of kinetic to internal energy. When the balance equations in their flux form are integrated over the fluid volume, one obtains the balance equations for the total mass, total momentum, and total internal energy.

The balance can alternatively be written in the advective form

$$\frac{D}{Dt}v = v\nabla_x^T u \tag{C.6}$$

$$\frac{D}{Dt}u^T = v\nabla_x^T \sigma - \nabla_x^T \Phi_{ext} \tag{C.7}$$

$$\frac{D}{Dt}e = -v\nabla_x^T q + vtr(\sigma^T(\nabla u^T)) \tag{C.8}$$

where

$$\frac{D}{Dt} := \partial_t + u^T \nabla_x \tag{C.9}$$

is the advective derivative. The advective form describes changes along fluid trajectories. The dependent variables are now specific quantities, quantities per unit mass.

Phenomenological flux laws. The balance equations contain the unknown stress tensor σ and heat flux q. They consist of reversible and irreversible (or molecular) contributions, $\sigma = \sigma^{rev} + \sigma^{mol}$ and $q = q^{rev} + q^{mol}$. There is no reversible heat flux, $q^{rev} = 0$. A basic linear theory assumes these irreversible fluxes to be proportional to the gradients of the velocity and temperature field. Assuming an isotropic fluid, one arrives at the specific flux laws or constitutive relations for the stress tensor and heat flux

$$\sigma^{rev} = -pI \tag{C.10}$$

$$\sigma^{mol} = 2\rho\nu S + 3\rho\nu' N \tag{C.11}$$

$$q^{mol} = -\rho c_p \lambda \nabla T \tag{C.12}$$

where p is the thermostatic pressure, T the temperature, ν the kinematic shear viscosity, ν' the kinematic expansion viscosity, $c_p = T\partial\eta(p, T)/\partial p$ the specific heat at constant pressure, λ the thermal conduction coefficient, and where

$$S := D - \frac{1}{3}tr(D)I$$

$$N := \frac{1}{3}tr(D)I \tag{C.13}$$

are the shear and normal components of the rate of deformation tensor $D := [\nabla u^T + (\nabla u^T)^T]/2$. It is only the symmetric part of the velocity gradient tensor ∇u^T that matters. In the anti-symmetric part, the vorticity tensor does not induce stresses since it represents a uniform rotation of an infinitesimal fluid element. The flux laws can be viewed as the first term in an expansion about the state of thermal equilibrium. The thermostatic pressure p, the temperature T, and the specific heat c_p can be derived from the specific entropy function $\eta(e, v)$. The coefficients ν, ν' and λ are the new coefficients that need to be determined phenomenologically. They must also be assumed to depend on η and v, though in practice one determines them as functions of the pressure p and the temperature T. If the molecular (irreversible) fluxes are neglected ($\nu = \nu' = \lambda = 0$), then one obtains the ideal or perfect fluid equations.

The variables $\rho(v)$, u, and e for which one has time evolution equations are called prognostic variables, and the variables η, T, p diagnostic variables. One may choose different sets of prognostic and diagnostic variables. A useful choice is to use u, p, and T as prognostic variables. Then the density ρ becomes a diagnostic variable which is determined by the equation of state.

C.2 Navier–Stokes and Euler Equations

Numerous approximations are applied to the basic fluid dynamical equations. One such approximation assumes a homogeneous fluid, $\rho = \rho_0 =$ constant, and $\nu = \nu_0 =$ and results in the Navier–Stokes equations

$$\partial_t u + (u^T \nabla_x) u = -\frac{1}{\rho_0} \nabla_x p + \nu_0 \Delta_x u - \nabla_x \Phi_{ext} \tag{C.14}$$

$$\nabla_x^T u = 0 \tag{C.15}$$

where $\Delta_x := \nabla_x^T \nabla_x$ is the Laplace operator. The dependent variables are the fluid velocity u and the pressure p. The constant density ρ_0 and the constant viscosity coefficient ν_0 enter as parameters. Equation (C.14) represents the momentum balance of the fluid. Equation (C.15) states the conservation of volume and is the consequence of mass conservation for a homogeneous fluid. This equation is also referred to as the incompressibility condition. The thermostatic specification and the energy equation are not needed because of the assumption of a constant density. The ideal fluid version ($\nu_0 = 0$) is called the Euler equations.

The Navier–Stokes or Euler equations have to be solved within a domain Ω subject to appropriate boundary conditions. Studies have been made in both two and three dimensions and for infinite, finite, and periodic domains. In the case of an infinite domain, one has to impose conditions on the solution as $|x|$ approaches infinity. For rigid boundaries, one generally imposes the no-slip boundary condition $u = 0$ on the boundary $\partial\Omega$ of the domain in the viscous case and the free-slip condition $u^T n = 0$

in the inviscid case, where n is the normal vector of the boundary. A free boundary is generally assumed to be a material surface, moving with the fluid velocity.

Whether or not there exist dynamical systems that corresponds to the Navier–Stokes equations or Euler equations is a complicated and not yet fully resolved issue. The complicating fact is that the pressure p enters the Navier–Stokes equations as a Lagrange multiplier to satisfy the incompressibility constraint (C.15). Assuming for simplicity no external forces and a domain fully enclosed by rigid surfaces, the pressure becomes determined by the Neumann problem

$$\begin{aligned} \Delta_x p &= -\rho_0 tr((\nabla_x u^T)^T(\nabla_x u^T)) \text{ in } \Omega \\ (\nabla_x p)^T n &= \rho_0 \nu_0 (\Delta_x u)^T n \text{ on } \partial\Omega \end{aligned} \tag{C.16}$$

The Poisson equation follows from the divergence of the momentum Eq. C.14 taking the incompressibility condition (C.15) into account. The boundary condition follows from the normal component of the momentum equation on the boundary taking the no-slip condition $u = 0$ into account. The pressure is at every time instant determined by the velocity field. This implies in particular that only the initial velocity field can be freely specified. The initial pressure field is then determined. Denoting the solution of the Neumann problem by $p = A[u]$, the momentum equation takes the form

$$\partial_t u + (u^T \nabla_x) u = -\frac{1}{\rho_0} \nabla_x (A[u]) + \nu_0 \Delta_x u \tag{C.17}$$

The time derivative of $u(x, t)$ is not determined locally by values of u in the neighborhood of x but globally by the values of u in the whole domain. Equation (C.17) is thus a functional equation, complicating the analysis.

C.3 Eulerian Versus Lagrangian Description

The fluid dynamical equations have so far been formulated in what is called the Eulerian description of the fluid flow. In the Eulerian description, the position x and time t are the independent variables, and the specific volume $v(x, t)$, the fluid velocity $u(x, t)$, and the specific internal energy $e(x, t)$ are the dependent variables. The Eulerian description is a field description in physical space. Variables are considered at points in physical space, independent of which fluid particles occupy this point.

An alternative description is the Lagrangian description. In the Lagrangian description, the fluid flow is described by the position $x(s, t)$ of a fluid particle as a function of its label s and time t. Usually, the label s is taken to be the initial position. The Lagrangian description follows the paths of (marked) fluid particles. It is a field description but a field description in label space. The independent variables are s and t.

The connection between the Eulerian $f_E(x,t)$ and Lagrangian fields $f_L(s,t)$ is given by

$$f_L(s,t) = f_E(x(s.t),t) \text{ and } f_E(x,t) = f_L(s(x,t),t) \tag{C.18}$$

where $s(x,t)$ is the inverse of the mapping $x(s,t)$. Especially, the Eulerian fluid velocity is given by $u(x,t) = \dot{x}(s(x,t),t)$ where the dot denotes the time derivative at constant s. Lagrangian and Eulerian time derivatives are related by

$$\dot{f}_L(s,t) = [(\partial_t + u^T\nabla_x)f_E](x(s,t),t) = \frac{Df_E}{Dt}(x(s,t),t) \tag{C.19}$$

where ∂_t is the time derivative at constant x. Thus, the time derivative at time t and constant label s equals the advective derivative at position $x(s,t)$ and time t. The gradients in label and physical space are related by

$$\nabla_s = F\nabla_x \text{ and } \nabla_x = G\nabla_s \tag{C.20}$$

where the deformation or displacement gradient tensors $F(s,t) := \nabla_s x^T(s,t)$ and $G(x,t) = \nabla_x s^T(x,t)$ have components $F_{ij}(s,t) = \partial x_j(s,t)/\partial s_i$ and $G_{ij}(x,t) = \partial s_j(x,t)/\partial x_i$. Trivially, $G = F^{-1}$.

In the Lagrangian description, the momentum balance becomes

$$\ddot{x} = \frac{1}{\rho}\nabla_x\sigma - \nabla_x\Phi_{ext} \tag{C.21}$$

where the dots denote partial differentiation with respect to time at constant s and ∇_x the gradient operator in physical space. Note that it is the gradient in physical space that appears in this momentum balance. One would have to substitute the expression $\nabla_x = G\nabla_s$ to make Eq. C.21 a proper partial differential equation with respect to the independent variables s and t. The Lagrangian description thus becomes awkward because it is the gradient in physical space that matters and not the gradient in label space. And there is another gradient in physical space hidden in (C.21) since the irreversible stress depends on the velocity gradient in physical space. The specification of boundary conditions is also more complicated in the Lagrangian description. Again, the gradients at the current position of the fluid particle matter, not the gradients at their original position. For these reasons, the Lagrangian description is hardly used to study fluid processes, especially irreversible ones. The preferred choice is the Eulerian description. There the fluid flow can be obtained without calculating the fluid trajectories. If one is interested in the fluid trajectories, one can obtain them from the Eulerian velocity field a posteriori by solving the three ordinary differential equations $\dot{x}(s,t) = u(x(s,t),t)$ with initial conditions $x(s,t=0) = x_0(s)$.

For perfect fluids, there is no production of entropy. For this reason, one usually replaces the specific internal energy e by the specific entropy η as a dependent variable and replaces Eq. C.8 by

$$\frac{D}{Dt}\eta = 0 \tag{C.22}$$

These dynamical equations have then to be augmented by an equation of state $p = p(v, \eta)$ which characterizes the fluid.

In the Lagrangian description, Eqs. C.6, C.7, and C.22 become

$$v(s, t) = v_0(s)J(s, t) \tag{C.23}$$

$$\ddot{x} = -\frac{1}{\rho}\nabla_x p - \nabla_x \Phi_{ext} \tag{C.24}$$

and

$$\eta(s, t) = \eta_0(s) \tag{C.25}$$

where $J(s, t) = \partial(x(s, t))/\partial(s)$ is the Jacobian. These equations have again to be augmented by an equation of state $p = p(v, \eta)$.

C.4 Variational Principles for Perfect Fluids

The neglect of irreversible molecular fluxes makes perfect fluids susceptible to a Hamiltonian formulation. One must distinguish between the Lagrangian and Eulerian descriptions.

Hamilton's principle for the Lagrangian description. The Lagrangian equations of motion (C.24) for a perfect fluid can be obtained from Hamilton's principle applied to continuous systems. One has to specify the Lagrangian density $\mathcal{L}$ which is the kinetic minus potential energy expressed in terms of the position x, the velocity $\dot{x}$, the displacement gradient tensor F, and the independent variables s and t. Hamilton's principle then states that

$$\delta \int dt \int\int\int \mathcal{L}(x, \dot{x}, F; s, t) d^3 s = 0 \tag{C.26}$$

for variations $\delta x(s, t)$ that vanish on the boundary. The Euler–Lagrange equations are

$$\frac{\partial}{\partial t}\frac{\partial \mathcal{L}}{\partial \dot{x}_i} + \frac{\partial}{\partial s_j}\frac{\partial \mathcal{L}}{\partial F_{ji}} - \frac{\partial \mathcal{L}}{\partial x_i} = 0 \tag{C.27}$$

and are derived in the same way as the ones for a discrete particle system.

For a perfect fluid, the appropriate Lagrangian density is

$$\mathcal{L} = \rho_0(s)\left[\frac{1}{2}\dot{x}^T\dot{x} - e(v, \eta) - \Phi_{ext}(x, t)\right] \tag{C.28}$$

where $e(v, \eta) = e(v_0(s)J(s,t), \eta_0(s))$ is the specific internal energy. The acceleration term and the external force term in (C.24) are straightforward to derive. To derive the pressure term, one needs to notice that the only dependence on the displacement gradient tensor F is through $v(s,t) = v_0(s)J(s,t)$. Since $p = -\partial e(v,\eta)/\partial v$, one finds

$$\frac{\partial}{\partial s_j}\frac{\partial \mathcal{L}}{\partial F_{ji}} = \frac{\partial}{\partial s_j}\left[\rho_0(s)pv_0(s)\frac{\partial J}{\partial F_{ji}}\right]$$
$$= J(F^{-1})_{ij}\frac{\partial p}{\partial s_j} = J\frac{\partial p}{\partial x_i} \tag{C.29}$$

where we have used (see Appendix E.20)

$$\partial J/\partial F_{ji} = J(F^{-1})_{ij} \tag{C.30}$$

$$\partial^2 J/\partial F_{ji}\partial s_j = 0 \tag{C.31}$$

and where sums over repeated indexes are implied.

This continuous form of Hamilton's principle can easily be rationalized as the continuum limit of a discrete particle system. To obtain the Lagrangian density (C.28), one needs to replace in the Lagrangian of an N-particle system the $\sum_n m_n$ by an integral $\int\int\int \rho_0(s)d^3s$, $q^{(n)}$ by x, $p^{(n)}/m_n$ by $\dot{x}$, and associate the pair interaction potential energy with the internal energy.

Ertel's potential vorticity. Hamilton's principle encodes all the dynamics in one scalar field, the Lagrangian density $\mathcal{L}$. This abridgment has many important consequences. One such consequence is formulated in Noether's theorem which states that infinitesimal symmetries of the Lagrangian density imply conservation laws. Thus, as in classical particle physics, conservation of energy and momentum is assured if the Lagrangian density is invariant under infinitesimal translations δt and δx. There is, however, another much more interesting application of Noether's theorem in fluid dynamics.

The Eulerian and Lagrangian descriptions differ in the number of prognostic variables. The time evolution in the Lagrangian description is described by a set of 3 second-order differential equations (equivalent to 6 first-order equations), whereas the time evolution in the Eulerian description is described by only 5 first-order differential equations. This reduction is possible since the dynamical evolution at a point in physical space depends only on the velocity, specific volume, and specific entropy at that point and not on the particular fluid particle that occupies this point. Specific volume and specific entropy constitute a two-dimensional manifold, whereas the label coordinates constitute a three-dimensional manifold. This fact implies that the Lagrangian density is invariant under a one-dimensional infinitesimal particle relabeling transformation that leaves the specific volume and specific entropy unchanged. This invariance implies, via Noether's theorem, the material conservation of a quantity called potential vorticity (Müller 1995). This conservation law is absent in classical particle physics since there the particle label is discrete and not continuous and an infinitesimal particle relabeling cannot be performed, as required by Noether's theorem.

Hamilton's principle for the Eulerian description. Hamilton's principle can also be formulated in the Eulerian frame. To transform the Lagrangian density (C.28) to the Eulerian frame, one has first to change the independent variables from (s, t) to (x, t) and to introduce the inverse mapping $s(x, t)$ instead of the forward mapping $x(s, t)$. Hamilton's principle (C.26) then becomes

$$\delta \int dt \int\int\int d^3x \rho \left(\frac{1}{2} u^T u - e(\rho, \eta) - \Phi_{ext}\right) = 0 \tag{C.32}$$

Here, the velocity has to be determined from the Lin constraint

$$\frac{D}{Dt} s = 0 \tag{C.33}$$

which implies

$$u_j = (G^{-1})_{ij} \frac{\partial s_i}{\partial t} \tag{C.34}$$

where G is the displacement gradient tensor of the inverse mapping with components $G_{ij} = \partial s_j / \partial x_i$. The density and specific entropy are given by $\rho(x, t) = M \rho_0(s)$ and $\eta(x, t) = \eta_0(s)$ where $M = \partial(s)/\partial(x)$ is the Jacobian of the inverse mapping. The Lagrangian density in (C.32) then depends on $(s, \partial_t s, G; x, t)$ as required for field variational principles. Variations $\delta s(x, t)$ yield the momentum equation in its Eulerian form.

More elegant forms and Hamiltonian formulations with pairs of canonically conjugate variables can be obtained by adding the Lin constraint (C.33) and the continuity equation to the variational principle (C.32) with associated Lagrange multipliers. Details can, for example, be found in Müller (1995). All these Eulerian versions require non-physical auxiliary variables with associated gauge groups.

Hamilton's principle, whether in the Lagrangian or Eulerian frame, only applies to the dynamics of perfect fluids. Most problems in fluid dynamics involve dissipative processes that defy such an approach. Nevertheless, Hamilton's principle can be a powerful tool, especially for theoretical studies. By applying analyses, abridgments, and approximations directly to the Lagrangian or Hamiltonian density, one assures that the symmetries and covariances of the underlying physical problem are respected.

Appendix D
Some Proofs and Explicit Calculations

D.1 Derivation of the Lorenz Equations

The Lorenz equations (2.42) are supposed to describe some essential features of Rayleigh–Bénard convection. This type of convection occurs in a planar layer of fluid heated from below and is characterized by regular patterns of convection cells, called Bénard cells. The Lorenz equations can be derived from the fluid dynamical equations by the following assumptions and operations which follow Schuster (1995) who in turn follows Saltzmann (1962):

(1) The starting points are the fluid dynamical equations for an incompressible fluid in the form

$$\nabla_x^T u = 0 \tag{D.1}$$

$$\partial_t u + (u^T \nabla_x) u = -\frac{1}{\rho_0} \nabla_x p - \frac{\rho - \rho_0}{\rho_0} g + \nu_0 \Delta_x u \tag{D.2}$$

$$\partial_t T + (u^T \nabla_x) T = \lambda_0 \Delta_x T \tag{D.3}$$

$$\rho = F(T) \tag{D.4}$$

Here, u is the velocity vector, ρ the density, T the temperature, and p the pressure field, all fields being functions of spatial position x and time t, and ρ_0 a constant reference density, ν_0 a constant viscosity coefficient, λ_0 a constant thermal conduction coefficient and $g = (0, 0, g_0)$ the constant gravitational acceleration vector, and $\Delta_x = \nabla_x^T \nabla_x$ the Laplace operator. A justification of these equations can be found in textbooks on fluid dynamics such as Müller (2006) but is somewhat irrelevant here. The major point is that these equations form a set of partial differential equations, a somewhat convoluted set since the pressure p has to be determined from a Poisson equation obtained by taking the divergence of the momentum balance and imposing the incompressibility constraint $\nabla_x^T u = 0$.

P. Müller, *Handbook of Dynamics and Probability*,
https://doi.org/10.1007/978-3-030-88486-4

(2) Now adopt the geometry depicted in Fig. D.1. We write $x = (x, y, z)$ and $u = (u, v, w)$ and assume no variation in the y-direction, $\partial_y = 0$. The volume conservation $\nabla_x^T u = 0$ thus reduces to

$$\partial_x u + \partial_z w = 0 \tag{D.5}$$

One can thus introduce a stream function ψ such that $u = -\partial_z \psi$ and $w = \partial_x \psi$.

(3) Next consider the temperature deviation $\theta(x, z, t)$ from a linear temperature gradient such that

$$T(x, z, t) = T_0 - \frac{z}{h}\Delta T + \theta(x, z, t) \tag{D.6}$$

where ΔT is the temperature difference between the bottom and top of the container separated by a distance h. The temperature equation then becomes

$$\partial_t \theta + J(\psi, \theta) = \frac{\Delta T}{h}\partial_x \psi + \lambda_0 \Delta_h \theta \tag{D.7}$$

where $J(a, b) = \partial_x a \partial_z b - \partial_z a \partial_x b$ is the Jacobian and $\Delta_h = \partial_x \partial_x + \partial_z \partial_z$ the two-dimensional Laplace operator. The boundary conditions at the top and bottom are $\theta = 0$ at $z = 0$ and $z = h$.

(4) Taylor expansion of the equation of state $\rho = F(T)$ gives to first order

$$\rho = \rho_0 - \alpha_0 \rho_0 (-\frac{z}{h}\Delta T + \theta) \tag{D.8}$$

where $\rho_0 = F(T_0)$ and $\alpha_0 \rho_0 = -(\partial F / \partial T)(T_0)$ is the thermal expansion coefficient.

(5) Taking the curl of the momentum equation, one then arrives at

$$\partial_t \Delta_h \psi + J(\psi, \Delta_h \psi) = \alpha_0 g_0 \partial_x \theta + \nu_0 \Delta_h^2 \psi \tag{D.9}$$

The boundary conditions at the bottom and top are assumed to be $\psi = \Delta\psi = 0$ at $z = 0$ and $z = h$, corresponding to $w = 0$ and the free-slip condition $\partial_z u = 0$.

All the approximations and simplifications so far are reasonable and reduced the somewhat convoluted set of four coupled PDEs for the three velocity components and temperature to a more benign set of two coupled PDEs for the stream function and the temperature deviation. The substantial reduction is accomplished by the next step.

(6) To account for the observed structure of the convection cells, the spatial structure of the temperature deviation θ and the stream function ψ is assumed to be of the form

$$\frac{\pi R}{R_c \Delta T}\theta(x,z,t) = \sqrt{2}x_2(t)\cos\left(\frac{\pi a}{h}x\right)\sin\left(\frac{\pi}{h}z\right) - x_3(t)\sin\left(2\frac{\pi}{h}z\right) \quad \text{(D.10)}$$

$$\frac{a}{1+a^2}\frac{1}{\lambda_0}\psi(x,z,t) = \sqrt{2}x_1(t)\sin\left(\frac{\pi a}{h}x\right)\sin\left(\frac{\pi}{h}z\right) \quad \text{(D.11)}$$

This representation may be viewed as an expansion of the vertical structure with respect to the basis functions $\sin(n\pi z/h)$, retaining only the lowest order term $n = 1$ for the stream function and the two lowest order terms $n = 1, 2$ for the temperature deviation, and a similar expansion for the horizontal structure. The variables $x_1(t)$, $x_2(t)$, and $x_3(t)$ are the time-dependent expansion coefficients. The coefficient x_1 characterizes the convection velocity, x_2 the temperature difference between upward and downward moving fluid, and x_3 the deviation of the vertical temperature gradient from its equilibrium value. These expansion coefficients will turn out to be the state variables of the Lorenz equations. The parameter a is the ratio of the vertical to the horizontal scale of the rolls shown in Fig. D.1, $R := g_0\alpha_0 h^3 \Delta T/\lambda_0\nu_0$ the Rayleigh number, and $R_c := \pi^4(1+a^2)^3/a^2$ a critical Rayleigh number.

(7) When the representations (D.10) and (D.11) are substituted into Eqs. D.7 and D.9, the nonlinear advection terms generate higher order harmonics proportional to $\sin(n\pi z/h)$. When only the harmonics $n = 1$ and $n = 2$ are retained in the equation for the temperature deviation and only the harmonic $n = 1$ in the equation for the stream function, then these equations turn into the Lorenz equations

$$\dot{x}_1 = -\sigma x_1 + \sigma x_2 \quad \text{(D.12)}$$

$$\dot{x}_2 = r x_1 - x_2 - x_1 x_3 \quad \text{(D.13)}$$

$$\dot{x}_3 = x_1 x_2 - b x_3 \quad \text{(D.14)}$$

where the dot denotes differentiation with respect to the normalized time $\tau := \pi^2(1+a^2)\lambda_0/h^2$, $\sigma := \nu_0/\lambda_0 > 0$ is the Prandtl number, $b = 4/(1+a^2) > 0$, and $r = R/R_c \propto \Delta T$. The Prandtl number is a property of the fluid, the constant $b > 0$ characterizes the geometry of the convective rolls, and the Rayleigh number is a control parameter which characterizes the magnitude of the applied temperature difference ΔT that drives the convection. Miraculously, the horizontal structure drops out of the equations, testifying to the ingenuity of the 'Ansatz' (D.10) and (D.11).

This derivation of the Lorenz equations follows the common procedure: First, expand the spatial structure of a PDE in terms of a set of basis functions. This turns the PDE into a hierarchy of ODEs for the time-dependent expansion coefficients. Second, truncate this hierarchy to obtain a manageable but still faithful representation of the original problem. It is surprising that in this case, three coefficients seem to suffice to capture some essential features of Rayleigh–Bénard convection.

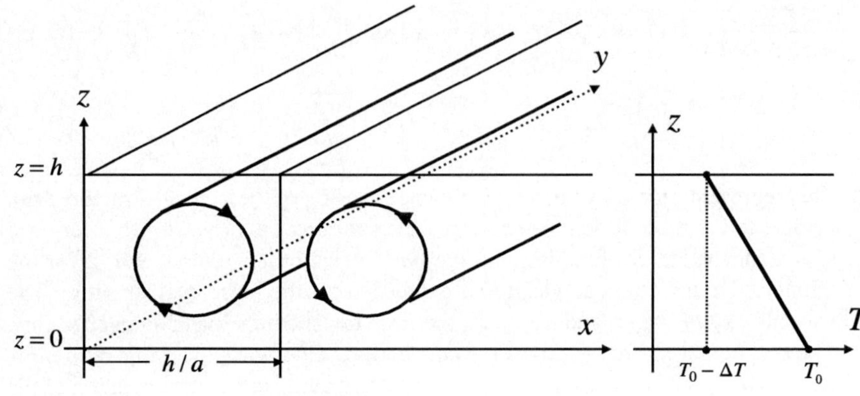

Fig. D.1 Geometry of Lorenz system

D.2 Inverse Scattering Transform of the KdV Equation

The Korteweg–De Vries (KdV) equation

$$\partial_t \rho(x, t) - 6\rho(x, t)\partial_x \rho(x, t) + \partial_x^3 \rho(x, t) = 0 \tag{D.15}$$

is a nonlinear partial differential equation that was originally derived as an approximation to the equation for surface waves. In that context, the variable ρ would be a surface elevation, the second term would describe the nonlinear interaction among surface waves, and the third term the dispersion of surface waves. If the nonlinearity is neglected, the KdV equation has wave solutions $\rho(x, t) \sim \exp\{i(kx - \omega t)\}$ with wavenumber k, frequency ω, and dispersion relation $\omega = k^3$. Since the phase speed $c := \omega/k = 3k^2$ depends on the wavenumber, any initial condition disperses. However, if the nonlinearity is included, the KdV equation has solutions of the form $\rho(x, t) = f(x - ct)$ where

$$f(z) = -\frac{1}{2}c\frac{1}{\cosh^2(\frac{1}{2}\sqrt{c}(z - z_0))} \tag{D.16}$$

for every $c \in \mathbb{R}$. These solutions represent disturbances that propagate at constant speed c without changing their shape. These solutions are called solitons. They are a depression for $c > 0$ and an elevation for $c < 0$. Their amplitude becomes larger, and the width the smaller, the faster the soliton travels. The discovery of these soliton solutions has been quite important since it shows that nonlinear PDEs can have 'simple' solutions. If one equates solitons with particles, then the KdV equation is an example that particles can be extracted out of field theories.

The KdV equation can also be written in the flux form

$$\partial_t \rho + \partial_x(-3\rho^2 + \partial_y^2 \rho) = 0 \tag{D.17}$$

implying a local conservation. There are actually additional conservation laws. For example, one finds

$$\partial_t \rho^2 + \partial_x(-4\rho^3 - (\partial_x \rho)^2 + \rho \partial_x^2 \rho) = 0 \tag{D.18}$$

Indeed, the KdV equation has an infinite number of such conservation laws, and this is one of the intrinsic reasons that it is amenable to a solution technique called inverse scattering transform. This technique utilizes a judicious choice of two linear operators, one defining an eigenvalue problem and the other a time evolution operator. The field ρ enters the eigenvalue problem as a potential. One has to determine the 'scattering data' for the given potential $\rho(x, t)$ in the eigenvalue problem, advance these scattering data by the time-evolution operator, and reconstruct the potential $\rho(x, t')$ at a later time. Explicitly, the inverse scattering transform employs the following four steps:

1. Finding a Lax pair of operators. Consider two linear operators L and M acting on a field $\psi(x, t)$ in the following way:

$$L\psi(x, t) = \lambda\psi(x, t) \ , \quad \frac{\partial}{\partial t}\psi(x, t) = M\psi(x, t) \tag{D.19}$$

The equation with the operator L constitutes an eigenvalue problem with eigenvalue λ. The operator M describes a time evolution. Such a pair of operators constitutes a Lax pair for a given nonlinear PDE if

$$\partial_t L + LM - ML = 0 \tag{D.20}$$

represents the nonlinear PDE. The operator on the left-hand side of (D.20) is thus not a differential operator but just a multiplier. Crucial for further development is the fact that Eq. D.20 is also the condition that the eigenvalue λ does not depend on time, as is easily demonstrated by taking the time derivative of the eigenvalue problem in (D.19).

In the case of the KdV equation, the operators

$$L = -\partial_x^2 + \rho \tag{D.21}$$

$$M = -4\partial_x^3 + 6\rho\partial_x + 3\partial_x \rho \tag{D.22}$$

form a Lax pair, as can be shown by explicit calculation. The coefficients in front of all differential operators ∂_x^n, $n = 1, \ldots, 5$ vanish.

2. Determining the scattering data. The physicist realizes that Eq. (D.21) is the time-independent Schrödinger equation for the quantum-mechanical energy eigenstates in a potential $\rho(x,t)$. A major point to note is that the time dependence in the potential $\rho(x,t)$ is just a parametric dependence. As far as the Schrödinger equation is concerned, it is just a parameter that describes the potential. The Schrödinger equation is, of course, well-studied. If $\rho(x,t)$ belongs to a certain class of functions, then the Schroedinger equation admits two types of solutions, scattering solutions and bound solutions. These solutions come in pairs. The scattering solutions $\psi_k^{L,R}(x,t)$ correspond to positive energies and have a continuum of eigenvalues $\lambda = k^2$ with $k \in \mathbb{R} \setminus \{0\}$. Asymptotically, they consist of linear combinations of e^{ikx} and e^{-ikx}. Their asymptotic behavior can be written in the form

$$\begin{aligned} \lim_{x\to+\infty} \psi_k^L(x,t) &= T_k(t)e^{-ikx} + R_k(t)e^{ikx} \\ \lim_{x\to-\infty} \psi_k^R(x,t) &= T_k(t)e^{ikx} + L_k(t)e^{-ikx} \end{aligned} \tag{D.23}$$

defining the reflection coefficients $R_k(t)$ and $L_k(t)$ and the transmission coefficients $T_k(t)$. The details are given in books on quantum mechanics.

The bound states $\psi_n^{L,R}(x,t)$ correspond to negative energies and have a finite set of discrete eigenvalues $\lambda_n = -\kappa_n^2, (n = 1, \ldots, N), (\kappa_n \in \mathbb{R})$. These bound states are square integrable and can be characterized by the normalization constant

$$c_n^{L,R}(t) := \left[\int_{\infty}^{\infty} dx \psi_n^{L,R}(x,t)\right]^{-1/2} \tag{D.24}$$

The coefficients $L_k(t)$, $R_k(t)$, $T_k(t)$, $c_n^{l,R}(t)$ are called the scattering data. Determining these scatterings for a given potential $\rho(x,t)$ is called the direct scattering problem. Not all of these scattering data are independent. Even more so, there is the remarkable result that the scattering data $R_k(t)$ and $c_n^L(t)$ are sufficient to reconstruct the potential $\rho(x,t)$.

These results suggest the following solution strategy: Given the initial condition $\rho(x,t=0)$, solve the direct scattering problem and determine the scattering data $R_k(t=0)$ and $c_n^L(t=0)$. Then advance the scattering data in time to determine $R_k(t)$ and $c_n^L(t)$ and use these to reconstruct $\rho(x,t)$. So the next step is to derive the evolution equations for the scattering data.

3. Calculating the time evolution of scattering data. The time evolution of the eigenfunctions is governed by the operator M, given by (D.22), and is fairly complicated. The evolution of the scattering data is, however, quite simple, and given by

$$\frac{d}{dt}R_k(t) = 8ik^3 R_k(t) \tag{D.25}$$

$$\frac{d}{dt}L_k(t) = -8ik^3 L_k(t) \tag{D.26}$$

$$\frac{d}{dt}T_k(t) = 0 \tag{D.27}$$

$$\frac{d}{dt}c_n^L = 4k_n^3 c_n^L \tag{D.28}$$

$$\frac{d}{dt}c_n^R = -4k_n^3 c_n^R \tag{D.29}$$

The eigenvalues do not change in time by construction, $d\lambda/dt = 0$. The evolution equations are linear, reflecting the linearity of the operator M, and are easily solved. One can thus determine the scattering data at time t from the scattering data at time t=0. It remains to invert the scattering problem

4. Inverting the scattering data. The potential $\rho(x,t)$ can be reconstructed from the scattering data $R_k(t)$ and $C_n^L(t)$ as follows.

Define the Marchenko kernel

$$g(x,t) := \sum_{n=1}^{N} (c_n^L)^2 e^{-\kappa_n x} + \frac{1}{2\pi}\int_{-\infty}^{\infty} dk R_k(t) e^{ikx} \tag{D.30}$$

and then solve the linear integral equation (Gelfand–Levitan–Marchenko)

$$f(x,y,t) + g(x+y,t) + \int_{-\infty}^{\infty} dz g(x+z,t) f(z,y,t) = 0 \;\; x < y, \infty \tag{D.31}$$

Then the potential $\rho(x,t)$ is given by

$$\rho(x,t) = -2\frac{d}{dx} f(x,x) \tag{D.32}$$

Though solving the Gelfand–Levitan–Marchenko integral equation may not be an easy task, its solution is greatly facilitated by the fact that it is a linear integral equation.

The potential $\rho(x,t)$ constructed from the linear operators L and M is a solution of the KdV equation since L and M satisfy Eq. D.20 which is the KdV equation for ρ.

D.3 Adiabatic Elimination of Fast Variables

Consider the system[1]

$$\begin{aligned}\dot{x}(t) &= \frac{1}{\epsilon}u(x(t), y(t)) \\ \dot{y}(t) &= v(x(t), y(t))\end{aligned} \tag{D.33}$$

where ϵ is a small parameter. It describes the coupled evolution of a fast variable x and a slow variable y. The fast variable can formally be eliminated by a two-time-scale expansion. Introduce a fast time scale

$$\tau := t/\epsilon \tag{D.34}$$

and assume an expansion

$$\begin{aligned}x(t;\epsilon) &= x^{(0)}(\tau,t) + \epsilon\, x^{(1)}(\tau,t) + \epsilon^2\, x^{(2)}(\tau,t) + \dots \\ y(t;\epsilon) &= y^{(0)}(\tau,t) + \epsilon\, y^{(1)}(\tau,t) + \epsilon^2\, y^{(2)}(\tau,t) + \dots\end{aligned} \tag{D.35}$$

where each $x^{(i)}$ and $y^{(i)}$ depends both on the slow time scale t and the fast time scale τ. Expansion of the time derivatives then yields

$$\begin{aligned}\tfrac{d}{dt}x &= \tfrac{1}{\epsilon}\, x_\tau^{(0)} + (x_\tau^{(1)} + x_t^{(0)}) + \epsilon(x_\tau^{(2)} + x_t^{(1)}) + \dots \\ \tfrac{d}{dt}y &= \tfrac{1}{\epsilon}\, y_\tau^{(0)} + (y_\tau^{(1)} + y_t^{(0)}) + \epsilon(y_\tau^{(2)} + y_t^{(1)}) + \dots\end{aligned} \tag{D.36}$$

where subscripts denote differentiation. Insert this expansion into the original equations and collect equal powers of ϵ. At order $O(1/\epsilon)$, one finds

$$\begin{aligned}x_\tau^{(0)} &= u(y^{(0)}(\tau,t), x^{(0)}(\tau,t)) \\ y_\tau^{(0)} &= 0\end{aligned} \tag{D.37}$$

From the first equation, one concludes that $y^{(0)}$ does not depend on the fast time scale τ, so that

$$y^{(0)} = y^{(0)}(t) \tag{D.38}$$

Then the first equation reads

$$x_\tau^{(0)} = u(y^{(0)}(t), x^{(0)}(\tau,t)) \tag{D.39}$$

The variables (τ, t) are *independent* arguments of the expansion functions $x^{(i)}(\tau,t)$ and $y^{(i)}(\tau,t)$. As a consequence, Eq. D.39 is a closed first-order ordinary differential equation for the dependence of $x^{(0)}$ on τ *at fixed* t. Given the function u, this ordinary

[1] The text for this section was to a large extent provided by Rupert Klein, Berlin.

differential equation must now be solved. This is in general an easier problem than the original one.

The result is, if this solution has been obtained successfully, the functional dependence of $x^{(0)}(\tau, t)$ on the fast variable τ, with t as a parameter. Denote this partial solution by $X(\tau; t)$ and consider it given.

At order $O(1)$, we then find

$$x_\tau^{(1)}(\tau, t) = -X_t(\tau; t) + x^{(1)} \frac{\partial g}{\partial x}(X(\tau; t), y^{(0)}(t)) + y^{(1)} \frac{\partial u}{\partial y}(X(\tau; t), y^{(0)}(t))$$
$$y_\tau^{(1)}(\tau, t) = -y_t^{(0)}(t) + v(Y(\tau; t), y^{(0)}(t)) \tag{D.40}$$

The evolution equation for $y^{(0)}$ is now obtained by the so-called 'sub-linear growth' condition: Taking again into account that t and τ are *independent* arguments of the expansion functions, we integrate the second equation with respect to τ from, say, τ_0 to τ_1 at fixed t. This yields

$$y^{(1)}(\tau_1, t) - y^{(1)}(\tau_0, t) = -(\tau_1 - \tau_0)\, y_t^{(0)} + \int_{\tau_0}^{\tau_1} d\tau v(X(\tau; t) x^{(0)}(t)) \tag{D.41}$$

For the asymptotic expansion to remain meaningful even for ranges of the fast time of the order $|\tau_1 - \tau_0| = O(1/\epsilon)$, we must require that $|y^{(1)}(\tau_1, t) - y^{(1)}(\tau_0, t)|$ grows slower than linear with $|\tau_1 - \tau_0|$. Otherwise, the perturbation assumption $|\epsilon y^{(1)}| \ll |y^{(0)}|$ would be violated and the whole approach of collecting 'terms of equal order of magnitude' by grouping expressions multiplied by the same power of ϵ would be flawed. This statement is equivalent to requiring that

$$\lim_{|\tau_1 - \tau_0| \to \infty} \left(\frac{y^{(1)}(\tau_1, t) - y^{(1)}(\tau_0, t)}{\tau_1 - \tau_0} \right) = 0 \tag{D.42}$$

which is achieved in Eq. D.41 by setting

$$y_t^{(0)} = \lim_{|\tau_1 - \tau_0| \to \infty} \frac{1}{\tau_1 - \tau_0} \int_{\tau_0}^{\tau_1} d\tau v(Y(\tau; t), y^{(0)}(t)) \tag{D.43}$$

provided the limit does exist.

This is the classical result of the 'method of averaging', which states that one should

- solve the 'fast problem' for fixed slow coordinates (here the result is $X(\tau; t)$, which solves Eq. D.39 for fixed t),
- substitute the result into the right-hand side of the next order equation (here this means substitution into the right-hand side of Eq. D.40), and
- average the resulting equation over the fast time scale which results in the slow time derivative of the leading order function via the sub-linear growth condition (here this result is Eq. D.43).

D.4 Balance Equations

One of the important ingredients of fluid dynamics is the balance equations for mass, momentum, and energy. Here, we give an explicit derivation of the macroscopic balance equation from the microscopic equations of motion. We do this in order to pinpoint the essential assumptions and problems.

Microscopic dynamics. For simplicity, we consider a N-particle mono-atomic gas with no internal degrees of freedom. The state space X (often denoted by Γ in this context) is then a $6N$-dimensional space with elements $x = (q, p)$ where $q = (q^{(1)}, \ldots, q^{)N)})^T$ and $p = (p^{(1)}, \ldots, p^{(N)})^T$ are $3N$-dimensional vectors and $q^{(n)}$ and $p^{(n)}$ are elements of the three-dimensional position space and three-dimensional momentum space of the nth particle. We assume that the microscopic dynamics of this system is governed by the Hamilton function

$$H(q, p) = \sum_{n=1}^{N} \frac{1}{2m} p^{(n)T} p^{(n)} + \frac{1}{2} \sum_{\substack{n,m=1 \\ n \neq m}}^{N} m\Phi_{int}(q^{(nm)}) + \sum_{n=1}^{N} m\Phi_{ext}(q^{(n)}, t) \quad \text{(D.44)}$$

where $q^{(nm)} := |q^{(n)} - q^{(m)}|$. Here, $\Phi_{int}(r)$ is an isotropic (spherically symmetric) pair interaction potential as, for example, sketched in Fig. 17.1 and $\Phi_{ext}(y, t)$ an external potential that may explicitly depend on time.

Hamilton's equations give

$$\dot{q}_t^{(n)} = \frac{1}{m} p_t^{(n)} \quad \text{(D.45)}$$

$$\dot{p}_t^{(n)} = \sum_{\substack{m=1 \\ n \neq m}}^{N} F_t^{(nm)} - m \frac{\partial \Phi_{ext}(q_t^{(n)}, t)}{\partial q_t^{(n)}} \quad \text{(D.46)}$$

where

$$F_t^{(nm)} := -m \frac{\partial \Phi_{int}(q_t^{(nm)})}{\partial q_t^{(nm)}} \frac{(q_t^{(n)} - q_t^{(m)})}{q_t^{(nm)}} \quad \text{(D.47)}$$

is the force exerted by particle m on particle n. The partial derivative $\partial \Phi_{int}(q_t^{(nm)})/ \partial q_t^{(nm)}$ has to be interpreted as evaluating the derivative $\partial \Phi_{int}(r)/\partial r$ at position $r = q_t^{(nm)}$ and similarly for the partial derivative of Φ_{ext}. The solutions of Hamilton's equation are denoted by $q_t^{(n)}(q, p)$ and $p_t^{(n)}(q, p)$ where (q, p) are the initial conditions.

Densities in physical space. We now introduce microscopic mass, momentum, and energy densities in the volume V to which the gas is confined by the definitions

$$m_{y,t}(q,p) := \sum_{n=1}^{N} \delta(y - q_t^{(n)}(q,p))m \tag{D.48}$$

$$p_{y,t}(q,p) := \sum_{n=1}^{N} \delta(y - q_t^{(n)}(q,p))p_t^{(n)}(q,p) \tag{D.49}$$

$$h_{y,t}(q,p) := \sum_{n=1}^{N} \delta(y - q_t^{(n)}(q,p))h_t^{(n)}(q,p) \tag{D.50}$$

where $y \in V$ and where

$$h_t^{(n)} := \frac{1}{2m} p_t^{(n)T} p_t^{(n)} + \frac{1}{2} \sum_{\substack{m=1 \\ n \neq m}}^{N} m\Phi_{int}(q_t^{(nm)}) + m\Phi_{ext}(q_t^{(n)}, t) \tag{D.51}$$

is the energy of the nth particle which changes according to

$$\dot{h}_t^{(n)} = \frac{1}{2m} \sum_{\substack{m=1 \\ n \neq m}}^{N} F_t^{(nm)T}(p_t^{(n)} + p_t^{(m)}) + m\frac{\partial \Phi_{ext}(q_t^{(n)}, t)}{\partial t} \tag{D.52}$$

The densities can be interpreted as the values of a family of observables indexed by the position $y \in V$. Note that the densities depend on the initial conditions (q, p), as indicated. In the following, we suppress the arguments (q, p) and replace the subscript notation by the argument notation, writing the densities as $m(y, t)$, $p(y, t)$, and $h(y, t)$.

Microscopic balance equations. The densities are the product of a delta function and a property $\varphi_t^{(n)} = (m, p_t^{(n)}, e_t^{(n)})$. The partial time derivative of the densities at fixed y (and fixed (q, p)) thus takes the general form

$$\partial_t \sum_{n=1}^{N} \delta(y - q_t^{(n)})\varphi_t^{(n)} = -\nabla_y^T \left[\sum_{n=1}^{N} \delta(y - q_t^{(n)})\dot{q}_t^{(n)}\varphi_t^{(n)} \right] + \sum_{n=1}^{N} \delta(y - q_t^{(n)})\dot{\varphi}_t^{(n)} \tag{D.53}$$

The first term on the right-hand side is the divergence of a flux. Explicitly, one gets the following expressions for these contributions to the mass, momentum, and energy flux

$$j_m(y,t) := p(y,t) \tag{D.54}$$

$$j_p^{(1)}(y,t) := \sum_{n=1}^{N} \delta(y - q_t^{(n)}) \frac{p_t^{(n)}}{m} p_t^{(n)T} \tag{D.55}$$

$$j_h^{(1)}(y,t) := \sum_{n=1}^{N} \delta(y - q_t^{(n)}) \frac{p_t^{(n)}}{m} h_t^{(n)} \tag{D.56}$$

The mass and energy fluxes are vectors; the momentum flux is a symmetric second-order tensor.

The second term in (D.53) yields no contributions to the mass equation. The contributions to the momentum and energy equation are twofold, one arising from the interaction potential and one from the external potential. The contribution from the interaction potential contains the force $F_t^{(nm)}$ which is anti-symmetric in the indexes n and m. That is, $F_t^{(nm)} = -F_t^{(mn)}$. This anti-symmetry arises from our choice of an isotropic interaction potential. Because of this anti-symmetry, we can write

$$\sum_{\substack{n,m=1 \\ n\neq m}}^{N} \delta(y - q_t^{(n)}) F_t^{(nm)} = \frac{1}{2} \sum_{\substack{n,m=1 \\ n\neq m}}^{N} [\delta(y - q_t^{(n)}) - \delta(y - q_t^{(m)})] \tag{D.57}$$

and use the formula

$$\delta(y - y_1) - \delta(y - y_2) = -\nabla_y^T \int_0^1 d\lambda (y_1 - y_2) \delta(y - \lambda y_1 - (1 - \lambda) y_2) \tag{D.58}$$

which can be verified by direct evaluation. The contribution from the interaction potential can therefore also be written as the divergence of a flux. Explicitly, these contributions to the fluxes are given by

$$\begin{aligned} j_p^{(2)}(y, t) := &\int_0^1 d\lambda \frac{1}{2} \sum_{\substack{n,m=1 \\ n\neq m}}^{N} (q_t^{(n)} - q_t^{(m)}) F_t^{(nm)T} \\ &\cdot \delta(y - \lambda q_t^{(n)} - (1 - \lambda) q_t^{(m)}) \end{aligned} \tag{D.59}$$

$$\begin{aligned} j_h^{(2)}(y, t) := &\int_0^1 d\lambda \frac{1}{4m} \sum_{\substack{n,m=1 \\ n\neq m}}^{N} [F_t^{(nm)T} (p_t^{(n)} - p_t^{(m)})](q_t^{(n)} - q_t^{(m)}) \\ &\cdot \delta(y - \lambda q_t^{(n)} - (1 - \lambda) q_t^{(m)}) \end{aligned} \tag{D.60}$$

The momentum flux is again a symmetric tensor. The contributions from the external potential become source terms which take the form

$$Q_p(y, t) = -m(y, t) \nabla_y \Phi_{ext}(y, t) \tag{D.61}$$

$$Q_h(y, t) = m(y, t) \partial_t \Phi_{ext}(y, t) \tag{D.62}$$

In summary, the balance equations take the form

$$\partial_t m(y, t) + \nabla_y p(y, t) = 0 \tag{D.63}$$

$$\partial_t p(y, t) + \nabla_y j_p(y, t) = -m(y, t) \nabla_y \Phi_{ext}(y, t) \tag{D.64}$$

$$\partial_t h(y, t) + \nabla_y j_h(y, t) = m(y, t) \partial_t \Phi_{ext}(y, t) \tag{D.65}$$

where $j_p = j_p^{(1)} + j_p^{(2)}$ and $j_h = j_e^{(1)} + j_h^{(2)}$. If there is no external potential, the balance equations take the form of conservation laws. These balance equations are just a continuous formulation of the microscopic balance equations for mass, momentum, and energy. Nothing has been lost (or gained) by this continuous formulation of the discrete particle dynamics. The equations are still invariant under time reversal.

Averaged balance equations. The densities and fluxes depend on the initial conditions (q, p). One now assumes that these initial conditions are a random variable with probability density $\rho_0(q, p)$ and takes the expectation of the balance equations which is trivial since taking the expectation commutes with differentiation, both with respect to the spatial coordinate y and the time coordinate t. The result is

$$\partial_t\langle m(y,t)+\rangle\nabla_y\langle p(y,t)\rangle = 0 \tag{D.66}$$

$$\partial_t\langle p(y,t)+\rangle\nabla_y\langle j_p(y,t)\rangle = -\langle m(y,t)\rangle\nabla_y\Phi_{ext}(y,t) \tag{D.67}$$

$$\partial_t\langle h(y,t)+\rangle\nabla_y\langle j_h(y,t)\rangle = \langle m(y,t)\rangle\partial_t\Phi_{ext}(y,t) \tag{D.68}$$

The expectation of any of the quantities $f_{y,t}(q, p)$ might be taken either in the Heisenberg or the Schrödinger representation

$$\langle f(y,t)\rangle = \int_X dq dp \rho_0(q,p) f_{y,t}(q,p) \tag{D.69}$$

$$= \int_X dq dp \rho_t(q,p) f_{y,0}(q,p) \tag{D.70}$$

In the Heisenberg representation, one has to calculate $f_{y,t}(q, p)$ by solving Hamilton's equation with initial condition (q, p). In the Schrödinger representation, one has to calculate $\rho_t(q, p)$ by solving the Liouville equation

$$\partial_t\rho_t + u_q\nabla_q\rho_t + u_p\nabla_p\rho_t = 0 \tag{D.71}$$

with initial condition ρ_0. In both cases, one has to specify the initial distribution ρ_0 and solve the dynamical equations. The hope is, of course, that the averaged equations can somehow be closed and one gets away without solving the microscopic dynamics. This is the task of non-equilibrium statistical mechanics.

Fluid dynamical equations. To arrive at the fluid dynamical equations as given in Sects. 18.2 and C.1, we have to identify the expectation with the fluid dynamical quantities as follows:

$$\rho(y,t) := \langle m(y,t)\rangle \quad \text{mass density} \tag{D.72}$$

$$u(y,t) := \frac{\langle j_p(y,t)\rangle}{\langle m(y,t)\rangle} \quad \text{fluid velocity} \tag{D.73}$$

$$e(y,t) := \frac{\langle h(y,t)\rangle}{\langle m(y,t)\rangle} \quad \text{specific total energy} \tag{D.74}$$

The balance equation then take the form

$$\partial_t \rho + \nabla^T(\rho u) = 0 \tag{D.75}$$

$$\partial_t(\rho u) + \nabla^T(\rho u u^T - \Pi) = -\rho \nabla \Phi_{ext} \tag{D.76}$$

$$\partial_t(\rho e) + \nabla^T(\rho e u - \Pi u + q) = \rho \partial_t \Phi_{ext} \tag{D.77}$$

where the stress tensor $\Pi = \Pi^{(1)} + \Pi^{(2)}$ and the heat flux vector $q = q^{(1)} + q^{(2)}$ are given by

$$\Pi^{(1)}(y,t) := -\langle \frac{1}{m} \sum_{n=1}^{N} \delta(y - q_t^{(n)}) \delta p_t^{(n)} \delta p_t^{(n)} \rangle \tag{D.78}$$

$$\Pi^{(2)}(y,t) := -\langle j_p^{(2)}(y,t) \rangle \tag{D.79}$$

$$q^{(1)}(y,t) := \langle \frac{1}{2} \sum_{\substack{n,m=1\\ n\neq m}}^{N} \delta(y - q_t^{(n)}) \delta p_t^{(n)} U_{int}(q_t^{(nm)}) \rangle$$

$$+ \langle \frac{1}{2m^2} \sum_{n=1}^{N} \delta(y - q_t^{(n)}) \delta p_t^{(n)} \delta p_t^{(n)T} \delta p_t^{(n)} \rangle \tag{D.80}$$

$$q^{(2)}(y,t) := +\langle \int_0^1 d\lambda \frac{1}{4m} \sum_{\substack{n,m=1\\ n\neq m}}^{N} [F_t^{(nm)T}(\delta p_t^{(n)} - \delta p_t^{(m)})](q_t^{(n)} - q_t^{(m)})$$

$$\cdot \delta(y - \lambda q_t^{(n)} - (1-\lambda) q_t^{(m)}) \rangle \tag{D.81}$$

with

$$\delta p_t^{(n)}(y,q,p) := p_t^{(n)}(q,p) - m u(y,t) \tag{D.82}$$

The equations for the kinetic, potential, and internal energies can be obtained by straightforward manipulations.

We have thus derived the phenomenological fluid dynamical equations from the microscopic dynamics and have explicit expressions not only for the mass, momentum, and energy densities but also for the stress tensor and the heat flux. The drawback is that these expressions still require the solution of the N-particle problem and an initial probability distribution.

If one smooths the microscopic flux laws with a smoothing function $H(y)$ in physical space, one follows pretty much the same steps. The crucial formula (D.58) is then replaced by

$$H(y - y_1) - H(y - y_2) = \nabla_y^T[(y_2 - y_1) H(y - y_1) \tag{D.83}$$

which is the first term of a Taylor expansion with respect to the ratio of the scale of the intermolecular forces and the scale of the smoothing function.

D.5 Reduced Densities

In Hamiltonian dynamics, one makes use of the fact that particles are identical. Consider the N-particle Hamiltonian system. The N-particle density $\rho_t^{(N)}(q^{(1)}, p^{(1)}, \ldots, q^{(N)}, p^{(N)})$ in the $6N$-dimensional phase space Γ can be assumed to be symmetric in the indexes counting the particles since the particles are identical. Next introduce the reduced or marginal densities

$$\rho_t^{(s)}(q^{(1)}, p^{(1)}, \ldots, q^{(s)}, p^{(s)}) := \int dq^{(s+1)}dp^{(s+1)} \ldots dq^{(N)}dp^{(N)} \rho_t^{(N)}(q^{(1)}, p^{(1)}, \ldots, q^{(N)}, p^{(N)}) \tag{D.84}$$

These reduced densities often suffice for the calculation of expectation values of observables. Consider the expectation of the mass density

$$\begin{aligned}\langle m(y,t)\rangle &= \int_\Gamma dq^{(1)}dp^{(1)} \ldots dq^{(N)}dp^{(N)} \sum_{n=1}^{N} \delta(y - q_t^{(n)}(q,p)) m \rho_0^{(N)}(q^{(1)}, p^{(1)}, \ldots, q^{(N)}, p^{(N)})\\ &= \int_\Gamma dq^{(1)}dp^{(1)} \ldots dq^{(N)}dp^{(N)} \sum_{n=1}^{N} \delta(y - q^{(n)}) m \rho_t^{(N)}(q^{(1)}, p^{(1)}, \ldots, q^{(N)}, p^{(N)})\\ &= N \int_\Gamma dq^{(1)}dp^{(1)} \delta(y - q^{(1)}) m \rho_t^{(1)}(q^{(1)}, p^{(1)})\\ &= N \rho_t^{(1)}(y)\end{aligned} \tag{D.85}$$

where $\rho_t^{(1)}$ is the 1-particle density. Similarly, observables that are expressible as double sums only require the 2-particle density.

The time evolution of these reduced densities can be derived from the Liouville equation

$$\frac{D}{Dt}\rho_t^{(N)} = 0 \tag{D.86}$$

Consider specifically the N-particle system governed by the Hamilton function (D.44), with a (spherically symmetric) pair interaction potential and without external forcing for simplicity. Then

$$\begin{aligned}\frac{D}{Dt} &:= \partial_t + \sum_{n=1}^{N} \dot{q}^{(n)T} \nabla_{q^{(n)}} + \sum_{n=1}^{N} \dot{p}^{(n)T} \nabla_{p^{(n)}}\\ &= \partial_t + \sum_{n=1}^{N} \frac{1}{m} p^{(n)T} \nabla_{q^{(n)}} + \frac{1}{2} \sum_{\substack{n,m=1\\ n\neq m}}^{N} F^{(nm)T} (\nabla_{p^{(n)}} - \nabla_{p^{(m)}})\end{aligned} \tag{D.87}$$

where $F^{(nm)}$ is the force that particle m exerts on particle n. Since this interaction force is assumed to arise from a (spherically symmetric) pair interaction potential, it is anti-symmetric in the particle indexes, $F^{(nm)} = -F^{(mn)}$. Integrating the Liouville equation over the remaining $(N-s)$ variables $(q^{(s+1)}, p^{(s+1)}, \ldots, q^{(N)}, p^{(N)})$, taking into account that the density $\rho_t^{(N)}$ is symmetric and the interaction force $F^{(nm)}$ anti-symmetric in the particle indexes and assuming that all densities vanish at the boundaries, one obtains the celebrated BBGKY hierarchy

$$\frac{D}{Dt}\rho_t^{(s)} = -(N-s)\sum_{n=1}^{s}\int dq^{(s+1)}dp^{(s+1)}\,F^{(n(s+1))T}\nabla_{p^{(n)}}\rho_t^{(s+1)} \tag{D.88}$$

where

$$\frac{D}{Dt} := \partial_t + \sum_{n=1}^{s}\frac{1}{m}p^{(n)T}\nabla_{q^{(n)}} + \frac{1}{2}\sum_{\substack{n,m=1\\ n\neq m}}^{s} F^{(nm)T}(\nabla_{p^{(n)}} - \nabla_{p^{(m)}}) \tag{D.89}$$

The term on the right-hand side accounts for the interaction with the remaining $(N-s)$ particles. Because there are only pair interactions, only the density $\rho_t^{(s+1)}$ is required. The evolution of the s-particle density is coupled to the $(s+1)$-particle density.

Solving the BBGKY hierarchy is equivalent to solving the Liouville equation, unless one can close the hierarchy at some level. Closure at sth order means that all densities of order $s+1$ and higher are expressed in terms of densities of order s or smaller. A first-order closure thus expresses $\rho_t^{(2)}$ in terms of $\rho_t^{(1)}$. The most famous such closure is the molecular chaos assumption leading to the Boltzmann equation for $\rho_t^{(1)}$.

Appendix E
Mathematical Tools, Conventions, and Notation

E.1 Sets

We use the following standard symbols, names, and meaning/definition that occur in set theory:

- {}, set , collection of elements.
- $a \in A$, a is an element of the set A.
- $A \cap B$, intersection, the set of elements that belong to A and B.
- $A \cup B$, union, the set of elements that belong to A or B.
- $A \subseteq B$, subset, set A has fewer elements than B or is equal to B.
- $A \subset B$, strict subset, set A has fewer elements than B.
- $\emptyset$, the empty set, set containing no elements.
- $\mathcal{P}(X)$, power set of X, set of all subsets of X, the empty set belongs to the power set.
- $A \setminus B$, set difference, all elements that belong to A but not to B. The set $X \setminus A$ is called the complement of the set A.

The operations intersection and union are idempotent, commutative, associative, and distributive. A power set $\mathcal{P}(X)$ together with these operations thus forms a Boolean algebra. The set difference satisfies de Morgan's laws

$$A \setminus (B_1 \cap B_2) = (A \setminus B_1) \cup (A \setminus B_2) \tag{E.1}$$

$$A \setminus (B_1 \cup B_2) = (A \setminus B_1) \cap (A \setminus B_2) \tag{E.2}$$

A countable set of subsets A_i, $i = 1, \ldots$ is called a countable *partition* of the state space if $X = \cup_{i=1}^{\infty} A_i$ and $A_i \cap A_j = \emptyset$ for $i \neq j$.

Furthermore, we denote by $\mathbb{Z}, \mathbb{N}, \mathbb{Q}, \mathbb{R}$, and $\mathbb{C}$ the integers, natural numbers, rational numbers, real numbers, and complex numbers. $\mathbb{R}_+$ denotes the positive real numbers.

P. Müller, *Handbook of Dynamics and Probability*,
https://doi.org/10.1007/978-3-030-88486-4

Open intervals of the real axis are denoted by $]a, b[$ with $a < b \in \mathbb{R}$ and closed intervals by $[a, b]$. The convention for the open interval avoids confusion with the pair (a, b).

E.2 Functions

We use the words 'function' and 'map' interchangeably. A function f from a set X to a set Y is denoted by

$$\begin{aligned} f : X &\mapsto Y \\ x &\mapsto f(x) \end{aligned} \tag{E.3}$$

The following definition and theorems apply:

- The *composition* of two functions $g : X \mapsto Y$ and $f : Y \mapsto Z$ is the function $f \circ g : X \mapsto Z$ defined by $(f \circ g)(x) := f(g(x))$ for all $x \in X$.
- The *image* of a set $A \subset X$ under the map f is defined by

$$f(A) := \{f(x) : x \in A\} \subseteq Y \tag{E.4}$$

- The *pre-image* of a set $B \subset Y$ by

$$f^{-1}(B) := \{x : f(x) \in B\} \subseteq X \tag{E.5}$$

  If $h = f \circ g$ then $h^{-1} = g^{-1} \circ f^{-1}$.
- A function f is called *one-to-one* or an *injective* function if $f(x_1) = f(x_2)$ implies $x_1 = x_2$.
- A function f is called *onto* or a *surjective* function if $f(X) = Y$.
- A function f is called a *bijection* if it is one-to-one and onto. Then the *inverse map* $f^{-1} : Y \mapsto X$ exists and satisfies $(f^{-1} \circ f)(x) = x$ and $(f \circ f^{-1})(y) = y$ for all $x \in X$ and all $y \in Y$.

Note that we follow the customary practice and denote by the same symbol f^{-1} the inverse of a function which is a function from Y into X and the pre-image which is a function from the power set of Y into the power set of X. The pre-image always exists and satisfies $f(f^{-1}(B)) \subseteq B$ and $A \subseteq f^{-1}(f(A))$. The equality signs hold if f is invertible.

The *indicator function* of a subset A of X is a function

$$\chi_A : X \mapsto \{0, 1\} \tag{E.6}$$

where

$$\chi_A(x) = \begin{cases} 1 \; if \;\; x \in A \\ \\ 0 \; if \;\; x \notin A \end{cases} \tag{E.7}$$

It satisfies

$$\chi_{A \cup B} = \chi_A + \chi_B - \chi_{A \cap B} \tag{E.8}$$

$$\chi_{A \cap B} = \chi_A \chi_B \tag{E.9}$$

$$\chi_{A^c} = 1 - \chi_A \tag{E.10}$$

E.3 Groups

A set G with an operation

$$G \times G \mapsto G \tag{E.11}$$

$$(a, b) \mapsto a \circ b \tag{E.12}$$

is called a group if

(1) the operation is associative, $a \circ (b \circ c) = (a \circ b) \circ c$,
(2) there exist a neutral element e such that $a \circ e = e \circ a = a$, and
(3) there exists for each element $a \in G$ an inverse element a^{-1} such that $a^{-1} \circ a = a \circ a^{-1} = e$.

The operation '$\circ$' is sometimes called the product. If in addition $a \circ b = b \circ a$, then the group is called commutative or Abelian.

A semi-group is a set G with an operation that is associative but does not have inverse elements. There might be a neutral element. A semi-group might be commutative.

The integers $\mathbb{Z}$ and the real numbers $\mathbb{R}$ are Abelian groups under addition with the neutral element 0. The set $\mathbb{Z} \setminus \{0\}$ and $\mathbb{R} \setminus \{0\}$ are Abelian groups under multiplication with neutral element 1. The natural numbers and the non-negative real numbers $\mathbb{R}_+$ are semi-groups under addition and multiplication.

Two groups G and G' are *isomorphic* if there exists a map $f : G \mapsto G', a \mapsto a'$ such that $f(a \circ a') = f(a) \circ f(a')$.

E.4 Vector Spaces

To avoid notational and technical complications, this appendix restricts itself to vector spaces of countable dimension, finite or countably infinite. There are, of course, vector spaces with a dimension of higher cardinality and some of our definitions and statements apply to them, but theory is simplest for countable dimensions.

Vector spaces. A complex (real) vector space is a set X with two operations:

(1) An addition

$$\begin{aligned} X \times X &\mapsto X \\ (x, y) &\mapsto x + y \end{aligned} \tag{E.13}$$

(2) A scalar multiplication

$$\begin{aligned} \mathbb{R}(\mathbb{C}) \times X &\mapsto X \\ (a, x) &\mapsto ax \end{aligned} \tag{E.14}$$

If the scalar a is a real number the vector space is called a *real* vector space, and if it is a complex number it is called a *complex* vector space. Addition must satisfy the following rules:

(1) $(x + y) + z = z + (y + z)$ (associativity).
(2) $x + y = y + x$ (commutativity).
(3) there exists a null vector 0 such that $x + 0 = x$.
(4) for each $x \in X$, there exists a negative vector $-x \in X$ such that $x + (-x) = 0$.

The scalar multiplication must satisfy

(1) $a(x + y) = ax + ay$ (distributivity).
(2) $(a + b)x = ax + bx$ (distributivity).
(3) $a(bx) = (ab)x$ (associativity).
(4) $1x = x$ and $0x = 0$.

Hamel basis. A finite set $e^{(1)}, \dots, e^{(N)}$ of vectors is called linearly independent if

$$\sum_{n}^{N} x_n e^{(n)} = 0 \tag{E.15}$$

implies $x_1 = x_2 = \dots = x_N = 0$.

Given a set of vectors, $B = \{e^{(i)} | i \in I\}$ where the counting set I may have any cardinality. One can form the *finite linear combinations*

$$x = \sum_{k=1}^{K} x_{i_k} e^{(i_k)} \tag{E.16}$$

If every vector $x \in X$ can be written as such a finite linear combination, the set B of vectors is called *complete*. If B is complete and if all finite subsets of B are linearly independent, then the set B is called a *basis*. The representation (E.16) is then unique and the expansion coefficients x_{i_k} are called the *components* of the vector x. Note the adjective 'finite' in the above definition. Every vector space has such a basis. All bases of a vector space have the same cardinality, called the *dimension* of the vector space. There exist vector spaces of finite, countably infinite, and uncountably infinite dimensions. The proof of the theorem that every vector space has a basis requires the axiom of choice. Thus, there is in general no constructive way to find or construct a basis for vector spaces of infinite dimension. This specific concept of a basis using finite linear combinations is thus only useful for finite dimensions but not for infinite dimensions where such bases are called Hamel bases. A more useful definition of a basis of infinite-dimensional vector spaces is obtained when one allows for infinite linear combinations. This requires the vector space to be a Euclidean or unitary vector space.

Scalar product. Given a complex (real) vector space X, a scalar (inner) product is a map

$$\begin{aligned} X \times X &\mapsto \mathbb{C}(\mathbb{R}) \\ (x, y) &\mapsto \langle x, y \rangle \end{aligned} \tag{E.17}$$

that satisfies the following axioms for all vectors $x, y, z \in X$ and all scalars $a, b, \in \mathbb{C}(\mathbb{R})$

(1) Symmetry

$$\langle x, y \rangle^* = \langle y, x \rangle \tag{E.18}$$

where * denotes complex conjugation.

(2) Linearity

$$\langle x, ay + bz \rangle = a\langle x, y \rangle + b\langle x, z \rangle \tag{E.19}$$

(3) Positive definiteness

$$\langle x, x \rangle \geq 0 \tag{E.20}$$

with strict equality holding only for $x = o$.

The scalar product satisfies the *Schwarz inequality*

$$|\langle x, y \rangle|^2 \leq \langle x, x \rangle \langle y, y \rangle \tag{E.21}$$

Euclidean and unitary vector spaces. A real vector space X endowed with a scalar (inner) product is called a Euclidean vector space. A complex vector space endowed with a scalar product is called a unitary vector space. Often, the definition of a Euclidean vector space includes the requirement that the vector space is of finite dimension.

Having a scalar product allows one to define the length of a vector x by $|x| := \sqrt{\langle x, x \rangle}$, the orthogonality of two vectors x and y by $\langle x, y \rangle = 0$, and the angle φ between two vectors by $\cos^2 \varphi = \langle x, y \rangle \langle y, x \rangle / (|x||y|)$.

The scalar product induces a *norm*

$$\| x \| := (\langle x, x \rangle)^{1/2} \tag{E.22}$$

and a *metric*

$$d(x, y) := \| x - y \| \tag{E.23}$$

Having a norm allows one to define convergence of vectors. A sequence of vectors $x^{(1)}, x^{(2)}, \ldots$ in a Euclidean or unitary vector space is said to *converge strongly* to a vector x if

$$\lim_{n\to\infty} \| x^{(n)} - x \| = 0 \tag{E.24}$$

A sequence is said to be a *Cauchy sequence* if for every $\epsilon > 0$, there exists a natural number N such that $\| x^{(n)} - x^{(m)} \| \leq \epsilon$ for $n, m > N$. The triangle inequality implies that every sequence that converges strongly is also a Cauchy sequence. The inverse is not necessarily true. If it is true, i.e., if every Cauchy sequence converges strongly to a vector in the space, then the vector space is called *complete*.

Basis. For Euclidean or unitary vector spaces of countably infinite dimensions, one can introduce a more useful definition of bases since the existence of a scalar product allows one to define a norm, a distance, and a topology, and one can hence assess whether infinite sums converge. Thus, given a countably infinite set $B = \{e^{(1)}, e^{(2)}, \ldots\}$ of vectors, the *infinite linear combinations*

$$x = \sum_{n=1}^{\infty} x_n e^{(n)} \tag{E.25}$$

may exist. When the set B is complete and linearly independent, meaning again that every vector $x \in X$ can be expressed as such an infinite linear combination and that $x = 0$ implies $x_n = 0$ for all n, then the set B is simply called a basis, distinct from the Hamel bases introduced above.

A countable (finite or infinite) basis is called *orthonormal* if $\langle e^{(i)}, e^{(j)} \rangle = \delta_{ij}$ for all $i, j = 1, \ldots$. Such basis vectors have length one and are mutually orthogonal. In terms of such orthogonal bases, the components of a vector are given by

$$x_n = \langle e^{(n)}, x \rangle \tag{E.26}$$

and the matrix of a linear operator by

$$A_{nm} = \langle e^{(n)}, \tilde{A}(e^{(m)}) \rangle \tag{E.27}$$

Change of basis in finite dimension: if one changes from a basis $e^{(1)}, \ldots, e^{(N)}$ to a basis $\hat{e}^{(1)}, \ldots, \hat{e}^{(N)}$ by a coordinate transformation $e^{(i)} = \sum_{j=1}^{N} B_{ji} \hat{e}^{(j)}$, then the coefficients in

$$x = \sum_{n=1}^{N} x_n e^{(n)} = \sum_{n=1}^{N} \hat{x}_n \hat{e}^{(n)} \tag{E.28}$$

transform as

$$\hat{x}_n = B_{nm} x_m \tag{E.29}$$

The scalar product in a finite-dimensional vector space X with basis $(e^{(1)}, \ldots, e^{(N)})$ is completely specified by the matrix S with components

$$S_{ij} := \langle e^{(i)}, e^{(j)} \rangle \tag{E.30}$$

such that for arbitrary vectors $x = \sum_{i=1}^{N} x_i e^{(i)}$ and $y = \sum_{i=1}^{N} y_i e^{(i)}$, the scalar product is given by

$$\langle x, y \rangle = \sum_{i=1}^{N} \sum_{j=1}^{N} x_i^* S_{ij} y_j = x^\dagger S y \tag{E.31}$$

The matrix S is a self-adjoint matrix that is positive definite and has hence positive eigenvalues. If one changes from a basis $e^{(1)}, \ldots, e^{(N)}$ to a basis $\hat{e}^{(1)}, \ldots, \hat{e}^{(N)}$ by a coordinate transformation $e^{(i)} = \sum_{j=1}^{N} B_{ji} \hat{e}^{(j)}$, then the matrix of the scalar product transforms as

$$\hat{S} = (B^\dagger)^{-1} S B^{-1} \tag{E.32}$$

Note that the matrix S of the scalar product transforms different from the matrix of a linear operator. If one chooses the special transformation

$$(B^{-1})_{nm} = \frac{1}{\sqrt{\lambda^{(m)})}} s_n^{(m)} \tag{E.33}$$

where $s^{(m)}$ is the mth eigenvector of the matrix S, then $\hat{S} = I$. The new basis is orthonormal, i.e., if $\langle e^{(i)}, e^{(j)} \rangle = \delta_{ij}$ for all $i, j = 1, \ldots, N$ and the scalar product is given by the identity matrix. Thus, one can always find a basis such that

$$\langle x, y \rangle = \sum_{i=1}^{N} x_i^* y_i = x^\dagger y \tag{E.34}$$

This scalar product is called the canonical scalar product. All scalar products of a finite-dimensional vector space can thus be transformed to the canonical scalar product by a suitable coordinate transformation.

A subset of a vector space is called a *(linear) subspace* when it is also a vector space. If $e^{(n_1)}, e^{(n_2)}, \ldots$ is a set of mutually independent vectors, then the subspace consisting of all their linear combinations is called the subspace spanned by these vectors. A subspace is called topologically closed if the limit vector of any convergent series of vectors also lies within the subspace.

Direct sum. The direct sum (or Cartesian product) of two vector spaces X and Y is a vector space written as $X \times Y$ or $X \oplus Y$ that consists of all pairs (x, y) with the operations

$$a(x, y) + b(x', y') = (ax + bx', ay + by') \tag{E.35}$$

If the vector spaces X and Y have dimensions N_X and N_Y, then the direct sum has dimension $N_X + N_Y$.

Tensor product. The tensor product of two vector spaces X of dimension N_X and Y of dimension N_Y is a vector space of dimension $N_X N_Y$, written as $X \otimes Y$. The basis vectors of the tensor product can be constructed as follows: If $e_X^{(n)}$, $n = 1, \ldots, N_X$ is a basis in X and $e_Y^{(m)}$, $m = 1, \ldots, N_Y$ is a basis in Y, then $e_X^{(n)} \otimes e_Y^{(m)}$ is a basis of $X \otimes Y$. Thus, every vector $x \in X \otimes Y$ can be written as

$$x = \sum_{n=1}^{N_X} \sum_{m=1}^{N_Y} x_{nm} e_X^{(n)} \otimes e_Y^{(m)} \tag{E.36}$$

with $N_X N_Y$ components x_{nm}.

If $x = \sum_{n=1}^{N_X} x_n e_X^{(n)}$ and $y = \sum_{m=1}^{N_Y} y_m e_Y^{(m)}$, then the tensor product of these two vectors takes the form

$$x = x \otimes y = \sum_{n=1}^{N_X} \sum_{m=1}^{N_Y} x_n x_m e_X^{(n)} \otimes e_Y^{(m)} \tag{E.37}$$

The tensor product of two vectors obeys the rules

$$c(x^{(1)} \otimes x^{(2)}) = (cx^{(1)}) \otimes x^{(2)} = x^{(1)} \otimes (ce^{(2)}) \tag{E.38}$$

and

$$(ax^{(1)} + by^{(1)}) \otimes x^{(2)} = a(x^{(1)} \otimes x^{(2)}) + b(y^{(1)} \otimes x^{(2)}) \tag{E.39}$$
$$x^{(1)} \otimes (ax^{(2)} + by^{(2)}) = a(x^{(1)} \otimes x^{(2)}) + b(x^{(1)} \otimes y^{(2)}) \tag{E.40}$$

Hilbert space. A Hilbert space is a complete unitary vector space. It is generally denoted by the symbol $\mathcal{H}$. A Hilbert space is called separable if it has a countable orthonormal basis.

E.5 Normed Spaces

A normed space is a vector space X together with a map

$$\begin{aligned} X &\mapsto \mathbb{R}_+ \\ x &\mapsto \| x \| \end{aligned} \tag{E.41}$$

that satisfies the axioms

$$\parallel x \parallel > 0 \ \text{ iff } \ x \neq 0 \tag{E.42}$$

$$\parallel cx \parallel = |c| \parallel x \parallel \quad \text{for } c \in \mathbb{C}(\mathbb{R}) \tag{E.43}$$

$$\parallel x + y \parallel \leq \parallel x \parallel + \parallel y \parallel \quad \text{triangle inequality} \tag{E.44}$$

A Euclidean or unitary vector space is a normed space by virtue of the definition

$$\parallel x \parallel := (\langle x, x \rangle)^{1/2} \tag{E.45}$$

E.6 Metric Spaces

A metric space is a set X together with the distance between two elements

$$\begin{aligned} X \times X &\mapsto \mathbb{R}_+ \\ (x, y) &\mapsto d(x, y) \end{aligned} \tag{E.46}$$

that satisfies the axioms

$$d(x, y) \geq 0 \tag{E.47}$$

$$d(x, y) = 0 \ \text{ iff } \ x = y \tag{E.48}$$

$$d(x, y) = d(y, x) \tag{E.49}$$

$$d(x, z) \leq d(x, y) + d(y, z) \quad \text{(triangle inequality)} \tag{E.50}$$

Normed spaces are metric spaces by virtue of the definition

$$d(x, y) := \parallel x - y \parallel \tag{E.51}$$

E.7 Limits

Limits are a well-defined concept in topology. A sequence $x_1, x_2, \ldots$ is said to converge toward x if for every neighborhood N of x, there exists a n_N such that $x_n \in N$ $\forall n > n_N$. If this is the case, one writes

$$\lim_{n\to\infty} x_n = x \tag{E.52}$$

When performing limits, some care must be taken. One has to realize that limits do not interchange. A counter-example is the double sequence

$$x_{nm} = \left(\frac{1}{n}\right)^{1/m} \tag{E.53}$$

When the limit with respect to n is carried out first, the double limit is zero. When the limit with respect to m is carried out first, the double limit is one.

The definition of the limit also applies to sums

$$\lim_{N\to\infty} \sum_{n=1}^{N} x_n = s \tag{E.54}$$

if the sequence of partial sums $s_N := \sum_{n=1}^{N} x_n$ converges toward s. Again, care must be taken. Limits do not interchange with summation. A counter-example is the double sequence

$$x_{nm} := \delta_{nm} \tag{E.55}$$

When the sum over n is first carried out and then the limit with respect to m, the result is one. When the limit with respect to m is first carried out and then the summation over n, the result is zero. Also, limits of sums are not infinite sums that can be manipulated like finite sums. Consider the limit s

$$s = \lim_{N\to\infty} \sum_{n=1}^{N} x_n \tag{E.56}$$

One might write

$$x_n = (s_n - a) - (s_{n-1} - a) \tag{E.57}$$

where a is an arbitrary number and $s_0 = 0$ for convenience. Thus, s can be written as an infinite

$$\begin{aligned} s = & \qquad\qquad (s_1 - a) + (s_2 - a) + \dots \\ & -(s_0 - a) \quad -(s_1 - a) - (s_2 - a) - \dots \\ = & -(s_0 - a) = a \end{aligned} \tag{E.58}$$

since all the terms cancel. Thus, every series can be made to converge to an arbitrary number. The recipe that guards against mistakes like these is to first perform any operations on the finite sum and then take the limit. Had we done this in this example, we would have obtained

$$\begin{aligned} \lim_{N\to\infty} \sum_{n=1}^{N} [(s_n - a) - (s_{n-1} - a)] = & \lim_{N\to\infty} [(s_1 - a) + \dots + (s_N - a) - (s_0 - a) \\ & - (s_1 - a) - \dots - (s_{N-1} - a)] = \lim_{N\to\infty} s_N \end{aligned} \tag{E.59}$$

the correct result.

E.8 Linear Operators

A map $A : X \mapsto X$ on a vector space X is called a *linear operator* if

$$A(a_1 x_1 + a_2 x_2) = a_1 A(x_1) + a_2 A(x_2) \tag{E.60}$$

for all $x_1, x_2 \in X$ and $a_1, a_2 \in \mathbb{C}(\mathbb{R})$.

A linear operator $\tilde{A}$ on a finite-dimensional vector space X with orthonormal basis $\{e^{(1)}, \dots, e^{(N)}\}$ can be represented by a square $N \times N$ matrix A with components A_{nm} such that

$$\tilde{A}(e^{(n)}) = \sum_{m=1}^{N} A_{mn} e^{(m)} \tag{E.61}$$

and $y = \tilde{A}(x)$ can be written as

$$y_n = \sum_{m=1}^{N} A_{nm} x_m \tag{E.62}$$

where x_m and y_n are the components of the vectors x and y in that basis. If one changes the basis by the coordinate transformation $e^{(i)} = \sum_{j=1}^{N} B_{ji} \hat{e}^{(j)}$, then the matrix transforms as

$$\hat{A} = BAB^{-1} \tag{E.63}$$

Among the linear operators, there are three important classes: self-adjoint, unitary, and projection operators.

Self-adjoint operators. The adjoint (or Hermitian conjugate) $A^\dagger$ of an operator A is defined by

$$\langle x, A^\dagger y\rangle = \langle Ax, y\rangle \tag{E.64}$$

for all $x, y \in X$. In matrix notation, $A^\dagger_{ij} := A^*_{ji}$.

The operator is called self-adjoint if $A = A^\dagger$. In matrix notation, $A^*_{ji} = A_{ij}$. For real vector spaces, self-adjoint operators can be represented by symmetric matrices. Self-adjoint operators have real eigenvectors. The eigenvectors corresponding to two different eigenvalues are orthogonal.

Unitary operators. An operator U is called *unitary* if

$$U^\dagger U = UU^\dagger = I \tag{E.65}$$

They are invertible and preserve the scalar product

$$\langle U(x), U(y)\rangle = \langle x, y\rangle \tag{E.66}$$

for all $x, y \in X$. The eigenvalues of unitary operators are complex numbers lying on the unit circle. The eigenvectors corresponding to two different eigenvalues are orthogonal. If A is a self-adjoint operator, then e^{iA} is a unitary operator.

Projection operators. Given a topologically closed subspace $W \subset X$, the orthogonal complement $W_\perp$ is the set of all vectors that are orthogonal to every vector in W. The orthogonal complement is also topologically closed. Any vector $x \in X$ can then be uniquely decomposed

$$x = x_W + x_{W_\perp} \tag{E.67}$$

where the vectors x_W and $x_{W_\perp}$ lie in W and $W_\perp$, respectively. The projection operator $\hat{P}_W$ is then defined by

$$\hat{P}_W x := x_W \tag{E.68}$$

$\hat{P}_W$ is a self-adjoint operator with eigenvalues 0 and 1. The eigenspaces are W and $W_\perp$. The qualifier 'topological closed' is necessary for the above definition since in a vector space of infinite dimension, not every subspace is topologically closed. The standard example is the vector space consisting of all sequences $x_1, x_2, \ldots$ that are square summable, i.e., for which $\sum_{n=1}^{\infty} x_n < \infty$. All sequences which only have a finite number of non-zero elements form a subspace that is, however, not closed. It does not contain all its limit points.

If $W_1, W_2, \ldots$ is a collection of closed subspaces, then X is said to be the direct sum of the W_n if the subspaces are mutually orthogonal and each vector $x \in X$ can be written as

$$x = \sum_{n=1}^{\infty} \hat{P}_{W_n} x \tag{E.69}$$

which implies $\sum_{n=1}^{\infty} \hat{P}_{W_n} = I$, the so-called resolution of the identity.

E.9 Matrix Notation

The tensor or index notation is the most appropriate notation when dealing with vectors and tensors. It can handle curvilinear coordinate systems and tensors of arbitrary order and can distinguish between contra-variant and covariant components in the case of non-orthogonal coordinate systems. For our purposes, it is sufficient to use matrix notation which applies to vectors and second-order tensors when represented in Cartesian coordinates. We use the following notation and conventions.

The Cartesian components of a vector x are represented by the *column vector*

$$x = \begin{pmatrix} x_1 \\ \cdot \\ \cdot \\ \cdot \\ x_N \end{pmatrix} \tag{E.70}$$

The *transposed vector* is given by the *row vector*

$$x^T = (x_1, \ldots, x_N) \tag{E.71}$$

A matrix is given by

$$A = \begin{pmatrix} A_{11} & \ldots & A_{1N} \\ \cdot & & \cdot \\ \cdot & & \cdot \\ \cdot & & \cdot \\ A_{N1} & \ldots & A_{NN} \end{pmatrix} \tag{E.72}$$

where the first subscript indicates the row and the second subscript the column number.

The following definitions and rules apply:

- The transpose A^T of a matrix A is given by

$$(A^T)_{ij} = A_{ji} \tag{E.73}$$

- The scalar product between two vectors x and y is a scalar given by

$$x^T y = \sum_{n=1}^{N} x_n y_n \tag{E.74}$$

- The tensor product xy^T of two vectors x and y is a matrix with components

$$(xy^T)_{ij} = x_i y_j \tag{E.75}$$

– The product between a matrix A and a column vector x is a column vector with components

$$(Ax)_i = \sum_{j=1}^{N} A_{ij} x_j \tag{E.76}$$

– The product between a row vector x^T and a matrix A is a row vector with components

$$(x^T A)_i = \sum_{j=1}^{N} x_j A_{ji} \tag{E.77}$$

with the general theorem $(Ax)^T = x^T A^T$.

– The product between two matrices A and B is a matrix with components

$$(AB)_{ij} = \sum_{k=1}^{N} A_{ik} B_{kj} \tag{E.78}$$

– The trace of a matrix A is a scalar given by

$$Tr(A) = \sum_{i=1}^{N} A_{ii} \tag{E.79}$$

– The trace of the product of two matrices A and B is the scalar given by

$$Tr(AB) = \sum_{i=1}^{N} \sum_{j=1}^{N} A_{ji} B_{ij} \tag{E.80}$$

E.10 Dirac Notation

The Dirac notation allows for a very elegant and economic formulation of linear algebra and is the notation is the notation of choice in quantum mechanics. The Dirac notation represents vectors $x \in \mathcal{H}$ ($\mathcal{H}$ being a Hilbert space) by ket-vectors $|x\rangle$ and the Hermitian conjugate $x^\dagger$ by the bra-vector $\langle x|$. The scalar product can then be written as

$$\langle x, y\rangle = x^\dagger y =: \langle x|y\rangle \tag{E.81}$$

The expansion theorem in a separable Hilbert space takes the form

$$|x\rangle = \sum_{n=1}^{\infty} \langle e^{(n)}|x\rangle |e^{(n)}\rangle \tag{E.82}$$

where $|e^{(1)}\rangle, |e^{(2)}\rangle, \ldots$ is an orthonormal basis.

If A is a self-adjoint operator, the scalar product can be written symmetrically as

$$\langle Ax, y\rangle = \langle x, Ay\rangle =: \langle x|A|y\rangle \tag{E.83}$$

Given an orthonormal basis, operators can be represented by their matrix elements

$$A_{ij} := \langle e^{(i)}|A|e^{(j)}\rangle \tag{E.84}$$

For self-adjoint operators, the matrix is Hermitian $A_{ij} = A^*_{ji}$.

The eigenvalues $a^{(i)}$ and eigenvectors $|a^{(i)}\rangle$ of an operator A are the solutions of the eigenequation

$$A|a^{(i)}\rangle = a^{(i)}|a^{(i)}\rangle \tag{E.85}$$

Self-adjoint operators have real eigenvalues $a^{(i)}$. If the eigenvalues are different, the eigenvectors are orthogonal. If the eigenvalue $a^{(i)}$ is degenerated with degeneracy $d(i)$, then there exists $d(i)$ orthogonal eigenvectors $|a^{(i,1)}\rangle, \ldots, |a^{(i,d(i))}\rangle$. The eigenvectors can be normalized such that the vectors $|a^{(1,1)}\rangle, \ldots, |a^{(1,d(1))}\rangle, |a^{(2,1)}\rangle, \ldots, |a^{(2,d(2))}\rangle, \ldots$ form an orthonormal basis.

The projection operator that projects onto the subspace spanned by the vector $|x\rangle$ can be written in Dirac notation

$$\hat{P}_x = |x\rangle\langle x| \tag{E.86}$$

If $\{\||e^{(1)}\rangle, |e^{(2)}\rangle, \ldots\}$ is an orthonormal basis, the resolution of the identity reads

$$I = \sum_{n=1}^{\infty} |e^{(n)}\rangle\langle e^{(n)}| \tag{E.87}$$

an identity which is useful in proofs.

E.11 Eigenvectors and Eigenvalues

Let A be a complex $N \times N$ matrix. Its eigenvectors $e^{(s)}$ and eigenvalues $\lambda^{(s)}$ are the solutions of the equation

$$Ae^{(s)} = \lambda^{(s)} e^{(s)} \tag{E.88}$$

The eigenvalues are the N zeros of the *characteristic polynomial* $det(A - \lambda I) = 0$. If d roots coincide, the corresponding eigenvalue λ is said to be degenerate and to have the algebraic multiplicity d. The geometric multiplicity $K \leq d$ of an eigenvalue λ is the dimension of the null space of the matrix $\lambda I - A$. There exist K independent eigenvectors in this null space. If there is at least one eigenvalue whose geometric multiplicity is smaller than its algebraic multiplicity, the matrix is called defective.

The eigenvector $\hat{e}^{(s)}$ of the adjoint matrix are defined by

$$A^{\dagger}\hat{e}^{(s)} = \kappa^{(s)}\hat{e}^{(s)} \tag{E.89}$$

which can also be written as

$$\hat{e}^{(s)\dagger}A = \kappa^{(s)*}\hat{e}^{(s)\dagger} \tag{E.90}$$

For this reason, the eigenvectors $\hat{e}^{(s)}$ are also called the *left eigenvectors* and the eigenvectors $e^{(s)}$ the *right eigenvectors* of the matrix A. Their eigenvalues are related by $\lambda^{(s)*} = \kappa^{(s)}$.

If the matrix is non-defective, the eigenvectors form a basis and the left and right eigenvectors can be normalized to satisfy the completeness relation

$$\sum_{s=1}^{N} e^{(s)}\hat{e}^{(s)\dagger} = I \tag{E.91}$$

and the orthogonality relation

$$\hat{e}^{(s)\dagger}e^{(s')} = \delta_{ss'} \tag{E.92}$$

The completeness relation can also be written as $E\hat{E}^{\dagger} = I$ where E is the matrix whose columns are the right eigenvectors and $\hat{E}^{\dagger}$ the matrix whose rows are the left eigenvectors. The matrix $\hat{E}^{\dagger}$ is thus the inverse of E.

Expansion of any vector x with respect to the eigenbasis gives

$$x = \sum_{s=1}^{N} a^{(s)}e^{(s)} \tag{E.93}$$

where the components or amplitudes

$$a^{(s)} = \hat{e}^{(s)\dagger}x \tag{E.94}$$

are the projection of x onto the left eigenvectors.

If the matrix is normal, $A^{\dagger}A = AA^{\dagger}$, then the left and right eigenvectors are identical and can be normalized to form an orthonormal basis.

A normal matrix is Hermitian if $A = A^{\dagger}$. The eigenvalues are then real. A Hermitian real matrix is a symmetric matrix.

If the matrix is real, the eigenvalues and eigenvectors come in complex conjugate pairs.

E.12 Topological Sets

A topological set is a set X together with a family $\mathcal{T}$ of subsets $T \subset X$ called *open sets* that satisfy the axioms

$$\emptyset, X \in \mathcal{T} \tag{E.95}$$

$$\bigcup_{i\in I} T_i \in \mathcal{T} \text{ if } T_i \in \mathcal{T} \tag{E.96}$$

$$T_1 \cap T_2 \in \mathcal{T} \text{ if } T_1, T_2 \in \mathcal{T} \tag{E.97}$$

where I is an arbitrary index set. The empty set and X are open. Arbitrary unions of open sets are open. The finite intersection of open sets is open.

Further useful definitions and theorems are as follows.

- The complement of an open set is a *closed set*. The empty set and X are closed. The arbitrary intersection of closed sets is closed. The finite union of closed sets is closed.
- A set N is called a *neighborhood* of a point $x \in X$ if there exists an open set T such that $x \in T \subseteq N$.
- Given an arbitrary set B, a point $x \in X$ is called an *interior point* of B if B is a neighborhood of x and an exterior point if $X \setminus B$ is a neighborhood of x.
- A point $x \in X$ is a *limit point* of B if every open set containing x contains at least one point of B different from x itself. The set of all limiting points of B is denoted by $L(B)$. A closed set contains all its limit points, $L(B) \subseteq B$.
- The *interior* $\mathring{B}$ of a set B consists of all interior points. The *closure* $\overline{B}$ of a set B consists of all points that are not exterior points of B. The closure $\overline{B}$ is a closed set. The closure is given by $\overline{B} = B \cup L(B)$.
- A set A is called *dense* in X if $\overline{A} = X$. Equivalently, a set A is dense if any neighborhood of any point $x \in X$ contains at least one point of A. Rational numbers $\mathbb{Q}$ and the irrational numbers $\mathbb{R} \setminus \mathbb{Q}$ are dense in $\mathbb{R}$.
- A set A is called *nowhere dense* in X if its closure has an empty interior.
- A set is called *separable* if there exists a countable dense sequence of points such that every non-empty open set contains at least one point of the sequence. The real line is separable with the rational numbers being one possible dense subset.
- A family of sets $(A_i)_{i\in I}$ is called a *cover* if $X \subseteq \cup_{i\in I} A_i$.
  A topological set is called *compact* if there exists a finite sub-cover for every open cover of X.
- A topological set is called a *Hausdorff space* if distinct points have distinct neighborhoods. If $x \neq y$, there exists neighborhood N of x and a neighborhood M of y such that $N \cap M = \emptyset$. The real line with the standard topology is a Hausdorff space.

- A sequence $(x_n)_{n \in \mathbb{N}}$ is said to *converge* to a point x if for every neighborhood N of x, there exists a n_0 such that $x_n \in N$ for $n \geq n_0$.
- A function $f : X \mapsto Y$ between two topological spaces is called *continuous* if the pre-image of every open set is open or, equivalently, if $\lim_{n\to\infty} f(x_n) = f(x)$ for all sequences x_n that converge to x.

An important theorem concerning continuous functions is as follows: If $f : X \mapsto Y$ is a continuous function from a separable space X with a countable dense subset Q into a Hausdorff space Y, then the function is completely determined by the values of the function on the countable dense subset Q. For each $x \in X$, there exists a sequence $(x_1, x_2, \ldots)$ with $x_n \in Q$ and $\lim_{n\to\infty} x_n = x$ such that $\lim_{n\to\infty} f(x_n) = f(x)$.

If the underlying set X is a metric space with a distance d, some of these definitions may be reformulated in terms of *open balls* (spheres) that are defined by

$$B(x, \epsilon) := \{y : d(x, y) < \epsilon\} \tag{E.98}$$

for all $\epsilon > 0$.

A set A is open if for every $x \in A$, there exists an $\epsilon > 0$ such that $B(x, \epsilon) \subseteq A$.

A set $N \subset X$ is a neighborhood of a point $x \in X$ if there exists an $\epsilon > 0$ such that $B(x, \epsilon) \subseteq N$.

A point x is a limit point of A if $B(x, \epsilon) \cap A \neq \emptyset$.

A sequence $(x_n)_{n \in \mathbb{N}}$ converges toward x if $\lim_{n\to\infty} d(x_n, x) \to 0$.

A function $f : X \mapsto X$ is continuous if for every $\epsilon > 0$, there exists a $\delta > 0$ such that $d(x, y) < \delta$ implies $d(f(x), f(y)) < \epsilon$.

E.13 Measure Spaces

A measure space is a triple $\{X, \mathcal{A}, \mu\}$ consisting of a basic set X, a σ-algebra $\mathcal{A}$, and a measure μ.

Sigma-algebras. Let $A \subseteq X$ be a subset of the basic set X. A family $\mathcal{A}$ of such subsets is called a σ-algebra if

$$\emptyset \in \mathcal{A} \tag{E.99}$$

$$A \in \mathcal{A} \Rightarrow X \setminus A \in \mathcal{A} \tag{E.100}$$

$$A_n \in \mathcal{A}, n \in \mathbb{N} \Rightarrow \bigcup_{n=1}^{\infty} A_n \in \mathcal{A} \tag{E.101}$$

This definition of a σ-algebra ensures that $X \in \mathcal{A}$ and that one can perform all countable set operations without leaving the family though one may leave the family by uncountable operations. Two trivial examples of σ-algebras are $\mathcal{A} = \{\emptyset, X\}$ which is the smallest σ-algebra of X and $\mathcal{A} = \mathcal{P}(X)$, the power set of X which is largest σ-algebra of X. There are usually additional families of subsets that form a σ-algebra.

For any $A \subset X$, the set $\mathcal{A} = \{\emptyset, A, X \setminus A, X\}$ forms a σ-algebra. σ-algebras are thus not given but must be chosen or assigned.

There is an important *extension theorem* for σ-algebras. If $\mathcal{E}$ is an arbitrary family of subsets of X, then there exists a smallest σ-algebra $\mathcal{A}(\mathcal{E})$ of X which contains all sets of $\mathcal{E}$. $\mathcal{A}(\mathcal{E})$ is called the σ-algebra generated by $\mathcal{E}$.

Borel set. The σ-algebra generated by the finite half-open intervals $]a, b]$ of the real line is called the σ-algebra of Borel sets and denoted by $\mathcal{B}$. The Borel set is larger than the set of all countable unions of half-open intervals and smaller than the power set of $\mathbb{R}$, though it is hard to explicitly construct non-Borel sets. The analogous construction in $\mathbb{R}^N$ leads to the Borel set $\mathcal{B}^N$.

Measures. Given a σ-algebra $\mathcal{A}$, a measure is a function $\mu : \mathcal{A} \mapsto [0, \infty]$ with properties

$$\mu(\emptyset) = 0 \tag{E.102}$$

$$\mu\left(\bigcup_{n=1}^{\infty} A_n\right) = \sum_{n=1}^{\infty} \mu(A_n) \tag{E.103}$$

for all families $(A_n)_{n \in \mathbb{N}}$ of pairwise disjoint members of $\mathcal{A}$. The second property is called σ-additivity.

A measure μ thus assigns a non-negative number, including infinity, to each element A of the σ-algebra $\mathcal{A}$.

A measure is called finite if $\mu(X) < \infty$. Such measures can be normalized to 1 and then become probability measures. A measure is called σ-finite if there exists disjunct subsets $A_1, A_2, \ldots$ such that $X = \cup_{n=1}^{\infty} A_n$ and $\mu(A_n) < \infty$. It is a weaker condition and usually sufficient in proofs. A statement is called almost true or true almost everywhere if the set A for which it is not true has measure zero.

There are many possible measures on a given σ-algebra:

Discrete measures. Take a point $x_0 \in X$ and an arbitrary σ-algebra, then

$$\delta_{x_0}(A) := \chi_A(x_0) \tag{E.104}$$

defines a measure concentrated on x_0, called the Dirac measure. Here χ_A is the indicator function of A. If $x_i \in X, i = 1, 2, \ldots$ is a countable number of points, then

$$\mu = \sum_{i=1}^{\infty} \mu(x_i)\delta_{x_i} \tag{E.105}$$

with $0 \leq \mu(i) \leq \infty$ defines a measure on $\mathcal{P}(X)$. It is called a discrete measure with support $x_1, x_2, \ldots$. If X has a countable number of elements, then all measures on $\{X, \mathcal{A}\}$ are of this form. A particular discrete measure is then $\mu(A) = |A|$ where $|A|$ denotes the number of elements in A. It is called the counting measure.

As for σ-algebras, there is also an extension theorem for measures: Let $\mathcal{E} \subseteq \mathcal{P}$. If
(i) ν is a σ-additive measure on $\mathcal{E}$,

(ii) $E_1 \cap E_2 \in \mathcal{E}$ if $E_1, E_2 \in \mathcal{E}$, and
(iii) $X = \cup_{n \in \mathbb{N}} E_n$, $E_n \in \mathcal{E}$ and $\nu(E_n) < \infty$,
then there exists a unique measure μ in $\mathcal{A}(\mathcal{E})$ with $\mu(E) = \nu(E)$ for $E \in \mathcal{E}$.

Two important applications of the extension theorems for σ-algebras and measures are the product measure and the Lebesgue measure.

Product measure. Let $\{X_n, \mathcal{A}_n, \mu_n\}$ $n = 1, \ldots, N$ be a set of measure spaces. The product space $\{X, \mathcal{A}, \mu\}$ then consists of the direct sum (Cartesian product) $X := X_1 \times \ldots \times X_N$, the σ-algebra $\mathcal{A} := \mathcal{A}_1 \times \ldots \times \mathcal{A}_N$ which is the smallest σ-algebra generated by $A_1 \times \ldots \times A_N$, and the product measure μ which is the unique extension of $\nu(A_1 \times \ldots \times A_N) = \nu_1(A_1) \cdot \ldots \cdot \nu_N(A_N)$. If all the measures are identical, the measure is called the Bernoulli measure.

Lebesgue measure. The measure that assigns to all half-open intervals $]a, b]$ on the real line the length $b - a$ satisfies the requirements of the extension theorem and can thus uniquely be extended to a measure on the Borel set $\mathcal{B}^1$. This extended measure is called the Lebesgue measure and denoted by μ_L^1. As stated, the Borel set $\mathcal{B}^1$ is smaller than the power set of $\mathbb{R}$. There exist sets which are not measurable. If these sets are subsets of measurable sets of measure zero, then one can assign measure zero to these sets as well. This extension of the Lebesgue measure is called the Borel measure. The Lebesgue measure μ_L^N in N dimensions is defined as the unique extension of the measure for which $\mu_L^N(B_1 \times \ldots \times B_N) = \mu_L^1(B_1) \cdot \ldots \cdot \mu_L^1(B_N)$.

Measurable functions. A function f from one measure space $\{X, \mathcal{A}, \mu\}$ into another measure space $\{Y, \mathcal{B}, \nu\}$ is called measurable if $f^{-1}(B) \in \mathcal{A}$ for all $B \in \mathcal{B}$. Often $Y = \mathbb{R}$.

Lebesgue integral. Every measure induces a generalized integral called the Lebesgue integral. Starting points are a measure space $\{X, \mathcal{A}, \mu\}$ and a measurable function $f : X \mapsto \mathbb{R}$. Consider pairwise disjoint sets $A_n \in \mathcal{A}$, $n = 1, \ldots, N$ and the function

$$g(x) := \sum_{n=1}^{N} a_n \chi_{A_n}, \quad a_n \in \mathbb{R} \tag{E.106}$$

where χ_A is the indicator function of the set A. Such a function is called a simple function. Its Lebesgue integral is defined by

$$\int_X \mu(dx) g(x) := \sum_{n=1}^{N} a_n \mu(A_n) \tag{E.107}$$

To proceed, write $f = f^+ - f^-$ where $f^+(x) := max(0, f(x))$ and $f^-(x) := max(0, -f(x))$ and assume that there exist sequences $g_n^{\pm}$ of simple functions that converge uniformly to $f^{\pm}$. Then the Lebesgue integral of f is defined by

$$\int_X \mu(dx) f(x) := \lim_{n \to \infty} [\int_X \mu(dx) g_n^+(x) - \int_X \mu(dx) g_n^-(x)] \tag{E.108}$$

It can be shown that this integral exists and is independent of the sequences $g_n^{\pm}$ as long as they converge uniformly to $f^{\pm}$. Lebesgue integrals only require a measure space and no additional structure.

In the special case that the measure space is $\{\mathbb{R}, \mathcal{B}, \mu_L\}$ with μ_L being the Lebesgue measure, the Lebesgue integral is simply written as

$$\int_{\mathbb{R}} \mu_L(dx) f(x) = \int_{\mathbb{R}} dx f(x) \tag{E.109}$$

This Lebesgue integral (with respect to the Lebesgue measure) is more general than the Riemann integral since it allows for more general subsets A_n in its construction rather than just the intervals as in the construction of the Riemann integral.

The function spaces $\mathcal{L}^p$. Given functions $f : X \mapsto \mathbb{R}$, the function space $\mathcal{L}^p$ for $p > 0$ consists of all functions for which

$$\int_X dx |f(x)|^p < \infty \tag{E.110}$$

Such integrable functions represent an equivalent class. Two functions represent the same element if they only differ on a set of measure zero.

The space adjoint to $\mathcal{L}^p$ is $\mathcal{L}^q$ where $1/p + 1/q = 1$. For $p = 1$, the adjoint space consists of all bounded functions and is denoted by $\mathcal{L}^\infty$. If $f \in \mathcal{L}^p$ and $g \in \mathcal{L}^q$, then fg is integrable and defines a scalar product

$$\langle f, g\rangle := \int_X \mu(dx) f(x) g(x) \tag{E.111}$$

and a norm

$$\| f \| := [\int_X \mu(dx) |f(x)|^p]^{1/p} \tag{E.112}$$

If the measure is finite, one defines the expectation of f by

$$\langle f\rangle := \frac{1}{\mu(X)} \int_X \mu(dx) f(x) \tag{E.113}$$

often further assuming that the measure is normalized to one, $\mu(X) = 1$.

Densities. Measures operate on σ-algebras, which is usually awkward in practical calculations. Again, a theorem provides relief, this time the *Radon–Nikodym theorem*. Let ν and μ be two measures such that $\nu(A) = 0$ for all $A \in \mathcal{A}$ for which $\mu(A) = 0$. Then the Radon–Nikodym theorem states that there exists a density $\rho \in \mathcal{L}_1$ with $\rho(x) \geq 0$ such that

$$\nu(A) = \int_A \mu(dx) \rho(x) \tag{E.114}$$

for all $A \in \mathcal{A}$. These densities are much easier to deal with. In most of our examples, the measures μ will be given by such densities with respect to the Lebesgue measure, so we can write

$$\mu(A) = \int_A \mu_L(dx)\rho(x) = \int_A dx\rho(x) \tag{E.115}$$

If

$$\int_A \mu(dx)\rho_1(x) = \int_A \mu(dx)\rho_2(x) \tag{E.116}$$

for all $A \in \mathcal{A}$, then $\rho_1 = \rho_2$ almost everywhere, i.e., the set for which the equality does not hold has measure zero.

Discrete measures can also be represented by densities if we allow densities to be delta functions

$$\rho(x) = \sum_{i=1}^{\infty} \mu(x_i)\delta(x - x_i) \tag{E.117}$$

One then obtains

$$\mu(A) = \int_A dx\rho(x) = \sum_{i=1}^{\infty} \int_X dx\mu(x_i)\delta(x - x_i)\chi_A(x) = \sum_{i=1}^{\infty} \mu(i)\delta_{x_i} \tag{E.118}$$

as in (E.105). Note, however, that δ functions are not elements of $\mathcal{L}^1$. Nevertheless, we will make liberal use of the delta function, especially when measures are supported by lower dimensional subspaces of the state space.

E.14 Two Important Inequalities

Jensen inequality. A function $\phi : \mathbf{R} \mapsto \mathbf{R}$ is called convex if $\phi(\lambda x + (1 - \lambda x') \leq \lambda\phi(x) + (1 - \lambda)\phi(x')$ for $\lambda \in [0, 1]$ and $x, x' \in \mathbf{R}$. Convex functions are continuous. If they are differentiable, then a sufficient condition is $\phi'' \geq 0$. The exponential and the negative logarithm are convex functions. The Jensen inequality states

$$\langle \phi(f) \rangle \geq \phi(\langle f \rangle) \tag{E.119}$$

for random variables f where cornered brackets denote the expectation. This inequality has many applications. Consider in particular a discrete random variable f with probability mass function $p_n = p(\log x_n)$ where $x_n > 0$ and $\phi(x) = e^x$. Then

$$\sum_{n=1}^{N} p_n x_n \geq \Pi_{n=1}^{N}(x_n)^{p_n} \tag{E.120}$$

which states that the arithmetic mean is larger than the geometric mean.

Most importantly, Jensen's inequality implies *Gibbs' inequality*. Specifying $\phi(y) = -\log y$ and $f = p_2/p_1$ where p_1 and p_2 are two probability mass functions, one deduces

$$-\sum_{k=1}^{K} p_1(k) \log \frac{p_2(k)}{p_1(k)} \geq -\log \left\{ \sum_{k=1}^{K} p_1(k) \frac{p_2(k)}{p_1(k)} \right\} = 0 \tag{E.121}$$

or

$$-\sum_{k=1}^{K} p_1(k) \log p_2(k) \geq -\sum_{k=1}^{K} p_1(k) \log p_1(k) \tag{E.122}$$

The equality signs hold if and only if $p_1 = p_2$.

Chebyshev inequality. The Chebyshev inequality states that

$$P(|f - \langle f \rangle| \geq \epsilon) \leq \frac{\sigma_f^2}{\epsilon^2} \tag{E.123}$$

for all $\epsilon > 0$. To prove this inequality, define

$$A := \{\omega \in \Omega : |f(\omega) - \langle f \rangle| \geq \epsilon\} \tag{E.124}$$

Then using the Heaviside function θ, one finds

$$\begin{aligned} P(A) &= \langle \chi_A \rangle = \langle \theta(|f - \langle f \rangle| - \epsilon) \rangle \\ &= \langle \theta(\frac{|f - \langle f \rangle|}{\epsilon} - 1) \rangle \leq \langle (\frac{|f - \langle f \rangle|}{\epsilon})^2 \rangle \end{aligned} \tag{E.125}$$

since $\Theta(\xi - 1) \leq \xi^2$.

E.15 Permutations and Combinations

Combinatorics studies the arrangements of objects into patterns. One important example are permutations and combinations which count the number of ways in which one can draw n objects from N different objects. The result depends on whether the order matters and on whether one can draw an object more than once. One thus distinguishes between permutations (the order matters) with and without repetition and combinations (the order does not matter) with and without repetition. With no repetitions, n must be smaller or equal to N.

(1) The number of permutations without repetition is given by

$$P(n; N) = \frac{N!}{(N-n)!} = N(N-1)\cdots(N-n+1) \tag{E.126}$$

This formula can be rationalized as follows: the first time one can draw from N objects, the second time from $(N-1)$ objects, and the nth time from $(N-n+1)$ objects. Multiplying these numbers gives the formula. The special case $n = N$ gives $P(N; N) = N!$.

(2) The number of permutations with repetitions is

$$\hat{P}(n; N) = N^n \tag{E.127}$$

and it can be rationalized that one can draw n times from N objects.

(3) The number of combinations without repetition is

$$C(n; N) = \frac{N!}{n!(N-n)!} = \binom{N}{n} \tag{E.128}$$

It is $P(n, N)$ divided by $n!$ since the $n!$ permutation of the chosen objects count as one when the order does not matter.

(4) The number of combinations with repetition is

$$\hat{C}(n; N) = \binom{N+n-1}{n} \tag{E.129}$$

which is the number of combinations that you can draw n objects from $(N+n-1)$ different objects without repetition.

Note that the symbol 'P' stands here for **p**ermutation, not for probability.

E.16 Calculus of Variation

Functional. A functional F is a function from a function space into the real numbers

$$\begin{aligned} F : \mathcal{F} &\mapsto \mathbb{R} \\ f &\mapsto F[f] \end{aligned} \tag{E.130}$$

A functional assigns to every function $f \in \mathcal{F}$ a real number. It is somewhat idiosyncratic to introduce a new name, namely functional, for a function that has functions as its arguments, but the name is being used. As always, we assume that $\mathcal{F}$ is made up of 'well-behaved' functions $f : \mathbb{R} \mapsto \mathbb{R}$, so that all operations we perform on them are well-defined.

Functional derivative. In mathematics, the functional derivative is a distribution, denoted by $\delta F/\delta f$, such that

$$\langle\frac{\delta F}{\delta f}, g\rangle = \lim_{\epsilon\to 0}\frac{F[f+\epsilon g]-F[f]}{\epsilon} \tag{E.131}$$

holds for all test functions g. For a complete definition, the function space needs to be defined. Here, we use the physicist's definition of the functional derivative which is given by

$$\frac{\delta F}{\delta f(x_*)} := \lim_{\epsilon\to 0}\frac{F[f+\epsilon\delta(x-x_*)]-F[f]}{\epsilon} \tag{E.132}$$

The physicist's definition differs from the mathematical definition that it chooses a specific function as a test function, namely the δ-function, which is not a valid test function in the mathematical definition. The mathematical definition describes how the entire functional varies when disturbed whereas the physicist's definition only describes variations to disturbances near a specific argument x_*. The physicist's definition is thus a generalization of the partial derivative of a function $F = F(x_1, \ldots, x_i, \ldots)$ of many arguments which is defined by

$$\frac{\partial F}{\partial x_i} = \lim_{\epsilon\to 0}\frac{F(x_1,\ldots,x_i+\epsilon,\ldots)-F(x_1,\ldots,x_i,\ldots)}{\epsilon} \tag{E.133}$$

Lagrange multipliers. If one wishes to maximize the functional $F[f]$ under the constraint that $G[f] = 0$, one may proceed as follows: First consider the completely equivalent problem of maximizing the functional $H[f] = F[f] + \lambda G[f]$ under the same constraint for any constant λ. Second, find for every λ, the maximum of H without constraint and call the solution $f(x;\lambda)$. When a λ^*, called a Lagrange multiplier, can be found such that $G[f(\lambda^*)] = 0$, then this $f(\lambda^*)$ also maximizes the constrained F and is hence a solution to the problem.

E.17 Legendre Transformation

The Legendre transform of a convex function $f : \mathbb{R} \mapsto \mathbb{R}, x \mapsto f(x)$ is defined by

$$g(u) := \sup_x[ux - f(x)] \tag{E.134}$$

When the function f is differentiable, then the maximum is obtained for that value of x where

$$u = \frac{df}{dx} \tag{E.135}$$

Inverting this relation to obtain $x = x(u)$ and substitution into the definition gives the more explicit form of the Legendre transform

$$g(u) = ux(u) - f(x(u)) \tag{E.136}$$

Fig. E.1 Truth table for the conjunction, disjunction, and implication

A	B	$A \wedge B$	$A \vee B$	$A \rightarrow B$
T	T	T	T	T
T	F	F	T	F
F	T	F	T	T
F	F	F	F	T

The Legendre transformation thus produces a function where the independent variable is changed from x to the derivative $u = df/dx$. Furthermore, one finds

$$\frac{dg}{du} = x \tag{E.137}$$

E.18 Classical Propositional Logic

Classical propositional logic follows the rules of binary logic whose formal structure is given by a Boolean algebra.

Boolean algebra. Consider a set $\mathcal{E}$ of propositions $A, B, C, \ldots$ that can either be true or false but not both. Thus, every proposition can be assigned a truth value in the set $\{F, T\}$ where F stands for false and T for true. Three basic connections are considered within the set $\mathcal{E}$:

(1) The conjunction $A \wedge B$ is true if A is true and B is true and is false otherwise.
(2) The disjunction $A \vee B$ is false if A and B are false and is true otherwise.
(3) The negation $\neg A$ is true if A is false and is false if A is true.

These connections are explicated in truth Table E.1

The connections are

$$\text{idempotent} \begin{cases} A \wedge A \equiv A \\ A \vee A \equiv A \end{cases}$$

$$\text{commutative} \begin{cases} A \wedge B \equiv B \wedge A \\ A \vee B \equiv B \vee A \end{cases}$$

$$\text{associative} \begin{cases} A \wedge (B \wedge C) \equiv (A \wedge B) \wedge C \equiv A \wedge B \wedge C \\ A \vee (B \vee C) \equiv (A \vee B) \vee C \equiv A \vee B \vee C \end{cases}$$

$$\text{distributive}\begin{cases} A \wedge (B \vee C) \equiv (A \wedge B) \vee (A \wedge C) \\ A \vee (B \wedge C) \equiv (A \vee B) \wedge (A \vee C) \end{cases}$$

$$\text{dual}\begin{cases} \neg(A \wedge B) \equiv \neg A \vee \neg B \\ \neg(A \vee B) \equiv \neg A \wedge \neg B \end{cases}$$

where the equivalence sign means that the propositions on both sides have the same truth value. The set $\mathcal{E}$ together with the connections $\wedge$, $\vee$, and $\neg$ constitutes a Boolean algebra. The power set of any set also constitutes a Boolean algebra where $\wedge$ corresponds to the intersection of sets, $\vee$ to the union, and $\neg$ to the complement.

A Boolean algebra corresponds to the Boolean ring $\mathbb{Z}mod2 = \{0, 1\}$ with the multiplication and addition

$$\begin{gathered} 0 \cdot 0 = 0; 0 \cdot 1 = 0; 1 \cdot 0 = 0; 1 \cdot 1 = 1 \\ 0 + 0 = 0; 0 + 1 = 1; 1 + 0 = 1; 1 + 1 = 0 \end{gathered} \tag{E.138}$$

If one substitutes $F = 0$ and $T = 1$ in the case of binary logic and introduces the index function χ_A in the case of set theory, the conjunction $\wedge$ and the intersection $\cap$ correspond to multiplication. The disjunction $\vee$ and union $\cup$ do not correspond to addition. Rather, addition corresponds to the symmetric connection $(A \wedge \neg B)(B \wedge \neg A)$ or $(A \setminus B) \cup (B \setminus A)$.

Implication. An additional important connection in binary logic is the implication $A \rightarrow B$ which is false if A is true and B is false and is true otherwise. The truth table of the implication is also given in Table E.1.

The implication $A \rightarrow B$ is equivalent to the implication $\neg B \rightarrow \neg A$ and to the connection $A \wedge \neg B$ since these generate the same truth table. These equivalences imply especially that if $A \rightarrow B$ is true then $A \wedge \neg B$ must be false.

The implication does not imply that either A or B is true. It does imply the following: if A is true then B must be true; if B is false then A must be false. It does not say anything about the other two cases: if B is true then A can be true or false; if A is false then B can be true or false.

The logical implication $A \rightarrow B$ does not imply that one can deduce or derive B from A. First, any derivation would have to specify the set of allowed intermediary implications. Second, the logical implication is all-encompassing. If A is true, then $A \equiv A \wedge B$ is equivalent to $A \rightarrow B$ which is true if B is true. Thus, every true proposition implies any other true proposition. If A is false, then $A \equiv A \wedge B$ is also equivalent to $A \rightarrow B$ which is true for B being either true or false. Thus, every false proposition implies all propositions. The logical implication also does not imply a causal or temporal relation between the two propositions.

E.19 Fourier Transforms and Series

Fourier transforms and series are representations of a complex-valued function of the real line, of a periodic complex-valued function of the real line, and of natural numbers.

Fourier transform. If $f : \mathbb{R} \mapsto \mathbb{C}$ is an integrable function, its Fourier transform defined by

$$\hat{f}(\omega) := \int_{-\infty}^{\infty} dt f(t) e^{-i\omega t} \tag{E.139}$$

exists, and the original function f can be expressed by the inverse formula

$$f(t) = \frac{1}{2\pi} \int_{-\infty}^{\infty} d\omega \hat{f}(\omega) e^{i\omega t} \tag{E.140}$$

There are other conventions regarding the factor $1/2\pi$ in front of the integral and the sign. One such convention is to symmetrize the formulas and have a factor $1/\sqrt{2\pi}$ in front of both integrals. Also, the formulas are often expressed in terms of the circular frequency $\nu := \omega/2\pi$ instead of the angular frequency ω. By writing $e^{i\omega t} = \cos \omega t + i \sin \omega t$, the Fourier transform can also be expressed in terms of cosine and sine transforms. If f is a real-valued function, then $\hat{f}(-\omega) = \hat{f}^*(\omega)$.

In a linear algebra context, the transformation formulas state that the functions $e^{i\omega t}$, $\omega \in \mathbb{R}$ form a non-countable orthogonal basis of the vector space $\mathcal{L}^1$. The completeness and orthogonality relation may be written as

$$\int_{-\infty}^{\infty} d\omega e^{-i\omega(t-t')} = 2\pi\delta(t - t') \tag{E.141}$$

$$\int_{-\infty}^{\infty} dt e^{(\omega-\omega')t} = 2\pi\delta(\omega - \omega') \tag{E.142}$$

Fourier series. If f is a complex-valued function on the interval $[-T/2, T/2[$, periodically extended to the real line, then its Fourier series is defined by

$$\tilde{f}(t) = \sum_{n=-\infty}^{+\infty} c_n e^{i\omega_n t} \tag{E.143}$$

where the frequencies are given by

$$\omega_n := n\frac{2\pi}{T} \tag{E.144}$$

and the coefficients by

$$c_n := \frac{1}{T} \int_{-T/2}^{+T/2} dt f(t) e^{-i\omega_n t} \tag{E.145}$$

The frequency $\Delta\omega := 2\pi/T$ is called the fundamental frequency. Indeed one finds $\tilde{f}(t+T) = \tilde{f}(t)$.

In the limit $T \to \infty$, one obtains

$$\lim_{T\to\infty} \sum_{n=-\infty}^{+\infty} c_n e^{i\omega_n t} = \frac{1}{2\pi} \int_{-\infty}^{\infty} d\omega \hat{f}(\omega) e^{i\omega t} \tag{E.146}$$

and recovers the Fourier transform. The Fourier series is thus the Fourier transform of a periodic function. The periodicity causes the discreteness of the Fourier coefficients.

If f is a real-valued function, then $c_n = c^*_{-n}$ and the decomposition of the exponential function into sine and cosine functions yields the representation

$$\tilde{f}(t) = \frac{a_0}{2} + \sum_{n=1}^{\infty} (a_n \cos\omega_n t + b_n \sin\omega_n t) \tag{E.147}$$

where

$$a_n = c_n + c_{-n} = \frac{2}{T} \int_{-T/2}^{+T/2} dt f(t) \cos\omega_n t \qquad n \geq 0 \tag{E.148}$$

$$b_n = i(c_n - c_{-n}) = \frac{2}{T} \int_{-T/2}^{+T/2} dt f(t) \sin\omega_n t \qquad n \geq 1 \tag{E.149}$$

There exist various conditions under which the Fourier series $\tilde{f}$ converges to the original function f. One such condition requires that f be piecewise smooth in the interval $[-T/2, T/2[$, meaning that f is continuous and differentiable except at a finite number of points. Then $\tilde{f}(x)$ converges to $f(x)$ at the points x where f is continuous and to $(f_-(x) + f_+(x))/2$ at points x where f has a jump discontinuity, which includes the points $x = T/2 + nT$ if $f(-T/2) \neq f(T/2)$. Here, $f_\pm$ are the right and left limits at the discontinuity.

Orthogonality relations. In a linear algebra context, the Fourier series state that the functions $\cos\omega_n t$, $n = 0, 1, 2, \ldots$ and $\sin\omega_n t$, $n = 1, 2, \ldots$ with $\omega_n = 2n\pi/T$ form an orthogonal basis of the associated function space. The orthogonality relations are

$$\frac{2}{T} \int_{-T/2}^{+T/2} dt \cos\omega_m t \cos\omega_n t = \delta_{nm} \tag{E.150}$$

$$\frac{2}{T} \int_{-T/2}^{+T/2} dt \sin\omega_m t \sin\omega_n t = \delta_{nm} \tag{E.151}$$

$$\frac{2}{T}\int_{-T/2}^{+T/2} dt \sin\omega_m t \cos\omega_n t = 0 \tag{E.152}$$

Note that the functions $\sin\omega_n t, n = 1, 2, \ldots$ and $\cos\omega_n t, n = 0, 1, 2, \ldots$ are orthogonal on the interval $[-T/2, +T/2[$, not on the interval $[0, T/2[$ where we also obtain

$$\frac{2}{T}\int_{0}^{+T/2} dt \cos\omega_m t \cos\omega_n t = \delta_{nm} \tag{E.153}$$

$$\frac{2}{T}\int_{0}^{+T/2} dt \sin\omega_m t \sin\omega_n t = \delta_{nm} \tag{E.154}$$

but

$$\frac{2}{T}\int_{0}^{\pi} dt \sin\omega_m t \cos\omega_n t = \begin{cases} 0 & \text{for } n-m \text{ even} \\ \frac{1}{\pi}\frac{2m}{m^2-n^2} & \text{for } n-m \text{ odd} \end{cases} \tag{E.155}$$

The functions $\sin\omega_n t$, $n = 1, 2, \ldots$ and $\cos\omega_n t$, $n = 0, 1, 2, \ldots$ with $\omega_n = 2n\pi/T$ provide an orthonormal basis for the interval $[-T/2, +T/2[$ but not for the interval $[0, +T/2[$.

Discrete Fourier transform. The discrete Fourier coefficients of a signal consisting of N given numbers $x_0, \ldots, x_{N-1}$ are defined by

$$\hat{x}_k := \frac{1}{N}\sum_{n=0}^{N-1} x_n e^{-i\omega_n k} \quad k = 0, \ldots, N-1 \tag{E.156}$$

where again $\omega_n = n\Delta\omega$ and the fundamental frequency is $\Delta\omega = 2\pi/N$. The inverse formula is given by

$$x_n = \sum_{k=0}^{N-1} \hat{x}_k e^{i\omega_k n} \tag{E.157}$$

The signal is a superposition of a constant term plus terms with the fundamental frequency and a finite number of higher harmonics. The signs in the exponent and the placement of the factor $1/N$ are again conventions.

The numbers x_n can be periodically extended such that $x_{n+N} = x_n$ for $n \in \mathbb{N}$. Similarly, formula (E.156) can be used to define $\hat{x}_k$ for all $k \in \mathbb{N}$. The resulting sequence $\hat{x}_k$ is also periodic with period N, $\hat{x}_{k+N} = \hat{x}_k$, as is $\exp\{i\omega_{n+N}k\} = \exp\{i\omega_n k\}$. Because of these periodicities, the summation range can be shifted by M points as long as it contains N consecutive indexes and one obtains alternatively

$$x_n = \sum_{k=-M}^{N-1-M} \hat{x}_k e^{i\omega_k n} \tag{E.158}$$

Often, one chooses a summation range that is symmetric around zero. These formulas are used in signal processing.

If the numbers x_n are real, then $\hat{x}_0$ is real and $\hat{x}_k = \hat{x}^*_{N-k}$. If N is even, the signal can then be written as

$$x_n = \frac{a_0}{2} + \sum_{k=1}^{N/2} (a_k \cos \omega_n k + b_k \sin \omega_n k) \tag{E.159}$$

where

$$a_k = \hat{x}_{N-k} + \hat{x}_k = \frac{2}{N} \sum_{n=0}^{N-1} x_n \cos \omega_n k \quad 0 \le k \le \frac{N}{2} \tag{E.160}$$

$$b_k = i(\hat{x}_{N-k} - \hat{x}_k) = \frac{2}{N} \sum_{n=0}^{N-1} x_n \sin \omega_n k \quad 1 \le k \le \frac{N}{2} \tag{E.161}$$

If N is odd, the summation extends to $(N-1)/2$.

One may represent the periodically extended infinite sequence of numbers x_n by the function

$$f(t) = \sum_{k=-\infty}^{\infty} x_n \delta(t-n) \tag{E.162}$$

Formula (E.145) then gives the Fourier coefficients

$$c_n := \frac{1}{N} \sum_{k=-N/2}^{N/2-1} x_k e^{-i\omega_n k} \tag{E.163}$$

which is equivalent to the expression (E.156) for $\hat{x}_n$. The discrete Fourier transform may thus be viewed as the Fourier transform of a periodic function of discrete support. The periodicity causes the discreteness of the Fourier coefficients, the discrete support the periodicity of the coefficients. The different cases are summarized in Fig. E.2.

Spectrum. Assume $f(t)$ to be a stationary stochastic process with covariance function $C(\tau) = \langle f(t) f(t+\tau) \rangle - \langle f(t) \rangle \langle f(t+\tau) \rangle$ where τ is the time lag. The Fourier transform of the covariance function

$$S(\omega) := \int_{-\infty}^{\infty} d\tau C(\tau) e^{-i\omega\tau} \tag{E.164}$$

is called the spectrum. Since the covariance function is a symmetric function of τ, the spectrum is a symmetric function of the frequency, $S(-\omega) = S(\omega)$. Introduce the Fourier components of the stochastic process

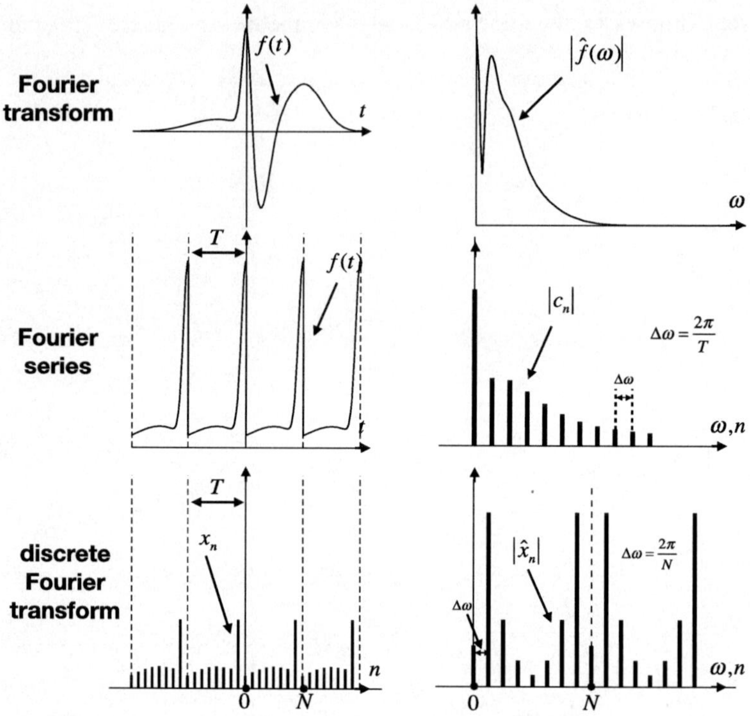

Fig. E.2 The different Fourier transformations. If time t is continuous then the frequency ω is continuous. If t is continuous and $f(t)$ a periodic function with period T, then the frequency is discrete with spacing $\Delta\omega = 2\pi/T$. Conversely, if time t is discrete with spacing Δt, then ω is continuous and $\hat{f}(\omega)$ is periodic with period $\Omega = 2\pi/\Delta t$. If t is discrete with spacing Δt and $f(t)$ periodic with period $T = N\Delta t$, then ω is discrete with spacing $\Delta\omega = 2\pi/T$ and $\hat{f}(\omega)$ periodic with period $\Omega = 2\pi/\Delta t$, here shown for $\Delta t = 1$

$$\hat{f}(\omega) := \int_{-\infty}^{\infty} d\tau f(t) e^{-i\omega t} \tag{E.165}$$

These Fourier components constitute a family of random variables with the frequency ω as the index set. Using the rules for Fourier transformations, one obtains for these random variables

$$\langle \hat{f}(\omega) \rangle = 0 \tag{E.166}$$

$$\langle \hat{f}(\omega) \hat{f}^*(\omega') \rangle = 2\pi S(\omega) \delta(\omega - \omega') \tag{E.167}$$

The Fourier components of a stationary process are uncorrelated. Furthermore

$$\int_{-\infty}^{\infty} d\omega S(\omega) = 2\pi C_0 = 2\pi\sigma_0^2 \tag{E.168}$$

The spectrum thus describes how the variance is distributed over the different frequencies. The factor of 2π may be avoided by a different normalization of the Fourier transform. The formulas might be further simplified to account for the fact that the covariance function and spectrum are symmetric functions and that $f(t)$ is a real function. If time is discrete $t_n = n\Delta t, n \in \mathbb{Z}(\mathbb{N})$, then the spectrum is periodic with period $2\pi/\Delta t$. If the process has a finite duration T (or is periodic with period T), then the spectrum is discrete with frequency spacing $\Delta\omega = 2\pi/T$.

E.20 Jacobian

Let $f : \mathbb{R}^N \mapsto \mathbb{R}^N$, then the Jacobian matrix is defined by

$$F := \nabla_x f^T \tag{E.169}$$

with components

$$F_{ij} = \frac{\partial f_j}{\partial x_i} \tag{E.170}$$

Its determinant is called the Jacobian determinant or just the Jacobian and variously written as

$$J(x) = \frac{\partial(f(x))}{\partial(x)} = det F = |F| \tag{E.171}$$

The Jacobian determines how infinitesimal volume elements change. The cofactor matrix is defined by

$$C_{ij} := \frac{\partial J}{\partial F_{ij}} \tag{E.172}$$

and has the properties

$$(F^{-1})_{ij} = \frac{C_{ji}}{J} \tag{E.173}$$

$$\frac{\partial C_{ij}}{\partial x_i} = 0 \tag{E.174}$$

E.21 Stirling's Formula

The Euler–Maclauren sum formula gives for the logarithm of $n!$ the series

$$\ln n! = n \ln n - n + \frac{1}{2} \ln 2\pi n + \frac{1}{12n} - \frac{1}{360n^3} + \cdots + \frac{B_k}{(2k-1)2k} \frac{1}{n^{2k-1}} + \ldots \tag{E.175}$$

where B_k is the kth Bernoulli number. Taking only the first few terms of this series into account gives various versions of Stirling's formula or approximation.

References

Müller, P.: Ertel's potential vorticity in physical oceanography. Rev. Geophys. **33**(1), 67–97 (1995)
Schuster, H.G.: Deterministic Chaos: An Introduction . Wiley-VCH (1995)
Saltzmann, B.: Finite amplitude free convection as an initial value problem - I. J.atmos. Sci. 19, 329-341 (1962)
Müller, P.: The Equations of Oceanic Motions. Cambridge University Press (2006)

Index

P. Müller, *Handbook of Dynamics and Probability*,
https://doi.org/10.1007/978-3-030-88486-4